approximate conversion tables

ENGLISH TO METRIC

English	×	= Metric
inches	2.54	centimeters
feet	0.305	meters
yards	0.91	meters
miles	1.61	kilometers
square inches	6.45	square centimeters
square feet	0.09	square meters
square yards	0.84	square meters
acres	0.40	hectares
square miles	2.6	square kilometers
cubic inches	16.4	cubic centimeters
cubic feet	0.028	cubic meters
cubic yards	0.76	cubic meters
cubic miles	4.19	cubic kilometers
ounces	28.3	grams
pounds	0.45	kilograms
tons	0.9	tons
fluid ounces	30	milliliters
quarts	0.95	liters
gallons	3.8	liters

METRIC TO ENGLISH

Metric	×	= English
centimeters	0.39	inches
meters	3.28	feet
meters	1.09	yards
kilometers	0.62	miles
square centimeters	0.15	square inches
square meters	11	square feet
square meters	1.20	square yards
hectares	2.47	acres
square kilometers	0.38	square miles
cubic centimeters	0.06	cubic inches
cubic meters	35.31	cubic feet
cubic meters	1.31	cubic yards
cubic kilometers	0.24	cubic miles
grams	0.04	ounces
kilograms	2.20	pounds
tons	1.1	tons
milliliters	0.033	ounces
liters	1.06	quarts
liters	0.26	gallons

ENERGY

1 barrel of crude oil = 42 gallons
7 barrels of crude oil = 1 metric ton = 40 billion BTUs
1 metric ton of coal = 28 million BTUs
1 gram U_{235} = 2.7 metric tons of coal = 13.7 barrels of crude oil
1 BTU (British Thermal Unit) = 252 calories = 0.0002931 kilowatt-hour
1 kilowatt-hour = 860,421 calories 3412 BTU

Metric conversions in the text are approximate.

GEOLOGY OF CALIFORNIA

GEOLOGY OF CALIFORNIA

Second Edition

ROBERT M. NORRIS
Professor of Geology, Emeritus
University of California, Santa Barbara

ROBERT W. WEBB
Late Professor of Geology
University of California, Santa Barbara

JOHN WILEY & SONS, INC.

New York   Chichester   Brisbane   Toronto   Singapore

Library of Congress Cataloging in Publication Data:

Norris, Robert M. (Robert Matheson), 1921–
Geology of California/Robert M. Norris, Robert W. Webb—2nd ed.
    p.   cm.

    Includes bibliographical references.
    ISBN 0-471-50980-9
    1. Geology—California.   I. Webb, Robert Wallace, 1909–84, joint author.
  II. Title.

QE89. N67   1990
557.94—dc20

89-36302
CIP

Printed in the United States of America

10 9 8 7 6 5 4 3

Cover Photo: The spirelike topography results from weathering along widely
spaced joints in the White Fang Quartz Monzonite. In boulder-pile mountains
of this type, much of the weathering is thought to have occurred beneath a cover
of soil during a more moist climate. The bulk of the soil cover was stripped off
as the climate became increasingly arid. Granite Mountains C, eastern Mojave
Desert, San Bernardino County. (Photo by R. M. Norris)

*To Ginny Norris and Elaine Webb
whose encouragement and patience
made this book possible*

preface

Much has happened to enhance our understanding of California geology during the past decade, providing ample reason for preparing a complete revision of this textbook. In addition, many users of the first edition have offered valuable suggestions for improving the book and its usefulness. Finally, the new edition offers an opportunity to correct errors and to attend to omissions in the first edition.

Because an unexpectedly large number of college courses using this textbook are open to students with no previous training in geology, we have elected to add a new introductory chapter in which some of the basic principles of geology, particulaly relevant to California, are covered in an abbreviated manner. It is expected that this chapter will also provide a useful review for any students whose introductory course was completed some years ago.

Just as work on the second edition was beginning, my colleague, Professor Robert W. Webb, died quite unexpectedly while we were on a field trip to the Mono Basin. This delayed the revision somewhat and made it necessary for me to rely more heavily on others. I especially appreciate the assistance of Professor Gregory R. Wheeler of California State University, Sacramento, who read the entire manuscript, wrote part of the Sierra Nevada chapter, and prepared the chapter on the Klamath Mountains. Others whose comments on individual chapters have been very helpful include Bennie W. Troxel, University of California, Davis; William P. Irwin, U.S. Geological Survey; York T. Mandra, San Francisco State University; Richard A. Schweickert, University of Nevada, Reno; Kent Murray and Marlon Nance, California State University, Sacramento; and my colleagues at Santa Barbara, John C. Crowell, Stanley M. Awramik, and Arthur G. Sylvester. Steven Spear, Palomar Community College, David Schwartz, Cabrillo College, and Ted Herman of West Valley College reviewed the entire manuscript and offered many cogent suggestions. Ellie Dzuro skillfully typed the manuscript, and Dianne Griffin drafted most of the maps and diagrams. Any errors that survive editorial scrutiny and find their way into print in the second edition are my responsibility, and I trust the users of the book will not hesitate to let me know about them.

ROBERT M. NORRIS
SANTA BARBARA

July 1988

contents

one

BASIC CONCEPTS OF GEOLOGY—A BRIEF REVIEW

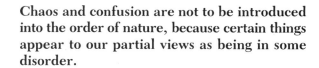

Chaos and confusion are not to be introduced into the order of nature, because certain things appear to our partial views as being in some disorder.

—*James Hutton*

Geology is both a laboratory and a field science, but much of its subject matter is, with the aid of a few basic concepts, easily understandable to the intelligent layperson. In a sense, geology is a *third-order* science. In this scheme, a *first-order* science would be physics, which concerns itself with the most fundamental aspects of matter and energy. Chemistry would be a *second-order* science concerned with the interactions of various combinations of matter and energy, and sciences such as geology, astronomy, and biology are *third-order* sciences dependent upon the more basic sciences. Geology, for example, could be considered the physics and chemistry of the earth over time, or perhaps the biology of the earth. Although the scope of geology is very broad and its problems often quite complex, much of the science deals with things that can be seen and readily understood by a reasonably alert layperson. For example, it is obvious that floods cause erosion of hillsides and deposition in the lowlands; these concepts are much easier to grasp than some of the more abstract or unobservable phenomena dealt with by physics or even chemistry.

SCALE OF GEOLOGIC PHENOMENA

Although many geologic phenomena are easily observed and readily understood, geology is also the most historical of the sciences and it is necessary to appreciate the vastness of geologic time in order to grasp the full importance of many seemingly minor events. We know, for example, that sea level has changed dramatically in the past 20,000 years, and is likely to change a great deal in the future as well, but if we make observations throughout the world and discover that sea level is rising at a rate of 5 to 6 millimeters per year (about the thickness of an ordinary pencil), we may be unimpressed. This same rate multiplied by 1,000 years amounts to 5 or 6 meters, or about 20 feet. If the average level of the sea were to rise that much with the high tides added to it, the effects on our coastal areas would be catastrophic.

We could provide examples of slow geologic processes that loom quite large when they persist for considerable periods of time. For example, slip on the San Andreas fault averages about 5.5 centimeters (2 inches) per year. In a million years this amounts to 55 kilometers (32 miles). The Sierra Nevada appears to have been elevated at a rate of about 0.1 centimeter (0.04 inches) per year for the past 3 million years, a total uplift of about 3,000 meters (10,000 feet) for the period. Another consequence of the vastness of geologic time is the increasing inevitability of the unlikely event. Put another way, if a given, but rather unlikely phenomenon is scientifically possible and reasonable, given enough time the event becomes probable, or indeed almost inevitable; it is governed by the laws of chance. To illustrate, we might think of a newly formed volcanic island totally lacking vegetation and animal life. We know that insects, seeds, birds, and so on may be carried long distances by the wind. Floating logs may house a few lizards or perhaps a pregnant female mouse. Sooner or later the winds and the sea currents will deliver the first plants and animals to our newly minted island. Many of these new immigrants will fail to survive or reproduce, but given enough time, our new volcanic island will acquire a flora and a fauna. If the island is sufficiently isolated and the new immigrants arrive infrequently enough, evolution may result in the development of unique forms such as the moas of New Zealand or the dodos of Mauritius.

Recorded history shows us that severe earthquakes may be associated with an abrupt horizontal slip along the San Andreas fault in California. These slips may occur once every 100 years or so, and the distance moved may vary from a few to perhaps 10 meters (30 feet). Given a history of recurrent activity over a period of perhaps 5 million years, we are able to show that one part of California has slid by the other many kilometers and in so doing has produced a number of strong earthquakes (Figure 1-1). In a million years' time, if strong earthquakes are produced by the San Andreas only once in a century, we can expect 10,000 of them! Geologists then, must pay close attention to any small but repetitive events, for given enough time, these small events can produce enormously important and far-reaching changes.

The earth as seen from a human being's perspective is a large place and the spatial scale of geologic phenomena may be very difficult to grasp. For instance, standing on a shoreline looking out to sea, most of us would be forgiven if we considered the sea to be a significant portion of the whole earth. Yet if we could reduce the earth to a sphere a couple of meters in diameter, the sea would then be only a very thin film several millimeters thick; actually it is only a minor part of the whole earth, albeit an important part.

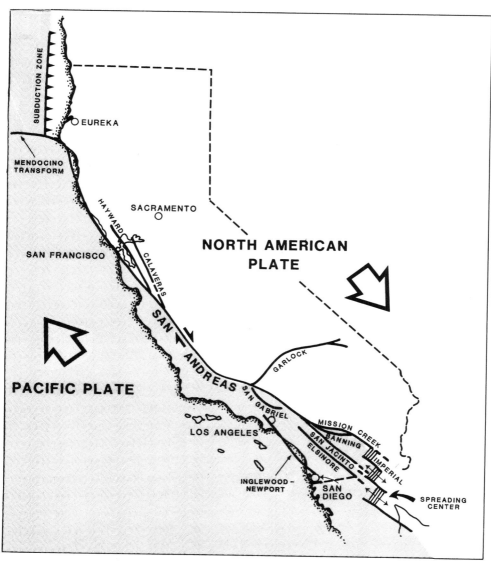

FIGURE 1-1 The San Andreas fault system and its principal branches.

We have long known that immense blocks of the earth's crust, such as the Sierra Nevada, may be slowly raised thousands of meters above the surrounding lowlands, requiring stupendous forces acting over a long time. More recently, we have learned that entire continents move slowly across the face of the earth like ponderous rafts, sometimes colliding and other times splitting apart, or sliding by one another, all in very slow motion. These examples illustrate that some geologic phenomena are difficult to appreciate because they are so big, so slow, and so inconspicuous in our daily life that their ultimate significance is easily overlooked. Human history is so short, in geological terms, that it represents barely more than an instant in geologic time, leading us to confer on our landscape an unwarranted permanence and to miss almost entirely the continual change that affects the earth we live on.

Although sudden events, some of truly catastrophic dimension, do occur, such as explosive volcanic activity, severe earthquakes, and destructive landslides, most geologic changes go on gradually and inconspicuously and produce dramatic results only over long periods of time.

GEOLOGIC MATERIALS

The rocky outer crust of the earth is the chief concern of geology and it is from the crust and uppermost part of the mantle that virtually all the visible physical features of California are formed. Thus it is important that we know something of crustal materials and how they behave.

The basic building blocks of the crust are called *minerals*, which are naturally-occurring crystalline chemical compounds. Crystals are geometrically ordered arrangements of chemical compounds or elements. For example, the external form of a quartz crystal is merely the obvious reflection of the orderly and always consistent internal arrangement of the silicate tetrahedron, where four oxygen atoms are linked to one silicon atom. Each chemical compound has its characteristic internal pattern or order that confers on minerals a predictable and characteristic external form (Figure 1-2). Not all mineral grains in nature exhibit external crystal form, because many have been formed in confined spaces and have been forced to take on the shape defined by their surroundings. However, even these irregularly shaped grains possess the orderly internal nature characteristic of all other specimens of the particular mineral involved. Some hornblende grains have no external crystal faces, but all hornblende grains have the same sort of internal geometry, and where space was available during growth, all have the same external geometry as well.

Although several thousand mineral species have now been identified, named, and described, most are quite rare, and a group of about two dozen minerals—called *rock-forming* minerals—make up most of our crustal rocks. The reason so few minerals are so abundant is not hard to understand; it is related to the relative abundance of the various chemical elements in the earth's crust. The most abundant are oxygen, silicon, and aluminum. The commonest minerals are the silicates (silicon and oxygen) and the aluminum silicates. The most abundant single mineral, not surprisingly, is quartz, silicon dioxide. The most abundant group of closely allied minerals is the feldspars, aluminum silicates of calcium, sodium, and potassium for the most part. Table 1-1 lists the 10 most common elements in the earth's crust.

Virtually all the rock-forming minerals are composed of these 10 elements. Interestingly, many metallic ores of zinc, lead, copper, tin, and silver represent rather rare minerals that are only locally concentrated in the earth's crust. Unfortunately, of the most important metals in modern civilization, only iron and aluminum are abundant and widespread in crustal rocks. We have emphasized the great importance of rock-forming minerals and mentioned a few examples, but it may be useful to list the major groups here because you will encounter these names frequently as your study of geology continues.

We have thus far not made a clear distinction between *minerals* and *rocks*, although it has probably become evident that most rocks are made up of various combinations

FIGURE 1-2 Shiny black crystals are the mineral neptunite, a titano-silicate of iron, manganese, potassium, and sodium. The stubbier, glassy crystals are the pale blue mineral benitoite, a barium titano-silicate. Benitoite is California's state gem, and crystals as large as the ones in this specimen have so far been found at only one locality at the head of the San Benito River in southern San Benito County. The very fine-grained white matrix is natrolite, one of the zeolite group of minerals. (Photo by R. M. Norris)

of minerals. This is generally true, but some rocks are composed of single minerals. We call these occurrences rocks rather than minerals because they occur in large masses or thick layers in the manner of rocks, although a geologist, given a small hand specimen of any one of them would surely identify it as a mineral. An example of such a rock is marble or limestone. These rocks are often composed entirely of the mineral calcite or partially or entirely of a closely related mineral, dolomite. Rock salt is composed entirely of the mineral halite. On the other hand, a familiar rock like granite is composed of quartz and two kinds of feldspar, usually with minor amounts of other rock-forming minerals such as amphibole, garnet, and the micas.

Unlike minerals, which by definition must always be inorganic, some rocks are made up of organic materials. Coal and limestone are two examples. Coal is not common in California and limestones are not abundant in either the coastal mountains or the Sierra Nevada. However, a rather unusual rock of organic origin, composed of the remains of billions of tiny silica-secreting marine plants called diatoms, is a prominent rock in the southern Coast Ranges. This rock is called *diatomite* or *diatomaceous earth*. In short, rocks like granite are aggregates of minerals. Other rocks form from single minerals, and some form from organic materials like diatoms or plants.

TABLE 1-1 The Most Abundant Elements in the Earth's Crust, Expressed as Percent

Oxygen	46.6
Silicon	27.7
Aluminum	8.1
Iron	5.0
Calcium	3.6
Sodium	2.8
Potassium	2.6
Magnesium	2.1
Titanium	0.4
Phosphorus	0.1
All others	0.1

Igneous Rocks

The usual classification of rocks is a threefold division into *igneous, metamorphic,* and *sedimentary.* Igneous rocks make up almost all of the earth's crust, whereas most of the surface of the earth is covered by sediment and sedimentary rocks. Igneous rocks are those that solidified from a molten material called *magma.* All the various kinds of lava erupted from volcanoes are igneous rocks. Similarly, rocks solidified from magma below the surface of the earth, such as granite, are igneous rocks, specifically *plutonic* igneous rocks as contrasted with *volcanic* igneous rocks cooled at or very close to the surface (Figure 1-3). Dozens of different igneous rocks have been recognized and named. These various kinds exist for two main reasons. First, the original molten material, the magma, can vary widely in composition depending upon where and from what materials it was formed (see rock tables in Appendix 2). Second, the rate of cooling or solidification may vary a great deal. If the magma cools very slowly at depth, the resulting rock is coarsely crystalline, like granite. The same granitic magma, if erupted by a volcano, cools very quickly, producing a fine-grained crystalline volcanic rock called rhyolite, glassy, noncrystalline varieties of which may be called pumice if they were frothed by a high gas content, or obsidian if they had little gas.

Before we consider some of the varied kinds of volcanic materials we find on the surface of the earth, recall that the liquid parent of all volcanic products is called

TABLE 1-2 Common Rock-forming Mineral Groups

Common oxides such as quartz.
Feldspars, aluminum silicates chiefly of sodium, calcium, and potassium.
Amphiboles, chiefly silicates of sodium, calcium, magnesium, and iron.
Pyroxenes, also chiefly silicates of sodium, calcium, magnesium, and iron.
Micas, thin, laminated aluminum silicates.
Clays, chiefly hydrated aluminum silicates.
Garnets, equidimensional silicates, mostly of calcium, magnesium, and iron.
Olivine, silicates of iron and magnesium.
Carbonates, mainly carbonates of calcium and magnesium.

FIGURE 1-3 An exposure of the White Fang quartz monzonite, a course-grained plutonic igneous rock, Granite Mountains C, eastern Mojave Desert, San Bernardino County. (Photo by R. M. Norris)

magma. Some magma erupts as fluid lava in the form of flows, but much is erupted explosively as a fine-grained material called *ash*. Coarser solid materials are given other names; particles from about the size of peas to walnuts are called *cinders* or *lapilli*. Larger masses are called *blocks*. Inevitably we find some materials that are intermediate between our categories. One of these types is called *volcanic bombs*. Bombs are fluid blobs that are hurled out of volcanic vents and cool to some degree as they fly through the air. Larger ones often take on a spindle shape, some look like cow pies, and others resemble twisted and burned strips of bacon. Some have cores of granite ripped off the throat of the volcano, whereas others have picked up cores of deep crustal rocks like peridotite (olivine). Such cored bombs are found at Dish Hill in the Mojave Desert (Figure 1-4). Bombs vary from the size of an almond to monsters a meter or more across.

Because the gas content of a magma varies considerably, we find some volcanic products that show the effects of a high gas content and others that tell us that the magma contained little dissolved gas. If a parent magma is gassy and rich in silica, a frothy, light-colored pumice is erupted. Pumice may be so full of gas bubbles that it will float on water. Great quantities of pumice cover much of the ground between Lake Crowley and Mono Lake on the eastern side of the Sierra Nevada. Rounded

FIGURE 1-4a A group of spindle-shaped volcanic bombs shaped by spinning during flight before hardening had occurred. The larger bomb is about 25 centimeters (10 inches) long. (Photo by R. M. Norris)

FIGURE 1-4b An olivine-cored volcanic bomb fragment from Dish Hill, Mojave Desert. The core is believed to have been derived from peridotites in the upper mantle and brought to the surface by rising basaltic magma. Specimen is $7\frac{1}{2}$ centimeters (3 inches) across. (Photo by R. M. Norris)

pumice beach pebbles are common along the southern shores of the Salton Sea, and pumice grains mantle the slopes of Cinder Cone in Lassen Volcanic National Park.

If the gassy lava happens to be a dark-colored basaltic material, the resulting rock is called *scoria*. Scoria looks something like pumice, but is generally black or dark red. Fewer gas pockets in scoria make it heavier and it will not float like some pumice.

It is apparent that differences in magma composition give rise to differences in the kind of appearance of solids erupted from volcanic vents, so it is not surprising that fluid lavas also differ because of composition. Not all lava flows look alike. We see short, stubby, steep-sided flows at Mono Craters, which have been called *coulees*, and the widespread, much thinner, dark flows characteristic of the Modoc plateau. The short flows at Mono Craters are silica-rich rhyolites and the more fluid, dark lavas seen in the Modoc country are silica-poor basalts (Figure 1-5). Lavas of intermediate composition are not uncommon; Mount Shasta is composed of such rocks.

Volcanic cones and vents also show much variation in style and geometry, again reflecting differences in the chemistry of the magma and the geologic setting in which the eruptions took place. A few examples will be given here, but others will be described in later chapters as specific features are discussed.

Very short-lived eruptions of solids and gases may produce pitlike craters surrounded by layers of cinders such as Ubehebe Crater in northern Death Valley. Fire fountains of red-hot solid and pasty lava blobs may produce cinder cones around a vent in just a few weeks or months. This is a very common type of eruption in Hawaii and in the recent past in California as well. Hundreds of cinder cones are found in

FIGURE 1-5 A horizontal basalt flow of Quaternary age showing crude columnar jointing. Near Fort Bidwell, Modoc County. (Photo by R. M. Norris)

FIGURE 1-6 Pisgah Crater, a Quaternary cinder cone in the Mojave Desert with associated basalt flows. The photo was taken in 1973 before extensive mining had altered the cone. (Photo by R. M. Norris)

California from the Oregon border south to the Mojave Desert. These cinder cones frequently are associated with the extrusion of fluid basaltic lava flows, which often surround the cinder cone. Fine examples can be seen at Pisgah and Amboy Craters in the Mojave Desert (Figure 1-6).

The short, stubby rhyolite flows from the Mono Craters often issue from low cones composed of pumice. Some of the Mono Craters are little more than pumice rings, whereas others have a rough, bulbous mass of rhyolite in the center, a plug that cooled before extruding enough material to fill the crater and ooze out as a coulee.

California's most recently active volcano, Lassen Peak, is made up of pasty lava flows and plugs of dacite, a somewhat darker, less silicic lava than the rhyolites of the Mono Craters. The dacite flows of Lassen Peak did not extend far beyond the vent, giving the mountain a somewhat rounded, blunt shape.

Mount Shasta is the most prominent volcano in California and a magnificent peak by any standards. It is called a *composite* cone because it is built up from layers of fragmental material like cinders, blocks, and ash, interbedded with lava flows of andesitic composition. Andesites, like dacites, have compositions between the very siliceous rhyolites and the much less silicic, dark-colored basalts, but are closer to the basalts. Most of the world's large volcanic cones are andesitic and all are products of alternating explosive episodes in which solids are erupted, and quieter periods in which fluid lavas are produced. These eruptive styles do not follow one another in a predictable, reliable pattern and may in fact occur together.

Although this brief summary provides a rough guide for classifying and understanding California volcanoes and their products, it is incomplete. Volcanoes erupt liquid, solid, and gaseous products, and the proportion of these produced will greatly influence the resulting landform.

Metamorphic Rocks

The second large category of rocks is a group we call *metamorphic*. These rocks have complex histories and have been altered, sometimes substantially, by heat, pressure, or chemical solutions in various combinations. Alteration may be very slight or very extensive, but it is limited always by melting. Once a rock melts, magma is formed and when it cools, we have an igneous rock. Unfortunately, this distinction is not always easy to make when studying rocks.

Metamorphism can affect rocks of any type. Sedimentary, igneous, and even metamorphic rocks may be metamorphosed. Some rocks are quite susceptible to metamorphism and quickly develop new characteristics; sedimentary shales and mudstones are two such rocks. Conversely, some rocks like sandstone and some volcanic rocks resist metamorphism, and any changes produced are likely to be quite subtle. In metamorphism of sedimentary shale, even moderate amounts of heat and pressure will recrystallize clays to fine-grained micas and alter a shale to a slate. Additional metamorphism can alter a slate to a mica schist in which crystals of new minerals become evident. Further metamorphism may convert a mica schist to an even coarser-grained rock with prominent bands of quartz and dark minerals called gneiss.

The alert reader will have noted that metamorphism, like igneous activity, involves the formation of new crystalline minerals, but the crucial difference is that crystallization in a metamorphic setting takes place in a solid state, whereas it occurs in a fluid or melt (i.e., magma) in igneous rocks.

Many metamorphic rocks have almost the same chemical composition as their unmetamorphosed predecessors. The atoms and molecules are simply rearranged in response to the heat or pressure. For example, coral limestone, a porous sedimentary rock composed of limy remains of organisms, may be metamorphosed to a dense, coarsely-crystalline marble in which all trace of the original organisms has vanished, but in which the basic chemical composition is unchanged. Both rocks are composed entirely of calcium carbonate.

Before we leave metamorphism, we should know about certain minerals that tell us the temperature and pressure at which they formed, and hence reveal much about the metamorphic environment. One such mineral is glaucophane, common in parts of the California Coast Ranges. On the basis of laboratory experiments this mineral is known to be a product of metamorphism in which the pressure was high—implying deep burial, but in which the temperature was relatively low, suggesting perhaps that despite the deep burial, the rocks were cooler than might have been expected at that depth. Another metamorphic mineral, sillimanite, forms only under conditions of high pressure and high temperature. We know that glaucophane requires pressures equivalent to depths of 8 to 10 kilometers (5–6 miles) in the crust, but temperatures more like those usually found only 2 to 3 kilometers (1–2 miles) deep. Glaucophane schists found in the Franciscan formation of the California Coast Ranges suggest that relatively cool materials, probably from the sea floor, have been rapidly carried to great depths in what are called *subduction zones*. These materials, however, were soon expelled before they had time to reach equilibrium with the higher temperatures generally found at these great depths. From such evidence, we have put together the rather astonishing history of our Coast Ranges in which some of the rocks had brief sojourns at depths of 8 to 10 kilometers (5–6 miles) before being returned to near surface conditions.

Sedimentary Rocks

Sedimentary rocks cover about three-quarters of the earth's land surface, and they are the rocks we are likely to see in most parts of California. They are a varied lot, which were deposited by water, wind, and even ice. Most of them are composed of grains derived from all types of rock by erosion and weathering, and moved and deposited by the wind, running or standing water, and ice.

The majority of our *sedimentary rocks* were formed from unconsolidated *sediments* such as mud, sand, and gravel that were transported to the sea by streams (Figure 1-7). Because sedimentary rocks form at the earth's surface at moderately low temperature, they often include remains or organisms (*fossils*) that reveal a great deal about the environment at the time of deposition. If our sedimentary rock is full of clam shells, we know it was deposited in shallow, near-shore, probably marine waters. If it contains abundant plant remains, leaf and stem impressions, and land snail shells, we would deduce that it was formed in a freshwater swamp.

Because life generally has evolved from the simple to the more complex and has

FIGURE 1-7 Tilted and folded late Cretaceous marine sedimentary rocks, Vaca Mountains at Lake Berryessa, Napa County. (Photo by Burt Amundson)

varied over geologic time, fossils have been used to set up a relative time scale (see Geologic Time Scale, inside front cover). If a rock contains reptilian remains, we know that it must be Pennsylvanian or younger. If it contains mammalian remains, it will not be older than Triassic. If fish are present, it is not older than Ordovician, and so on. Numerous details have been added as our understanding improves, and we now know of many *index fossils* whose stay on earth is sufficiently well-documented that their presence indicates the rocks are Cretaceous in age, or Miocene, or Silurian, or some other age.

Sedimentary rocks, because of their fossil remains and because their agent of transportation often can be recognized, tell us more about conditions on the earth's surface at the time they were formed than any other rocks. If we find layers of rock salt or gypsum in a sequence of sedimentary rocks, we can be reasonably sure that very arid conditions prevailed at the time of deposition. If, conversely, we find peat or coal beds containing the remains of ferns and succulent, broad-leafed plants, we would suspect very moist conditions.

Some sedimentary rocks are composed of former glacial deposits and, of course, tell us that the climate was cold when they formed. If we find fossil reef-forming corals, we can postulate tropical marine conditions.

Classifications that divide rocks into igneous, metamorphic, and sedimentary types or igneous rocks into volcanic or plutonic varieties are synthetic conveniences designed to facilitate communication and understanding. But like all such human inventions, they are imperfect. Nature cares nothing for our classifications and boundaries. The separation of rock types usually proves to be gradational rather than sharp, and we sometimes find rocks that are neither clearly metamorphic nor clearly igneous, but display features characteristic of both. Other rocks lie on the boundary between sedimentary and igneous and we are hard-pressed to assign them to one of our categories. We must remember that sharp boundaries exist mostly in our minds and that nature's boundaries are usually fuzzy and gradational.

THE DYNAMIC CRUST: CONSTRUCTIVE AND DESTRUCTIVE PROCESSES

As soon as people began to find rational explanations for what they saw in the rocks of hillside and shore, they began to realize that the land was changing and impermanent. But because the changes were so slow or of such apparent day-to-day insignificance, understanding of the dynamic nature of our planet came slowly. Nevertheless, discovery of marine shells in rocks high in the mountains showed either that the land had been raised out of the sea or that the sea had receded at some time in the prehistoric past. The early visitors to rivers such as the Grand Canyon of the Colorado saw the sediment-laden reddish stream flowing in the canyon bottom, but it was not until the nineteenth century that geologists learned that rivers such as the Colorado were slowly carving their own valleys and transporting the resulting debris to the sea (Figure 1-8).

A number of dramatic geologic processes have long been observed, such as landslides, dust storms, and volcanic eruptions, but geologists were slow to appreciate the importance of even these impressive events in shaping the earth's surface, because the events seemingly occurred infrequently and affected only widely separated areas.

FIGURE 1-8 Colorado River just upstream of Parker Dam at the mouth of the Bill Williams River. View from San Bernardino County side. (Photo by Mary Hill)

As we have come to appreciate the vast stretch of geologic time, we have begun to understand more clearly the cumulative importance of a multiplicity of small events as well as the less frequent occurrences of large-scale events.

Weathering

Among the important but generally inconspicuous processes at work on the surface of the earth is *weathering*. We are all familiar with some forms of weathering, such as the rusting of steel tools left outdoors in the rain, but rock weathering is less evident and less noticeable, even though it is enormously important in understanding geologic change.

If you have ever walked through an old cemetery looking at the dates on the tombstones, chances are you have seen some good examples of weathering. In such places, soft, porous rocks like sandstone may have crumbled to such a degree that the inscriptions are barely visible, and easily soluble rocks may be deeply etched.

In areas where the temperature regularly swings back and forth across the freezing point of water, porous rocks crumble and fall apart as the water, which has soaked into the rock, freezes and wedges the grains apart. Frost-wedged angular blocks of fresh rock often litter the surface of high mountains. Water has entered tiny cracks in such rocks and freezing has widened and lengthened these fractures, ultimately shattering the rock. The formation of ice, a very common natural phenomenon, is one important agent of weathering.

All weathering does not depend upon the freezing of water, however. We have already mentioned rusting of steel tools as a form of weathering. Rusting is caused by the combined attack of moisture and oxygen in the air on the iron. Atmospheric gases and moisture also attack rocks, or more properly the minerals in the rocks (Figure 1-9).

Because minerals have a wide range of chemical composition, they also differ greatly in their response to the gases and moisture in the atmosphere. The common mineral quartz is highly resistant to weathering and is also hard and tough so it tends

FIGURE 1-9 Granitic boulder weathering by exfoliation, Indio Hills, Riverside County. (Photo by R. M. Norris)

to outlast the other minerals with which it occurs in rocks. The end result is that most stream and beach sands are dominately quartz. The feldspar group of minerals is even more abundant than quartz in igneous rocks, but they are much more susceptible to weathering and are less resistant in a physical sense, so are more readily altered than quartz and tend to be less abundant than quartz in most sands.

As a general rule, minerals are most stable in the environment in which they first formed, and less stable in different settings. The green mineral olivine is found in rocks derived from deep within the earth's crust. Exposed to the atmosphere, olivine is quite unstable and soon alters to other minerals more stable at the surface. Pyrite, or fool's gold, forms in many ore bodies at depths where oxygen is in short supply. Exposed to surface conditions, pyrite breaks down into iron oxides (rust) and sulfuric acid.

Some minerals are soluble in rainwater and dissolve fairly quickly when exposed to the atmosphere. The mineral calcite, which makes up most limestones and marbles, is such a mineral. Rainwater charged with carbon dioxide from the air and from decayed material in the soil produces a weak acid (carbonic acid), which is an effective agent for dissolving limestone. All natural rain is slightly acid because of carbon dioxide, but in the case of acid rain as a pollutant, the acidity has been increased above natural levels by such substances as sulfur dioxide from industrial flue gases. These gases react with atmospheric moisture to form sulfuric acid, which is far more corrosive than carbonic acid. Limestone buildings and monuments in cities with polluted air quickly succumb to weathering. Limestone monuments that endured almost unchanged for thousands of years in the clean, dry air of the Egyptian desert were rapidly corroded when taken to cities such as New York and London.

Weathering then, may be considered the sum total of natural processes that act to break down or alter rock and mineral materials at or near the surface of the earth. As soon as these materials are moved, we enter the realm of erosion; weathering does not involve any transportation of earth materials.

It is hard to overestimate the importance of weathering. It goes on everywhere on the land surface; it is enormously important in producing our soils, limestone caverns, clays for ceramics, and even our chief aluminum ore deposits to cite just a few examples.

Apart from volcanism, most of the agents of change we have mentioned are destructive and act to wear away the land surface and reduce it to a low plain.

Structural Processes

Those who live in California and similar mountainous areas should realize that there is more to the geologic story than destruction. A few measurements and calculations would show that the present mountains, even the highest ones, would be reduced to low plains, given the persistence of present erosional rates, in just a few million years. We still have mountains—impressive ones, too—so something else is going on, canceling out the effects of erosion and weathering.

A moment's reflection reveals that these other processes, soon to be examined, have in fact exceeded the effects of destruction, otherwise we would have no mountains at all. There is much that we do not understand about the ultimate causes of mountain building, but we are making progress.

Most Californians have not had a chance to see mountains growing in any meaningful way, but a fortunate few, possibly frightened out of their wits, have seen a mountain range uplifted with respect to the adjacent valley. In 1872, during the great Owens Valley earthquake, some Lone Pine residents saw that the Sierran block had been raised with respect to the Owens Valley nearly 5 meters (15 feet) just north of the town. This sort of ground rupture results from movement along a *fault*. During the 1971 San Fernando earthquake, visitors to Lopez Canyon could see a place where the San Gabriel Range was raised with respect to the adjacent valley by about a meter (3 feet) (see Figure 10-9). These kinds of events seldom occur in human terms, but geologic study of many of our mountain ranges shows quite clearly that repeated movements along the boundary faults of mountains have occurred frequently enough to raise the mountains much faster than erosion can tear them down (Figure 1-10).

We should not conclude that all faulting involves raising or depressing one block with respect to the adjacent one. The great San Andreas fault system is an example of another type called *strike-slip* faults. Fractures of this sort are nearly vertical, and blocks on either side move horizontally with little or no up-and-down motion. Land-

FIGURE 1-10 Deep Spring Valley, a typical closed Great Basin valley. Note especially the fault-controlled linear base of the Saline Range. The position of the playa lake next to the range front shows that the valley floor has been tilted downward toward the scarp during faulting. (Photo by R. M. Norris)

scapes cut by such faults are characterized by stream valleys with kinks in them or by stream valleys in which the headwater portions have been completely detached from the lower reaches. The persistence of such movement over 5 million years or so has separated formerly adjacent places by several hundred kilometers (Figure 1-11). Still other faults are nearly flat-lying surfaces where one block is shoved nearly horizontally over the other, sometimes for considerable distances. These are called *thrust* faults. Some thrusts are nearly flat-lying, but others are much steeper and when inclined more than about 30 degrees are called *reverse* faults, in which compression of the crust pushed one block upward and somewhat over the adjacent block. Reverse faults are common in the Transverse Ranges, and some of the faults that mark the western and southern sides of the San Gabriel Range near Los Angeles are of this type. Many of the major rock masses in the Klamath Mountains and the western Sierra Nevada are separated by reverse faults. Other fine examples are found in the southern San Joaquin Valley where it meets the Transverse Ranges. Conversely, inclined faults in areas undergoing tension or stretching result in one block sliding down the face of the adjacent block, producing what is called a *normal fault*. Many of the range-front faults in the Basin and Range province are of this sort, and the prominent fault at the east base of the Sierra Nevada is primarily a normal fault, though it does have appreciable strike slip (Figure 1-12).

Geologists have recently learned that some normal faults, steep at the surface, curve and flatten at depth to become *listric normal* faults. Where a number of these listric normal faults merge at depth, they may form a *detachment fault* or *detachment*

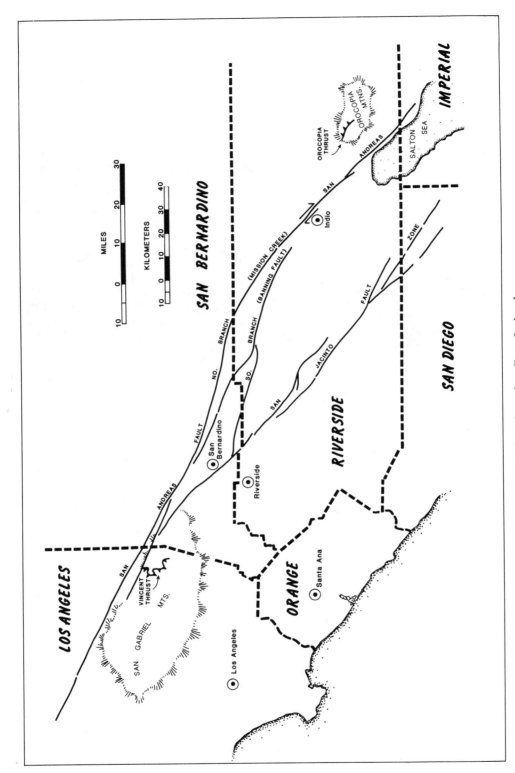

FIGURE 1-11 Right slip displacement on the San Andreas fault is shown by offset of related rock sequences and structural features in the Orocopia Mountains northwest of the Salton Sea and the San Gabriel Mountains in Los Angeles County, a distance of about 200 kilometers.

FIGURE 1-12 Eastern face of the Sierra Nevada south of Big Pine, Inyo County. The scarp is steep and linear, but much eroded. Although the Sierra Nevada frontal fault is dominantly a normal fault, it has a significant strike-slip component. Owens River is in the middle distance. The high peak is North Palisade (4,344 meters or 14,242 feet). The floor of the valley is about 1,200 meters or 4,000 feet, and the view is from the Inyo Range at an elevation of about 2,400 meters or 8,000 feet. (Photo by R. M. Norris)

surface. The term *low-angle normal* fault is also used for such faults whose dip is less than about 30 degrees or so.

Finally, it should be remembered that types of faults, like types of rocks, can and do grade into one another, and there are many examples of faults that show both horizontal and vertical slip.

Great compressive forces sometimes develop in parts of the earth's crust. Blocks of rock the size of mountain ranges are often squeezed into wrinkles or welts by these forces, and many of our mountains show such a history; their rocks, originally deposited as flat or nearly flat sheets on the sea floor, are bent into great arches or so crumpled that some layers stand on end or are even overturned. Such folding has been prominent in the development of most California mountains. For example, archlike folds or *anticlines* are common in the Coast and Transverse ranges. Examples are illustrated in Figures 1-7 and 10-5. Downward bends or *synclines* are less often exposed but underlie many of our valleys such as the Salinas, San Joaquin, and Eel River. A small, well-exposed example in the Mojave Desert is illustrated in Figure 7-5.

Although we can find some examples of mountains or hills mostly caused by faulting, even the simplest fold features like Signal Hill near Long Beach (see Figure 10-14) upon close examination turn out to be affected by faulting. What appear to be examples of pure fault-blocks in the Modoc Plateau include some folding (see

Figure 5-17). We are reminded once again that nature prefers gradation, whereas humans seem to like clear-cut categories best. As we have come to a better understanding of our mountains, we find that most are complex mixtures of faulting, folding, and intrusion with one process dominant at a given time and the others at different times. Folding and faulting often occur simultaneously.

Occasionally, large areas of land are elevated to form high plateaus with only minor folding and faulting. To some degree this is characteristic of the high, flat Modoc Plateau of northeastern California, but the best American examples lie to the east where Utah, Colorado, Arizona, and New Mexico meet.

In light of the previous discussion, the steep mountains and deep valleys of California should make even the most skeptical person suspect that the part of the earth's crust called California is in the throes of rapid geological change. A few examples may serve to dispel any remaining doubts.

Perhaps the most compelling evidence for the assertion that California is undergoing rapid geological change is provided by earthquakes, which forcibly direct the attention of even the most blasé to matters geological.

Even though much remains to be discovered about earthquakes, most would now agree that they are but one manifestation of geologic activity and are much more frequent in young mountainous areas like California than in the apparently more stable continental interior. For example, California and western Nevada together account for approximately 80 percent of all earthquakes in the 48 contiguous states.

We will be examining the underlying causes of active mountain belts later in this chapter, but at this point we need to consider earthquakes and their direct causes in more detail.

Conventional geological wisdom holds that earthquakes result when abrupt movement occurs along a fault. To better appreciate what happens we need to understand the theory of *elastic rebound*, originally proposed by Harry Fielding Reid following his analysis of the 1906 San Francisco earthquake. This theory holds that between earthquakes on a fault such as the San Andreas, stress gradually builds because of major crustal changes. During the period between earthquakes on the San Andreas, the forces that caused the western side to move northward as much as 6 meters (20 feet) during the 1906 quake, continue to act by bending and perhaps compressing the rocks on either side of the fault, but do not reach levels sufficient to overcome friction and cause the two blocks to shift. This slow application of energy is being stored in the rocks, just as energy is stored in a stretched rubber band or compressed spring. Eventually, the amount of elastic energy put into these rocks by as yet unspecified forces will exceed the frictional resistance, which keeps the two sides of the San Andreas in place, and one block will move suddenly by the other wherever stress is most effective in overcoming the frictional resistance. This response is the rebound part of the elastic rebound theory (Figure 1-13).

Not all faults behave the same, and even parts of the same fault show differences in behavior. The parts of the San Andreas where the trace of the fault changes direction seem to be characterized more by occasional large earthquakes, whereas some of the straighter portions, presumably where friction cannot reach such high levels, are prone to more frequent but smaller earthquakes as well as some *creep*, where the two sides slip slowly and more or less continually without producing any severe earthquakes (Figure 1-14).

Some earthquakes, of course, are produced in ways that do not seem to involve

A. ORIGINAL BLOCK

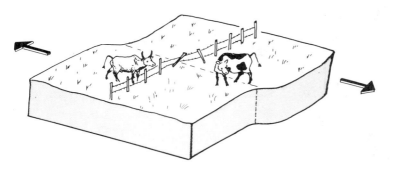

B. ELASTIC DEFORMATION (No slip on fault)
Fence and ground deformed

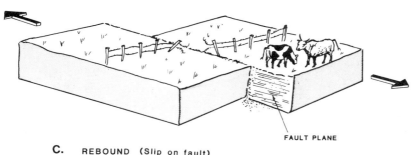

C. REBOUND (Slip on fault)
Fence and ground broken.

FIGURE 1-13 Development of elastic rebound in crustal blacks. The first block is undeformed, the second shows deformation without slip, and the third block shows that slip has occurred along the fracture and rebound of the distorted blocks to their original form.

faulting directly if at all. In recent years, swarms of earthquakes, some fairly strong, have shaken the Mammoth Lakes area in the central Sierra Nevada. Because this area is the site of much young volcanic activity, and because the quake activity was distinctive and very much like that occurring in other volcanic areas like Hawaii, these quakes were attributed to the movement of magma bodies at shallow depth, a disquieting notion to local residents who were concerned about the possibility of a new volcanic event in their backyard.

FIGURE 1-14 Distortion of an originally straight stone wall as a result of slow creep on the Calaveras fault, a member of the San Andreas system, Hollister, San Benito County. (Photo by J. C. Clark)

Geologists have learned also that changes in fluid pressure in rocks may activate faults, causing earthquakes. Though not proved, it has been suggested that the disastrous failure of the Baldwin Hills reservoir in Los Angeles in 1963 was caused in part by earthquakes and creep resulting from fluid pressure changes in the Inglewood oil field nearby. Similarly, a swarm of small earthquakes in the Santa Barbara Channel in 1968 and 1969 was believed by some geologists to be related to the removal of oil from some of the offshore fields and the reactivation of associated faults.

Erosional Processes

Siltation in reservoirs and the transported load in streams show that the California Coast Ranges from Santa Barbara north to the Oregon boundary are being eroded exceedingly rapidly. Some areas have been found to yield as much as 20,000 metric tons of sediment for every square kilometer of drainage basin per year. But the mountains are still present and are rugged with many steep canyons, so we know the rates of uplift have been more than a match for the rate of erosional destruction. Furthermore, we can find much evidence that this uplift is still in progress. Wave-cut benches, some formed about 40,000 years ago, from Santa Cruz south to the Mexican boundary, have been raised tens of meters in many places, and rates of uplift of a meter (3 feet) in 1,000 years are not unusual. At the mouth of the Eel

River in northern California, dated stream terraces document a rate of uplift of about 4 meters (13 feet) per thousand years at Scotia Bluffs (Figure 1-15).

Our understanding of the causes of mountain building including folding and faulting, is imperfect, but we do know some factors involved. One important factor is *isostasy*. The theory of isostasy assumes that the crust is made up of large blocks, which are supported by and float, so to speak, in a denser subcrustal material that can yield plastically. If two adjacent blocks of crust are of unequal density, the denser one will "float" lower than the lighter block. Continents, being made of rocks that, on average, are less dense than the ocean floor rocks, tend to stand at a higher level. Further, if one of the two adjacent blocks is thicker than the other, the thicker one will stand higher on the surface and project deeper into the plastic substratum as well. Geophysical measurements show that many mountain ranges like the Sierra Nevada stand above their surroundings because they have deep roots projecting far down into the subcrustal material.

Erosion tends to perpetuate these differences because it removes material from the higher blocks and adds it to the lower blocks. The higher-standing blocks are thereby lightened and tend to move upward in compensation, and the blocks receiving the materials are made heavier and sink. Such adjustment is never perfectly equal and the process does not continue indefinitely, but isostasy does help to explain why mountains persist much longer than might be expected from erosional rates alone.

Among the erosional processes that shape our landscape, erosion by running water

FIGURE 1-15 Tilted Pliocene marine sedimentary strata of the Wildcat Group, Scotia Bluff. Eel River is in the foreground. Near Fortuna, Humboldt County. (Photo by Don Norris)

is by far the most important. Some people may be surprised to learn that this is true even in the driest deserts where rain seldom falls, and with few exceptions the only permanent streams are those that arise in less arid areas and cross the desert. Although these very dry deserts are experiencing little stream erosion under present conditions, their landscapes are also the product of past events when more rain fell and streams were active. At the present moment in geologic time, wind may have the upper hand in the very driest deserts. California has no deserts as dry as the Atacama of northern Chile and adjacent Peru or parts of the Sahara and Arabian deserts, but the annual rainfall is very low, about 50 millimeters (1.9 inches) in Death Valley and around the Salton Sea. Despite the low rainfall in these California deserts, stream erosion remains the most active of the destructional processes.

Stream erosion, although infrequent in some California deserts, is particularly effective because it often occurs as intense summer thunderstorms, which are likely to drop sudden large amounts of water on barren and rocky terrain with minimal plant cover and thin soil. Without a continuous mat of vegetation to break the force of the rain, and thin and stony upland soils incapable of holding much water, a high proportion of the rain rushes down narrow canyons, sweeping mud, sand, and rocks out onto the valley floors below. The resulting rapid erosion of the uplands is matched by equally rapid and dramatic deposition in the valleys. Desert towns and roads can suddenly be inundated by slurries of mud and gravel with destructive consequences. In the late 1970s and early 1980s, Coachella Valley, the eastern Mojave and Death Valley have been struck repeatedly by such events.

The landscape of the much better-watered coastal mountains and the Sierra Nevada also is dominated by stream-produced features, and rates of erosion may be very high. From San Francisco southward, most streams are distinctly seasonal, many flowing only in winter and early spring. Summer and fall brush fires regularly remove large areas of vegetative cover, at least temporarily, exposing the soil to erosion and landsliding during the winter rains. In common with desert valleys, coastal lowlands can expect occasional destructive mud flows and flooding. North of San Francisco, rainfall is greater and forest cover is more continuous and less fire-prone, and numerous permanent streams grace the area.

But the Northern Coast Ranges, despite their thicker forests and more continuous cover of vegetation, are eroding rapidly. This results not only from the abundance of water, but because the northern Coast Ranges have numerous exposures of serpentine (mineral) or serpentinite (rock). Incidentally, serpentine is California's state rock, though it should be called serpentinite. These serpentinites are broken by myriad fractures and the material itself is slippery, so when hillsides underlain by this material get wet, landslides occur on a grand scale. Some geologists have claimed that in the northern Coast Ranges more hillside erosion is due to landsliding than to direct erosion by running water. The streams, of course, act like conveyer belts in carrying the slide debris downstream to the valleys and to the sea.

The form of stream valleys reveals much about the erosional history that shaped them. Erosion by running water and landsliding go hand in hand to shape most stream valleys and typically result in V-shaped cross-sections. The width of the V-shape varies from narrow, almost slotlike canyons to those that have broad, flaring form. Why the difference?

If you think about it, you can see that a narrow gorge indicates that the deepening processes are outrunning the widening processes (Figure 1-16). If the valley walls

FIGURE 1-16 Narrow, slotlike canyon cut in the soft, coarse-grained nonmarine Mecca Formation, Mecca Hills, Riverside County. (Photo by R. M. Norris)

are gently sloping and wide apart, the downcutting must be somewhat less dominant and the widening processes of greater importance. It is probably obvious that the stream that occupies the bottom of the V is the agent responsible for the downcutting, but it is probably less evident what processes widen valleys. We have mentioned one, landsliding; another is the activity of small tributary streams and gullies that join the main stream in the valley floor.

What might tip the scales in favor of the main stream and what might discourage downcutting by the stream? There are actually a number of factors, but we will discuss only a few of them. If we can contrive to steepen our stream, perhaps by elevating its headwaters or by lowering the area into which it flows, it will speed up and put more energy into downcutting. Some of California's mountains are quite young geologically and are still being raised with reference to sea level or the adjacent valleys. Streams coursing down rising mountains are continually steepened and as a result devote more of their energy to downcutting their channels to produce narrow, steep-sided gorges. Extreme examples are found in some of the mountains rimming Death Valley, where the valley floor has recently subsided with reference to the mountains.

FIGURE 1-17 Batequitos Lagoon, San Diego County. This estuary is the lower valley of San Marcos Creek, probably mainly the result of the post-Pleistocene sea level rise. (Photo courtesy Gerald Kuhn)

Conversely, some coastal stream valleys have had to adjust to gentler gradients because of the postglacial rise in sea level that occurred about 10,000 or 12,000 years ago. Sea level rose so rapidly it flooded the lower ends of these streams, forming *estuaries*. The rivers promptly responded; they could no longer cut downward—sea level interfered—so they began to fill in the estuaries, forming wide, flat-floored valleys. Some excellent examples can be seen along the San Diego County coast from San Onofre south to Del Mar (Figure 1-17).

We might be excused if we concluded from this that a steepening of gradient was the sole reason for slotlike canyons, and that widening of valleys results because slopes are flattened. It is not always so simple, and we must look for other independent evidence of uplift before deciding that a narrow canyon proves that uplift occurred.

Narrow and steep-sided canyons may develop largely independently of uplift and gradient change if the cover of vegetation and soil is reduced, either by a shift in climate or because of human practices such as overgrazing, burning, clear-cutting a forest, or even cultivation. Smooth, rounded grass-covered hills, particularly those underlain by soft, clay-rich rocks or soils, can quickly be converted to useless intricately gullied areas by removing or destroying the protective sod. When such areas lose most of their vegetation and original surface, they become dominated by closely spaced narrow gullies called *badlands*. Natural badlands occur in deserts. Zabriskie Point, near Furnace Creek in Death Valley, Red Rock Canyon near Mojave, and the Indio and Mecca hills in the Colorado Desert are good examples (Figure 1-18).

Manmade examples occur at a number of places in the southern Coast and Trans-

FIGURE 1-18 Intricate badland erosion produced in weakly consolidated nonmarine sedimentary rocks under arid conditions. Mecca Hills, Riverside County. (Photo by J. C. Crowell)

verse ranges where overgrazing or other ill-advised farming practices have destroyed the protective vegetation. Some well-known examples occur in the hills surrounding the Oxnard Plain in Ventura County. Off-road vehicles have created badlands where they have stripped away soil and vegetative cover near Gorman in Los Angeles County and at Ballinger Canyon in northeastern Santa Barbara County.

Stream Deposits

It is a truism that what a stream erodes in the uplands, it must deposit in the lowlands or in the sea. But it would be rare indeed for a stream to remove every loose particle from its entire drainage basin and transport it all to the appropriate lowland destination. Even after great floods, much loose material remains stored on the hillsides and in the stream channel, waiting for the next flood to move some of it farther along the way to the sea or valley below. Many stream deposits can be seen in stream valleys in the mountains, not just in the valleys below.

FIGURE 1-19 Large boulders moved out of Pickens Canyon, western San Gabriel Range during the 1934 New Year's Day flood. Montrose, Los Angeles County. (Photo courtesy U.S. Geological Survey)

Some people in the Los Angeles area remember the famous New Year's Day flood of 1934, which struck with particular severity at Montrose near Glendale. The events of that day showed graphic evidence of the size and volume of material that had been stored in Pickens Canyon, a short, steep gorge draining the western face of the rugged San Gabriel Mountains. In some places, the flood removed as much as 5 meters (16 feet) of loose material from the floor of Pickens Creek, exposing the bedrock channel. A huge volume of mud and coarse gravel was swept out into the town of Montrose, including some rocks as large as 3 meters (10 feet) across, causing severe damage to streets and buildings (Figure 1-19).

Sudden, large-scale stream deposits such as these attract our attention and may be very costly, but the enormous thickness of finer-grained stream deposits that floor all of the valleys and the continental shelf off the coastline are of far greater importance.

Alluvium is the term used for sediment deposited by streams. Alluvium and the soils that develop on it sustain California's number one industry, agriculture. All the best agricultural soils are found in our valleys and represent composites of the best mountain soils removed by erosion and transported by streams to the valleys.

Alluvium in broad, flat valleys like the Sacramento, San Joaquin, or San Fernando

FIGURE 1-20 Coalescing alluvial fans forming a bajada at the base of the Kingston Range, Inyo County. (Photo by R. M. Norris)

is likely to be fine-grained clay or silt or perhaps sand. Near the base of the mountains, particularly in the drier southern and eastern parts of the state, piles of coarse, gravelly stream deposits can be seen at the canyon mouths where the streams have room to spread out and dump their loads. Over time, a broad, generally low cone of gravel spreads out, fanlike from the mouth of each canyon, forming features called *alluvial fans*. Along the fronts of many desert ranges each canyon has its own fan and in time these fans coalesce to form a *bajada*, a continuous apron of sediment along the base of the mountains (Figure 1-20). Bajadas are characteristic features of the arid and semiarid valleys of the state. The cities along the steep south face of the San Gabriel and San Bernardino mountains from Pasadena east to San Bernardino are all built on coarse, alluvial fans.

In the drier parts of California, the Basin and Range Province, the Mojave and Colorado deserts, and even the southern San Joaquin Valley, rainfall is low, evaporation is high, and streams do not deliver enough water to fill valleys to overflowing. These valleys therefore contain permanent saline lakes if water is present year-round, or *dry lakes* or *playas* if stream flow is unreliable and irregular. Some playas have hard clay surfaces, whereas others are flooded with puffy saline clays, and some are shimmering white salt flats. Each of these is subject to occasional filling as a result of floods, but most remain dry. All are stream deposits, and some like Searles Lake in San Bernardino County are important commercial sources of clay or saline minerals.

Glaciers and Glaciation

In California one must go up into the high mountains to see the effects of glacial erosion, but even there, glaciers are active today in only a few places. Several small glaciers persist in favorable spots in the Sierra Nevada and on the upper slopes of Mount Shasta, but during the Ice Ages of the Pleistocene epoch, which began about 1.8 million years ago and ended only about 10,000 or 12,000 years ago, large glaciers covered the high Sierra, Klamath, and Cascade mountains. As you will discover in later chapters, there wasn't just one episode of glaciation during the Pleistocene, but several, so glacial erosion has affected our California landscape repeatedly.

Among the various destructive processes that shape our landscape, none produces more dramatic and spectacular results than glaciation. This is true not only in California, but in the Alps, the fjords of Norway, and the Canadian Rockies to mention just a few examples.

Glaciers have ground and polished hard rocks in the Sierra, leaving behind towering Matterhorn-like peaks (*horns*), shining rounded rocky domes, thousands of rock-floored basins now occupied by sparkling lakes (*tarns*), and hugh glacial stairways over which modern waterfalls plunge (Figure 1-21).

The most dramatic of the glacially produced landscapes in California is the Yosemite Valley, but hundreds of lesser examples occur on both the eastern and western

FIGURE 1-21 Glacial stairway at Yosemite Valley. Vernal Falls (lower) and Nevada Falls (upper) are on the Merced River. Liberty cap (2,158 meters or 7,076 feet) to the left of Nevada Falls. View from Glacier Point. Mariposa County. (Photo by R. M. Norris)

slopes of the Sierra Nevada. Glaciated valleys like the Yosemite have developed because ice streams fed by ice fields or ice caps near the mountain crest have repeatedly followed former stream valleys toward lower elevations. During this movement, influenced by fracture and joint patterns the ice has greatly deepened these former stream valleys, producing dramatic U-shaped, steep-walled, glaciated valleys. In the higher parts of the mountains, these U-shaped valleys often have a stairlike long profile; the eroding glaciers cut deeper at some places than at others, perhaps because of inequalities in rock resistance or because cutting was enhanced below junctions of two ice streams where valleys joined.

In the lower parts of these glacially produced valleys, damlike deposits (*moraines*) have often been left where the glacier terminated. When the ice melted, lakes occupied some valleys for a time, until they were drained by downcutting of their outlets or were filled by sediment from above. As you will learn later, these occurrences explain the flat floor of Yosemite Valley near the village.

Like all erosional agents, glacial ice must deposit what it erodes. The form of glacial deposits differs in some respects, depending upon whether the glaciers were parts of large continental ice sheets or confined to mountainous terrains. In California, all glaciers of Pleistocene age were mountain glaciers and hence we need not be concerned with the types of glacial deposits particularly characteristic of continental glaciation.

Erosional features of grand dimensions such as Yosemite Valley are more noticeable than moraines, ridges of glacial material called *till*, left along the margins of the ice or at its terminus. We recognize a number of kinds of moraines. At the lower end of a glacier—its terminus—the melting ice releases any rock, gravel, or other debris incorporated in the ice. If the melting end remains relatively stationary for a time, a ridge of glacial till will accumulate. Should the glacier again advance, some of the moraine may be bulldozed ahead of the advancing ice, but most will be smeared out beneath the glacier, and when the ice advance stops once again, a new moraine will be built. The moraine that marks the greatest advance is called the *terminal moraine*.

If the glacier melts back more rapidly than new ice can be delivered, the glacier is said to be retreating, although it never reverses its down-valley direction of flow. Following a period of glacial retreat, short-term advances may occur and each of these is likely to form a smaller moraine upstream from the terminal moraine. These upstream moraines are called *recessional moraines*, and they tell us that minor readvances of the ice occurred during a period of major retreat. Fallen Leaf Lake near the southwestern corner of Lake Tahoe is nestled in a fine set of recessional moraines (Figure 1-22).

Glaciers emerging from mountain valleys tend to spread out somewhat as they leave the confines of the mountain canyon, although they still form prominent tongues of ice extending sometimes a kilometer or so beyond the mountain front. Not only do such glaciers produce terminal moraines, but the melting sides of the ice tongues form ridges of till called *lateral moraines*, which are just the sides of continuous loops and merge with the terminal moraine. Because the edges of these glacial tongues are likely to have accumulated large loads of rock and gravel from the walls of the canyons down which they flowed, lateral moraines may be of impressive size. Some on the eastern side of the Sierra are several hundred meters high, indicating

FIGURE 1-22 Multiple morainal loops around Fallen Leaf Lake southwest of Lake Tahoe, El Dorado County. Moraines are chiefly those of the Tioga and Tahoe glacial stages. Ice advanced northward toward top of photo. (Photo courtesy U.S. Department of Agriculture)

that the ice tongues that formed them were at least that thick and carried an immense volume of rock debris. Striking examples can be seen at the mouth of Pine Creek west of Bishop, at McGee Creek, and below Convict Lake in southern Mono County (Figure 1-23).

Erratic boulders also are very common indicators of the former presence of a glacier and are strewn by the thousands on polished glaciated bedrock surfaces in the higher parts of the Sierra. In most instances, these erratics are composed of the same type of rock as that on which they rest, but many show some degree of rounding and an occasional one is ground, polished, or even scratched (*striated*) as it was dragged over hard rocks by the enclosing ice. These erratics may have been carried 10 or 20 kilometers (6–12 miles) from sources higher in the mountains, but as long as erratics are not distinctive rocks, we can only guess how far the ice carried them. Occasionally, however, we are lucky, and find the erratic is composed of a distinctive rock quite unlike the rock on which it rests, but whose source can be pinpointed. In such a fortunate event, we can tell how far the transporting ice moved, and in what direction (Figure 1-24). We can, of course, use this same method whenever we discover a distinctive boulder in a moraine.

FIGURE 1-23 Lateral moraine about 200 meters (640 feet) high at McGee Creek, Mono County. Mt. Baldwin (3,847 meters or 12,614 feet) is on the skyline. (Photo by R. M. Norris)

CONTINENTAL DRIFT AND PLATE TECTONICS

In recent years we have learned that the earth's crust is divided into a number of continent-sized blocks we call *plates*. It has been found that the oceanic plates form along great sutures such as the Mid-Atlantic Ridge and East Pacific Rise, features we now call *spreading centers*. Crustal material wells up along these spreading centers

FIGURE 1-24 Schist erratic boulder transported by a glacier and left stranded on an exposure of granodiorite. Near Mono Hot Springs, Fresno County. (Photo by R. M. Norris)

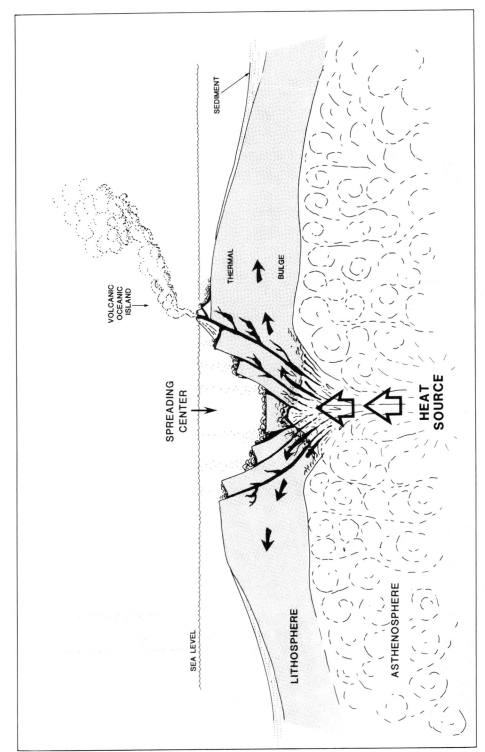

FIGURE 1-25 Diagrammatic view of a sea-floor spreading center or divergent plate boundary. Spreading centers are commonly offset along transform faults and characterized by crestal valleys. The sea floor typically decreases in elevation away from the central rift as the cooling rocks shrink and subside.

and is added to the edges of the plates, which move away from one another on opposite sides of the spreading center. At present, the spreading of the oceanic plates in the Atlantic is widening that ocean by carrying the continents farther apart (Figure 1-25).

A moment's consideration will show that on an earth of finite size, this same widening cannot also be going on in the other oceans unless some compensating shortening is occurring somewhere. As we have come to understand this process of *sea-floor spreading* and the associated process of continental drift, we have discovered several other kinds of plate boundaries in addition to the spreading centers, and we have also found that these boundaries may change from time to time.

Before we consider the various types of plate boundaries it is worthwhile to consider the supporting evidence for continental drift and sea-floor spreading. As improved maps of the Atlantic Ocean became available in the seventeenth century, a number of sharp-eyed observers noted that the east and west shores of this ocean had a remarkably similar orientation as if the continents on either side had once been joined. But it wasn't until the twentieth century that strong evidence was assembled to show that the Americas had indeed been joined with Europe and Africa until near the end of the Triassic Period, when a developing rift resulted in the slow westward drift of the Americas away from the Old World and the formation of the new Atlantic Ocean.

Until about 1960, most northern hemisphere geologists refused to support the theory of continental drift despite the large amount of supporting evidence assembled by the German geographer Alfred Wegener, beginning about 1910. Part of the reason for this rejection was the absence of any plausible explanation for the driving force, but it was also true that much of the strongest supporting evidence was found in the southern hemisphere lands that northern hemisphere geologists saw only occasionally, if at all. Southern hemisphere geologists, far fewer in number, had to face the evidence on a regular basis, and proportionally far more of them believed Wegener was correct, notwithstanding the absence of a suitable explanation for the driving mechanism.

During the 1960s, new evidence supporting drift was found by geophysicists studying the east Pacific floor off the western coast of United States and Canada. This evidence and the development of the theory of plate tectonics and sea-floor spreading caused the rapid collapse of most of the opposition to continental drift by 1970.

Before considering what this evidence was, it will be useful to sketch the general nature of rock magnetism and learn what it can tell us about rock history.

Volcanic rocks particularly, but some sedimentary rocks as well, record in their iron-bearing minerals the orientation of the earth's magnetic field at the time the volcanic rocks cooled or the sediments were deposited. The iron-bearing minerals in these rocks behave like little compass needles locked into place when the enclosing rocks formed. Not only do these minerals reveal magnetic north and south by their orientation, but they also show that the earth's magnetic field has frequently reversed its polarity. We say that magnetic compass needles point north today; during the last reversal, about 700,000 years ago, these same compass needles would have pointed south instead. Periods of normal and reversed magnetization are of irregular length and uncertain cause, but the record has been worked out and dated for about the past 80 million years, providing an extremely useful, increasingly well-defined magnetic chronology.

These magnetic minerals not only align themselves parallel to the generally north-south magnetic lines of force, but also record a vertical component, from which magnetic latitude can be determined. It should be remembered that a compass needle points straight up and down (vertically) over the magnetic poles and horizontally over the magnetic equator. In other areas, its alignment is intermediate. This component is called *magnetic dip.*

In the early 1960s, R. G. Mason, A. D. Raff, and V. Vacquier discovered stripelike magnetic patterns in the Pacific sea floor off western North America; some of the stripes showed stronger magnetic fields caused by normal magnetization parallel to present magnetization, but others showed weaker fields caused by a reversed pattern. These stripes at first were not understood, although they provided strong evidence for considerable horizontal offset on some of the Pacific Ocean sea-floor fracture zones (Figure 1-26).

By the mid 1960s the magnetic sea-floor stripes, the concept of sea-floor spreading proposed by H. H. Hess and R. S. Dietz, and the magnetic chronology developed by A. Cox and his fellow workers, were brought together by F. J. Vine and D. H. Mathews. They were able to show that as lava wells up in the spreading centers and cools, it records the north-south orientation of the earth's magnetic field as well as its polarity and latitude. Whenever the earth's magnetic field reverses, these lavas record it and eventually produce a series of magnetic stripes parallel to the spreading center and arrayed in opposite order on each side of the center. The sea-floor stripes have been likened to a sort of enormous tape recorder that reveals the rate and history of sea-floor spreading when it is matched with the magnetic chronology largely developed and dated from continental lavas.

Many independent lines of confirmatory evidence have now been found that have convinced most geologists that continental drift has occurred and that it is linked to sea-floor spreading. Some of these evidences are paleontological, some come from matching glacial sequences and patterns on widely separated land areas, and some are based on rock and age relationships.

We have already mentioned one type of plate boundary, the spreading center, where plates are generated by the addition of new material moving up from depth. The second type of boundary occurs where an oceanic plate collides with a continental plate. Because the oceanic plate is denser and thinner, it bends down and is shoved beneath the edge of a continent. We call this process *subduction,* and variations on the same theme occur when two oceanic plates collide or when two continental plates are forced together. This appears to have happened when India was shoved into the southern edge of the Asiatic continent, resulting in the abnormally elevated Tibetan plateau. As might be expected, the leading edge of a continent so affected is likely to show the effects of this sort of collision, and intensely folded and faulted coastal mountains are formed and often raised to considerable elevation.

As an oceanic plate is subducted beneath a continental plate, heat is generated and some melting may occur, giving rise to igneous activity in the coastal mountains. As the same time, of course, the downgoing oceanic plate is likely to produce a trench or depression on the sea floor next to the continent. At present, this type of boundary does not exist off California except north of Cape Mendocino. We believe, however, that subduction was active in the not too distant past and accounts, at least

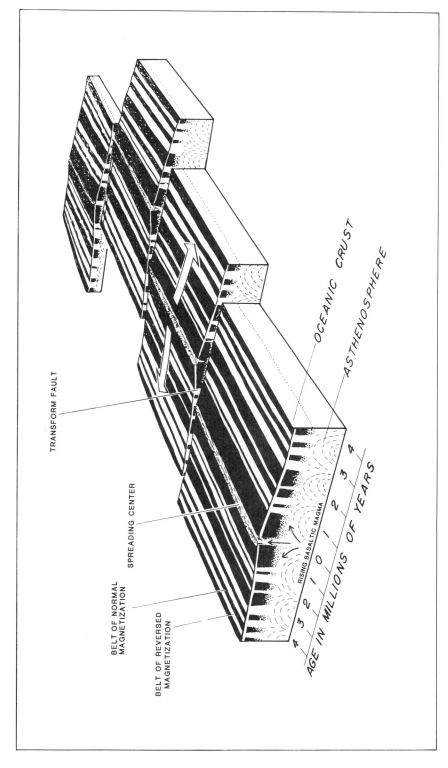

FIGURE 1-26 An idealized view of magnetized stripes in the sea-floor volcanic rocks on either side of the spreading center. Note the symmetrical pattern. Because periods of magnetization are of unequal length and opposite polarity, it is possible to match blocks on opposite sides of the transform fault and establish the amount of displacement that has occurred.

in part, for California's Coast Ranges. The west coast of South America with the high Andes and the deep Peru-Chile Trench is a fine example of an active subduction zone.

As we will see when we discuss the Sierra Nevada and Klamath Mountain provinces, subduction of oceanic crust beneath coastal California has occurred a number of times in the past, dating back at least as far as Paleozoic time more than 200 million years ago. During these periods of subduction, the oceanic crustal material did not glide smoothly under the continental edge, but was subjected to intense friction, which not only raised coastal mountains as noted previously, but often scraped off the sediment cover and some of the underlying oceanic crustal rocks against the leading edge of the continent. These scrapings have been plastered onto the continental margin, adding to the width of the continent. Not infrequently, the scrapings consist of slope and deep-water sedimentary materials, sea-floor volcanic rocks, and sea-floor basement rock all jumbled together to form a *melange*. The melanges are separated from each other and from the continental rocks by faults and until the theory of plate tectonics was assembled, the melanges of the California Coast Ranges were enigmatic in the extreme and had perplexed geologists for many years.

On a larger scale, especially in the Coast Ranges, the Sierra Nevada, and in the Klamath Mountains, fault-bounded blocks many kilometers long and broad are found, which appear to consist of rocks that formed long distances from their present location because they are often very unlike the neighboring blocks. These appear to have been delivered to their present location by sea-floor spreading or strike-slip faulting. In those cases where the origination of these blocks can be ascertained with some confidence, the blocks or *microplates* (according to some geologists) are referred to as *exotic terranes*. In those instances where information about their nature and origin is less certain, they are called *suspect terranes*.

The reasons for labeling these microplates as exotic or suspect terranes are numerous and may include rock types, fossils, stratigraphic sequence, and magnetic characteristics among other things. Some combinations of such evidence has been sufficient to convince geologists that these blocks are distinctive from their neighbors in some manner and that they ought to be designated as exotic or suspect terranes.

A number of such terranes have been proposed for the Sierra Nevada and Klamath Mountains, but probably the best known is the slice of Coast Ranges known as the Salinian Block, which appears to have moved hundreds of kilometers northward along the San Andreas fault from its ancestral home far to the south.

The San Andreas fault represents the third major type of plate boundary, one in which the plates slide by one another along a major fracture system. The area of California lying west of the San Andreas, plus the Baja California peninsula of Mexico, is regarded as being a portion of the Pacific Plate. Most of the state and the rest of North America belongs to the North American Plate, which under present circumstances extends all the way east to the Mid-Atlantic Ridge. At present, the continental materials of North America appear to be firmly joined to the oceanic rocks of the North Atlantic Ocean and no movement is occurring on this continent-ocean boundary.

Our understanding of the causes of mountain building remains incomplete, but the moving plates (plate tectonics) are very helpful in explaining the origin of many of our mountains. Isostasy is useful too, because it helps us to understand how mountains, once produced, persist in the face of erosion.

Before we leave the subject of mountain building, it may be useful to emphasize once again the temporary nature of features such as mountains when long spans of geologic time are considered. The rock record shows plainly that areas now mountainous were once lowlands or shallow sea bottom, and some areas now almost featureless plains were once the site of prominent mountains.

The oldest sedimentary rocks in the Sierra Nevada show us that the site of the present mountains was once a shallow continental shelf on which a considerable thickness of sediment accumulated. This changed in later time to an outer shelf and slope environment in which materials of volcanic origin became increasingly important. Eventually, when this pile of sediment had reached a thickness of perhaps 9 or 10 kilometers (5–6 miles), compression, associated with subduction, converted the pile of marine sediment into a belt of folded mountains. It is quite possible that the thick accumulation of sedimentary materials in the Great Valley of California will one day, when the Sierra and Coast Ranges are worn down, be converted into a new belt of folded mountains.

UNDERGROUND WATER

Running water, as we have seen, is the most important of the sculpturing agents that act on the surface of the earth. We have also seen that solid water—ice—is similarly an important erosional agent. Water, of course, also makes up the oceans and in this form is responsible for erosion of our coasts and deposition of sedimentary materials over the vast extent of the world's oceans. Largely unseen water contained in the rocks, soils, and alluvium, *ground water*, is our next concern in this review of general geology.

It has always been easy to see where the water in rivers came from when it was raining, or when the river was fed by a lake, snowbank, or glacier, but it wasn't until the seventeenth century that any sort of rational explanation was available to account for water flowing in streams when it hadn't rained for a while and when there were no lakes, snowbanks, or glaciers to act as sources. A surprising number of writers nonchalantly declared that rivers emerged full-blown from large springs somewhere in the headwaters. Some suggested that these springs were fed by the sea through some mysterious underground system, presumably with pumps and desalination equipment.

An inquisitive Frenchman named Perrault was dissatisfied with these explanations and decided to find out what was really happening in one small branch of the river Seine. He constructed a weir, a measured notch in a low dam, to calculate the total outflow (*discharge*) of this tributary, and installed a number of simple rain gauges in the drainage basin of the tributary in order to find out how much water entered the system. By these simple devices he was able to measure the input in the form of rain and snow, and the output in the form of stream discharge. Much to everyone's surprise, he found that the river carried only about a third of the water that had fallen on the basin; the rest had soaked into the ground or evaporated back into the atmosphere. We now know that the proportions of runoff, evaporation, and percolation into the ground water vary from setting to setting, but we have compelling evidence that virtually all underground water comes ultimately from rain and snow.

There are few areas of geology in which the layperson's view is mixed with so much superstition and magic as the subject of ground water. Despite numerous scientific investigations since the day of Perrault, many people remain convinced that dowsers or water witches use supernatural powers to locate underground water by means of forked twigs or other such devices. Perhaps the main reason water witching has maintained its popularity in the face of great scientific advance is because, in many cases, it appears to work, and water *is* found in the place the dowser predicted. It is hard to argue with success, but success does not demonstrate cause and effect, and just because the dowser is convinced that the behavior of his wand was controlled by the presence of underground water, this does not prove that there was such control.

Why then do dowsers often succeed? There are several good reasons. First, many work in areas where conditions are fairly uniform and ground water will be found at about the same depth anywhere. Second, in more complex areas experienced dowsers eventually acquire a good practical knowledge of the nature of the ground water occurrence. And, finally, there is at least some element of luck involved; locating a buried water-bearing fracture system in an area with little ground water is an example.

The geologist, on the other hand, dispenses with magical paraphernalia such as witch-hazel twigs, and seeks instead a rational and detailed understanding of the rocks, sediments, and structures of the area. The geologist wants to investigate the porosity of the rocks, the amount of open space between grains, and the permeability of the earth materials, the degree to which pore spaces are connected with one another. Some rocks, perhaps surprisingly, are quite porous but not very permeable. Most shales fit this description; they have high porosity, but the spaces are so tiny that thin films of water cling tightly to the shale grains so that water will not flow through the rock. Such a rock is quite properly described as *impermeable*. Conversely, a coarse sandstone or cavernous limestone may have large pores with very good interconnections, even though the percentage of pore space is less than in a shale. These latter rocks are both porous and permeable.

As a rule sands and sandstones are good *aquifers* or water-bearing layers, and shales are usually good *aquicludes* or poor source rocks. Shales, in fact, often separate aquifers from one another and act as seals to confine ground water to certain horizons.

If the upturned end of an aquifer is exposed in the foothills of a mountain range, the aquifer may receive water from streams that flow across its exposure, and the water will migrate downward into the aquifer to its termination under the adjacent valley. In an undisturbed natural setting such an aquifer may fill with water and the water within it may be under considerable pressure, provided there is an impermeable cap rock—usually a shale—keeping the water confined (Figure 1-27).

If a well in the valley is drilled into such a pressurized aquifer, an *artesian well* will result because the water in the well will rise toward the surface and may even flow strongly at the surface. Artesian wells are much less common today in California because so many wells have been drilled and so much water has been produced that pressures in our confined aquifers are reduced almost everywhere.

Much of our ground water occurs in *unconfined aquifers* in the unconsolidated alluvium in our valleys. This is water that sinks more or less directly downward

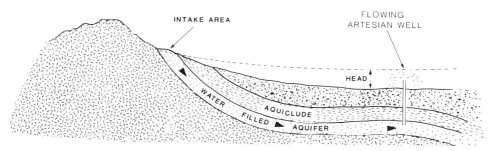

FIGURE 1-27 Cross-section of a confined aquifer or artesian system. If the aquifer is filled with water, the system is under pressure and artesian wells result when the aquifer is drilled.

rather than percolating along buried aquifers. It is fed by every rain- and snowstorm and usually lies only a few meters below the surface if it is not pumped heavily. As the water drains downward between rains, the upper part of the soil or alluvium dries, but at some depth, all the pore spaces are filled with water, and this level is described as the water table. In hilly country, the water table generally follows the land surface closely during and immediately following rains, but the water table gradually flattens under the influence of gravity and because some of the water leaks out onto the surface in low places to form springs, seeps, ponds, or marshy places. This is what feeds streams between rains.

In much of California, particularly in the southern half of the state, strongly seasonal rains cause many streams to flow only in the winter and early spring because the winter rains raise the water table, which then leaks slowly into water courses. By summer, the water table often falls below the bottom of the streams and they dry up until the next rainy period (Figure 1-28). In wetter winters, more water is added to the underground reservoir and streams may flow until midsummer or even later.

Fortunately for California, a lot of the high mountains receive more rain over a longer period than the valleys, and often snow as well. The greater amount of mountain precipitation ensures that more permanent streams develop as does the slow melting of snow at higher elevation, but even the permanent streams and rivers respond to the summer drought that affects much of California.

Limestone is not a particularly common rock type in California and most of the occurrences are in the dry Basin and Range and Mojave Desert provinces, but minor occurrences are found in the Coast Ranges, Sierra Nevada, the Klamath Mountains, and elsewhere. Limestones, because of their solubility, are greatly affected by the presence of ground water, which may dissolve the rock to form large and beautiful caverns (Figure 1-29).

Although there remains some disagreement about the way in which ground water produces limestone caverns, most authorities believe that the open spaces are dissolved below the water table by slowly circulating ground water. Unless the ground water moves through the rocks, it quickly becomes saturated with lime and rapidly loses its ability to dissolve enough rock to produce even a small cave.

The features that make limestone caves interesting are collectively called *dripstone*, or *speleothems*, to use a more elegant term. There are many named types of

FIGURE 1-28 Big Sandy Creek, a typical intermittent water course in the southern Coast Ranges. Streams such as this flow only during and shortly after winter rains. Monterey County. (Photo by R. M. Norris)

dripstone, but the most common are the *stalactites*, which hang from the ceiling, and *stalagmites*, which project upward from the floor; these are called *pillars* or *columns* if they join. Sometimes, sheetlike forms develop, which are called *curtains*. Stalactites form on the ceiling because drops of lime-laden water may cling to the ceiling long enough to deposit a tiny amount of lime. The lime deposition results from decreased solubility when some of the water evaporates, or some carbon dioxide escapes from the water, or due to a slight temperature change. If the dripping persists at the same place long enough, a hollow tubular structure—the stalactite—will develop. As the excess water drips off the lower end of the stalactite, it falls to the floor and repeats the process to form a somewhat stubbier structure with no internal tube—a stalagmite. Dripping water forms a number of other cave features as well, all of which have been given special names, but the process involved is basically the same as that described for the formation of stalactites and stalagmites.

California's best-known limestone caves are found on the western slopes of the central Sierra Nevada and in the Mojave Desert. Mitchell Caverns in the Providence Mountains of the eastern Mojave Desert is administered by the California State Park System and is a small but particularly nice example.

Ground water is a major natural resource in California. An important part of the domestic and agricultural water supply is obtained from wells; ground water allows some flexibility in meeting high demands for water during the dry summer and fall months, when surface-water flows are at a minimum. Furthermore, underground storage of water has some clear advantages over storage in surface reservoirs, chief

FIGURE 1-29 Dripstone features in Mitchell Caverns, Providence Mountains State Recreational Area, San Bernardino County. The caverns are developed in the Permian Bird Spring Formation. (Photo courtesy John Kelso-Shelton, California State Parks)

of which is the prevention of loss by evaporation. It is also generally much less expensive to store water underground than it is to build surface reservoirs, but there are limits to the amount and location of suitable underground storage sites, and considerable energy may be required to pump underground water. Utilization of both underground and surface storage is necessary to support both California's great agricultural enterprise and its burgeoning population.

As we will see in the chapters dealing with the various provinces of California, management of ground-water resources in California is far from perfect; legal requirements and scientifically rational management are sometimes at odds, and excessive pumping has depleted and degraded some underground reservoirs. For example, pumping has led to subsidence of the land surface above the ground-water reservoir in some parts of the San Joaquin Valley and Mojave Desert.

One of the most insidious sorts of damage to California's ground-water resources has come from improperly handled toxic wastes. Today's modern industrial society requires a vast array of natural and synthetic chemical compounds, many of which are dangerous. As the state's population and industrial activity have increased, more toxic chemicals are needed and the demand for ground water increases. It is much harder to find safe, remote areas to store and dispose of toxic substances, and the risk of contaminating ground water is increasing accordingly.

Ground water has been degraded in some coastal areas such as Ventura County because poorly managed pumping has allowed salt water from the sea to replace the fresh water pumped out.

Threats to the quality and quantity of ground water are numerous in California and can be expected to continue into the forseeable future, but the enormous importance of our ground water demands that our stewardship be as good as our understanding and technical skills allow; short-term costs and needs should not be substituted for long-term benefits.

THE SEA AND MARINE PROCESSES

The waters of the Pacific Ocean, its waves, its tides, and its currents have a profound effect on California both directly and indirectly. While acknowledging that the sea exerts great influence, and the climate in turn has a marked effect on geologic processes, we will confine our attention here to the more direct occurrences along the California coast, and on the sea bottom immediately offshore.

Marine processes are responsible for erosion, transportation of eroded debris, and for deposition, and once again water is the medium.

Most of the California coast from the Oregon boundary south to Point Conception at the western end of the Santa Barbara Channel is an open, exposed coast unprotected by offshore islands. South and east of Point Conception, the coast changes direction and no longer faces into the weather. Moreover, from Point Conception to the Mexican border, eight mountainous offshore islands provide some storm protection for mainland shores. As a consequence, the impact of wave energy tends to be greater on the northern and central coasts than on the southern coast.

The sea attacks the shore somewhat like a horizontal saw, confining its direct impact to a rather narrow zone in a vertical sense. If sea level remains stable and tectonic activity neither raises nor depresses the coastal regions, marine erosion eventually produces a broad, nearly flat rocky platform just below sea level. The vertical range of marine erosion depends upon three things. Perhaps the most obvious are the daily tides. The higher the tidal range, the greater the area of wave attack. In some parts of the world, with tidal ranges of 10 meters (30 feet) or more, a fairly broad band along the shore is subjected to daily attack by the sea. In California, where the tide range is about 2 or 3 meters (6–10 feet), there is a narrower band exposed to wave erosion and there are no marked differences along the California coast.

A second factor influencing the effectiveness of wave erosion is the degree to which the coast is exposed to the open Pacific. Large storm waves can greatly increase the extent of the shore coming under direct wave attack. Storm beaches sometimes lie as many as 5 meters (15 feet) or more above the highest tides and may be affected by only a few severe winter storms each year.

The third factor, usually quite inconspicuous because it occurs very slowly, is a change in *relative* sea level. This change may come about because the land is elevated or depressed with respect to the sea. Or, it may be caused by an actual worldwide change in sea level. California coastal landforms that reveal a relative drop in sea level are the numerous marine terraces or benches that occur along the coast. Good

FIGURE 1-30 Broad, elevated marine terraces west of Santa Cruz, Santa Cruz County. (Photo by R. M. Norris)

examples are seen on some of the offshore islands such as San Clemente and San Nicolas, and near the city of Santa Cruz on the central coast (Figure 1-30).

Coastal subsidence or a rise in sea level is demonstrated by numerous coastal marshes and estuaries like San Francisco and Humboldt bays.

Coastal erosion takes place in much the same manner as stream erosion. Rocks delivered to shore by streams or derived directly from the sea cliffs, which edge much of the California coast, are rolled back and forth or hurled against the base of the sea cliff by the waves. These rocks enable the waves to cut a notch in the base of the cliff and also help to grind and abrade the rocky shore platform (Figure 1-31).

Most marine deposits are materials laid down on the shelves and slopes offshore and are not particularly obvious to land dwellers. One kind of marine deposit, the beach, is very evident to coastal dwellers or visitors. Most California beaches are sandy, but some range from muddy beaches in some sheltered bays to boulder or cobble beaches on some very exposed coasts.

Before considering beach behavior in more detail, it is worthwhile to ask where the sandy or gravelly beach materials came from. An obvious source is the nearest coastal cliff. Cliffs do indeed make some contribution, but the most important source of beach materials is sand from the coastal streams. If the coast was flat or gently sloping like that of Louisiana, many streams would have built small deltas at their mouths. But California's coast is mostly steep and mountainous and deep water lies close inshore, so there are few deltas of any size. Nonetheless, streams when in flood may carry large amounts of sediment to the sea. Much of this sediment is deposited as a sort of submarine delta in the surf zone off the stream mouths. During the intervals between floods, waves work over these piles of sediment and add the sandier portions to the beaches and inshore areas; the fine portions are carried out into deeper and quieter waters to be deposited. Coarser materials may remain off the stream mouths or form small bouldery deltas. The Ventura River has such a small bouldery delta.

FIGURE 1-31 Wave-cut bench in steeply-dipping Monterey Formation, a thin-bedded sili-cified shale of Miocene age. Near Gaviota, Santa Barbara County. (Photo by R. M. Norris)

Anyone who has closely observed the beaches for a few years knows that numerous changes occur, some quite suddenly. Energetic winter waves flatten and widen beaches that lose sand into deeper water, often exposing parts of the underlying rocky platform. Gentler summer waves bring the sand back in, making it somewhat steeper and covering the rocks, a fortunate arrangement because most people prefer to use the beaches when the weather is warm. Summer waves, in short, sweep up the sand from the surf zone and bank it up against the coastal cliffs or dunes.

We have also learned by costly experience that waves transport sand along the coast. Because most Pacific waves come from the west or northwest, and the California coast trends generally somewhat east of south, waves strike the beaches at an angle. For most parts of the Californian shoreline, this means that a stream of beach sand is driven slowly in a southerly and easterly direction. It is not true, however, that a sand grain starting on the beach near Crescent City in the far north will ever reach the beaches of San Diego, because the shoreline is divided into a series of more or less self-contained *cells*. A typical cell is limited at both ends by rocky headlands around which little sand passes. Most of the sand supply in the cell is delivered by streams directly tributary to the cell.

It is logical to wonder why California beaches don't get wider and wider as streams in flood keep adding new sand to the beaches. Although there are some local exceptions, under natural circumstances beaches tend to maintain about the same configuration from year to year, so there must be some means for disposing of the surplus sand. The loss is accomplished by three means.

1. Some minor amount of sand is carried offshore beyond the reach of summer waves by particularly severe winter storms.
2. On some coasts, such as northern Santa Barbara and southern San Luis Obispo counties, or near Bodega Bay in Sonoma County, appreciable sand is blown inshore from the beach by the wind to form tracts of coastal dunes beyond the reach of waves (Figure 1-32).
3. Most important, however, are the large submarine canyons reaching close to shore. These not only are disposal systems for surplus beach sand, which is periodically flushed down them into deep water far beyond the reach of waves, but the canyons themselves are largely due to the abrasive action of these underwater sand streams.

   Monterey Bay, roughly between Santa Cruz on the north and Monterey on the south, is one such cell. Sand delivered to this cell by the Pajaro and Salinas rivers and other streams, finds its way northward from Monterey to the head of Monterey submarine canyon just offshore from Moss Landing, and south from Santa Cruz to the same place where the surplus sand is sluiced down this large canyon to a depth of about 3,000 meters (10,000 feet) below sea level, where a large submarine fan has developed (Figure 1-33).

FIGURE 1-32 Small coastal dunes supplied by sand from the beach. The form of coastal dunes is much more dependent upon and influenced by vegetation than is the case in desert dunes. Jalama Beach County Park, Santa Barbara County. (Photo courtesy John Lentz, California Coastal Commission and University of California Press, *California Coastal Access Guide*, 1982)

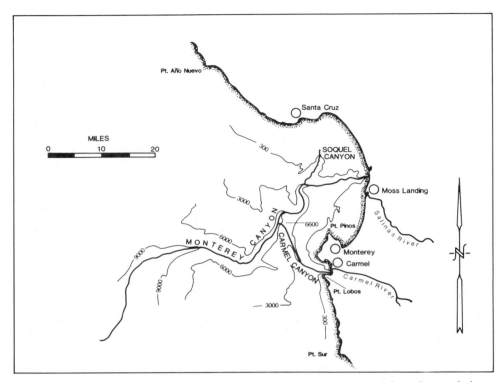

FIGURE 1-33 Map of Monterey submarine canyon showing the main branches and their relationship to land features. (Topography after U.S. Coast and Geodetic Survey)

Where artificial obstructions to this system are installed, such as dams on streams, or breakwaters on the shore, problems of beach erosion or sand depletion follow. Several specific examples are discussed in later chapters.

Marine deposits include beaches, of course, but more important are the land-derived sediments that are deposited on the continental shelves or in the deep offshore basins off southern California. These are important because they are the modern examples of the depositional environments in which many of the marine sedimentary rocks of California formed, particularly those of the younger sedimentary rocks found in the various coastal ranges. The marine environments in which many of the older sedimentary rocks of the Basin and Range and Mojave Desert provinces were formed were quite unlike anything we now see off the coast of California. Those old marine environments were more akin to conditions now prevailing in the Bahamas or other shallow tropical areas.

GEOLOGIC TIME

As was noted earlier, one of the characteristics that most sharply distinguishes geology from the other physical sciences is its strong concern with time. From the days when James Hutton and John Playfair set geology on its course as a distinct science, it became apparent to many students of the earth that unless one was prepared to attribute many geologic features and phenomena to sudden catastrophic events without clear historical precedent, a vast amount of time was needed to account for earth features. Admittedly, some geologic processes are dramatically sudden—volcanic eruptions, earthquakes, great floods, and such. Trying to assemble the entire rock record of the earth, including its evidence of numerous past climatic changes, advances and retreats of the sea and glaciers, the appearance and disappearance of a vast number of plant and animal species of steadily increasing complexity, and the formation and destruction of mountain ranges requires a vast stretch of time in order to concede that these changes occurred by orderly and observable processes.

From the beginnings of geology as a science about 200 years ago, this long view of time has been criticized by those whose view of biblical chronology convinces them that the earth is only a few thousand years old. Some of the most severe criticism came from within the scientific community. In 1862, the eminent physicist Lord Kelvin, on the basis of his calculation of the time needed for the earth to cool from a molten state, declared that the earth was between 98 and 400 million years old. Geologists, who had by then become accustomed to having unlimited time at their disposal, were shocked. Kelvin, for the next 30 years refined his estimates, supported by seemingly unshakable evidence from thermodynamics, and by 1897 declared that 24 million years was the best estimate of the earth's age. This had a healthy if profoundly disquieting effect on geologists because it forced them to reexamine their evidence and their assumptions. But even after they had done so, many of them could see that the rock record simply did not agree with Kelvin's restrictive 24 million years. It was a major dilemma and something was missing.

The missing evidence wasn't long in coming. Henri Becquerel discovered radioactivity in the spring of 1896, and in 1903 Pierre Curie and his assistant Albert Laborde announced that radium salts constantly released heat. It quickly became evident that the earth had not cooled steadily from its birth, but included a source of enormous amounts of heat, and as a result was vastly older than Kelvin thought. It wasn't long before the steady radioactive decay that changes one element into another was quantified, allowing rocks that contained the parent radioactive elements and their decay products to be dated. Radiometric dating has been greatly improved during the present century and now permits us to estimate with confidence an age for the earth of about 4,600 million years.

A geologic time scale much like the one in use today had been assembled by the end of the nineteenth century, based largely on the fossil record that shows life became more complex and diverse from the beginnings of the record to the present time. The time scale is constantly undergoing refinement and at present one of the most lively disputes concerns the boundary between the Phanerozoic and Proterozoic (see Time Scale, inside front cover). Since the first edition of this book, there is increasing support for the addition of a new period at the beginning of the Phanerozoic—the Ediacarian or Vendian—when multicellular organisms appeared, but before they had developed hard parts.

The geologic time scale is a *relative* system based on the order rather than the precise duration of events. For example, rocks containing the first fish remains (Ordovician) rest on rocks containing nothing more advanced than trilobites (Cambrian). The amphibian record begins before the reptilian record, and conifers appeared before the flowering plants. The periods and eras were designated to reflect this evolutionary succession and as a result are of unequal lengths.

Because radiometrically dated mileposts in the geologic record are increasingly numerous and well defined, they now permit us to calibrate the geologic time scale and in some instances to measure the rates of geologic processes whether they be the time needed to raise and destroy a mountain range, the length of time the dinosaurs roamed North America, or the time needed for the Colorado River to excavate the Grand Canyon.

Finally, the great span of geologic time makes rare events in a human lifetime commonplace and probable instead of improbable. From the human perspective, there is little possibility that a female lizard could successfully ride a floating log to an island hundreds of miles from land and there establish a new colony of lizards. Yet over millions of years such an unlikely event becomes probable if not virtually certain. The violent volcanic eruption that produced the Bishop Tuff in eastern California hasn't happened again in our state in the last 710,000 years, yet the geologic record shows events of this sort have occurred numerous times in North America during the Tertiary period alone.

REFERENCES

Barnes, Charles W., 1980. Earth, Time, and Life. John Wiley & Sons, 582 p.

Larson, Edwin E., and Peter W. Birkeland, 1982. Putnam's Geology. Oxford University Press, 789 p.

Robinson, E. S., 1982. Basic Physical Geology. John Wiley & Sons, 661 p.

Zumberge, J. H., and C. A. Nelson, 1972. Elements of Physical Geology. John Wiley & Sons, 395 p.

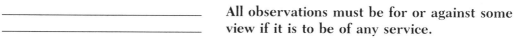

two

GEOLOGIC SETTING

All observations must be for or against some
view if it is to be of any service.

—*Charles Darwin*

California is a state of geologic contrasts. Of the 48 contiguous states, it contains the highest and lowest elevations only 130 kilometers (80 miles) apart, plus a variety of rocks, structures, mineral resources, and scenery equaled by few areas in the world. Furthermore, because much of the state is arid or nearly arid, its rock sequences are exposed with unusual clarity.

California's rocks vary from Proterozoic to presently forming sediments, and several of the state's formations are type examples for North America and the world. The geologic map of California (Figure 2-1 and in pocket) shows general rock ages, and Figure 2-2 locates the geomorphic provinces. The extent and relationships of the provinces should be compared with the data of the geologic map.

The oldest rocks found in California are Proterozoic gneisses and schists as much as 1,800 million years old, which occur in the San Gabriel and San Bernardino mountains and in the Mojave and Basin and Range provinces. Only in the Basin and Range are these old crystalline rocks overlain by relatively unaltered Proterozoic sedimentary rocks. The Proterozoic sea in which these rocks were deposited certainly covered much more of California than the Basin and Range, but in other areas these ancient sediments have either been much altered by metamorphism or stripped off by erosion.

The ensuing Paleozoic was a time during which much of eastern California was a broad, shallow marine shelf on which great thicknesses of limestone were deposited. Mountain-building episodes, chiefly to the north and east, interrupted this pattern from time to time, resulting in either erosional gaps in the record, or influxes of sand, mud, and gravel from the highlands.

The outer edge of the Paleozoic shelf lay somewhere along the eastern side of the

51

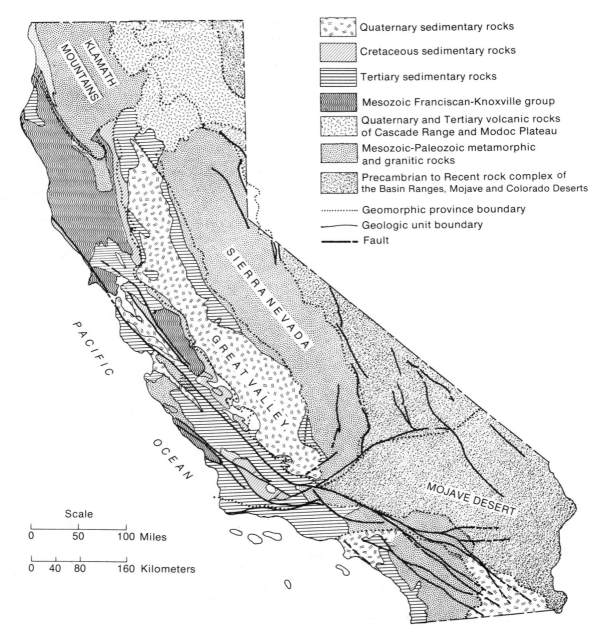

FIGURE 2-1 Geologic map of California, showing principal faults and generalized geologic units. (Source: California Division of Mines and Geology)

Sierra Nevada and Klamath mountains. Paleozoic marine sedimentary rocks are particularly common in the northern Sierra and in the eastern Klamaths, but for the most part they are not shelf sediments, but deeper-water deposits composed of volcanic detritus from offshore islands mixed with oceanic crustal materials, a subduction complex. Paleozoic rocks are also present in the eastern Transverse Ranges

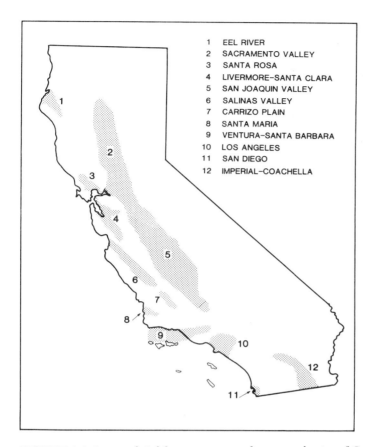

1 EEL RIVER
2 SACRAMENTO VALLEY
3 SANTA ROSA
4 LIVERMORE–SANTA CLARA
5 SAN JOAQUIN VALLEY
6 SALINAS VALLEY
7 CARRIZO PLAIN
8 SANTA MARIA
9 VENTURA–SANTA BARBARA
10 LOS ANGELES
11 SAN DIEGO
12 IMPERIAL–COACHELLA

FIGURE 2-2 Principal California marine sedimentary basins of Cenozoic age.

where they appear to be mainly shelf deposits, and in the Peninsular and Coast ranges where their origins are more obscure because of intense metamorphism.

Marine conditions persisted into early Triassic time in much of California, but with the onset of active subduction along the continental margin in the late Triassic, the sea withdrew to the west, and mountains developed in eastern California accompanied by some granitic intrusive activity.

Active subduction continued during much of the Jurassic, bringing island arcs, atolls, and other scraps of land from perhaps as far away as the western Pacific. By late Jurassic time, these collisions resulted in the Nevadan orogeny, during which time extensive mountain building occurred and vast amounts of granitic magma were generated. With some pauses or interruptions, subduction continued through much of the ensuing Cretaceous period when granitic intrusions were again voluminous and very widespread. Granitic rocks of Nevadan age are by far the most abundant plutonic igneous rocks in California and are exposed in most of the provinces.

As the Nevadan Mountains were raised during the Jurassic and Cretaceous, huge volumes of material were stripped off the mountains and carried to the deep marine basins lying to the west. The Jurassic Great Valley beds, best displayed in the northern Coast Ranges, make up the first, deeper-water part of this sequence and

FIGURE 2-3   The geomorphic provinces of California.

as the basins filled in the Cretaceous, shelf deposits succeeded the deep-water deposits. Late Cretaceous shelf sediments are abundant in much of California west of the Sierra Nevada.

At the same time, to westward, probably near a chain of volcanic islands lying offshore, subduction continued and what we know as Franciscan rocks, composed of a mixture of deep-water sediments, oceanic crustal material, and volcanic rocks were scraped off the sea bottom, mixed together, and plastered against the edge of the continent.

Uplift associated with development of the Nevadan Mountains once again blocked the sea from reaching eastern California, and younger rocks in the Basin and Range and Mojave Desert provinces are limited to nonmarine sedimentary rocks and to volcanic and intrusive rocks.

Marine Cenozoic rocks are voluminous and widespread from the slopes of the Sierra westward to the sea, however, particularly in the San Joaquin Valley, the Transverse and Coast ranges. Smaller amounts occur in the Peninsular Ranges and in the Colorado Desert. In the Los Angeles, Ventura, Santa Maria, and San Joaquin basins, the middle and late Cenozoic marine rocks are thousands of meters thick. The Ventura Basin contains a sedimentary section 15,000 meters (50,000 feet) thick, forming one of the most complete middle and late Cenozoic sedimentary records in the world. Thick marine sections are also present in the Eel River embayment near

Eureka. California's principal Cenozoic marine basins are shown in Figure 2-3. These geographic features are better understood when compared with the geologic map.

In the Mojave and Colorado deserts and in the Basin and Range, restricted continental basins, including those occupied by extinct lakes Searles, Owens, and Manly (Death Valley), reveal the sedimentary climatic, faunal, and floral history of the late Cenozoic.

Active faulting is an important feature of California's structural pattern. Examination of the geologic map shows the dominance of the San Andreas fault and its branches. This fault, of right-slip movement, has been crucial in California's geologic history since at least the Miocene and in some opinions since the Jurassic. Cumulative displacement has been estimated to be as much as 560 kilometers (350 miles). Other important faults are the Calaveras and Hayward in the San Francisco Bay area, the Nacimiento in the southern Coast Ranges, the San Jacinto of the Peninsular Ranges, the Sierra Nevada in eastern California, and the Garlock, which separates the Mojave Desert from the Sierra Nevada and Basin and Range.

California's landscapes are extremely varied, ranging from the broad, nearly flat floor of the Great Valley to the jagged, glaciated Sierra Nevada. Twelve geomorphic provinces are recognized (Figure 2-3): the Sierra Nevada, the Klamath Mountains, the Cascade Range, the Modoc Plateau, the Basin and Range, the Mojave Desert, the Colorado Desert, the Peninsular Ranges, the Transverse Ranges, the Coast Ranges, the Great Valley, and the Offshore area. The geomorphic provinces are topographic-geologic groupings of convenience based primarily on landforms and late Cenozoic structural and erosional history.

GEOMORPHIC PROVINCES

Sierra Nevada

When the forty-niners moved west across the interior desert, the formidable barrier of the 640-kilometer (400-mile)-long Sierra Nevada blocked their path. Those who arrived in the colder months and who made frontal assaults on the Sierran wall endured hardships of snow and storm; those choosing to circumvent it were parched by the arid wastes of the deserts of the south and southwest. The Sierra Nevada, the highest and most massive of California's topographic features, plays a key role in making California habitable. It squeezes moisture from clouds moving inland from the Pacific, stores winter snows and rainfall in its soils, lakes, and forests, and provides large rivers that water the agriculturally rich Central Valley, the San Francisco Bay area, and the city of Los Angeles.

Parenthetically, the name Sierra Nevada is derived from Spanish: *sierra* meaning "jagged range" and *nevada* meaning "snowed upon." The plural form "Sierras" is a common but redundant usage as is Sierra Nevada Mountains.

Great Valley

Following descent from the crest of the Sierra Nevada through deep canyons and foothills of the Mother Lode gold belt, a traveler enters the flat-floored trough known as the Great Valley or the Central Valley. Drained by the Sacramento and San Joaquin rivers, the Great Valley extends nearly 800 kilometers (500 miles) north and south, separating the Sierra Nevada from the Coast Ranges by an average of 65 kilometers (40 miles). From Red Bluff to Bakersfield, the monotonous floor is interrupted only by Sutter (Marysville) Buttes. Much of the valley's elevation is close to sea level (Sacramento, 10 meters [30 feet]); even Bakersfield (comparatively high) has an elevation of only about 120 meters (400 feet). Beneath the valley's silt and gravel cover, a thick sedimentary sequence carries valuable petroleum and natural gas deposits.

Northern Provinces

Three major geologic units occur in northern California. The broad, largely forest-covered volcanic Modoc Plateau and the volcanic peaks of the Cascade Range are the southern and western extensions of topographies belonging more typically to provinces of northwestern United States. The third northern province, relatively inaccessible and the least known geologically, is the Klamath Mountains. All three provinces close the Great Valley to the north.

Southern Provinces

Terminating the Great Valley on the south are several units of the Transverse Ranges, which are composed of many overlapping mountain blocks of nearly east-west trend in contrast to the northwest-southeast lineation of the Coast Ranges and the Sierra Nevada. The Transverse Ranges consist of several major parallel and subparallel ranges and intervening valleys. Chief among the ranges are the Santa Ynez, Santa Susana, Santa Monica, San Gabriel, and San Bernardino, with intervening sediment-filled valleys or basins such as the Santa Ynez, Ventura, Ojai, Santa Clara, Simi, San Fernando, San Gabriel, and Santa Barbara Channel. Between the Transverse Ranges and the Peninsular Range lies the Los Angeles Basin, with thousands of meters of post-Jurassic sediment overlying crystalline basement rocks. Southward, the Peninsular Ranges incorporate the San Jacinto, the Santa Ana, and other ranges in the hinterland of San Diego and are dominated by rock types prevalent in the Sierra. Much of the Peninsular Range province lies outside California, continuing south nearly 1,300 kilometers (800 miles) as the peninsula of Baja California.

Southeastern Provinces

The Mojave Desert, the Colorado Desert, and the Basin and Range province constitute California's desert regions. The Basin and Range extends from Utah's Wasatch Mountains to the Sierra Nevada. In these desert regions, borates and saline minerals are derived from modern playa and ancient lake basins, and many of the mountains have mineral deposits. Elevations range from 4,345 meters (14,246 feet) at White Mountain in the Inyo-White mountains to −86 meters (−283 feet) near Badwater in Death Valley. Great differences in relief provide some spectacular landscapes, for instance, where San Jacinto Peak (3,296 meters or 10,805 feet) overlooks Palm Springs (145 meters or 475 feet) and Salton Sea (−72 meters or −235 feet). Such varied features as Death Valley National Monument, Salton Sea, and the Colorado River delta are included in the southeastern provinces.

Coast Ranges

West of the Great Valley the Coast Ranges extend 880 kilometers (550 miles) north and south and are divided by the San Francisco Bay system, a network of waterways and straits produced by drowning of river-cut and block-faulted valleys. The Coast Ranges show strong northwest-southeast trends, induced by folds and faults of the same trend. The chain contains dominately sedimentary rocks underlain by two unlike kinds of basement rocks that are mostly of middle Mesozoic age. One of these, the Franciscan Formation, is widespread and figures prominently in today's interpretations of the roles of sea-floor spreading and plate tectonics in developing the continental margin. The other type is granitic with associated metasediments. These rocks are exposed in several areas, and the granitic rocks probably correlate with granitic units of the Sierra Nevada or Peninsular Ranges. Geologic history of the Coast Ranges is intricately interwoven with tectonics of the San Andreas and other major faults, particularly those of the western part of the state. Sedimentary units include the Great Valley sequence, ancient submarine fan and shelf deposits exposed along the boundary between the Coast Ranges and the Great Valley, plus Cenozoic basin deposits of thousands of meters of clastic sediments. In some places a narrow, discontinuous coastal plain faces the Pacific. More often, however, especially in the south, the Coast Ranges rise abruptly from the sea to almost 1,800 meters (6,000 feet), forming a western barrier almost as continuous as the Sierra Nevada to the east.

Offshore

The offshore province is composed of two main regions. North of Point Arguello, the province appears to be related to the Coast Ranges and is characterized by continental shelf and slope topography. East and south of Point Arguello, the province may correlate with the Transverse and Peninsular ranges. This southern section contains elevated blocks and ridges that are occasionally expressed as islands, with deep intervening closed basins.

CALIFORNIA IN THE CONTINENTAL FRAMEWORK OF NORTH AMERICA

California's geologic setting is only one aspect of the complex patterns of rocks, structure, and history composing the western Cordillera of the United States and the geologic features circumscribing the Pacific Basin. The mountains of southern Alaska, western Canada, western continental United States, western Mexico, Central America, the Andes, and parts of Antarctica all seem to share roughly similar geologic histories. Since the renascence of the concept of continental drift and development of the ideas of plate tectonics and sea-floor spreading, it is important to understand the eastern Pacific. For example, we now appreciate that the building of the Sierra Nevada during the Mesozoic Nevadan orogeny was only a minor incident in the major event of Cordilleran orogeny, which involved most of western North America.

Sophisticated isotopic age dating now permits correlating such California events as the Nevadan and Coast Range orogenies with episodes like the Basin and Range

Sevier and Rocky Mountain Laramide orogenies. Consequently, geologists have established that such local deformations were merely parts of a major event of the eastern Pacific-western North American margin, with tectonic pulsations and intervals of quiescence.

In recent years, as our understanding of California's complex geology has improved, we have discovered that much of the basement rock from the Sierran foothills westward consists of oceanic rocks scraped off against the western edge of the continent during several periods of subduction. This basement is also believed to include some slices and blocks of rock that have been transported hundreds or even thousands of kilometers by strike-slip faulting or by rafting on the subducting oceanic plates. These are the exotic or suspect terranes referred to in Chapter 1.

CURRENT GEOLOGIC STUDIES

Geologic maps are the principal products of geologic field work, and thorough understanding of a region's geology is possible only when a geologic map can be examined. Accordingly, most countries have organized geologic surveys. The U.S. Geological Survey, which employs several thousand geologists, is responsible for all facets of the federal government's geologic mapping program. In addition, most states have their own geologic surveys and cooperate with the federal survey in undertaking many geologic studies, including map publication. In California, the responsible agency is the Division of Mines and Geology, a branch of the California Department of Natural Resources.

Although geologic mapping of California is proceeding systematically, much remains to be done. Large areas are incompletely studied, and sometimes pertinent geologic maps are unavailable or outdated. Surface geology was summarized originally as the Geological Map of California, published by the Division in 1916 on a scale of approximately 1 to 750,000. It was an extremely generalized representation, drawn from vague reconnaissance studies that appeared before World War I. Up to that time the Division's primary concern had been mining and mineral resource problems.

The Division's emphasis changed gradually, however, with the 1929 appointment of Olaf P. Jenkins. Jenkins initiated a second geologic map, which appeared in 1938 on a scale of approximately 1 to 500,000. The geologically unknown areas of California were glaringly apparent, since more than one-third of the state appeared as "unmapped." Jenkins was joined in 1948 by Gordon B. Oakeshott. Together they sponsored the detailed mapping required as the base for solving California's geologic problems. The necessity of educating the public and the government to the importance of the geologic environment was also apparent. As Deputy Chief Geologist, Oakeshott focused on geologic education and for almost a quarter of a century was one of the state's leading earth science educators. The first two sheets of the present geologic atlas appeared as Jenkins closed his administration in 1958. Ian Campbell then was appointed State Geologist and Chief of the Division, and under his direction the third geologic map was completed by 1969. It is an atlas of 27 sheets, on a scale of 1 to 250,000 (Figure 2-4). During this period, the orientation of the state program shifted toward detailed studies of environmental problems while still encompassing research on mineral commodities and mining potential. This new series of geologic maps is continually updated and several are in the second or even third edition.

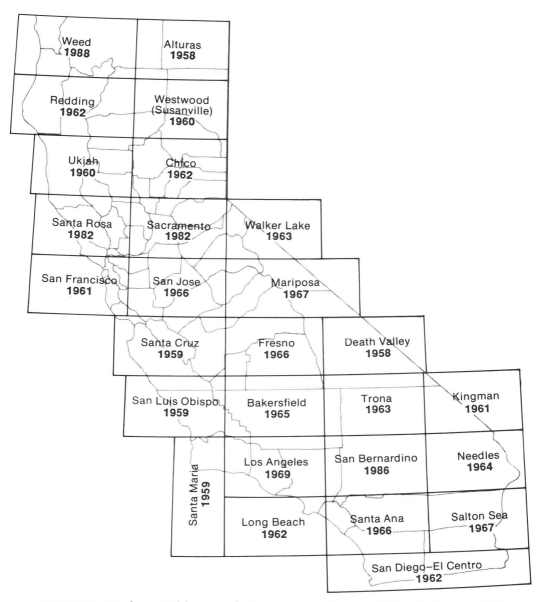

FIGURE 2-4 Index to California geologic map sheets. (Source: California Division of Mines and Geology)

SYSTEMATIC STUDY OF CALIFORNIA GEOLOGY

The geology of California is exceedingly complex and becomes more so as the consequences of plate tectonics are evaluated. Where, then, to start, and why start at a particular place in developing a logical, interesting, and yet necessarily limited coverage of this fascinating geology?

We will begin with the Sierra Nevada because it provides a basis for understanding the bedrock geology of the entire state. Other regions differ from the Sierra primarily in events and rocks subsequent to the middle Mesozoic, though the earlier record is best preserved in the Basin and Range, Mojave, and Transverse Range provinces. The plutonic rocks throughout the state are overwhelmingly products of granitic magmatic invasions that accompanied the Nevadan orogeny. In addition, the Sierra is the topographic backbone of the state and has shed debris since the middle Mesozoic, forming many of the state's younger rock units. The northern provinces (Klamath Mountains, Cascade Range, and Modoc Plateau) then are discussed, followed by the Basin and Range, the Mojave Desert, the Colorado Desert, the Peninsular Ranges, the Transverse Ranges, the Coast Ranges, the Great Valley, and the Offshore, with a special section on the San Andreas fault.

REFERENCES

Whenever possible, publications of the California Division of Mines and Geology and the U.S. Geological Survey have been cited, because they are generally more widely available in local libraries. Lists of available geologic publications may be obtained from: California Division of Mines and Geology, Publication Sales, P.O. Box 2980, Sacramento, California 95812; U.S. Geological Survey, Room 555 Battery Street, San Francisco, California 94111; U.S. Geological Survey, Room 7638, 300 North Los Angeles Street, Los Angeles, California 90012. The authors recommend that readers who are interested in an up-to-date detailed wall map of the geology of California purchase the 1:750,000 edition from the Division of Mines and Geology in Sacramento.

The following four references are especially pertinent and are frequently suggested as collateral reading.

California Geology (formerly Mineral Information Service). California Division of Mines and Geology. A monthly digest of new data on mining, plus general geology and announcements of publications pertinent to California geology.

Geologic Maps of California. California Division of Mines and Geology. Usually referred to by name of sheet and date of publication.

Bailey, Edgar H., ed., 1966. Geology of Northern California. California Division of Mines and Geology Bulletin 190. 508 p. A series of papers by different authors, organized into 10 chapters, each with extensive bibliographies. (Still important, although some parts are outdated by newer studies.)

Jahns, Richard H., ed., 1954. Geology of Southern California. California Division of Mines and Geology Bulletin 170. A series of papers and maps by different authors, organized into 10 chapters, with 34 map sheets and 5 geologic guides. (Again, outdated in some respects, but still a very important reference.)

Other general references are:

Alt, David D., and Donald W. Hyndman, 1975. Roadside geology of northern California. Mountain Press Publishing Co., Missoula, Montana. 244 p.

Anderson, F. M., 1932. Pioneers in the Geology of California, California Division of Mines and Geology Bulletin 104, p. 1–24.

Geotimes. Published monthly by the American Geological Institute, 4220 King Street, Alexandria, Virginia 22302.

Harbaugh, John W., 1974. Geology: Field Guide to Northern California. William C. Brown Co., Dubuque, Iowa, 123 p.

Hinds, Norman E. A., 1952. Evolution of the California Landscape. California Division of Mines and Geology Bulletin 158, 240 p.

Oakeshott, Gordon B., 1971. California's Changing Landscape. McGraw-Hill Book Co., New York, 379 p.

Sharp, Robert P., 1972. Geology: Field Guide to Southern California. William C. Brown Co., Dubuque, Iowa, 181 p.

————, 1978. Coastal Southern California. Kendall/Hunt Publishing Co., Dubuque, Iowa, 268 p.

three

SIERRA NEVADA

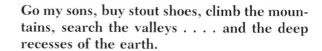

Go my sons, buy stout shoes, climb the mountains, search the valleys . . . . and the deep recesses of the earth.

—*Peter Severinus*

The Sierra Nevada is California's topographic backbone. Consider how different the state's geography and human resources would be if the Sierra were lower or non-existent. For instance, the Sierra squeezes moisture from Pacific storms that must rise above its crest. Another example is the presence of mineral resources near the surface because ancient rocks have been exposed by erosion in the Sierra Nevada's deep canyons and on its slopes. Placer gold derived from Sierran lodes was the main attraction for thousands of pioneers, many of whom reached the gold country only after an arduous crossing of the range.

Sierran grandeur has evoked both wonder and inspiration. One of its first and greatest interpreters was the famous naturalist John Muir, who spent much of his life exploring and extolling this stupendous landform. Muir recorded his observations meticulously, and his journals provide probably unexcelled portrayals of Sierran majesty. Furthermore, Muir's explorations resulted in corrections of earlier, erroneous ideas regarding the origins of the natural features in the Range of Light, as he called the Sierra.

Differences of opinion regarding the Sierra, exemplified by the extended controversy about the origin of Yosemite Valley, resulted in the initiation of a comprehensive study of the valley by the U.S. Geological Survey. The project was undertaken by Francois E. Matthes, who prepared the first large-scale topographic map of Yosemite, and F. C. Calkins. The result was the classic publication *Geological History of the Yosemite Valley*. Although it did not appear until 1930, this study established Matthes as a leader in Sierran geology. Like Muir, Matthes was a persuasive and eloquent though a less prolific writer. Several posthumous papers, based on Matthes's voluminous field notes, were prepared by Fritiof Fryxell primarily for the layperson. Two of these papers were published by the University of California Press in 1950 on the occasion of the state's centennial. The U.S. Geological Survey published two more in 1960 and 1965.

GEOGRAPHY

The Sierra Nevada extends about 640 kilometers (400 miles), terminating in the north near Lassen Peak in the Cascade Range province. The rock and structural patterns of the Sierra disappear beneath the lavas of the Cascades and the volcanic and sedimentary cover of the northern Sacramento Valley. They reappear in the Klamath Mountains, where rock units and faults have geologic histories similar to those of the Sierra. (The relationships between the Sierran and Klamath provinces are shown in the geologic map [Figure 2-1].) Some geologists believe that the same basement patterns extend north into Oregon, curving northeast and disappearing again beneath the volcanics of eastern Oregon and southern Washington.

To the south, the Sierra grades into the Tehachapi Mountains, which enclose the southern end of the Great Valley. The Garlock fault, separating the Sierra-Tehachapi from the Mojave Desert, is the southern geologic boundary. The Great Valley's Cenozoic gravels (derived primarily by erosion of the Sierra) overlap and conceal the western extensions of the Sierra. The Sierran basement terminates near the western margin of the Great Valley, presumably in contact with the Franciscan Formation. The exact relationship between these two major rock units, Sierran and Franciscan, is debated, but the contact is probably a major fault.

The Sierra is from 65 to 160 kilometers (40–100 miles) wide. Its elevations vary from 120 meters (400 feet) at the Great Valley boundary to summits of more than 4,250 meters (14,000 feet) adjacent to the Basin Ranges. Extensive vertical movement on the Sierra Nevada fault system has produced an almost unbroken eastern wall for more than 150 kilometers (100 meters), making descents to the adjacent valley extraordinarily precipitous. The Sierra bifurcates near Lake Tahoe, which occupies a down-dropped fault block between the Sierra Nevada proper on the west and the Carson Range on the east. Important geographic names are given in Figures 3-1, 3-2, and 3-3.

ROCKS: SUBJACENT SERIES

An important section of the earth's outer crust is exposed in the Sierra Nevada. The rocks are mainly igneous and metamorphic units of diverse composition and age, including volcanic and metasedimentary interlayered rocks. In the central and southern Sierra, plutonic igneous rocks, mostly silicic (granitic), form the multiple intrusions of the Sierra Nevada batholith and constitute 60 percent of the exposed rock. Early geologists called the metamorphic and igneous rocks (basement) the *subjacent series*. Sedimentary and volcanic rocks that overlie the basement are most prominent in the central and northern Sierra. These are known as the *superjacent series* and are considered later in the chapter.

Although the bulk of the plutonic rocks in the Sierra Nevada are late Jurassic to late Cretaceous in age, some Triassic and even Paleozoic age plutons are present in

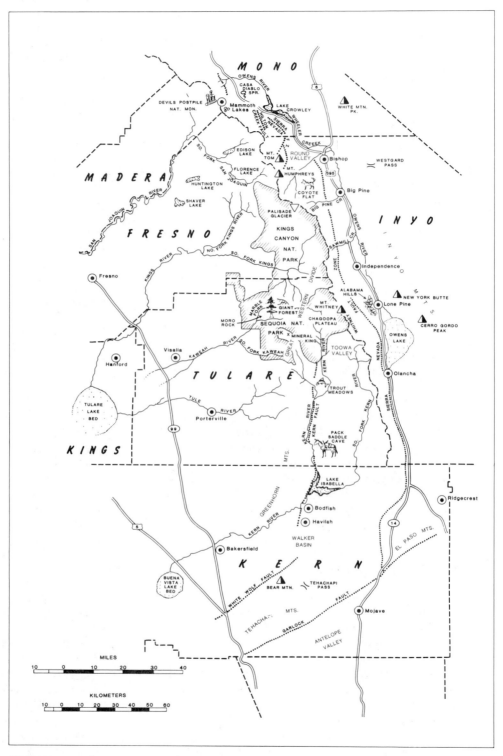

FIGURE 3-1 Place names, southern Sierra Nevada.

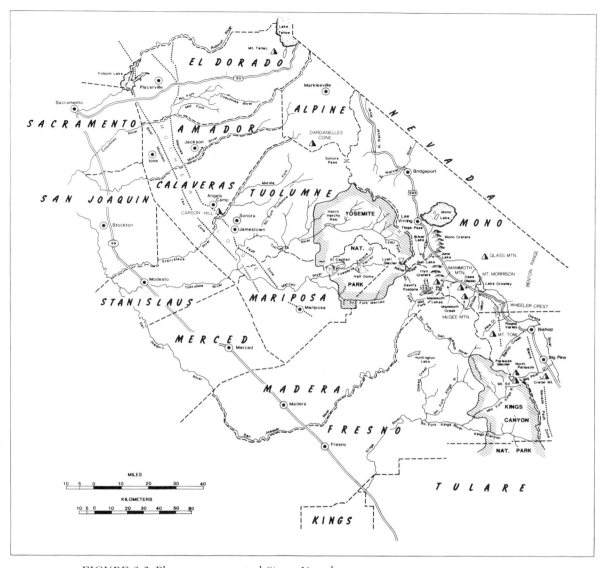

FIGURE 3-2 Place names, central Sierra Nevada.

the eastern and northern part of the range. The Sierra Nevada batholith is a composite of more than 100 distinct plutons, some of which have surface exposures of nearly 1,500 square kilometers (570 square miles).

The late Jurassic to late Cretaceous plutons represent intrusive events that are a part of what is known as the Nevadan Orogeny, which also included intense deformation and metamorphism of the older sedimentary rocks that covered the developing batholiths. The Nevadan Orogeny produced the Nevadan mountains—site of the present Sierra Nevada. The modern Sierra presents a rejuvenation of the deeply eroded Nevadan mountains, mainly by faulting.

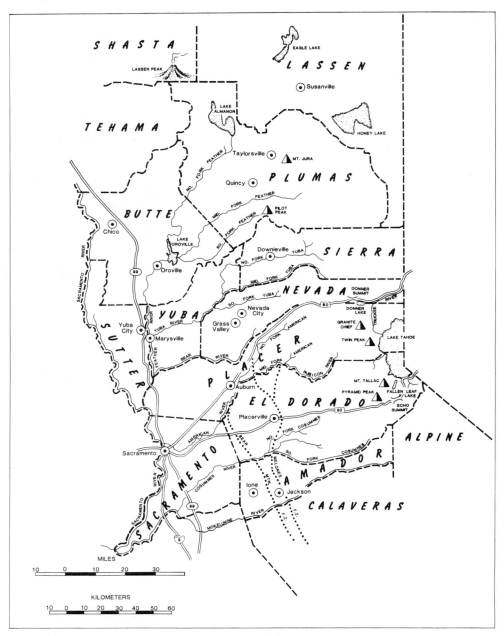

FIGURE 3-3 Place names, northern Sierra Nevada.

Basement Metasedimentary Rocks

For many years the oldest known Sierran rocks were the Ordovician metasediments found in the Mount Morrison roof pendant near Crowley Lake[1] but in 1980, Lower Cambrian archaeocyathids were discovered in the Big Pine pendant about 60 kilometers (40 miles) to the south. Roof pendants from near Tioga Pass southward to Big Pine include rocks form all or almost all Paleozoic periods from Cambrian to Permian, though no one roof pendant is known to have a complete sequence (Figure 3-4 and 3-5). These roof pendants, particularly the southern ones, are clearly the westward more metamorphosed continuation of the Paleozoic rock sequence so well exposed in the White and Inyo mountains directly to the east. In the vicinity of Mount Morrison, about 9,700 meters (32,000 feet) of hornfels, chert, marble, slate, and quartzite reflect nearly continuous deposition during the Paleozoic era. The metamorphic rocks in the west-central and northwestern Sierra, particularly north of the Mother Lode, are often less deformed and metamorphosed than those in the southern portion of the range. The Shoo Fly complex in the northwestern Sierra is truncated by late Devonian volcanic rocks like the Sierra Buttes Formation. The lithologically diverse Shoo Fly complex is thus pre-late Devonian, and parts may be much older. Approximately 7,600 meters (25,000 feet) of marine deposited pyroclastic debris, lava flows, chert, and tuffaceous siltstone make up the seven formations of the Paleozoic island arc sequence that overlies the Shoo Fly complex with marked unconformity and ranges from late Devonian to Permian in age. Although no completely continuous Paleozoic section is known, all Paleozoic periods are represented in the province. The Paleozoic rocks show a higher percentage of volcanic and volcaniclastic rock in the northern Sierra than in the central and eastern parts of the range.

In addition to the metasedimentary roof pendants definitely established as Paleozoic by fossils or radiometric dating, extensive but apparently unfossiliferous pendants occur in the southern Sierra. These are composed of greenstones and other metamorphosed volcanic rocks, with subordinate amounts of metasediments, and are usually designated as Paleozoic. Although high percentages of volcanic rock are typical of roof pendants definitely established as Mesozoic (like those at Mineral King and in the Ritter Range), thick volcanic sections alone do not necessarily reflect Mesozoic age.

Mesozoic sedimentary rocks are especially widespread in the west-central and northwestern Sierra. Triassic sequences up to 3,000 meters (10,000 feet) thick consist of metavolcanics and sedimentary rocks derived by weathering of volcanics. (Isotopic dates from the volcanic source suggest a Permian age.) The Hosselkus limestones and Swearinger slate have yielded Triassic fossils from their exposures at Taylorsville at the head of the North Fork of the Feather River in Plumas County. Jurassic metamorphic rocks and fossiliferous marine sediments are found from the Ritter Range northwest about 450 kilometers (280 miles) to Mount Jura. The Mount Jura section near Taylorsville is nearly 4,600 meters (15,000 feet) thick and consists of at least 14 named formations.

At least two major unconformities are recognized within the roof pendant rocks,

---

[1]Roof pendants are remains of the older covering rocks into which the granitic magmas intruded.

Composite Geologic Column, Northern Sierra Nevada

HOLOCENE	Dredge tailings, alluvium	
PLEISTOCENE	Glacial deposits, alluvium	
PLIOCENE	Mehrten, Table Mountain Latite	
MIOCENE	Valley Springs (nonmarine), various volcanics	
EOCENE	Ione (nonmarine), River channel deposits	
CRETACEOUS	Chico (marine)	
JURASSIC	Mariposa { Copper Hill (marine) / Salt Spring (marine) / Logtown Ridge (volcanic) }	Intrusive rocks of the Sierran batholith
TRIASSIC	Swearinger (marine) Hosselkus Limestone (marine)	
PERMIAN	Robinson (marine) Reeve (marine)	Melange and Ophiolite
PENNSYLVANIAN	Goodhue (marine)	
MISSISSIPPIAN	Peale (marine) Taylor (volcanic) Arlington (marine)	Calaveras Group
DEVONIAN	Sierra Buttes Taylorsville (marine)	
SILURIAN	Bullpen Lake (marine) Montgomery (marine) Grizzly (marine)	Shoo Fly Complex
CAMBRIAN-ORDOVICIAN	various rocks	

Stratigraphic Column, Southern Sierra Nevada

QUATERNARY	Glacial and stream deposits	
PLIOCENE	Basalt — Andesite, rhyolite, latite, basalt	
MIOCENE		
EOCENE	Ione	
CRETACEOUS	Sierran batholithic intrusives Kings River roof pendant rocks	
JURASSIC		
TRIASSIC	Kern River and Kaweah River roof pendants	Kernville Series
PERMIAN	Ophiolites	
CARBONIFEROUS		

FIGURE 3-4 Selected geologic columns, Sierra Nevada.

FIGURE 3-5 Split Mountain roof pendant with granitic rocks (light) intruding metasedimentary units (dark). Note other roof pendants in the ridges beyond. A typical view of Sierran rocks, this picture was taken looking west across the Sierran crest in the vicinity of Big Pine. Arcuate glacial moraines resulting from ice recession are also shown. (Photo by Spence Air Photos, courtesy of Department of Geography, University of California, Los Angeles)

demonstrating that several periods of tectonic disturbance interrupted the deposition and accretion of the Sierran basement before the onset of the Mesozoic Nevadan orogeny. In the Mount Morrison pendant, an unconformity (Antler orogeny) separates Pennsylvanian from older rocks, and in many places late Triassic metasediments rest unconformably on Permian and older Paleozoic rocks (Sonoma orogeny).

It is currently acknowledged that early to middle Jurassic sediments were the youngest rocks involved in the Nevadan orogeny. Upper Cretaceous sediments lie by transgressive unconformity on the basement in the northern and north-central parts of the range's west flank. This suggests that the Nevadan orogeny developed prominent mountains from the opening of the middle Jurassic through the middle Cretaceous.

Pre-Cenozoic sedimentation on the site of today's Sierra formerly was attributed to the presence of a (Cordilleran) geosyncline during much of the Paleozoic and most of the Mesozoic. This shallow trough was presumed to have extended from the Gulf of Mexico or eastern Pacific north to the Arctic, periodically submerging much of western North America. This view has been superseded by the concept of a borderland with sedimentation occurring along continental margins in subsiding miogeoclinal (shelf) and eugeoclinal (slope) belts. The continental margin of North America in the Sierra-Klamath area is thought to have grown oceanward by tectonic accretion during the Phanerozoic, an idea consistent with continental drift and plate tectonic theory. The sedimentary prisms that accrue on the forward and trailing margins of plates are commonly described as accruing in geoclinal environments.

In summary, thick Paleozoic sediments in the south and in the Shoo Fly complex on the west were followed by equally thick Mesozoic volcanics in the northwest. Early Mesozoic volcanic and volcaniclastic deposits thicken from southeast to northwest along the Sierran axis. In the south, only fragments of the cover survive today, as roof pendants. It appears that during the Paleozoic miogeoclinal conditions dominated in the Ritter Range and south Sierra, but oceanic crustal material was accreted to the Sierra along the Calaveras-Shoo Fly Thrust and the northern part of the Melones fault in the Mesozoic. The rocks west of the Melones fault are so unlike those east of the fault that they are thought to be exotic blocks attached to North America in the early Triassic.

Basement Igneous Rocks

Plutonic igneous rocks form more than 100 plutons in the Sierra Nevada. The earliest intrusives are small, mafic, often coarse-textured with hornblende phenocrysts in a plagioclase matrix, and generally dioritic and gabbroic. Remnants of these intrusives became inclusions in later granitic rocks, or became part of the metamorphic sequences. The dark plutonic intrusives currently are considered forerunners of the widespread plutonic invasion, during which most of the Sierran batholith was constructed. Some workers have emphasized the dioritic and gabbroic plutonic phases, suggesting that perhaps they represent remnants of a larger magmatic episode preceding the Jura-Cretaceous granitic invasion. Isotopic dates from gabbroic-dioritic bodies yield ages 10 to 15 million years older than those of the major granitic bodies.

Such age differences are inconclusive for establishing an earlier mafic intrusive episode. Devonian plutons, like the Bowman Lake batholith near the Middle Fork of the Yuba River, have been recently identified in the northern Sierra, but may be parts of an accreted exotic block.

Sierra Nevada Batholith

The Sierra Nevada batholith is composed of granitic rocks variously described as granite, quartz-monzonite, granodiorite, and quartz diorite. These rocks were intruded into pre-Jurassic roof rocks as a series of overlapping plutons and have ages of about 88 to 206 million years. Plutons of similar granitic rocks in the White Mountains east of the Sierra have been dated at 70 to 225 million years. Both sets of plutons are considered part of the general Cordilleran plutonic episode of western North America. As the study of batholithic rocks has progressed, the Sierra Nevada

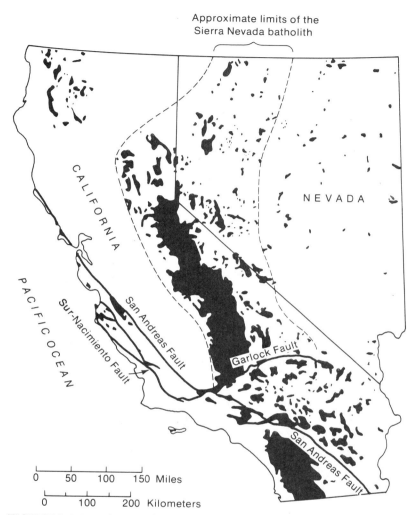

FIGURE 3-6 Distribution of granitic rocks in the Sierra Nevada batholith. (Source: Geological Society of America)

batholith itself has come to embrace the expanse of silicic plutonic rocks in the White-Inyo Mountains and western Nevada, plus the granitic rocks of southern California (Figure 3-6).

Several dozen granitic plutons have been mapped. Each pluton has specific temperature-pressure composition characteristics that determine the final rock product. The span of ages currently established suggests that the plutons were emplaced from early Triassic in the east-central Sierra and White Mountains to late Cretaceous west of the Whitney Crest. A study of these isotopic dates indicates that the plutons were emplaced during a span of approximately 150 million years in bursts of plutonic activity with intervening periods of quiescence or readjustment. These plutonic episodes constitute the minimum age spread of the Nevadan orogeny, assuming, as

TABLE 3-1 Some Dates of Nevadan Pluton Emplacement in the Sierra

Chronologic Age (maximum in m.y.)	Geologic Age	Location
85–70 (?)	Latest Cretaceous	White Mountains
90–79	Late Cretaceous	Cathedral Range (south of Yosemite)
98	Late Cretaceous	Donner Lake
121–104	Middle Cretaceous	Huntington Lake
148–132	Early Cretaceous	Yosemite
180–160	Middle and Late Jurassic	Inyo Mountains
210–195	Early and Middle Jurassic	Lee Vining
225 (?)	Early Triassic	White Mountains

is customary, that the orogeny began when the plutonic activity first affected the roots of the mountains and that it closed when plutonic activity ceased. Plutonic activity is summarized in Table 3-1.

The batholithic granites display varied grain size and mineral content. Dark-colored minerals occur in high proportion in some rocks, but are absent in others (alaskite). Consequently, the rate of weathering and the forms of the weathered rocks vary substantially.

A common granite is a porphyritic rock in which feldspar crystals, often $2\frac{1}{2}$ centimeters (1 inch) on a side are embedded in a light-colored, finer-grained matrix. This granite may occupy many square kilometers and its crystals are often so abundant that a tenth of a square meter (1 square foot) of outcrop may show several dozen. Porphyritic plutons and plutons of even grain with typical "salt-and-pepper" appearance often are closely associated. These differences in granite have been described in detail in the Yosemite Valley region where careful mapping has delineated many variations in the rocks.

Other differences distinguish the granitic plutons. For example, large areas frequently occur where thousands of fine-grained dark blebs are contained in the parent rock. These inclusions sometimes are derived by the fragmentation of rocks of the older country rock cover into which the plutons were intruded (xenoliths). In other cases, the blebs represent segregations crystallized from the magma itself before final congealment (autoliths).

Plutons may be separated from one another by septa or screens of older metamorphic rocks, or by younger dike rocks, but in many cases they abut one another directly without separation and are distinguished by their contrasting mineral compositions and contact relationships.

Dikes, sills, and veins—usually lighter in color than the enclosing rocks— may represent late-stage intrusions into the parent. This process is thought to occur near the final consolidation point of the magma. Some plutons have gneissic textures with foliation and lineation where minerals are aligned before complete consolidation of the magma. Such rocks should not be confused with metamorphic gneisses.

How do plutonic bodies form? What combinations of energies are at work? Are all plutons of similar origin? Is there such a thing as a parent magma? Are some plutons formed in place by heating but not melting of older rocks, thus producing nonmagmatic (granitized) plutons? What theories of origin best explain Sierran rock bodies? Commonly offered models are as follows.

1. Magma is generated in the lower crust by melting of crustal basement. Sedimentary and volcanic materials that were deposited on the margins of continental boundaries as the continents were depressed by subduction into a zone of melting may be added to the magma from lower crust melting.
2. Magma forms in subduction zones where oceanic plates are carried beneath continental plates. Magmatic differentiation occurs as magma rises through the crust toward final emplacement.
3. Magma is mantle-generated, pulsating upward from deep-seated "hot spots" into the crust during periods of increased temperature. (This is the interpretation for magmatic origin in areas like the Hawaiian Islands.)
4. Various combinations of 1, 2, and 3.

The origin of Sierran intrusive rock is debated vigorously. Worldwide, granitic batholiths are restricted to areas underlain by continental plates. All lines of evidence strongly suggest crustal sources for the magmas that produced Sierran plutons.

Most research on the origin of intrusive rock of western North America has been conducted by the U.S. Geological Survey. A major contributor to Sierran petrogenesis is Paul C. Bateman, who has mapped and interpreted Sierran geology for more than 30 years. His studies have been complimented by the work of such senior scientists as L. D. Clark, J. B. Eaton, N. K. Huber, Anna Heitanen, Warren Hamilton, C. D. Rinehart, Richard Schweickert, and Jason Saleeby.

In 1967, Bateman proposed that magma was generated by depression and downfolding of a geosynclinal accumulation of Precambrian and early Paleozoic sediments whose compositions produced granitic magma. Emplacement accompanied and followed compression in the tectonic episode initiating the Nevadan orogeny. Generally speaking, he suggested that the plutons formed 3–8 kilometers ($1\frac{1}{2}$–5 miles) below the surface.

By 1974, new field data had prompted Bateman to present an alternate model. He proposed that magma was generated in an east-dipping subduction zone, where the Pacific plate slid beneath the North American plate. The magma then migrated upward to fuse lower crustal material. This process generated magmas with compositions intermediate between those derived primarily from mantle sources and those derived from crustal sources.

Studies of the rubidium and strontium isotopic components of Sierran plutons reveal a high incidence of rubidium in the trace-element chemistry, with highest percentages in the range's eastern axis. This supports the view that granitic magmas were derived primarily from crustal materials, contaminated only minimally by increments of mantle. Studies in the northern Sierra reinforce the position that granitic magmas are crustally derived, whereas the older gabbroic and dioritic plutons are probably from mantle sources.

Structure

Faults. Prebatholithic rocks are complexly faulted, especially in the northwest and north-central Sierra. Faults of the Mother Lode both predate and postdate the batholithic invasions (Figure 3-4). Fault patterns were partially responsible for the idea advanced by earlier workers that mountain building occurred in the Permian or late Pennsylvanian (Sonoman orogeny). This view is being revived because investigators have reported evidence in the Klamath Mountains for late Devonian thrust faulting (Antler orogeny), late Paleozoic faulting (Sonoman orogeny), and a later Jurassic episode of mountain building (Nevadan orogeny). These episodes also seem to be recorded in the stratigraphy of the northwestern Sierra, where pre-Jurassic faults are regionally significant (Figure 3-7). Although correlation between prebatholithic faulting in the northwestern Sierra and the Klamath Mountains cannot

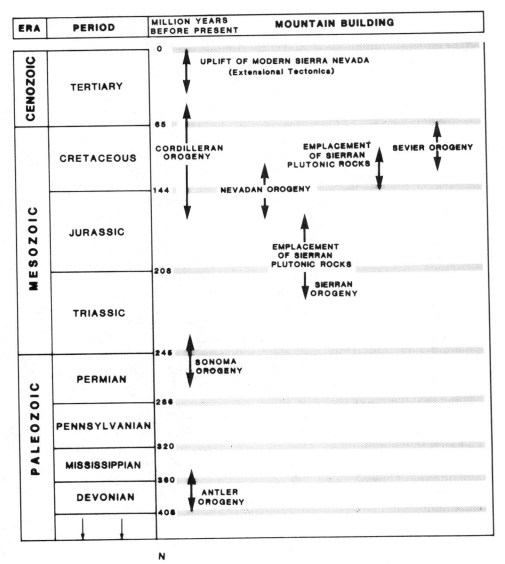

FIGURE 3-7 Major orogenic events in the history of the Sierra Nevada.

be established unequivocally, the preponderance of evidence from both structural relations and rock sequences provides a very strong case for the similarity of history in these two provinces.

The western Paleozoic and Triassic plate rocks of the Klamath province are quite similar to the Calaveras complex of the northwest Sierra. The Northern Melones fault is an east-dipping thrust fault along which older rocks thrust westward over the Calaveras complex. The multiple plate fragments, which include oceanic crust and overlying volcanic and sedimentary material, are discussed more fully in the Klamath Mountains discussion in Chapter 4.

Some rocks, like those of the Culbertson Lake Terrane (North Yuba River) were added to the continent 410 million years ago in the late Devonian. The youngest plate or piece of associated rock joined the northwest Sierra at the beginning of the early Jurassic, 185 million years ago. Each successive docking of new material moved the continental boundary westward until, as the Nevadan Orogeny began, the coast was located along the current Sierran foothills and jutted westward from the northern Sierra to encompass the Klamath Mountain province.

The easternmost of the prominent faults that cut the western slope of the central Sierra Nevada is the Calaveras-Shoo Fly thrust, a steeply east-dipping fault of middle Jurassic or older age, extending from near Auburn south to the vicinity of Mariposa. This fault is regarded as pre-Nevadan because it is folded with its enclosing rocks near Sonora and is cut by a plutonic rock dated at 170 million years. Some geologists believe that this fault extends northward to the North Fork of the Feather River along the east branch of the northern Melones fault.

The most prominent fault in the Sierran foothills is the Melones, which has been traced from the northernmost Sierra near the North Fork of the Feather River southward beyond the Merced River near Mariposa. Some geologists extend it even farther southward to the vicinity of the Tule River in Tulare County. This fault zone is a major structural boundary between older Sierran rocks to the east and presumed exotic terranes that arrived from the west during the early Triassic Sonoma Orogeny. Because the suites of rocks on either side of the Melones fault zone are so dissimilar, it is not certain whether this steeply-dipping fault has reverse or normal slip. But because it may be correlated with faults in the Klamath Mountains, which are clearly thrusts, reverse slip is regarded as likely by most geologists, and it may be an old subduction zone.

Westernmost of the three Sierran foothill fault zones is the Bear Mountain fault, separating two subparallel belts of rock that may be parts of single Triassic island arc or two separate assemblages accreted to the continent (Figure 3-8).

All three faults are marked by elongate slivers of serpentinite and other sheared metamorphic rocks, which give rise to greenish-black shiny exposures in road cuts in the Mother Lode belt. In many places these fault zones are closely associated with the gold-bearing quartz dikes that occur in the Mother Lode along the Melones fault. South of Placerville, these faults are reasonably well-defined, but to the north they appear to divide into a complex of smaller faults.

Significant postbatholithic faulting does occur in the basement complex, but it is older than present Sierran block faulting. Prominent older faults are the Kern Canyon fault and the sheared and lineated margins of Dinkey Creek and other plutons. Additional older faults are the lineaments along subordinate crests of the Sierra like

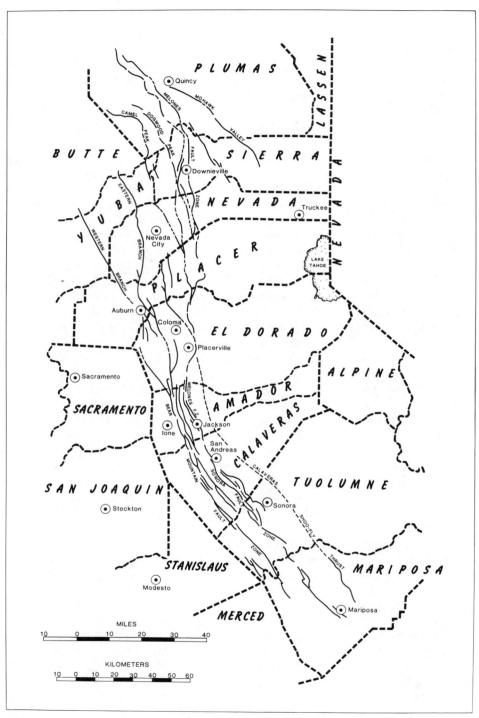

FIGURE 3-8 Principal faults of the Sierra Nevada province.

the Great Western Divide, where aligned scarps have been interpreted as fault-line features.

The Kern Canyon fault is one of the larger structures entirely within the Sierra. Located in the north-south trending upper Kern River drainage, the fault parallels the river for 120 kilometers (75 miles). Movement on the Kern Canyon fault must predate the formation of an elevated erosion surface that truncates the fault. Movement could not have occurred in the last 3.5 million years, because a basalt flow of that age covers the erosion surface. Some geologists have suggested a connection

FIGURE 3-9 Kern Canyon: view to the south from approximately Golden Trout Creek. This segment of the canyon is cut along the Kern Canyon fault, seen in the west wall (right foreground). (Photo courtesy of U.S. Geological Survey)

between the Kern Canyon fault and the White Wolf fault of the Tehachapi Mountains–Great Valley boundary, but little evidence supports this view. Figure 3-9 shows how strongly the Kern Canyon fault has influenced drainage direction and present topography. The Kern River follows the fault to produce a prominent fault-line valley, but Figure 3-10 shows that the fault and present-day drainage may not always coincide.

FIGURE 3-10 Kern Canyon fault: view to the south from south of Salmon Creek. Kern River follows irregular canyon to the right of the fault-line valley. (Photo courtesy of U.S. Geological Survey)

Folds. Folding has been continuous and complex since the early Paleozoic. Pre-Devonian deformation is now well-documented in the northwest Sierra. Folds in the basement Shoo Fly complex are cut by the late Devonian Bowman Lake pluton. At the culmination of the Jurassic Nevadan orogeny, the thick Sierran sediments had been transformed by progressive waves of deformation into a giant overturned syn-clinorium with steeply east-dipping beds on the western margin and equally steep east-dipping beds on the eastern margin. This gross structure includes many lesser folds of local importance. How many times folding deformed the subjacent formations is not known, but evidence for at least several episodes has been reported.

Joints. Sierran joint patterns, displayed mostly in plutonic rocks, are important in the modern landscape. Both vertical and horizontal directions can be seen. Many of these joints are sheetlike unloading fractures that are produced as erosion of rock overlying the plutons allows them to expand and crack. These joints are further enlarged by weathering. Other joints are produced by the mountain-building forces that have uplifted the Sierra. Of the vertical type, northwest-southeast systems are especially prominent. They are imposed on all basement rocks regardless of age and have been critical in developing giant domes like those of Yosemite Valley and Kings Canyon. The horizontal pattern is expressed by sheet joints that appear as horizontal or nearly horizontal structures in plutonic rocks. These provide the "layer-cake" rock patterns typical of many high ridges in the Sierran interior (Figure 3-11).

Many faultlike structures occur in plutonic igneous rocks, but they are not mapped

FIGURE 3-11 Sheet jointing in granitic rocks, Tioga Pass, Yosemite National Park. Slabs are 25 to 50 centimeters (1–2 feet) thick. (Photo by R. M. Norris)

FIGURE 3-12 Lineaments in Sierran granitic rocks. (Photo courtesy of U.S. Geological Survey)

easily because there are no readily available planes of reference for establishing displacement. Some of these structures are probably no more than joints, although others may be faults of large separation. High-altitude satellite photography has revealed hitherto unmapped lineaments in Sierran terrains. Figure 3-12 suggests rectilinear lineament patterns, prominent in the larger exposures of plutonic rock. The patterns probably reflect intersecting joints or ancient fault traces like the Kern Canyon. Fieldwork may eventually describe these linear expressions, but often vegetation, soil, and uniformity of rock type preclude field confirmation.

Unconformities. Many local and regional interruptions occurred in the history of Cordilleran sedimentation. There were periods of deformation, erosional interludes, and intervals of no deposition at all—preceded or followed by major and minor folding. This pattern has been established by detailed studies of the Proterozoic, Paleozoic, and Mesozoic beds exposed in the ranges from the Wasatch of Utah to the coastal mountains of California and Oregon.

ROCKS: SUPERJACENT SERIES

Overlying the Sierran basement are late Mesozoic and Cenozoic sediments and volcanics deposited after the long erosional interval that followed the Nevadan orogeny. Although the region generally had low relief during this interval, local relief may have reached about 1,000 meters (3,300 feet). The Cenozoic volcanic flows and volcaniclastics were deposited discontinuously. Nonmarine basins received continental, lacustrine, and fluviatile deposits, and marine sediments overlapped the western margin of the former Nevadan Mountains. These disconnected deposits have been labeled the *superjacent series* (Figure 3-4).

North of the Tuolumne River, the superjacent series covers ancient river channels. South of the Tuolumne, the superjacent cover is extremely fragmented, irregular, and discontinuous. Apparently volcanism was restricted or possibly Cenozoic erosion has removed much of a once-extensive cover. The limited volcanic cover in the southern Sierra is often quite young, sometimes even interbedded with Pleistocene glacial deposits. Prominent and extensive erosional surfaces are preserved under remnants of volcanic flows in the southern Sierra west of Whitney Crest, and in the Kern, San Joaquin, and Merced river drainages.

Northern Sierra

Cretaceous marine sediments overlap Sierran basement rocks along the Sacramento Valley margin. The sediments crop out discontinuously and include fossiliferous sandstones, shales, and basal conglomerates. Maximum exposed thicknesses are about 850 meters (2,800 feet). Well records have shown that Cretaceous fossiliferous sandstones occur on the Sierran basement as far south as the San Joaquin River.

Eocene sedimentation is represented by up to 300 meters (1,000 feet) of marine and fluviatile sediments resting on deeply weathered basement rock. The upper part of these Eocene beds is known as the Ione Formation, an interesting deposit containing a variety of clays, sands, and coals of economic value. The Ione is discontinuously exposed in the lower foothills from Butte County south to Madera County. The formation takes its name from the hamlet of Ione in western Amador County.

Ione clays are mined for making a variety of products from common red brick to whiteware and high-temperature ceramics. The sands, some of high purity, are mined for glass-making, and the lignites, formerly mined as a low-grade fuel, are now processed for the montan wax they contain. This wax is used for many purposes, including shoe polish and phonograph records. The Ione deposit has made the United States self-sufficient in montan wax.

The post-Eocene gravels and conglomerates distributed along the western third of the Sierra north of the Tuolumne River are younger than the Ione Formation but older than the initial volcanism characteristic of the superjacent series. Studies of the oldest gravels (extensively mined for gold) has permitted reconstruction of early Tertiary drainage patterns and gradients. Eocene and Oligocene streams drained west and southwest, forming a major network of five rivers (Figure 3-13). Subsequently, other gravels were deposited and interbedded with the volcanics whose eruption greatly altered the older drainage by shifting and burying former channels. In some places these interbedded gravels are gold-bearing, but they are less extensive and have less consistent values than the prevolcanic sequences.

There are three major episodes of Tertiary volcanism documented by the rocks of the northern Sierra. The Valley Springs Formation, which crops out east of Sacramento, is representative of the oldest (late Oligocene to early Miocene) volcanism. These rhyolitic flows and volcanic sediments are 19.9 to 25 million years old in the Valley Springs Formation and as old as 33.2 million years in other units. The Table Mountain latite that caps Table Mountain at Sonora is 9 million years old. This flow

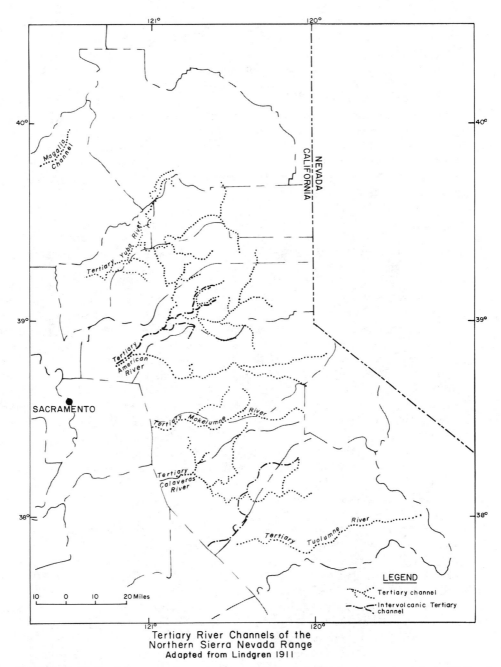

Tertiary River Channels of the
Northern Sierra Nevada Range
Adapted from Lindgren 1911

FIGURE 3-13 Buried stream channels concealed beneath Tertiary volcanic rocks (solid lines), and within the volcanic rock sequence (dashed lines). Most of these channels were discovered during gold prospecting as many of them yielded valuable placer deposits. (Source: California Division of Mines and Geology)

and other mid- to late Miocene andesites and basalts represent a second volcanic episode (see Figure 3-19). The youngest volcanism is Pliocene to Holocene. These young rocks are mostly basalts and range from 3.5 million years old to much younger. Curiously, one of the most extensive Tertiary units consists of volcanic mudflows whose ages span these three episodes. The Mehrten Formation crops out in the northern Sierra foothills. It ranges from 4 to 20 million years old.

Southern Sierra

Typically, southern Sierran volcanic rocks have been extruded from isolated vents, are limited areally, and are younger than most of those in the northern Sierra. An excellent example of this volcanic pattern occurs in the Owens Valley, north of Independence. Here the steep Sierran escarpment shows spectacular, deep V-shaped canyons from which streams are building huge alluvial fans that form aprons extending almost the width of the valley. In the canyon cut by Sawmill Creek, a basaltic lava flow emerged from vents high on the escarpment, flowed down Sawmill Canyon, and displaced a stream. Subsequently the stream recut its canyon through the lava flow. Field study has established that the lava erupted between glacial stages. Morainal deposits are preserved above and below the lava, which congealed no more than 90,000 years ago (Figure 3-14).

Of the three volcanic episodes of the northern superjacent series, only the younger

FIGURE 3-14 Basalt flow at the mouth of Sawmill Canyon near Independence, Inyo County. This flow has been dated at between 60,000 and 90,000 years old and rests on a glacial till, presumably of Sherwin age. It is also overlain by a till of Tahoe age. Neither till can be seen in the photo although the stream-cut notch demonstrates appreciable erosion since the lava was emplaced. (Photo by R. W. Webb)

two are recognized in the south. Basalts characteristic of the second episode have been found in Miocene flows on the San Joaquin River and on Coyote Flat west of Bishop. Basalts of the third episode are represented by flows in the Kern River Canyon at Trout Meadows, in the South, Middle, and North forks of the San Joaquin, in the Owens River gorge, and at McGee Mountain in southern Mono County. Pleistocene to Holocene volcanism occurs at Mono Lake, Devils Postpile, Sherwin Hill, the volcanic fields of the Owens Valley, Toowa Valley volcanic field, Mammoth Mountain, and Inyo and Mono craters.

HISTORY OF THE NEVADAN MOUNTAINS

In the northern Sierra Nevada the basement rocks are the Shoo Fly complex. The oldest rocks are truncated by the 410-million-year-old Bowman Lake batholith on the North Fork of the Yuba River. A part of this complex is called the Bullpen Lake sequence. Study of this sequence indicates that the chemistry of the included pillow lavas and the structure and chemistry of rhythmically bedded cherts and some limestone present are typical of seamounts formed on oceanic plates. There are many modern seamounts, old volcanoes, some with eroded flat tops (guyots) now lying well below sea level, in the north Pacific. If this supposition is correct, then the Shoo Fly complex contains old ocean crust, and some geologists are convinced that the Shoo Fly complex is a pre-Devonian subduction complex, which has incorporated large slabs of sediment from land-derived deep sea fans as well as portions of the continental slope and rise rocks. These rock units were stacked upon one another during subduction and all are now overlain by the late Devonian Grizzly Formation.

The Grizzly Formation is the base of a series of Paleozoic island arc deposits that accumulated during the late Devonian to early Mississippian Antler orogeny. The late Paleozoic Calaveras complex appears to have joined the continent during the early Triassic Sonoma orogeny when it was thrust *beneath* the Shoo Fly complex. Evidence recorded in roof pendants and large country rock inclusions (described earlier) seems to indicate a dominately depositional environment for much of the central and southern Sierra throughout the Paleozoic. This combination of continental shelf accumulation and terrane accretion was profoundly altered, folded, and faulted by the Nevadan orogeny, which began early in the Jurassic. The dominately folded mountain belt produced during the Nevadan orogeny was extensively intruded by the granitic batholiths, which today dominate the higher parts of the range.

The western boundary of the Calaveras complex is the Melones fault zone. The Nevadan orogeny may have been caused by the arrival of a complex island arc at the Melones thrust boundary. The island arc collision added metavolcanic and metasedimentary rocks west of the Melones fault to the continent. Today these are known collectively as the Foothill Metamorphic Belt. By the end of Jurassic time, the Nevadan Mountains faced the sea along what is now the eastern edge of the Great Valley. Subsequent marine invasions have never extended farther east.

The development of the Nevadan Mountains poses some intriguing geologic problems. What is the magmatic source for the plutonic episode? How much rock of the Cordilleran depositional basin was originally available for deformation? How much (if any) was used in making the granitic magmas? What percentage of the original sediment do today's outcrops represent? How high were the Nevadan Mountains? How much erosion of cover (roof) rock occurred before the granitic bodies could be exposed?

How high were the Nevadan Mountains? Speculation is easy, but evidence regarding elevations of any former mountain range is usually obscure. Normally, at least three approaches are used: (1) thickness of deposits that were derived from the mountain mass; (2) slopes of reconstructed land surfaces preserved under later deposits; and (3) high mountain ridges, especially those with north-south trend, which block approaching weather systems and create climatic effects possibly reflected in nonmarine deposits in the lee of the range.

Evaluating the Nevadan Mountains under the first approach incorporates the fact that post-Jurassic marine sediments in the western Sacramento Valley are at least 7,600 meters (25,000 feet) thick. Sedimentary structures establish that the Cretaceous sediments were derived from sources to the east and northeast, presumably the Nevadan Mountains. If a 7.6 kilometer (5.75 mile) thickness of sediment accumulated in an adjacent marine basin, did the Nevadan Mountains concomitantly rise this amount? Probably not, although some unroofing of the mountain mass is likely. Rock fragmented by erosion occupies more volume when deposited than the original did, so a direct volume correlation between mountains and basin is unjustified.

The Cretaceous rocks involved are all shallow-water marine types, with fossils and conglomerates throughout the section. Consequently, the environment must have been primarily near shore and relief of the source area reasonably great. Since the conglomerates contain cobbles of moderate (fist) size, it is plausible also that there existed either large streams with extensive drainage areas or short streams on steep slopes, or both. Furthermore, many of the cobbles are granitic and can be traced mineralogically to the granitic plutons of the Nevadan batholith.

The second approach, reconstructing ancient slopes, does not usually yield many clues because most buried profiles have been altered significantly by tilting. Nevertheless, interest in locating gold-bearing channels has produced more data about stream slopes in the Nevadan Mountains than exists for most fossil ranges, a substantial height for the ancient mountains is inferred.

Under the third approach, both pollen grain studies and strictly geologic criteria have been used. Marked aridity definitely occurred east of the mountains, implying a persistent and prominent rain shadow. The rain shadow would be produced by a high mountain range in the path of rain-bearing storms from the Pacific.

Based on these data, speculation that the Nevadan Mountains were as high or higher than the modern Sierra seems justified, although estimates as low as 1,350 meters (4,500 feet) have been proposed. Certainly the Nevadan Mountains were more extensive geographically than the Sierra Nevada. Moreover, it is not unreasonable, in view of the isoclinal folding in some roof pendants, to suggest a significantly high elevation for at least some parts of the Nevadan Mountains and minimal crests of 3,000 meters (10,000 feet).

In summary, Nevadan orogenesis followed deposition of dominantly marine sediments in the east and dominantly submarine volcaniclastics in the west. Repeated disturbances both during and after deposition produced a major asymmetric synclinorium, characterized by extensive faulting on its west limb. Granitic magmas, generated below or within the synclinorium and probably heated by subduction, penetrated the fractured synclinal rocks. The resulting plutons were concentrated centrally in the syncline. Assimilation and differentiation during magmatic emplacement produced the compositional variety within and among plutons. With impetus from depth and concomitant compression, the Nevadan orogeny built a substantial range that contributed vast amounts of sediment primarily to the west, forming the Great Valley sedimentary sequence.

The erosion that stripped 11 kilometers (7 miles) or more of cover seems to have been largely completed by the close of the Eocene, leaving slowly moving streams draining an eastern crest. The overlying sequences of volcaniclastic and volcanic flows were deposited upon the eroded roots of the Nevadan Mountains. Subsequent elevation by faulting rejuvenated the region and produced today's Sierra Nevada.

GEOMORPHOLOGY OF TODAY'S SIERRA NEVADA

An impressive aspect of today's Sierra is the even skyline that stretches for miles across the summit and near the summit (Figure 3-15). Some geologists believe that these subdued summit uplands represent a fossil landscape now being destroyed by erosion. This surface was formed by earlier erosion across Nevadan roots and subsequently rejuvenated tectonically to present elevations. In this view, the undulating upland surface is the plain (peneplain) to which the Nevadan Mountains eventually were reduced. This landscape was a lowland with elevations possibly as high as 1,050 meters (3,500 feet), but with the crest line of the Nevadan roots near today's Great Western Divide, at an elevation of 600 to 750 meters (2,000–2,500 feet). The present landscape results from erosion in the late Tertiary and Quaternary and rejuvenation primarily due to movement on the Sierra Nevada fault. A new crest line, the Whitney or Muir Crest, was formed. Rapid uplift, aggregating more than 3,000 meters (10,000 feet) of vertical movement, mostly in the last 3 million years, produced the escarpment facing Owens Valley as the entire block was tilted west. Streams then rapidly cut new features into the old topography (Figure 3-16). The Sierra Nevada's elevation augmented the cold period dominating the earth during the Pleistocene, and glaciers developed, further modifying stream canyons.

Structure

Until pioneer geologist A. C. Lawson briefly studied the Kern River drainage in the early 1900s, the significance of Sierran high-level surfaces had not been noted. Lawson's classic papers on Kern River topography, followed by Francois Matthes's studies of Yosemite Valley, stimulated geologists to look more carefully.

Matthes systematically presented a generalized interpretation of the Sierra as a range, faulted extensively on the east and less so on the west, and containing rejuvenated streams that dissected an older subdued landscape. Matthes's simplified concept is reproduced in Figure 3-17. Furthermore, Matthes recognized that streams incise their valleys in response to changes in base level. This produces canyons within valleys and canyons within canyons, features that are particularly helpful in interpreting geomorphic history.

Evolution of the Sierra and its rate of uplift reflect the fault systems on which uplift is occurring. Evidence for boundary faults is prominent along the east side of

FIGURE 3-15 Even skyline summit of Sierra Nevada. Looking southwest from latitude of Bishop, view includes the entire summit, gently sloping west and south to the San Joaquin Valley seen in the distance. Note erosional contrast between the summit Sierra and the lower western flank. Farmlands in the foreground (center) are in Round Valley. Steep ridge at lower right is Wheeler Crest. Pine Creek drains into Round Valley through paired moraines of Tahoe-Tioga glaciation. Mount Tom is in right center foreground, south of Pine Creek. Tungsten Hills lie between Round Valley and farmlands in left center. The lower gorge (bottom center) of the Owens River has been cut through Bishop tuff. Bishop Creek (center) drains into the Owens Valley south of the Tungsten Hills. In the hazy far left is the canyon of the south-draining Kern River. Note subordinate flats like Coyote Flat southeast of Bishop Creek and the flat immediately southwest of Mount Tom. The black spot on this flat is Desolation Lake, largest natural lake in the southern Sierra. (Photo courtesy of U.S. Air Force and U.S. Geological Survey)

FIGURE 3-16 Sierra Nevada eroded fault scarp, as seen from the Owens Valley (east). Fresh trace of the fault is indicated by the abrupt change of slope and the fault facets terminating each ridge. Note that facets on the right do not align with those on the left. Note also the upland flats to the west, beyond the crest; the conical peak on the right is Monachee Mountain. (Photo courtesy of U.S. Geological Survey)

the range, but it is rarely seen along the west margin except in the Kern River-Tehachapi area.

Sierra Nevada Fault System. Like other faults, the Sierra Nevada fault is a zone of movement, not a line. The faults in this zone have an en echelon course and dominately dip-slip offset (Figure 3-18). The vertical displacement is as much as 1,500 meters (5000 feet) in the north and 3,350 meters (11,000 feet) in the south. Movement on the fault began in the middle Miocene and accelerated in the Pleistocene, as demonstrated by the overlap of glacial deposits, volcanics, and fault scarps. The Sierra Nevada fault resembles Basin and Range faults, of which it is probably the westernmost expression. The Sierran block is the largest of the north-south trending tilted fault blocks scattered across Nevada east to the Wasatch block of central Utah.

As noted in the caption for Figure 3-18, the eastern face of the Sierra Nevada is an eroded fault scarp; it does not represent the fault surface itself because the eastern face, as steep as it appears, seldom exceeds 30 degrees except for short distances, whereas the frontal fault has a dip of about 70 degrees.

Greenhorn Fault System. A total uplift of 600 to 2,200 meters (2,000–7,000 feet) has been suggested for the Greenhorn fault system, which bounds the western Sierra from the San Joaquin River south to the Tehachapi Mountains. Movement on the Greenhorn is thought to have accelerated the deformation initiated by large-scale doming of the southern Sierra. Prominent scarps occur irregularly. The Greenhorn presumably contributed to the development of the plateaulike form of the southern Sierra Nevada, which contrasts with the tilted-block pattern north of the San Joaquin River.

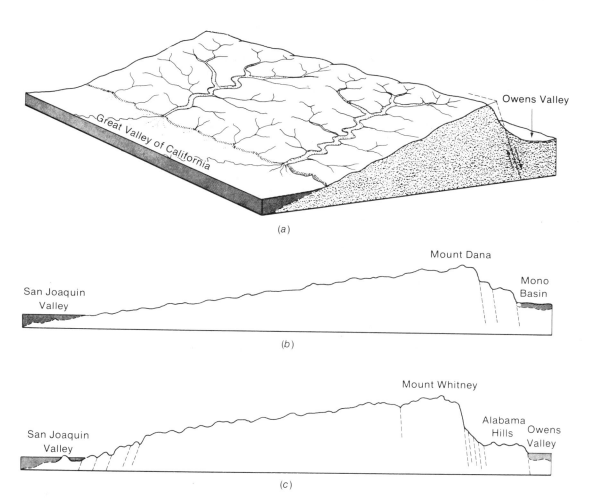

FIGURE 3-17 (a) Generalized diagram of part of tilted Sierran and Owens Valley blocks. Relative directions of movement are indicated by arrows. Height and slant of the Sierran block are much exaggerated. Streams are shown in characteristic arrangement, the main rivers flowing down the western slope with many of their tributaries at approximately right angles. No specific streams are represented. In front is a strip of the Great Valley, with thick layers of sand and silt derived from the elevated Sierran block. At the back is a strip of the Owens Valley, veneered with a thinner layer of sediment.

(b) Idealized section across the Sierra Nevada, representing the simple "textbook conception" of its tilted-block structure. This representation approximates the facts known about the range in the latitude of Yosemite Valley.

(c) Idealized section across the Sierra Nevada in the latitude of Sequoia National Park. Step faults exist at both western and eastern margins. Some distance from the western foothills, peaks of the down-faulted, buried fault block emerge as isolated hills above sediments of the San Joaquin Valley. (Source: U.S. Geological Survey)

FIGURE 3-18 En echelon patterns of the Sierra Nevada fault system seen from the north near Independence. The road to Onion Valley is in the foreground. Note the irregularity of the much-eroded Sierran escarpment, with its reentrants and small frontal hills. Compare with Figure 3-33, which shows the same escarpment near Lone Pine. The even skyline of the summit upland is easily seen. (Photo courtesy U.S. Geological Survey)

The existence of a western fault system has been questioned, however, because the scarps of the supposed faults are both discontinuous and irregularly aligned. Since the faults, if they exist, are presumably high-angle normal faults, the irregularity is almost inexplicable even if erosion modified initial escarpments. In addition, basalts of the San Joaquin River drainage were extruded before the stepped topography was produced and apparently are not displaced.

EVOLUTION OF THE SIERRAN LANDSCAPE

Most of the landscape or land surface seen in the Sierra Nevada is quite young, but in some places we can see scarps and remnants of landscapes that very likely date back to late Tertiary time. These include small, gently westward-sloping areas near the Sierran crest. Earlier it was believed that some of these surfaces were as old as Eocene, but most recent studies have shown that few, if any, are that old, and such old landscapes are preserved only where they have been buried by lava flows and lahars (volcanic mud flows). Only tiny bits and selvages of these old surfaces are preserved where modern erosion has removed the protective cover, but not quite removed the formerly buried surface. Nevertheless, we can see many of these older erosional surfaces in cross-section in numerous road cuts and on canyon walls. Miners followed some of these old surfaces by tunneling along gravel-filled channels beneath the volcanic rocks.

As was previously noted, in the headwaters of the Kern and San Joaquin rivers, lava flows fill some canyons cut into older erosional surfaces. By dating the flows, some of which are 9.5 million years old (late Miocene), we can demonstrate that the old erosional surface cut by the younger canyons must be at least 9.5 million years old. In the Kern River area, similar flows have been dated at 3.5 million years, demonstrating a Pliocene age for the erosional surface there. Similarly, the Table Mountain latite, dated at 9 million years, fills a stream channel and extends over an old erosional surface east of Sonora (Figure 3-19). Doubtless these Tertiary erosional surfaces or benches on either side of lava-filled canyons have been somewhat retouched by Quaternary erosion, but most geologists agree that they do indeed represent largely intact remnants of a Tertiary landscape.

Following the Nevadan orogeny in latest Jurassic time, the new mountains underwent a long period of erosion during which much of the covering metamorphic rock was stripped off, baring the underlying Nevadan plutonic rocks. There were almost certainly short-lived periods of renewed uplift, but by Eocene time (about 50 million years ago) the Sierra had been reduced to a relatively subdued range of low mountains, probably less dramatic than our central Coast Ranges are today. These mountains were flanked on the west by a broad, swampy coastal plain facing the sea, which lay near the location of our present Great Valley. The climate was warm and moist, perhaps nearly tropical as shown by the coal beds and deeply weathered clays and sands of the Ione Formation.

By the close of Oligocene time, the stream valleys and many intervening divides

FIGURE 3-19 Topographic inversion at Table Mountain: view to the northeast, from Jamestown. Light bands on the flow are unrelated to its course and may be the surface expression of northeast-trending regional jointing, propagated upward from the basement. (Photo by Spence Air Photos, courtesy of Department of Geography, University of California, Los Angeles)

and hills were buried under copious amounts of volcanic ash of rhyolitic composition, which forms a pinkish rock called tuff. The Valley Springs Formation is representative of this rock in the Mother Lode area. We do not know the location of the ancient volcanoes from which all this ash came, but it is likely that they lay somewhere to the east of the Sierra, perhaps in western Nevada.

Before these ash beds blanketed much of the northern Sierra, mainly from the Merced River north to the Feather River, streams had cut canyons deeply enough to expose gold-bearing quartz veins forming locally rich placer deposits in the stream gravels. During the nineteenth century when gold mining was at its height in the Sierra Nevada, miners followed some of these buried streams for long distances by tunneling under the tuffs. Although the tuffs themselves did not yield gold, they did provide an easily quarried and worked building stone that was used for construction; many historic buildings in Mother Lode belt towns were made from this rock.

About 20 million years ago, new streams had been established on the blanketing tuffs and when the next wave of volcanic activity began, basaltic lava flows and voluminous andesitic mudflows followed these newer channels, filling many and covering much of the older landscape. These deposits make up what we know today as the Mehrten Formation. Examples of this and closely related rocks can be seen along Donner, Carson, and Ebbets Pass roads. Grassy fields developed on the Mehrten, and related rocks are often little more than cobble fields nearly covered by thousands of melon-sized volcanic rocks, making them nearly impossible to cultivate.

FIGURE 3-20 Looking north across a glaciated section of the Chagoopa Plateau in the headwaters of the Big Arroyo, a tributary of the Kern River. (Photo courtesy of U.S. Geological Survey)

West of Jamestown and Sonora, in the Mother Lode foothills, is the flat-topped sinuous Table Mountain. This mesa is composed of the Table Mountain latite, up to 90 meters (300 feet) thick with a maximum width of 600 meters (2,000 feet) (Figure 3-19). The flows are 9 million years old. Superficially, Table Mountain resembles other mesalike landforms: a resistant lava cap burying weaker rock undercut by erosion to maintain steep cliffs on the more durable lava flow. Actually, however, Table Mountain is part of a larger lava flow that coursed down a river canyon, filling the channel and displacing the river. Subsequent erosion has inverted the topography so that the former lava-filled stream valley is now a ridge. The Table Mountain lava flows can be mapped discontinuously for more than 60 kilometers (40 miles). Such concise examples of reversed topography are rare.

In the upper Kern River Canyon is a prominent erosional surface known as the Chagoopa Plateau whose elevation is about 2,750 meters (9,000 feet). Lava flows that partly cover the Chagoopa surface are 3.5 million years old, showing that the Chagoopa surface is of Pliocene age; canyons 250 to 275 meters (800–900 feet) deep have been filled by these lava flows, indicating that the Chagoopa was considerably dissected before the lava was extruded (Figure 3-20).

A very controversial set of associated erosional surfaces has been recognized in the Kern River area. Some of these may represent fragments of old, nearly destroyed lowland landscapes, but many could have been produced by processes active at high altitude; the issue is not yet settled. One of these supposed old erosional surfaces, the Boreal Plateau, lies at an altitude of about 3,500 meters (11,500 feet) and is pictured in Figure 3-21.

FIGURE 3-21 Looking north from Cottonwood Lakes across the Whitney (Muir) Crest (left foreground) and the Boreal Plateau (right center). Note cirque embayments of the Whitney Crest. Each is a different size, depending primarily on the extent of headward stream erosion that preceded glacial excavation. (Photo courtesy of U.S. Geological Survey)

Most of the landscape in the higher parts of the Sierra shows the effects of Pleistocene glaciation. The later glacial stages have left the best and most extensive record because each glacial readvance tends to destroy or modify features produced during earlier advances. Because the different glacial events varied in intensity and length, however, minor readvances often failed to obliterate traces of earlier glacial events. Thus, the effects of the latest advances we call the Matthes and Recess Peak left behind their limited effects only in the higher parts of the range. Similarly, the Tioga, the last major advance, was unable to completely destroy the features left by the Tahoe glaciers, the most extensive of all Sierran Pleistocene glacial episodes.

Stream erosion has now replaced glaciation as the dominant erosional process almost everywhere in the Sierra Nevada. Streams are vigorously dissecting former glacial valleys, reducing the heights of waterfalls, filling glacial lakes with sediment, or draining them by deepening their outlets. Unless glaciation returns to the Sierra, streams eventually will erase almost every trace of today's dramatic glacial landscape.

The present landscape does not consist solely of erosional features, but includes many examples of constructional features produced by continuing volcanism. Although there is abundant evidence that volcanic activity has been waning in intensity since Pliocene time, it has continued right up to the present. Features like Mono and Inyo craters, Mammoth Mountain, Devils Postpile, Crater Mountain near Big Pine, the volcanic tableland near Bishop, and numerous lava flows and cones prove the recency of volcanic activity.

GLACIATION

A wide array of glacial features exists in the Sierra Nevada. Almost all of the landscape in the higher parts of the range is the product of glacial and periglacial processes. Nevertheless, despite the ubiquitous evidence of past glaciation, only about 60 small glaciers remain today. These are probably best described as glacierets, although the largest—Palisade Glacier—is about 1.5 kilometers (1 mile) long. Similarly, small patches of ice persist on the upper slopes of Mount Shasta and in a few sites in the Klamath Mountains. There are no longer any sizable glaciers anywhere in California. Moreover, the state's existing glaciers nearly disappeared during the warm interval that affected much of the world between about 1850 and 1950. Even a moderate warming would destroy California's remaining glaciers.

During the last main glacial stage (the Wisconsin), the Sierra Nevada had a discontinuous ice cap that extended about 450 kilometers (250 miles), from the Feather River to the upper Kern River. In several places, this ice cap was more than 48 kilometers (30 miles) wide. Few, if any, areas outside the polar regions sustain such a large ice cap today. The longest glaciers from this summit ice cap flowed about 100 kilometers (60 miles) down the gentle western slope toward the San Joaquin Valley. Glaciers on the eastern slope did not exceed 16 kilometers (10 miles) in length.

Glaciers have sculptured some of the world's grandest scenery, whether it be landscapes so well displayed in Glacier or Jasper national parks in the Rocky Mountains of United States and Canada, or the dramatic coastal scenery in the fjord country of Norway and New Zealand. Glacial landscapes were the main reasons for establishing Yosemite as a state park in 1866 and later as a national park, and for establishing Kings Canyon and Sequoia parks in the southern Sierra Nevada.

Evidence

Although signs of former glaciers are evident throughout the high Sierra, it has been only about 100 years since the influence of vanished ice sheets was accepted by the scientific community. The Sierra Nevada's numerous steep-sided, U-shaped valleys, including Yosemite (Figure 1-21) and the upper Kern River (Figure 3-9), all reflect glaciation. Cirques, often with almost vertical walls, frequently contain small lakes enclosed by bedrock and some by moraines deposited during the last glacial advance.

The varied effectiveness of the ice as it moved over the bedrock produced many depressions now occupied by small lakes called *tarns*. Sierran glaciated valleys often show distinctly stepped profiles. Gentle or even reversed slopes on the treads of these stairs are interrupted by steep, abrupt risers whose upper edges may form rims just high enough to contain small lakes, called *beaded* or *paternoster lakes* (Figure 3-22).

Rocks excavated at higher elevations by glacial ice frequently have been carried some kilometers downslope and then abandoned when the ice melted; these transported rocks are called *erratics*. It is usually apparent that these rocks are derived from distant sources, although they may be compositionally the same as the rocks on which they rest. Erratics commonly lie scattered over glacially planed, bare surfaces like Olmstead Point on Tioga Pass or west of Donner Summit on Interstate 80. If an erratic is distinctive enough to establish its source, direction and amount of ice movement can be calculated (Figure 3-23).

Other glacial geomorphic features are the numerous moraines that may form ridges 100 meters (330 feet) or higher. Generally, moraines are better developed on the

FIGURE 3-22 Beaded or Paternoster lakes on a glacial stairway. View east from near the summit of Piute Pass, elevation 3,484 meters or 11,423 feet, Bishop Creek, Inyo County. (Photo by R. M. Norris)

FIGURE 3-23 Glacial erratic boulder perched on a glacially eroded surface at an elevation of about 3,350 meters (11,000 feet), Piute Pass Fresno County. (Photo by R. M. Norris)

steep eastern slope of the Sierra than on its western side. Lateral moraines are conspicuous near the mouths of east-facing Sierran canyons between Tioga Pass and Bishop. In Figure 3-15, the two streams flowing into the cultivated areas are Bishop Creek (left) and Pine Creek (right). Both have prominent lateral moraines of Tahoe age. Those of Pine Creek extend from the base of the mountains onto the floor of Round Valley. Loop-shaped terminal moraines occur throughout the Sierra, but perhaps the best known are at Convict Lake north of Bishop. Here young moraines help form the basin that contains the lake. Another notable example is south of Lake Tahoe at Fallen Leaf Lake, where 10 to 15 arcuate moraines curve around the lake's lower end (see Figure 1-22).

Sequence

Although the existence of several stages of Sierran glaciation had been previously suspected, Francois Matthes proposed the first set in 1930. Based on his studies on the western slopes, he recognized three stages: from oldest to youngest, Glacier Point, El Portal, and Wisconsin. Assigned to the Wisconsin were the youngest and least weathered moraines, both lateral and terminal. El Portal deposits, on the other hand, were more deeply weathered and generally incorporated lateral moraines without terminal loops. El Portal was believed to reflect a more extensive glaciation than did the Wisconsin. The Glacier Point stage was thought to be preserved only as scattered erratics on the weathered bedrock that had escaped later glaciation. In 1931, as a result of his work on the eastern slopes, Eliot Blackwelder proposed a

TABLE 3-2 Sierran Glacial Sequence

Epoch	Eastern Slope tills	Western Slope tills	Age in years before 1950	Mid-Continent stage
HOLOCENE	Matthes till		0–700 (before existing glaciers)	
	Unnamed till	not yet recognized	1,000	
	Recess Peak till		2,000–6,000	
	Unnamed till		6,000–7,000	
			10,500	
PLEISTOCENE	Hilgard till		11,000	
	Tioga till	Wisconsin till	18,000–20,000	
	Tenaya till		26,000	Wisconsin
	Tahoe till II		65,000	
	Casa Diablo till		72,000	
	Mono Basin till		128,000	
	Donner Lake till	not yet recognized	145,000	Illinoian
	Tahoe till I			
	Hobart till		260,000–440,000	
	Reds Meadow till		590,000–700,000	Kansan
	Long Valley till		700,000	
	Sherwin till	El Portal till	750,000	Nebraskan
	McGee till	not yet recognized	1,600,000	
PLIOCENE			1,800,000	

Note: Matthes used "Wisconsin" for the latest glaciation in the Yosemite rather than a local name.

four-stage sequence: from oldest to youngest, McGee, Sherwin, Tahoe, and Tioga. Both sequences have been greatly expanded since Matthes and Blackwelder did their pioneer work. Table 3-2 gives the sequence of glacial tills now recognized, chiefly on the eastern slopes.

Most tills of 50,000 years and younger are dated by radiocarbon methods, but much glacial chronology remains relative. There are few radiometrically controlled dates older than 50,000 years, the limit of the radiocarbon method. Moreover, because of the uncertainties of the radiocarbon method, even the dates for younger glacial events do not necessarily equate with calendar years. The ages of older tills are even less well controlled, but some anchor points occur in the sequence.

Although considerable uncertainty persists in developing an acceptable glacial sequence for the Sierra Nevada, some success has been achieved in tying together worldwide sea level curves, which correlate with glacial advances and retreats, oxygen isotope records from the shells of marine organisms that reflect the cooling and warming of the seas, and the magnetic reversal chronology based mainly on volcanic rocks. Although perfection continues to elude us, the sequence of glacial events summarized in Table 3-2 represents a more modern refinement of the sequence given in the first edition of this book. Readers should expect additional refinements and corrections in the future.

One of the anchor points in the glacial chronology is the basaltic flow that nearly reaches the mouth of Sawmill Canyon near Independence (Figure 3-14). This flow

FIGURE 3-24 Bishop tuff resting on Sherwin till at Sherwin Grade summit on U.S. Highway 395. (Photo by R. M. Norris)

has yielded a potassium-argon date of 60,000 to 90,000 years. It rests on an older till, presumably of Sherwin age, and is in turn overlain by a Tahoe till. The Bishop tuff has a potassium argon age of 730,000 years and rests on Sherwin till in southern Mono County (Figure 3-24). The Sherwin till, in turn, rests on a basalt dated at 3.2 million years. Hence the age of the Sherwin till is older than 730,000 years and much younger than 3.2 million years; the best guess today is 750,000 years. The Casa Diablo till, also in southern Mono County, is capped by one basaltic flow and rests on another. The overlying basalt has a date of 64,000 ± 14,000 years and the underlying basalt an age of 129,000 ± 26,000 years. Basaltic flows dated at 2.6

million years lie beneath the McGee till. Both basalt and till have been elevated as much as 900 meters (3,000 feet) by faulting largely completed before the Bishop tuff was extruded. This substantial displacement emphasizes the accelerated movement on the Sierra Nevada fault system during the late Pleistocene. Unfortunately, this last example neither provides a tight age control for the McGee till nor allows us to distinguish it from the Sherwin till. Along Rock Creek in southern Mono County, however, undoubtedly Sherwin till rests on an older, more weathered till, possibly of McGee age.

Of the glaciations listed in Table 3-2, the Tenaya and Tioga are best displayed on the western slopes and the Tahoe on the eastern slopes. Older glacial deposits seldom have any topographic expression and are usually buried under younger materials. On the east side of the Sierra, the largest and most prominent moraines that extend from canyons are generally of Tahoe age, and the less prominent ridges and morainal loops within the Tahoe embankments are usually Tioga. Tahoe and Tioga moraines with this association are well displayed at Convict Lake. Tenaya and Tioga moraines have similar patterns and often are indistinguishable, but the older Tenaya moraines are more likely to be modifed by later erosion. Correlating the glacial deposits of the western and eastern slopes with those of the central and northern parts of the Sierra has been difficult, because most tills are confined to canyons and are not continuous for appreciable distances. Moreover, the older tills have been disrupted by subsequent glacial advances and by stream erosion.

Modern Glaciers

The 60 or more glacierets remaining in the Sierra Nevada are restricted to cool, moist, north-facing or northeast-facing slopes that receive minimal direct sunshine even in summer. The glaciers are limited to these sites because no part of today's Sierra is high enough to permit snow to lie on south-facing slopes indefinitely. The hypothetical elevation above which some snow would persist on south-facing slopes is called the *climatic firn limit*.

It is still widely believed that today's little glaciers are merely the shrunken remnants of grander Pleistocene forebears. Francois Matthes pointed out, however, that *all* present glaciers probably are less than 3,000 years old; his opinion is now sustained by most Sierran glaciologists. The relatively warm period that followed the final major glacial stage (Wisconsin) is known as the *climatic optimum* or *hypsithermal interval*; it probably began between 9,000 and 7,500 years ago and ended about 3,000 years ago. It is unlikely that the Sierra contained any glacial ice at all during most of the warmer hypsithermal.

The first of three recent ice advances known collectively as the Neoglaciation, or Little Ice Ages, began about 2,800 years ago. The smallest advance reached a maximum about A.D. 1350, and the last and probably largest about A.D. 1750. Marked shrinkage of all Sierran glaciers took place between 1850 and 1950, although some small ice advances occurred even then. If continued, the widely documented cooling tendency observed since 1950 undoubtedly will induce another advance of Sierran and other glaciers, if accompanied by incomplete snow melt in the summer. Conversely, global warming attributable to the "greenhouse effect" could easily counteract this cooling.

SUBORDINATE FEATURES

Tehachapi Mountains

The Tehachapi Mountains separate the Great Valley from the Mojave Desert and rise to 2,115 meters (6,934 feet) at Bear Mountain. The Great Valley-Tehachapi topographic boundary is primarily between sedimentary rocks in the valley and crystalline rocks in the mountains, expressed by a fault-line scarp eroded along the White Wolf fault. Movement on the White Wolf fault presumably triggered the 1952 Kern County earthquakes.

The rocks of the Tehachapi Mountains are compositionally like those of the Sierra Nevada. Moreover, the late Paleozoic or early Mesozoic sedimentary record and the middle Mesozoic intrusive record of the Nevadan orogeny is well exposed. Tehachapi roof pendants show the same northwest-southeast trend as those of the Sierra, but as a mountain block the Tehachapi trend is more nearly east-west. This appears to reflect left-slip on the Garlock fault, the master structure separating the Tehachapi Mountains from the Mojave Desert. The Garlock fault either cuts off the Sierra Nevada fault in the vicinity of the El Paso Mountains or the Sierra Nevada fault curves west to join the Garlock. In any case, the result is the major barrier of the Tehachapi Mountains, which separate northern and southern California. Since early Tertiary time, except for a brief invasion in Miocene time in the western Antelope Valley, this barrier has kept the sea from reaching any part of the Mojave Desert.

Within the Tehachapi Mountains are several flat-floored valleys described as grabens. Although unequivocal evidence of boundary faults is not available, the valleys either have internal drainage or show evidence of former internal drainage. Consequently, graben origin is essential.

Round Valley-Bishop Creek Embayment

Owens Valley is a narrow linear valley more than 240 kilometers (150 miles) long, usually fewer than 13 kilometers (8 miles) wide, and bounded by major faults. Near Bishop, a wide reentrant increases the width to about 30 kilometers (20 miles). This reentrant is Round Valley, bounded on the west by an extremely steep Pleistocene scarp that is the east face of the Wheeler Crest. (Wheeler Crest is sometimes called Wheeler Ridge and should not be confused with the oil-producing area of the same name in the southern San Joaquin Valley.) To the south, Round Valley rises about 1,300 meters (4,200 feet) across an alluvial fan to form a ramplike surface (the Coyote Warp) that culminates at the head of Bishop Creek at about 3,000 meters (10,000 feet). The Wheeler escarpment dissipates to the south, merging into the chain of Sierran summits of which Mt. Tom is the northern anchor. East of Bishop Creek, Coyote Flat drops steeply to the Owens Valley from 3,650 meters (12,000 feet). To the north, Round Valley is blocked by the Bishop volcanic tableland. Figure 3-15 places Round Valley in topographic perspective.

Movement and distribution of individual faults of the Sierra Nevada system are nowhere better demonstrated than in the Round Valley reentrant. The Sierra Nevada block faces east at Wheeler Crest. Here the major active Sierran fault is the Wheeler Crest fault, which emerges from beneath the volcaniclastic rocks and glacial debris of the Bishop volcanic tableland to become the major boundary between the Sierra Nevada and Round Valley for 15 to 25 kilometers (10–15 miles). To the south, the main Sierran escarpment is offset eastward about 16 kilometers (10 miles) by another Sierran fault, which bounds Owens Valley in the Bishop-Big Pine area. This en echelon pattern typifies the Sierra Nevada fault system throughout most of its 640 kilometer (400 mile) definable length.

Bishop Volcanic Tableland

Bounding Round Valley on the north is a major volcanic tableland that covers about 850 square kilometers (325 square miles) (Figure 3-25). It is composed of Bishop welded tuff, which is probably derived from the Long Valley-Lake Crowley area in southern Mono County. The tableland extends east and south to within a few kilometers of Bishop, and its tuff forms bold escarpments and other prominent landscape features in northern Owens Valley. The tuff was deposited across an irregular ridge-valley topography cut by both glacial and preglacial streams. As mentioned earlier, its association with glacial tills has provided a means for dating some Sierran glaciations (Figure 3-24).

The upper surface of the tuff is cut by numerous very young, well-preserved normal faults of north-south trend. These faults are often a kilometer or two long and show displacements of as much as 10 meters (30 feet) near the middle of the fault, dying away to nothing at both ends (Figure 3-26).

The Bishop tuff is not a single volcaniclastic unit, but is composed of many layers

FIGURE 3-25 Bishop volcanic tableland (upper right), with Wheeler Crest in the background. The Tungsten Hills are the low hills in the upper left. City of Bishop is in the foreground. The scarp at the edge of the tableland has been cut by the Owens River. (Photo by Spence Air Photos, courtesy of Department of Geography, University of California, Los Angeles)

FIGURE 3-26 Young fault scarp in the Bishop Tuff. Note how the 7-meter-high scarp dies out in the distance. This same effect can be seen on the scarp at the extreme left. These scarps developed after the tuff was deposited about 730,000 years ago. (Photo by R. M. Norris)

of tuff, ash, and pumice, all of which, however, are believed to have erupted in a very short time, probably days or weeks at most. Typical Bishop tuff is gray when fresh and weathers to pink, white, and brown. It includes glass and fragments of earlier-formed volcanic materials, plus granules of silicic minerals. The tuff's surface is irregular, with hummocks produced by gas that escaped during deposition, and shows prominent columns and radial joints. The rock is considered a welded tuff (ignimbrite), the product of incandescent, gaseous clouds of ash and pumice that showered the countryside, ultimately producing pasty flows of glowing material that moved quickly downslope into the Owens Valley as well as northward into a large area south of Mono Lake.

Because of its significance in glacial chronology (it is dated at about 730,000 years) and as an unusual volcanic deposit, the Bishop tuff has received considerable attention. It has been mined for building stone and pumice, and the picturesque gorges of the Owens River and Rock Creek are cut through the tuff into the older rocks below. A large, bizarrely shaped mass of Bishop tuff is on display in front of the Inyo County courthouse in Independence.

Long Valley and Crowley Lake

Long Valley, a major feature between Bishop and Mammoth Lakes, was occupied during glacial times by a lake (Long Valley Lake) as deep as 90 meters (300 feet); the shorelines, some examples of which are preserved on the southeast side of the basin, can easily be seen from Highway 395. This lake was drained by tectonic depression of its southeastern rim and by the downcutting of the Owens River. The valley is now occupied by the artificial Crowley Lake, formed by damming the Owens River. Long Valley occupies about 40 percent of what geologists refer to as *Long Valley caldera*. About 60 percent of the caldera is made up of a central group of hills called a *resurgent dome* and by hills around the rim of the caldera. The total area of the Long Valley caldera is about 450 square kilometers (170 square miles).

Long Valley is bounded on the west and southwest by the Sierra Nevada, on the north by Glass Mountain, on the east by the Benton Range, and on the south and southeast by the volcanic tableland composed of the Bishop tuff. It is drained by the Owens River whose main tributaries are Hilton, McGee, Convict, and Mammoth (Hot) creeks.

The Long Valley area has an extensive history of volcanism. Even before the caldera formed about 730,000 years ago, andesite and latite lavas were erupted between 3.6 and 3.2 million years ago. The rhyolites of Glass Mountain were extruded between 1.9 and 0.8 million years ago, and the large caldera we see today formed with the eruption of the Bishop tuff about 730,000 years ago. This eruption was an event of cataclysmic proportions and of incredible violence. Nothing on such a scale has been observed anywhere in the world during historic time, although the geologic record shows many events of the same sort, some much larger, have taken place in various parts of the world. Perhaps the most recent example is one that occurred in central North Island, New Zealand, about A.D. 120. On that occasion, about 21,000 square kilometers (8,000 square miles) were blanketed with at least 15 centimeters (6 inches) of ash and pumice, and huge areas of mature forest and countryside were buried by incandescent, steaming ash.

During eruption of the Bishop tuff, approximately 600 cubic kilometers[2] (144 cubic miles) of superheated ash and steam were erupted, covering an area of 1,500 square kilometers (570 square miles) in upper Owens Valley and Mono Basin. Lesser amounts of this same ash spread far and wide by winds; the ash has been detected in the soils east of the Rockies from El Paso to eastern Kansas and Nebraska. In the central part of the caldera, ash is as much as 1,500 meters (5,000 feet) thick.

While this volume of material was being erupted, the ground surface collapsed, probably along a series of ring-shaped faults forming the Long Valley caldera. Residual volcanic activity was sufficient, probably during the ensuing 100,000 years, to raise a bulge more or less in the center of the caldera, the resurgent dome. Today the dome rises about 450 meters (1,500 feet) above the adjacent caldera floor. In May 1980, four strong earthquakes shook the Mammoth Lakes area on the southern rim of the caldera, and by 1982 more than 3,000 earthquakes (most quite small) had been reported in the area. Since then, the earthquakes have continued, though at a lesser frequency. One earthquake of magnitude 6 occurred in the summer of 1986 east of the caldera at Chalfant Valley. These earthquakes have been accompanied by a rise in elevation of the resurgent dome and by spasmodic increases in the activity of the boiling springs at Hot Creek. Some of these springs changed from quietly boiling to furiously boiling, almost geyserlike for brief periods followed by more normal behavior.

These manifestations have convinced geologists familiar with the area that an active

---

[2]About 600 times the volume of material erupted from Mount St. Helens in 1980.

magma body has been formed between 4.9 and 7 kilometers (3–4½ miles) depth below the caldera floor. Given the continuing history of volcanic activity over the past 3 or 4 million years and the evidence of eruptions within the last 500 to 600 years, some sort of new eruption would not be surprising. No one to date, however, has been willing to make a specific prediction for the next eruption.

As noted earlier, recent studies suggest that the Long Valley caldera is bounded on all sides by faults that collectively form a circular or ring-shaped trace. Although such faults are difficult to establish because of the overlapping volcanic rocks, some of them are very obvious as well as very young. One of these, on the southwest side of Long Valley, probably doubles as a segment of the Sierran frontal fault system and as a bounding fault for the caldera. This is the Hilton Creek fault, which extends south from Long Valley into the Sierra, nearly paralleling Hilton Creek (U.S. Highway 395 follows the escarpment of this fault for a kilometer or two south of the Convict Lake crossing). This fault offsets a moraine at McGee campground about 12 meters (40 feet) vertically. The Hot Creek geothermal area appears to lie on the northern trace of this fault, so it is clearly a young feature.

Inyo and Mono Craters

Inyo and Mono Craters are two separate but aligned volcanic centers often seen by visitors to the High Sierra. The alignment results from eruption along a major northwest-southeast fissure (probably a major fault in the Sierra Nevada system) that extends about 65 kilometers (40 miles).

The northern domes of the chain are the Mono Craters (Figure 3-27), a group of primarily rhyolitic tuff rings and obsidian plug domes that constitute about 30 eruptive centers. The steep-sided rhyolitic flows have oozed out of the tuff rings and are visible south of Mono Lake from U.S. Highway 395. The plug-dome character is apparent because some younger effusions have pushed up older materials, as if floating them, with the younger material squeezing out underneath. Conversely, some younger plugs have moved up through the older material and overflowed the crater rims. These flows were very viscous when erupted and formed short, steep-sided features locally known as coulees.

Mono Craters have ages ranging from about 40,000 years to 500 years. If the islands of Mono Lake are included, the flows and cones on Paoha and Negit islands are probably younger than 500 years. A minor eruption is alleged to have occurred beneath the waters of Mono Lake in the 1890s. Some observers, however, consider this incident simply the result of a temporary increase in thermal spring activity.

Panum Crater, a nearly perfect pumice and ash (tephra) ring with a central plug of obsidian, is the northernmost of the Mono Craters located on the southern shore of the lake. Logs buried by ash of this crater have given a radiocarbon date of 640 years. Interestingly, after the 1980 eruption of Mount St. Helens in Washington, a feature similar to Panum Crater was formed in the vent area of that volcano.

Inyo Craters, about 20 kilometers (12 miles) to the south, also experienced their most recent eruptive episode between 600 and 500 years ago. Three of the vents seem to have been active at that time, one of which was Obsidian Dome, a prominent feature that can be seen a short distance west of Deadman Summit on U.S. Highway 395. A kilometer or so south of Deadman Summit, the highway passes close to Wilson Butte, the northernmost of the Inyo Craters (Figure 3-28). This steep-sided obsidian

FIGURE 3-27 Looking southwest over Mono Craters. The plug-in-crater pattern shows clearly in the largest crater (left center). The rounded mountain in front of the main Sierra in the upper right is Mammoth Mountain. (Photo by John S. Shelton)

plug is believed to have been last active about 1,200 years ago. The Inyo Craters also include two explosion pits containing small lakes. These pits are associated with flows of obsidian and pumice and are among the youngest volcanic features of the chain, yielding radiocarbon dates between 550 and 850 years before the present.

June Lake District

The June Lake area lies at the base of a particularly steep part of the Sierra Nevada's eastern face. The dominant topographic feature is a horseshoe-shaped trough whose toe abuts the steep escarpment. The depression contains four lakes: Grant in the northeast arm, Silver near the toe of the horseshoe, Gull and June in the southeastern arm (Figure 3-29). It is the drainage rather than the lakes themselves, however, that makes the area of special interest.

In 1889, I. C. Russell was struck by the peculiar course of Reversed Creek, which drains Gull Lake and flows directly toward the Sierra instead of away from it as is usual for Sierran streams. Before reaching Silver Lake, Reversed Creek is joined by Rush Creek at the toe of the horseshoe, and the combined stream (called Rush Creek) flows into Silver Lake and from there into Grant Lake, which drains into Mono Lake.

FIGURE 3-28 Wilson Butte, a prominent member of the Inyo Craters. This feature is an obsidian and pumice plug with a composition much like that of the Mono Craters to the north. Mono County. (Photo by R. M. Norris)

Since the extension of the Los Angeles aqueduct system into the Mono area, Grant Lake has been used as a reservoir, so now much of Rush Creek's flow is intercepted before it can reach Mono Lake. This is the chief reason the level of Mono Lake has fallen in recent years.

Russell was the first to offer an explanation for the anomalous course of Rush Creek. He suggested that the horseshoe-shaped trough was formed by a glacier forced to divide into two arms by Reversed Peak (2,891 meters, or 9,478 feet), which rises about 550 meters (1,800 feet) above June Lake. Russell contended that the lobe that excavated the June Lake arm left a terminal moraine across the valley just below June Lake's present position, and that a large block of glacial ice remained in the basin while the parent glacier melted back into the Sierran uplands. It was thought that the ice mass and the moraine produced meltwaters that subsequently cut a channel toward the Sierran front.

Half a century later, J. E. Kesseli examined the area and pointed out the difficulty of explaining why some of the glacier obligingly remained in the June Lake depression, whereas its bulk melted back into the higher mountains. He proposed that the horseshoe-shaped trough was an old, entrenched meander curve of a large river that flowed from the north through Grant, Silver, Gull, and June lakes (Figure 3-30). The subsequent reversal of flow was accounted for by tilting of the valley block to the northwest, probably by faulting. This explanation also necessitated vigorous stream erosion in the Grant Lake arm, which is almost 120 meters (400 feet) deeper than the June Lake arm. The weakness of Kesseli's interpretation is that little independent evidence substantiates either the large meandering river or the tilting and deep erosion of the Grant Lake arm.

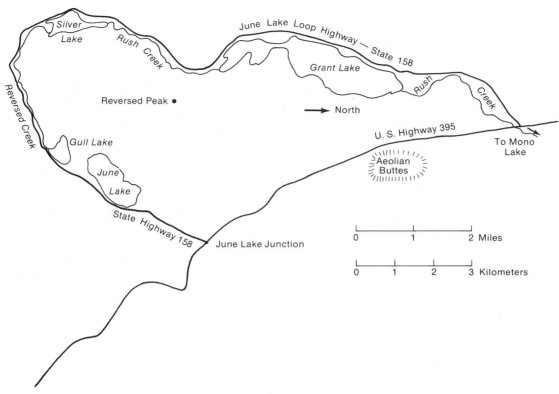

FIGURE 3-29 Drainage of Reversed Creek, June Lake area.

In 1949, W. C. Putnam concluded that the Rush Creek glacier did indeed divide into two lobes as it moved around Reversed Peak, but excavation was more effective in the Grant Lake area because bedrock there is less resistant than in the June Lake arm. The Grant Lake lobe of the glacier also was narrower and much thicker. The reversed slope in the June Lake arm merely expresses increased rock resistance along the arm. Furthermore, glaciers frequently cut valley floors with reversed slopes, because glacial motion depends upon the slope of the ice surface, not on the slope of the valley floor.

SPECIAL INTEREST FEATURES

Gold Mining

Most of California's gold came from the Mother Lode belt of the foothills of the western Sierra Nevada, but gold also was produced in quantity in the Klamath Mountains and to a lesser degree in the Basin and Ranges. Estimates of California's production vary, but it is probably a fair estimate that about 107 million ounces of gold have been extracted, mostly before 1900 and predominately from the Mother Lode. The Jamestown mine near Jamestown and the Carson Hill mine at Carson Hill, in Tuolumne and Calaveras counties, respectively, are producing mines in

FIGURE 3-30 June Lake, Mono County. Drainage from the lake flows away from the observer toward the Sierran front, and passes behind Reversed Peak just out of view to the right of the photo. (Photo by R. M. Norris)

disseminated gold deposits. Together they produce about 200,000 ounces of gold annually.

California gold is found as lode, placer, and disseminated deposits. Thick, gold-rich quartz veins were discovered by prospectors soon after the publicized discovery of placer gold at Sutter's Mill in 1848. The veins occur primarily in the Jurassic Mariposa slate of the Foothill Metamorphic belt.

The gold deposits of the Mother Lode belt are early Cretaceous (127–108 million years) in age. Most of the gold is found in a kilometer-and-a-half-wide (1 mile) band near the Melones fault. The mineralized quartz veins are less common in the southern Mother Lode, where deposits at some localities are large bodies of quartz and mariposite (a bright green, chrome-rich mica) with carbonate rock. These lower grade deposits of gold in country rock (disseminated deposits) have recently been the target of much exploration.

The gold found in Mother Lode rocks was mobilized by connate water released during metamorphism of the original marine sedimentary rocks and circulating in deep hydrothermal systems. The emplacement of plutons provided heat to drive the hot water systems. The water dissolved gold and sulfides and acquired carbon dioxide from the cooling granitic magmas. The pregnant solution precipitated gold and quartz in veins and altered other rocks to form the low-grade disseminated gold deposits. Many Sierran localities have characteristics like those of the enriched veins, but still lack significant gold content. The quartz veins of the Mother Lode belt are white and contain pockets, seams, and veinlets of native gold and pyrite (fool's gold) plus a few other precious and gangue (nonprecious) minerals.

Initially, gold was mined by placer methods, which involved working stream gravels and sands eroded from gold-bearing parent rock. Hydraulic mining (placer mining with high-pressure water) also was done, but the resulting destruction of the landscape and sediment choking of stream channels prompted the California legislature to outlaw hydraulic mining in 1884. Apart from hydraulic operations and gold dredging, most placer mining is a small-scale process employing sluice boxes, of which there are countless designs. A flow of water is directed across a confined channel (box) on which a series of cleats or riffles has been placed. The sluice box is inclined, the debris is passed down the channel, and the gold, if it is not too fine, is caught in the riffles because of its high specific gravity.

In recent years the high price of gold has made it profitable to reprocess some of the old mine tailings by leaching, and in a few instances to resume bedrock mining, most often by the open-pit process. The Carson Hill mine, between Angels Camp and Sonora is one of these. This mine is of special interest because it has produced more than 1,348,000 ounces (51,000 kilograms) of gold worth over $500 million. In 1854, this mine yielded the largest single mass of gold ever found in California, weighing about 88 kilograms (195 pounds, or 2,340 troy ounces), worth in those days a modest $110,000, but at 1989 prices about $930,000. The gold reserves at two active mines, Jamestown and Carson Hill, are at least $3\frac{1}{3}$ million and 825,000 troy ounces, respectively.

Gold mining in the Sierran foothill belt today is very much limited by environmental concerns. Modern operations are not the free-wheeling activity that was the norm during the nineteenth century.

Yosemite Valley

Because Yosemite National Park embraces a large part of the High Sierra already discussed, this section is confined to the characteristics and evolution of the valley itself and its immediate surroundings. The valley is the park's commanding feature and is world renowned for its imposing scenery (Figure 3-31). The verdant floor certainly contributes to Yosemite's beauty, but it is the sheer rock walls, the waterfalls, and the monolithic granitic peaks that make the valley special. Other Sierran valleys possess some of Yosemite's features, but only Yosemite unites them all in a single area. Figure 3-32 shows the location of Yosemite's important topographic features.

The portion of the valley seen by most visitors is only about 19 kilometers (12 miles) long. It lies between El Portal and the junction of the valley's two main branches, Little Yosemite Valley and Tenaya Canyon. Like Yosemite proper, these branches are fairly flat in their lower reaches, but are broken by abrupt risers above their junction. The south, or Merced River branch, is more spectacular because its channel is a glacial stairway of grand dimensions (see Figure 1-21). Vernal Fall plunges over the even, vertical cliff of the first riser as an unbroken curtain of water 97 meters (317 feet) high. A short distance upstream is the second riser, lying between Vernal Fall and Little Yosemite Valley above. The riser, the site of Nevada Fall, is higher (181 meters or 495 feet) but less regular than the first. Instead of the smooth curtain of Vernal Fall, Nevada Fall is a complex of dartlike jets of water that unite near the fall's base into a foaming cascade that pours into the Merced River below. Upstream, Little Yosemite Valley has no significant tributaries and so lacks notable waterfalls. Nevertheless, during the spring, when the snow melts rapidly in the high country, many cascades exist temporarily.

The park's highest falls plunge over the steep north wall of the main valley and

FIGURE 3-31 Yosemite Valley. (Photo by Fairchild Aerial Surveys, courtesy of Department of Geography, University of California, Los Angeles)

are fed by tributary streams that occupy hanging valleys abandoned by glaciers. Of these, Ribbon Fall is the highest, dropping 492 meters (1,612 feet). This is not a single leap, however, because much of the drop is confined to a narrow crevice cut by Ribbon Creek. Since it drains only a small area, once the snow has melted this creek dwindles to a trickle, and by late summer Ribbon Fall often dries completely.

Almost directly across the valley from Ribbon Fall is Bridalveil Fall, fed by Bridalveil Creek, another small stream that normally diminishes after the snow melts. Early in the season Bridalveil Fall often has impressive volume. The name reflects its early summer appearance, when the smaller volume of water breaks into a fine mist as it pours out of a deep notch in the valley wall and drops 260 meters (850 feet). About 189 meters (620 feet) of this is a clear leapover a vertical precipice.

The valley's most spectacular falls are the Yosemite Falls, two separate falls whose combined height is 782 meters (2,565 feet). The upper fall is one of the world's highest, 435 meters (1,430 feet). All but the upper 21 meters (70 feet) forms a clear, arc-shaped leap that carries the water so far from the cliff face that normally even the sloping base of the precipice is easily cleared. Lower Yosemite Fall drops 98 meters (320 feet) over the lowest cliff into the valley, a clear leap when water volume is high.

The hanging valleys have floors high above Yosemite Valley itself. Francois Matthes

FIGURE 3-32 Main topographic features of Yosemite Valley. (*Source*: University of California Press)

1	Basket Dome	13	Government Center	25	Mount Watkins
2	Bridalveil Fall	14	Half Dome	26	North Dome
3	Bunnell Point	15	Indian Creek	27	Quarter Domes
4	Cascade Cliffs	16	Leaning Tower	28	Ribbon Fall
5	Cathedral Rocks	17	Liberty Cap	29	Royal Arches
6	Cathedral Spires	18	Little Yosemite	30	Sentinel Dome
7	Clouds Rest	19	Merced River	31	Sentinel Rock
8	Dewey Point	20	Mirror Lake	32	Sunrise Mountain
9	Eagle Peak	21	Mount Broderick	33	Taft Point
10	Echo Peaks	22	Mount Florence	34	Tenaya Canyon
11	El Capitan	23	Mount Lyell	35	Washington Column
12	Glacier Point	24	Mount Maclure	36	Yosemite Falls (top of)

believed these valleys to be of two origins. Some, typical of glaciated regions everywhere, were left hanging when the main glaciers widened and deepened their channels. The more modest efforts of the tributary glaciers resulted in far less erosion, producing the discordant junctions seen today. The other type of hanging valley arises because some resistant, massive, unjointed granitic rocks effectively withstand the erosion of small, often intermittent streams. Consequently, the streams have been unable to keep pace with the downcutting of the large, perennial Merced River into which they drain.

The sheer, almost unjointed cliffs on both sides of the main Yosemite Valley are imposing features. Yosemite wouldn't be the same without El Capitan, Glacier Point, Half Dome, and the Washington Column. All were produced by widening of the valley as glaciers moved down the Merced River canyon. Actually, however, neither El Capitan, with its 884-meter-(2,898 foot)-high cliff, nor Half Dome, which rises 1,327 meters (4,352 feet) above the valley floor, was ever fully overridden by the ice. Their smooth sides away from the valley were sculptured by exfoliation rather than glacial scour. Furthermore, most of the other towering cliffs that rim Yosemite Valley are mainly the result of the El Portal glaciation and not the most recent Wisconsin glaciation, aided by vertical jointing. The last glacier occupying the valley, though certainly augmenting the widening done by earlier glaciers, was apparently no thicker than about 450 meters (1,500 feet). Nevertheless, it was this last glacier that produced the widespread, almost mirrorlike polished surfaces on the region's granitic rocks. Earlier glaciers certainly produced rock polish too, but none survives today. Indeed, though only about 10,000 years have passed since the last Wisconsin ice streams melted, weathering has roughened and destroyed considerable areas once highly polished. In such places, only the deeper gouges still persist.

Matthes divided the geologic history of Yosemite into three stages. These are still widely accepted, although the dates he assigned have been revised as improved dating methods have been developed. During the first or Broad Valley stage, the Sierra was of modest height with few peaks higher than 1,000 meters (3,300 feet). At this time the present drainage direction to the southwest was established. Matthes suggested an early Tertiary age for this stage—Pliocene is more likely. The Mountain Valley stage followed after some uplift that steepened the stream gradients and enhanced erosion. The ancestral Merced River responded by cutting a steep-sided gorge about 300 meters (1,000 feet) deep in the Yosemite region. During the Pliocene and perhaps on into the Quaternary, great uplift and tilting of the Sierran block occurred along eastern faults. Stream activity was greatly augmented, resulting in rapid deepening of the ancestral Merced River canyon to as much as 600 meters (2,000 feet). This period represents Matthes's Canyon (third) stage.

By the onset of glaciation about 2 million years ago, the Merced River occupied a deep, V-shaped, stream-cut canyon down which the ice eventually flowed. As indicated chiefly by moraines, earlier glaciations (particularly the El Portal) were characterized by greater extent and ice thickness than was the Wisconsin glaciation. Curiously, the evidence reveals that the ice in the source areas was about as thick during the Wisconsin as it was previously. Why, then, were the earlier ice sheets in the Yosemite region so much larger and thicker? Matthes's answer was that the earlier glacial stages lasted longer than the Wisconsin and so promoted greater accumulation of ice in peripheral areas. For example, during El Portal glaciation, ice in Yosemite Valley was probably about 1,800 meters (6,000 feet) deep, during which it excavated bedrock at least 600 meters (2,000 feet) below the present alluviated valley floor. In contrast, the Wisconsin glacier was never more than about 450 meters (1,500 feet) deep anywhere in the valley.

The El Portal glacier had a maximum length of 60 kilometers (37 miles) and extended 11 kilometers (7 miles) beyond the edge of the Sierran ice cap. It ended just below El Portal, where its terminal moraine can be seen today. During the El Portal stage, the ice covered Glacier Point to a depth of 160 meters (500 feet), and was more than 820 meters (2,700 feet) deep at Bridalveil Meadow where the later Wisconsin glacier terminated.

The smaller Wisconsin glacier ended in the main valley just above Bridalveil Meadow, where there is a well-preserved terminal moraine. This glacier received no ice from any of the canyons entering the main valley; it was dependent solely on the ice streams moving down Little Yosemite and Tenaya valleys. Maximum ice thickness of about 450 meters (1,500 feet) occurred near the junction of the two streams. At this time, Glacier Point towered more than 600 meters (2,000 feet) above the ice.

Once the Wisconsin glacier had melted and the main valley was free of ice, Lake Yosemite filled the basin upstream from the terminal moraine. The lake was 8.8 kilometers (5½ miles) long and occupied the main valley from wall to wall. The original depth of this lake is unknown, but the present valley floor is underlain by about 600 meters (2,000 feet) of lake deposits that rest on glacially scoured granite. Some of the lake beds may have been deposited in pre-Wisconsin lakes, however.

Lake Yosemite was destroyed by a combination of two processes. First, entering streams, mainly the Merced River and Tenaya Creek, quickly built large deltas into the lake. Smaller deltas were built by smaller streams entering the lake from the north. Second, the morainal dam at the lake's lower end was notched by the outflowing Merced River. This lowered the lake by at least 10 meters. The Merced River subsequently regraded and leveled the valley, leaving only small patches of former valley floor as low benches about 5 meters (15 feet) above present valley level.

Sequoia National Park

Sequoia National Park includes more than 1,560 square kilometers (600 square miles) of the highest part of the Sierra Nevada, from the Whitney crest on the east to the western foothills. It embraces the greatest range of altitude of any American national park or monument outside Alaska—4,100 meters (13,500 feet).

Topographically, the park is divided by the north-south Great Western Divide, which in any other setting would be a notable range in its own right. As it is, the Divide is actually a mountain range within a range. Seen from Moro Rock on the west, the Great Western Divide may seem to be the Sierran crest proper and has the craggy, alpine appearance of the main ridge. It is almost as high as the actual crest, with several peaks of more than 4,000 meters (13,000 feet).

East of the divide is the true high Sierra with its numerous lakes and relatively barren, heavily glaciated terrain. The area is drained by the Kern River and its gorge, which contrasts strongly with the Kaweah River drainage west of the Great Western

Divide. Not only does the partially fault-controlled Kern River follow an almost straight north-south course across Sequoia Park, but also the river's upper reaches occupy a pronounced U-shaped glacial valley (see Figure 3-9). In addition, the Kern Canyon has an openness of form that is unlike the deep gorges of the Kaweah system. This results mainly from the broad Chagoopa benchlands, a high-altitude erosional surface, fringing the Kern and the heavily glaciated uplands above the plateau. Bare rock and glacial lakes characterize much of this high country. Glaciers have stripped away almost all soil, further limiting tree growth already inhibited by the high-altitude climate. On the other hand, glaciers have left dozens of sparkling lakes that fill nearly every ice-scoured depression.

Rocks in the eastern section of the park are almost exclusively granitic. The older metamorphic rocks so prominent elsewhere in the Sierra are conspicuously missing; their few exposures are confined to the Mineral King area and the western foothills below Giant Forest village. Volcanic rocks are similarly absent from the eastern region, although they do occur a few miles south of the park's boundary.

Evidence of glacial scour prevails throughout the high eastern part of Sequoia, reflecting the existence of a long glacier that extended down the Kern Canyon to about 11 kilometers (7 miles) beyond the park's southern boundary. This glacier terminated at an elevation of 1,750 meters (5,700 feet), at latitude 36°14'N, and is the southernmost glaciation known in the Sierra Nevada. The southern 160 kilometers (100 miles) of the range were too low and too dry to sustain glaciers.

Unlike the Kern, the Kaweah River and its branches form dendritic patterns. They occupy particularly deep canyons, some lying almost 2,150 meters (7,000 feet) below the high peaks on either side. Usually, though, relief is between 1,200 and 1,500 meters (4,000–5,000 feet). The landscape of the western slope differs markedly from that on the eastern slope: Benchlands tend to be missing; the openness of form characteristic of the Kern River is not apparent; and glaciation was often confined to canyons, leaving uplands untouched and covered with thicker soil and forest. The deep valleys do provide considerable evidence of glaciation, however. In Tokopah Valley on the Marble Fork of the Kaweah, the canyon walls are nearly perpendicular and are spectacularly grooved and polished by latest Pleistocene glaciers. The polish seen here and at other places in the high country is almost without exception the result of the final ice advance.

Although most glacial ice has melted, Sierran rocks continue to be shattered and reduced to smaller pieces by frost action, which is probably the dominant weathering process in the high country. Some huge boulders look as if they had been split by a giant cleaver, with the discrete pieces still lying close to one another. Frost action causes this, attacking both large and small blocks with such indiscriminate vigor that whole valleys are choked with the resulting angular debris. The topography of the high country is thus gradually smoothed and rounded.

Limestone, or more properly marble caverns, are a minor but interesting feature of Sequoia. At least four are known, all in the Jurassic and Triassic metamorphic rocks of the Kaweah drainage. None is large, but all contain limestone cave features such as stalactites, stalagmites, and pillars. The most popular is Crystal Cave not far from Giant Forest. Others are Paradise Cave near Ash Meadow, and Palmer Cave and Clough Cave on the walls of the South Fork of the Kaweah River.

Moro Rock, overlooking the Middle Fork of the Kaweah River, is typical of the massive, bulbous, granitic domes throughout the Sierra Nevada. Usually glaciation is not involved in formation of these domes; instead they result from rock joints being more widely spaced than normal. This produces monolithic masses of rock with few avenues for weathering agents to enter. The normally jointed rocks surrounding the domes are quickly reduced to small blocks and gravel, leaving a projecting unjointed mass of rock that responds to weathering chiefly by exfoliation. In this process, thin shells are spalled off the exposed rock, gradually producing the rounded surfaces so characteristic of these domes.

Kern River—Lake Isabella Recreation Area

To control floodwaters, a dam was built across the Kern River at its confluence with the South Fork. As a result, Lake Isabella was formed and the area has been developed into a major recreational region.

The Kern River drains south from its headwaters at the west base of Mount Whitney (Figures 3-9 and 3-10), its course being somewhat determined by the Kern Canyon fault. The fault presumably poses little hazard since it apparently has been inactive for about 3 million years. Many geologic formations occur near the recreational area, including the pre-Nevadan Kernville series of metamorphic roof rocks and Nevadan granitic rocks. A typical Sierran limestone cavern, Packsaddle Cave, is developed in these roof rocks and is located about 24 kilometers (15 miles) north of the lake. Below the dam is the deep gorge of the Kern, where giant potholes are spectacularly displayed.

At the south end of the lake is the old mining town of Bodfish, once a stage stop for travelers passing from the gold-mining areas of old Kernville (now beneath Lake Isabella) to the antimony mines of Havilah, to the Walker Basin, and to the railroad that crosses the Tehachapi Mountains. The Big Blue gold mine, productive until about 1940, may still be seen on the west side of the Kern River north of Lake Isabella.

Mount Whitney

Mount Whitney (4,418.345 meters or 14, 495.881 feet) is the highest mountain in coterminous United States (Figure 3-33 and frontispiece). Its spire is unspectacular from lower elevations, however, because it is one of several high peaks forming Whitney Crest (Figure 3-34). Eighteen hundred meters (6,000 feet) below Mount Whitney is Whitney Portal, part of the steep Sierran scarp that is about 2,400 meters (8,000 feet) high in this latitude.

Mount Whitney was named in 1864 by Clarence King for J. D. Whitney, director of California's first geologic survey. (King was a member of the party organized by Whitney to explore the Sierra Nevada.) In recent years the peak has become a mecca for hiking enthusiasts. On a Fourth of July holiday, more than 1,000 people reportedly signed the register on the Whitney summit.

Mount Whitney patrol station near the timberline (elevation 3,270 meters or 10,720 feet) is one of the country's highest ranger stations and is only 13 kilometers (8 miles) west of the summit by trail. The federal government originally reserved Mount Whitney for weather observations, but the plan never materialized. It was later used as a base for solar radiation and Mars spectrum studies.

FIGURE 3-33 Mount Whitney. Iceberg Lake, elevation about (3,874 meters), 12,700 feet one of the higher lakes in the 48 coterminous states, is in the right foreground. This cirque lake does not thaw completely from one year to the next. The Great Western Divide is in the background. (Photo by Spence Air Photos, courtesy of Department of Geography, University of California, Los Angeles)

Alabama Hills

At the east base of the Sierra are the Alabama Hills, which extend about 16 kilometers (10 miles) and rise 90 to 120 meters (300–400 feet) immediately west of Lone Pine. The hills are a series of fault slivers raised in the Sierra Nevada fault zone (Figures 3-17 and 3-34). They are neither unique nor important in the total geology of the state, but they have achieved some prominence nevertheless. First, in the nineteenth century the persistent myth that the Alabama Hills are the oldest hills in the world was fabricated. Second, in 1872 an earthquake (possibly the strongest experienced in California history) destroyed Lone Pine, killed about 30 people, and produced significant surface ruptures at the base of the Alabama Hills and throughout the Owens Valley. Third, many motion pictures, particularly westerns, have been filmed here. The boulder-pile weathering along numerous joints in the Mesozoic granites makes an unusual striking backdrop.

The notion that the Alabama Hills are the oldest in the world results from two common misconceptions: that weathered features are of great age and that granite is always an old rock. Actually, Lone Pine and several other creeks flow across the hills (Figure 3-34, right center), indicating that the streams are antecedent or su-

FIGURE 3-34 Whitney (Muir) Crest, with Mount Whitney in almost the exact center. Crest is (4,200 meters) 14,000 feet in average elevation, with many peaks almost equal to Mount Whitney. Alabama Hills and Diaz Lake are in the foreground. (Photo courtesy of U.S. Geological Survey)

perimposed and that the hills are younger than the streams crossing them. This means uplift of the region is very young indeed—probably late Pleistocene. The myth of old age was also fostered by the peculiar appearance of the weathered granites and metamorphic rocks in contrast to the granites of the Sierra Nevada. Geologists recognize, however, that different climatic regimens produce different weathering patterns in similar rocks and would expect the Whitney summit climate and the arid environment of the Alabama Hills to produce different topographies.

Arid weathering as a cause of boulder-pile topography of the Alabama Hills has been questioned. It is suggested instead that the area was once under alluvial cover. Percolating ground water then promoted chemical decomposition of the covered granite, with decomposition being more rapid along joints. Subsequent shifts of erosional base level or a change to a more arid climate exhumed the granitic terrain, eroding the overlying alluvium and removing the decomposed material along the joints. This is an attractive explanation for the Alabama Hills, since they slope north and south beneath the giant alluvial fans derived from the Sierra and are crossed by streams older than the hills.

The 1872 earthquake corroborates the youth of the topography of the Alabama Hills. Surface displacements occurred along the base of the Sierra and the Alabama Hills for distance of 200 kilometers (120 miles), producing many sags in which intermittent lakes form. Vertical displacement was as much as 5 meters (17 feet), with a significant horizontal component. The resulting scarp shows littler erosional effect of the 100 or so years that have elapsed.

Palisade and Lyell Glaciers

The two largest modern glaciers in the Sierra are Palisade Glacier on upper Big Pine Creek and Lyell Glacier at the head of the Lyell Fork of the Tuolumne River. Palisade

FIGURE 3-35 Palisade Glacier, Inyo County near Big Pine. Note the moraines in the cirque below the glacier. Compare with Figure 3-5 (Photo by Spence Air Photos, courtesy of Department of Geography, University of California, Los Angeles)

Glacier is slightly larger, but neither glacier is more than about 1.5 kilometers (1 mile) long (Figure 3-35).

Palisade Glacier lies below a precipitous north-facing rock wall capped by three high peaks. On the east is Mount Sill (4,319 meters or 14,162 feet), in the center is North Palisade (4,344 meters or 14,242 feet), and on the west is Mount Winchell (4,199 meters or 13,768 feet). The southernmost true glacier in the United States today (latitude 37°04′N), Palisade is only about half the size it was in 1850. Though the steep cirque wall to the south shields the glacier from direct sun, a renewal of the 1850–1950 warming trend would bring it close to extinction relatively soon.

Like most Sierran glaciers, Palisade is rimmed by a fresh and youthful-looking moraine. During the latter part of the summer, a crescent-shaped crack (bergschrund) opens up near the head of the glacier as melting allows the ice to pull away from the cirque wall above. During the ensuing winter, the crack is filled with snow and ice, which are added to the glacier during the following years.

Lyell glacier is almost 3.2 kilometers (2 miles) across, but is shorter than Palisade Glacier. Lyell lies in a cirque on the north side of Mount Lyell (3,994 meters or 13,095 feet) at the Sierran crest in eastern Yosemite. In most respects Lyell Glacier resembles Palisade Glacier 96 kilometers (60 miles) to the south. They are about the same size, move at about the same rate, and lie at about the same elevation (3,650 meters or 12,000 feet) in comparable settings.

In 1933, two park naturalists were surprised to see a mountain sheep ram standing on the toe of Lyell Glacier. Their surprise was occasioned by the knowledge that mountain sheeep had been unknown in the Yosemite region since about 1880. Closer inspection revealed that the naturalists were looking at the ram's mummified remains, exposed as the glacier melted. Apparently, the ram had fallen into the bergschrund at the glacier's head about 250 years earlier. He had been frozen into the glacier as the bergschrund filled with snow and had been moved slowly through the advancing ice until melting released his body near the toe of the shrinking glacier. At the time of the discovery, two of the ram's feet were still frozen in the ice.

Mammoth Mountain and "Earthquake Fault"

Mammoth Mountain, a popular ski center, is a young silicic volcanic cone composed of glassy lavas. It was long thought to have begun about 400,000 years ago, but recent studies suggest it is much younger than originally supposed. It appears to have begun about 180,000 years ago and reached its present size after about 10 eruptions, the most recent of which took place approximately 40,000 years ago.

In the same area is a linear crack a few meters wide and several hundred meters long, which continues not as an open crack but as a fault trace. This feature is known locally as the "earthquake fault." Precise time of movement on the crack is not known, but the fault is definitely part of the Sierran series of north-south trending faults that cut the area's volcanic rocks. The view that the earth may open up in cracks like the earthquake fault is seldom supported by geologic observation. Such a feature is a rarity and is not the normal result of earth movement.

It has been suggested that the crack might have resulted from shrinkage of volume during cooling of the lava. The volcanic rock where the fissure is exposed is of Tertiary age, however. It is most improbable that a cooling crack could be preserved this long because weathering and deposition of sediment would long since have obliterated it. Finally, there is a possibility that the feature is part of the ring fractures that surround the Long Valley caldera.

Devils Postpile National Monument

Near the headwaters of the San Joaquin River is an example of columnar jointing known as the Devils Postpile. This feature developed in lava that was extruded from a fissure near Red's Meadow and subsequently flowed down river for several kilometers. The flow was part of the volcanic episode referred to earlier that began 3 to $3\frac{1}{2}$ million years ago. The columns of the Postpile are long, very regular, and numerous. Although it does not rank with Devils Tower (Wyoming) or Giant's Causeway (Ireland) for bulk, Devils Postpile is California's best display of columnar jointing (Figure 3-36).

Hot Creek

Mammoth Lakes drain into Mammoth Creek, which hugs the base of a low escarpment cut in young volcanic rocks as it flows past the town of Mammoth. At intervals, hot springs emerge along the creek from the base of the escarpment. As Mammoth Creek crosses the meadow toward Owens River, it becomes Hot Creek, along which the hot springs also occur (Figure 3-37).

These hot springs derive their heat from still-cooling young volcanic rocks. Sometimes small geysers have appeared, particularly after earthquakes, which seem often to alter the underground "plumbing." Some of the changes observed in hot spring

FIGURE 3-36 Devils Postpile, a particularly fine example of columnar jointing in an andesite flow about 600,000 years old. The curved columns are thought to reflect the irregular surface on which the flow cooled. Longest columns are about 18 meters (60 feet) long. Devils Postpile National Monument, Madera County. (Photo by Charles C. Plummer)

FIGURE 3-37 Looking west at the gorge cut by Hot Creek. The dissected tableland is rhyolitic. (Photo by Mary Hill)

activity may be due to movements in the magma body, which is present at relatively shallow depth (4.9–7 kilometers or 2–3 miles) beneath Long Valley caldera. Casa Diablo Hot Springs, an active fumarolic area a short distance northwest of Hot Creek, is the site of a new geothermal power development.

Mono Lake

Mono Lake occupies the bottom of a large valley at the east base of the steep Sierran escarpment near Lee Vining. It is a shallow water body, with no outlet and with a salinity exceeding that of seawater. Mono Lake is the remnant of larger, glacial Lake Russell, whose high stand is recorded by conspicuous shorelines around the basin (Figure 3-38). Lake Russell is presumed to have drained east, ultimately into Owens

FIGURE 3-38 Mono Lake from the east, July 1968. The shorelines of ancient Lake Russell and dwindling Mono Lake are apparent. The larger island is Paoha, the smaller immediately to the right is Negit. The third, light-colored patch between Negit and the shore is a shoal that in 1988 was largely connected to the shore. (Photo courtesy of U.S. Air Force and U.S. Geological Survey)

FIGURE 3-39 Tufa towers on the southwestern shore of Mono Lake. These structures are produced by lime-secreting cyanobacteria (blue-green algae) and tend to develop where springs bubble up from the lake bottom. The towers are exposed only when the lake level drops. (Photo by R. M. Norris)

Valley, prior to eruptions of the Bishop tuff. Mono Lake was much larger in 1940, when most of the entering streams were diverted to augment the water supply of the city of Los Angeles. Evaporation now usually exceeds inflow, and eventually Mono Lake will dry completely unless sufficient flow is provided to prevent this.

An unusual feature of Mono Lake is the display of tufa pinnacles that are forming in the briny waters where lake bottom springs encourage the growth of lime-secreting cyanobacteria (blue-green algae) (Figure 3-39). Most of the tufa towers are concentrated along an eroded fault scarp at the lake's southwestern edge and resemble those found in other Basin and Range lakes. Several new minerals have been described from the brines of Mono Lake. Its volcanic islands, white Paoha and black Negit, are younger than the lake itself and were formed by outpourings of lava and the construction of volcanic cones. They are even younger than the last high stand of Lake Russell.

Prospecting for geothermal power in the Mono Lake area has been discontinueed. Several wells were drilled, but low temperatures were encountered and the bedrock complex was reached at shallow depth.

Lake Tahoe

California's most famous lake, Tahoe (501 meters or 1,645 feet deep), is cradled in a downfaulted block in the central Sierra Nevada (Figure 3-40). It is bounded on the east by the Carson Range and on the west by the Mount Tallac ridge, part of the Sierra Nevada. Although the level of the lake is carefully regulated today, it has fluctuated widely in the past. There is evidence that it stood as high as 200 meters (600 feet) above its present level, and the caves in Cave Rock are believed due to

FIGURE 3-40 Looking east across Lake Tahoe, with South Tahoe (California) and Stateline (Nevada) clustered at the southern end. Washoe Lake is in the upper center. (Photo courtesy of U.S. Air Force and U.S. Geological Survey)

solution when the lake stood 43 meters (140 feet) above its present level of 1,900 meters (6,230 feet). In addition, submerged tree stumps show that the lake level dropped below the present level during a dry period thought to have occurred between 1750 and 1850.

The height of the outlet has been altered from time to time by the depositon and erosion of morainal material as well as by erosion of volcanic rocks, accounting for the natural fluctuations in level.

Unfortunately, urban growth threatens the future of this spectacular lake, and stringent regulations must be enforced if Lake Tahoe it to remain clean, clear, and beautiful.

REFERENCES

General

Bateman, Paul C., 1967. Sierra Nevada Batholith. Science, v. 158, pp. 1407–1417.

———, 1969. Geology of the Sierra Nevada. Mineral Information Service (now California Geology): Part 1, v. 22, pp. 39–42; part 2, v. 22, pp. 61–66.

_____, 1974. Model for the Origin of Sierran Granites. Calif. Geology, v. 27, p. 3–5.

_____and Clyde Wahrhaftig, 1966. Geology of the Sierra Nevada *In* Geology of Northern California. Calif. Div. Mines & Geology Bull. 190, pp. 107–171.

Calkins, F. C., 1930. The Granitic rocks of the Yosemite Region *In* F. E. Matthes, Geological History of Yosemite Valley. U.S. Geological Surv. Prof. Paper 160, pp. 120–129.

Clark, William B., 1979. Fossil River Beds of the Sierra Nevada. Calif. Geology, v. 32, pp. 143–149.

_____, 1987. Mother Lode Gold Mines. Calif. Geology, v. 40, pp. 57–58.

Hill, Mary, 1975. Geology of the Sierra Nevada. Univ. California Press Natural History Guide No. 37, 232 p.

Jenkins, Olaf P., 1981. Geological History of Sierran Gold Belt. Calif. Geology, v. 34, pp. 246–247.

Lindgren, Waldemar, 1911. The Tertiary Gravels of the Sierra Nevada, Calif. U.S. Geological Survey Prof. Paper 73, 226 p.

Rinehart, C. D., and D. C. Ross, 1957. Geology of the Casa Diablo Mountain Quadrangle, California. U.S. Geological Survey Map G0–99.

_____, 1964. Geology and Mineral Deposits of the Mount Morrison Quadrangle, Sierra Nevada, California. U.S. Geological Survey Prof. Paper 384, 106 p.

Special

Burnett, John L., 1971. Geology of the Lake Tahoe Basin, California Geology, v. 24, pp. *119– 127*.

_____, *and Robert A. Mathews, 1971. Geologic Look at Lake Tahoe. California Geology, v. 24, pp. 128–129.*

Buwalda, J. P., 1954. Geology of the Tehachapi Mountains, California. In Geology of Southern California, Calif. Div. Mines and Geol. Bull. 170, pp. 131–142.

Chakarun, John D., 1987. Tertiary Gold-bearing Gravels Northern Sierra Nevada. California Geology, v. 40, pp. 123–126.

Crippen, J. R., and B. N. Pavelka, 1970. The Lake Tahoe Basin, California-Nevada. U.S. Geological Survey Water Supply Paper 1972, 56 p.

Hill, Mary, 1975. Living Glaciers of California. California Geology, v. 28, pp. 171–177.

_____, ed., 1972. The Owens Valley Earthquake of 1872. California Geology, v. 25, pp. 51– 64.

_____, ed., 1975. Geologic Guide: Sierra Nevada—Basin and Range Boundary Zone. California Geology, v. 28, pp. 99–119.

Huber, N. K., and C. D. Rinehart, 1965. The Devils Postpile National Monument. Mineral Information Service (now California Geology), v. 18, pp. 109–118.

Jenkins, Olaf P., 1964. Geology of Placer Deposits. Mineral Information Service (now California Geology), v. 17, (passim). (Reprint of Calif. Div. Mines & Geology Bull. 135.)

_____, ed., 1948 and 1963. Geologic Guidebook Along Highway 49: The Sierran Gold Belt— The Mother Lode Country. Calif. Div. Mines and Geology Bull. 141, 164 p.

Kistler, R. W., 1966. Geologic Map of the Mono Craters Quadrangle, Mono and Tuolumne Counties, California. U.S. Geological Survey Map G0–462.

Matthes, F. W., 1930. Geologic History of Yosemite Valley. U.S. Geological Survey Prof. Paper 160, 137 p.

_____, 1950a. Sequoia National Park: A Geological Album. Univ. Calif. Press, 136 p.

————, 1950b. The Incomparable Valley—A Geological Interpretation of the Yosemite. Univ. Calif. Press, 160 p.

————, 1960. Reconnaissance of the Geomorphology and Glacial Geology of the San Joaquin Basin, Sierra Nevada, California. U.S. Geological Survey Prof. Paper 329, 62 p.

————, 1965. Glacial Reconnaissance of Sequoia National Park, California. U.S. Geological Survey Prof. Paper 504A, 58 p.

Oakeshott, Gordon B., ed., 1955. Earthquakes in Kern County, California in 1952. Calif. Div. Mines and Geology Bull. 171, 283 p.

————, ed., 1962. Geological Guide to the Merced Canyon and Yosemite Valley. Calif. Div. Mines and Geology Bull. 182, 68 p.

Putnam, William C., 1938. Mono Craters, California. Geographical Review, v. 28, pp. 68–82.

————, 1960. Origin of Rock Creek and Owens River Gorges, Mono County, California. Univ. Calif. Publ. Geol. Sci., v. 34, pp. 221–280.

Rapp, John S., 1974. Mammonton Dredge Field. California Geology, v. 27, pp. 201–202.

————, 1982. The Valley Springs Formation in the Sonora Pass Area. California Geology, v. 35, pp. 211–219.

Rinehart, C. D., and N. K. Huber, 1965. The Inyo Crater Lakes—A Blast in the Past. Mineral Information Service (now California Geology), v. 18, pp. 169–172.

Romanowitz, Charles M., 1970. California's Gold Dredges. Mineral Informatin Service (now California Geology), v. 23, pp. 155–168.

Schaffer, Jeffrey P., 1977. Pleistocene Lake Yosemite and the Glaciation of Yosemite Valley. California Geology, v. 30, pp. 243–248.

Smith, Genny, ed., 1976. Mammoth Lakes Sierra by C. D. Rinehart, E. Vestal and Bettie E. Willard. William Kaufmann, Inc., 147 p.

————, ed., 1982. Earthquakes and Young Volcanoes Along the Eastern Sierra by C. D. Rinehart, and Ward C. Smith. William Kaufman, Inc., 62 p.

Trent, D. D., 1983. California Ice Age Lost: The Palisade Glacier, Inyo County. California Geology, v. 36, pp. 264–269.

Turrin, Brent D., 1982. Potassium-Argon Dates and Stratigraphy of Pliocene Volcanic Domes near Mokelumne Hill, Calaveras County. California Geology, v. 35, pp. 220–222.

Twain, Mark, 1965. Islands of Mono Lake. Mineral Information Service (now California Geology), v. 18, pp. 173–180.

Woods, Mary C., 1977. Ice Age Geomorphology Middle Fork, San Joaquin River, Madera County, California. California Geology, v. 30, pp. 249–253.

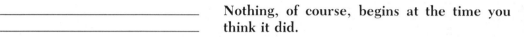

four

KLAMATH MOUNTAINS

Nothing, of course, begins at the time you think it did.

—*Lillian Hellman*

Drained by the tortuous, deep, and rock-filled Klamath River, the 30,300 square kilometer (11,800 square mile) mountains that straddle the California-Oregon border are collectively known as the Klamath Mountains. Ranges in the California portion of the province, from south to north, are the South Fork, Trinity, Trinity Alps, Salmon, Scott, Scott Bar, and Siskiyou mountains. The Klamath province adjoins the Cascade province on the east and is bounded by the northern Coast Ranges on the west and south (Figure 4-1).

The province is among the best watered in California, and the heavy rainfall results in many large streams and beautiful lakes. During the Pleistocene the combination of heavy precipitation and cooler climate resulted in the development of glaciers on the higher peaks. Upland summits in the Klamaths average between 1,500 and 2,100 meters (5,000–7,000 feet), but some peaks stand well above this old eroded plateau surface; Thompson Peak in the Trinity Alps is 2,746 meters (9,002 feet) high.

Geologic exploration began in this region when Major Pierson B. Reading discovered gold nuggets in the Trinity River near Weaverville in 1848. This gold discovery followed by only a few months the more famous find by John Marshall at Coloma in the Sierra Nevada.

J. S. Diller of the U.S. Geological Survey began mapping large portions of the Klamath Mountains around the turn of the century, but he and other geologists were long hampered by the lack of good topographic base maps and by difficult access, problems that persisted until almost 1960. During the last three decades W. P. Irwin (stratigraphy and structure), G. A. Davis (plutonism, metamorphism and structure),

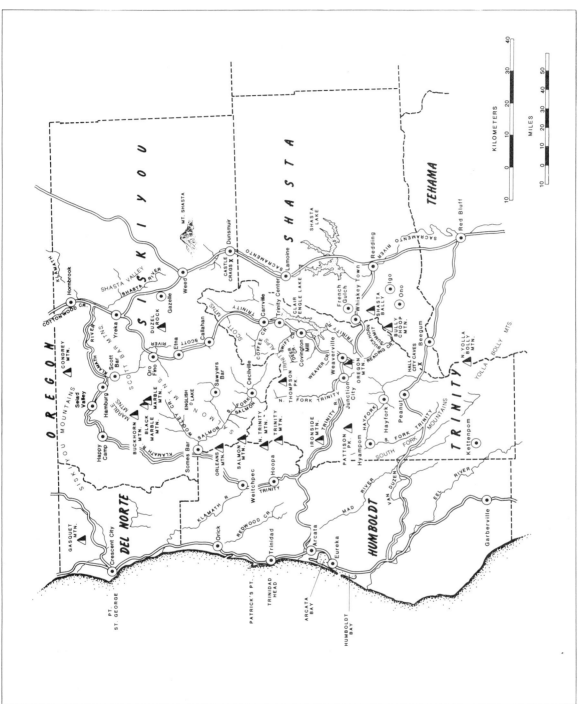

FIGURE 4-1 Place names, Klamath Mountains province.

127

and others have made major strides in explaining the geology of this rather remote area.

The Klamath Mountains are of considerable interest to geologists. In addition to their great scenic beauty, particularly in the Salmon-Trinity Alps Primitive area and the Marble Mountain Wilderness, the rocks include evidence of complex, multiple lithospheric plate interactions over a period from early Paleozoic to the present. The presence of accreted terranes, oceanic crustal slivers, and subduction complexes has made the Klamath region the center of many debates on California and global structural history.

ROCKS

The rocks of the Klamath province are generally divided into two groups; the older, subjacent units appear to have been accreted to the North American continent and deformed by numerous plate collisions beginning as far back as Devonian, and culminating in the massive deformation that affected much of the state during the Jurassic Nevadan orogeny. The younger superjacent rocks overlie the older rocks with marked unconformity and include rocks dating from early Cretaceous to the present. Although there are a few small patches of superjacent rocks in the center of the province, most lap up on the province margins and merge with the Great Valley sequence to the south and east, or with the volcanic rocks of the Cascade province to the east (Figure 4-2).

Subjacent Rocks

The subjacent rocks were divided into four belts or terranes by W. P. Irwin in 1960. These belts are distinguished by lithology, age, and structure. Several formations, metamorphic units, and plutonic bodies may thus be grouped as a related package. The four belts are all bounded by thrust faults (Figure 4-3). These four belts, which will henceforth be called *plates*, are from east to west, the Eastern Klamath plate, the Central Metamorphic plate, the Western Paleozoic and Triassic plate, and the Jurassic plate. These plates are all arcuate, open to the east. Each of the plates has ridden westward relative to the younger rocks beneath, so our discussion of the depositional and structural history of this part of California must logically begin with the Eastern Klamath plate. In this plate the Gazelle and Duzel formations of Paleozoic age overlie an oceanic basement rock called the Trinity Ophiolite (Figure 4-4). The Trinity Ophiolite is the largest exposure of oceanic crust and mantle anywhere in North America. It consists of iron- and magnesium-rich rocks that crystallized deep below the surface to form the plutonic rocks that are exposed from Covington Mill on the eastern shore of Clair Engle Lake to near Gazelle and Weed in southern Siskiyou County. Interstate Highway 5 and the Sacramento River follow the eastern edge of the Trinity Ophiolite from La Moine to Dunsmuir.

Eastern Klamath Plate. Duzel Rock, a prominent topographic feature about 30 kilometers (19 miles) northwesterly of Weed, is an exposure of the limestone in the Duzel Formation. In addition to limestone, the Duzel Formation includes a wide variety of sedimentary and volcanic rocks that are best seen between Yreka and

Composite Stratigraphic Column, Klamath Mountains

QUATERNARY	Alluvium Glacial deposits	
OLIGOCENE	Weaverville (nonmarine)	
CRETACEOUS	South Fork Mountain Schist	
JURASSIC	Galice (marine) Potem (marine) Bagley Andesite Arvison (marine)	Great Valley, Franciscan, and various plutonic rocks
TRIASSIC	Modin (marine) Brock Shale (marine) Hosselkus Limestone (marine) Pit (marine) Bully Hill Rhyolite Dekkas Andesite	
PERMIAN	Nosoni (metavolcanic) McCloud Limestone (marine)	
PENNSYLVANIAN	Baird (marine)	
MISSISSIPPIAN	Bragdon (marine)	
DEVONIAN	Balaklala Rhyolite, Kennett, Abrams Schist Copley Greenstone, Salmon Schist	
SILURIAN	Duzel, Gazelle (marine)	
ORDOVICIAN	Trinity Ophiolite	

FIGURE 4-2 Geologic column, Klamath Mountains.

Callahan. Radiolaria occur in cherts in the Duzel and have shown it to be Ordovician or Silurian in age (Figure 4-5).

The overlying Gazelle Formation contains more chert than the Duzel and considerably more volcanic material, both pyroclastic rocks and lava flows. Abundant invertebrate fossils date the Gazelle as early Silurian to early Devonian. These two formations seem to have distinctly different sources. The Duzel source rocks were metamorphic and plutonic for the most part, but the Gazelle was derived mainly from andesitic volcanoes. Perhaps an island arc had begun to grow on the old ocean floor now known as the Trinity Ophiolite.

The Trinity Ophiolite divides the Eastern Klamath plate in half. The southern part north of Redding is well exposed on the slopes of Shasta Lake. The oldest rock in this area is Devonian, and there is a nearly continuous sequence of sedimentary and volcanic rocks representing deposition up to middle Jurassic time. Evidence of volcanic activity first documented in the Gazelle is present in both the northern and

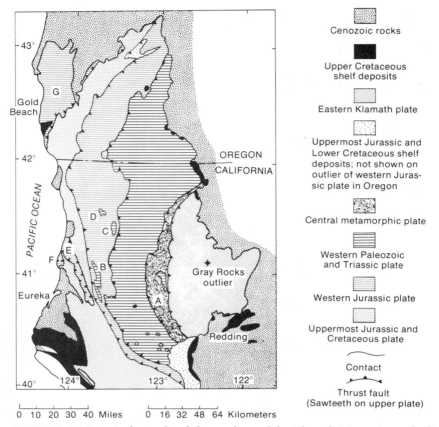

FIGURE 4-3 Principal postulated thrust plates of the Klamath Mountains and adjacent Coast Ranges. Thrust outliers are indicated by letter symbol: A, Oregon Mountain; B, Willow Creek; C, Prospect Hill; D, Flint Valley; E, Redwood Mountain; F, Patricks Point; G, southwestern Oregon. (*Source*: California Division of Mines and Geology)

southern parts of the plate. Although some of the later rocks exposed in the Klamath province seem to have originated far from California, the corals and brachiopods of the early to middle Devonian Kennett Formation and the paleomagnetic record of associated Permian rocks indicate that these rocks were deposited very close to their present location; they do not appear to be far-traveled exotic rocks.

Central Metamorphic Plate. The thrust fault that separates the Eastern Klamath plate from the Central Metamorphic plate is called the Bully Choop or Trinity thrust. It was active in the Devonian time, about 380 to 400 million years ago. As the Eastern Klamath plate rode westward over the adjacent oceanic plate, some of the older rocks near the bottom were scraped off. These scrapings, changed by heat and pressure in the subduction zone, may be the source of rocks in the Central Metamorphic plate. For example, lively debate has arisen over whether the Abrams schist is the metamorphic equivalent of the Duzel Formation. The quartz-mica schist, micaceous marble, and amphibolite found in the Abrams can be produced by me-

FIGURE 4-4 Pyroxenite layers in a peridotite, southeastern Trinity Alps. This rock is a part of a deep-sea ophiolite sequence. (Photo by W. P. Irwin, U.S. Geological Survey)

tamorphism of rocks like those of the Duzel. If the Abrams schist and the more iron-rich Salmon schist that underlines it were originally rocks belonging to the Eastern Klamath plate, the parent rocks of these two schists are as old as Ordovician. We know the metamorphism occurred about 380 to 400 million years ago because of dates obtained from metamorphically formed hornblende. Because metamorphism is believed to have taken place during underthrusting of the Central Metamorphic belt, the age of metamorphism is the age of thrust faulting. Much of the Trinity Alps area northwest of Weaverville is underlain by these schists.

Klamath Mountain plates once overlapped much more than they do today. Erosion has removed hundreds of meters of material, but because of greater resistance or structural location, some remnants of adjacent plates are found sitting on top of the younger, underthrust rocks. A "rootless" rock unit that has been thrust over other rocks and then isolated by erosion from its parent body, as rocks from different localities were shuffled together by thrusting, is called a *klippe*. Oregon Mountain, just south of State Highway 299 west of Weaverville, is an example of a large klippe of the Mississippian Bragdon Formation and a remnant of the Eastern Klamath plate.

Condrey Mountain $6\frac{1}{2}$ kilometers (4 miles) south of the Oregon boundary in the central Klamath Mountains may be the opposite sort of feature, a hole or window in the overthrust sheet that exposes the underlying rocks. Such a feature is called a *fenster*. This fenster is a 2,168-meter (7,112 feet) high mountain because the Condrey Mountain schist, which some geologists believe is metamorphosed Western plate rocks, is domed upward under the Western Paleozoic and Triassic plate, poking through these older rocks. The Condrey Mountain schist is also far more resistant to erosion than the older rocks that surround it. An alternative view is that Condrey Mountain is a klippe of the Central Metamorphic plate rocks.

FIGURE 4-5 Ordovician marble of the Duzel Formation at Duzel Rock, Eastern Klamath plate. Trinity Alps in the distance. (Photo by R. M. Norris)

Western Paleozoic and Triassic Plate. This plate is the most extensive of the structural units that make up the Klamath Mountains. It extends from well north of the California border south 96 kilometers (60 miles) to the South Fork of the Trinity River in southern Trinity County. The width of the belt varies from 40 kilometers (25 miles) west of Cecilville on the north slope of the Salmon Mountains in southern Siskiyou County, to almost 80 kilometers (50 miles) on the Oregon border. The plate is bounded on the east by the Siskiyou thrust zone, which deformed a middle to late Jurassic radiolarian chert about 133 to 158 million years ago, during the late Jurassic. It is unclear whether or not the Siskiyou fault is a subduction zone, though some geologists believe it is. The volcanic activity that produced rocks in the Eastern Klamath plate and some during the Jurassic, may indicate the occurrence of subduction along this fault.

Although the Klamath province as a whole has long frustrated geologic interpretation, it is probably this belt of rocks that continues to pose the knottiest problems. Its deposits include deep ocean crust, upper mantle rocks, and deposits that accumulated in deep ocean waters. The oceanic crust and mantle represented by dismembered or incomplete ophiolite sequences and the deep-sea deposits are volcanic and sedimentary rocks formed near an ancient island arc.

The southern portion of this plate, near Weaverville, is divided from east to west into the North Fork terrane, the Hayfork terrane, and the Rattlesnake Creek terrane. It is not yet clear whether these terranes are separated by major thrusts and represent major periods of accretion to the continent in middle to late Jurassic time, or if a single plate has been broken into separate parts during collision with the eastern plates. Geologists disagree as to whether we are looking at a single island arc evolving from late Triassic through Jurassic time with overriding basement rocks, or multiple arcs docking at North American shores one by one. The three-part division of the Western Paleozoic and Triassic plate mentioned previously is not recognized north of the Salmon Mountains. The northern part of the plate is more strongly metamorphosed and is intruded by many late Jurassic plutons.

The North Fork terrane crops out along a 1.6 to 10-kilometer (1–6-mile)-wide belt along the North Fork of the Trinity River. Its ultramafic (very iron- and magnesium-rich) rocks, mafic plutonic rocks, pillow basalts, and deep-sea sedimentary rocks like chert indicate that at least part of an ophiolite sequence is present. The upper portion of the North Fork terrane and the upper portion of the structurally adjacent Hayfork terrane are a mixture of rocks with both large and small blocks of seemingly unrelated origin. This mixed rock or *melange* is often referred to as an *olistostrome*. These chaotic deposits of heterogeneous materials accumulated as a semifluid body as a result of submarine gravity sliding or slumping. Some melanges, however, are produced by fragmentation and mixing during subduction; these are called *tectonic melanges*.

The lowest unit of the three units that make up the Hayfork terrane is called the Hayfork Bally meta-andesite, which is intruded by the Ironside Mountain batholith and the Wildwood pluton. Salmon Mountain, North Trinity Mountain, Ironside Mountain, and Pattison Peak, roughly along the Humboldt-Trinity County line, are carved from the granitic rocks of the Ironside Mountain batholith. Rocks stratigraphically between the meta-andesite and the melange are mostly chert, argillite, and minor limestone.

The age of the Western Paleozoic and Triassic plate depends upon one's point of view. Fossils as old as Silurian are found in limestone blocks embedded in an old submarine slide deposit in the Salmon Mountains, but elsewhere limestone blocks yield late Paleozoic, Triassic, and early Jurassic fossils. We know that the olistrostromal deposits incorporate rocks of many ages, and studies of the radiolaria in chert of the Western Paleozoic and Triassic plate show a similar range of dates.

The youngest known Permian fauna in the western hemisphere is found in the Hayfork terrane near Hall City caves and on the ridge west of Potato Creek. These localities are near the point where State Highway 36 crosses the Shasta-Trinity County line. Fusilinids in limestone blocks include a number of species found at only a few localities elsewhere along the Pacific coast of North America. Instead, they belong to a faunal assemblage represented in the rocks of China, Japan, and southeastern Siberia. How did they arrive in North America? They occur in a melange in the Western Paleozoic and Triassic plate, which must have formed far to the west of today's Klamath Mountains, perhaps carried by a plate that eventually was underthrust against North America. Other fossils, such as conodonts—small tooth like organic structures of unknown affinities—also indicate that the plate originated in the western Pacific. Rock ages in this plate span about 200 million years. We should remember that the age of the individual rock units in a melange is no more the age of the melange than the age of clay used to make bricks is the age of a brick wall. When were the pieces assembled? Probably the last units to be lithified did so just before the end of the Jurassic when all parts of the plate were deformed together during the Nevadan orogeny.

FIGURE 4-6 Galice Formation, a dark-colored slate and phyllite of late Jurassic age. Near the junction of Dillon Creek and the Klamath River, Western Jurassic plate, Siskiyou County. (Photo by Don Norris)

Western Jurassic Plate. This plate extends for 350 kilometers (220 miles) along the western edge of the Klamath Mountains. The youngest rock unit is the Galice Formation, which is at least thousands of meters thick (Figure 4-6). This rock is mostly slate and meta-graywacke with a few pyroclastic beds, but there are no distinctive marker units. The Galice seems to have been deposited in deep water near an island arc. Radiolaria and pelecypods give a Jurassic age that is confirmed by the dates of the Josephine ophiolite below, which is 157 million years old, and by a sill in the upper Galice that has been radiometrically dated at 150 million years.

Although the Galice rests directly on the Josephine ophiolite at many localities, in some places it rests on or interfingers with a thick pile (7,000 meters or 23,000 feet) of island arc volcanic rocks called the Rogue Formation. The nature of the contact between the Galice and Rogue is not clear, however. The two units may be about the same age and could have been brought into contact with one another during a period of subduction.

The western margin of this plate is the South Fork Mountain thrust, named for a prominent ridge that marks the west flank of the Klamaths for more than 150 kilometers (90 miles) (Figure 4-7). The western edge of the Klamaths along the east side of this thrust is marked by the South Fork Mountain schist, a well-foliated

FIGURE 4-7 South Fork Mountain near Mad River, Trinity County. The trace of the South Fork Mountain thrust fault lies along the crest of the mountain; the rocks in the photo are largely parts of the Franciscan complex, which here makes up the lower, overridden plate. (Photo by R. M. Norris)

quartz-mica schist with some meta-volcanic layers containing blue amphiboles (glaucophane and crossite). This schist was once thought to be the same age as the Rogue and Galice, but we know that the blueschist metamorphism took place during the Cretaceous. Moreover, some rocks in the Yolla Bolly Mountains in extreme southeastern Trinity County contain early Cretaceous pelecypods, and these rocks seem to grade, with increasing metamorphism, into normal South Fork Mountain schist. As a result, it now seems likely that the South Fork Mountain schist and its equivalents along the thrust are more closely related to the Franciscan Formation found in the Coast Ranges to the west.

Blueschist metamorphism takes place under conditions of high pressure and relatively low temperature. These conditions are believed to exist when one plate of the earth's crust is subducted beneath another. Hence, development of blueschist metamorphism is the South Fork Mountain schist indicates that these rocks were subducted along a fault. The metamorphic event is dated at 120 million years ago. Thus sometime in the early Cretaceous the Coast Ranges were attached to the Klamath province along the South Fork Mountain thrust.

Klamath Mountains Ophiolite Suite. A drill core of today's ocean floor would first penetrate the surface sediment accumulations that will one day form shale, limestone, and chert. Below these, the drill hold would encounter pillow basalts formed by extrusion of lava on the sea floor and cut by numerous feeder dikes. Still deeper, the core would include plutonic mafic rocks like gabbro and, ultimately, near the boundary of the crust and mantle, iron- and magnesium-rich peridotite. This assemblage of sea-floor crustal rocks is called an ophiolite suite. Fragments of such suites are found in the subjacent rocks of the Klamath Mountain province and are also present in the Sierra Nevada and Coast Ranges. In the Klamath Mountains, the most prominent examples are the Trinity ophiolite on the eastern side of the province and the Josephine ophiolite, which forms the basement for the western Jurassic plate along the western side of the province.

The Trinity ophiolite, called the Trinity ultramafic sheet by earlier workers, has most of the lithologic elements prescribed for ophiolites and has been extensively studied. It is one of the largest pieces of oceanic crust exposed on the North American continent. The age of crystallization has been radiometrically dated as Ordovician (Figure 4-4).

No ophiolite suite rocks are exposed in the Central Metamorphic plate, but many exposures of closely related rocks are present in the next plate to the west, the Western Paleozoic and Triassic plate. There are no areas in which a complete ophiolite sequence has been recognized in the Western Paleozoic and Triassic plate, but the North Fork ophiolite is nearly complete. This ophiolite is exposed 11 kilometers (7 miles) northeast of Cecilville near where Mathews Creek joins the South Fork of the Salmon River. These cherts contain Permian radiolaria, but most of the cherts associated with oceanic crustal rocks in this plate contain Triassic radiolaria.

The Josephine ophiolite, formerly known as the Josephine peridotite, is the basement for the volcanic and sedimentary rocks of the western Klamath Mountains. This nearly complete ophiolite is exposed extensively in the Smith River drainage north of the U.S. Highway 199 in Del Norte County, and outcrops extend southeasterly parallel to the South Fork Mountain thrust into northern Humboldt County.

Like most ultramafic rocks, the Josephine ophiolite contains deposits of chromium and nickel. Chromite has been obtained from mines near the head of the Smith River as well as elsewhere in the province, and nickel has long been produced from similar rocks in the Klamath Mountains of Oregon.

Granitic Rocks. Most of the intrusive rocks not part of ophiolite suites in the Klamath Mountains are granitic plutons similar to those of the Sierra Nevada. There are numerous exposures of these rocks in the province, but fewer examples occur in the Eastern Klamath plate than farther west (Figure 4-8). This may indicate that erosion of the eastern plate has not yet unroofed some plutons. For example, the nature of the roof pendant rocks surrounding the Shasta Bally batholith west of Redding at Buckhorn Summit, where State Highway 299 crosses the Trinity-Shasta County line, suggests that only a small part of this batholith has yet been exposed.

Another very large pluton is the Ironside Mountain diorite exposed for about 60 kilometers (37 miles) from Orleans Mountain lookout tower on the Humboldt-Siskiyou County line to 5 kilometers (3 miles) east of Hayfork on the south. For most of this distance, the outcrop is only 4 to 5 kilometers (2½–3 miles) wide, but at North Trinity Mountain it is 10 kilometers (6 miles) wide.

Most granitic plutons throughout the Klamath province range in age from 130 to 170 million years old. Because there is commonly some error range in radiometric dates, dates from plutonic rocks such as these are normally considered minimum dates. Although these radiometric dates are late Jurassic to early Cretaceous, most geologists who have seen the field evidence believe that all these plutons are truly

FIGURE 4-8 English Lake, a small cirque lake in the Marble Mountain Wilderness. Rocks in the foreground are diorites of the English Lake pluton within the Western Paleozoic-Triassic plate. The light-colored rocks on the skyline belong to a different pluton. Siskiyou County. (Photo by G. R. Wheeler)

Jurassic. For example, early Cretaceous dates (131–136 million years) come from the Shasta Bally batholith, which is overlain with prominent unconformity by early Cretaceous rocks of the Eastern Klamath plate. The prominence of the unconformity suggests a long period of erosion and hence an older date for the batholithic rocks. Alternatively, the dates may be correct and show that erosion was extremely rapid prior to deposition of the Cretaceous rocks.

A few diorite plutons give dates older than Jurassic, such as the Pit River stock dated at 250 to 260 million years (early Triassic) and the nearby Mule Mountain stock that is dated at 400 million years (Devonian). Both these plutons intrude the southern part of the Eastern Klamath plate. Apparently the magmatic pulses that formed so many plutons in the Sierra Nevada during the Cretaceous did not affect the Klamath Mountains.

Superjacent Rocks

The superjacent rocks are those that have been deposited since the Nevadan orogeny, and for that reason generally overlie the subjacent rocks with great unconformity. After the plate collisons and intrusions of magma that ended in the early Cretaceous, erosion began to level the new mountain range. The marine origin of the Cretaceous rocks in the Klamath province indicates that although some parts of the range may have remained above water, much of the province sank below sea level after the close of the Nevadan orogeny. Depositing of the thick pile of clastic material eroded partly from the newly uplifted Sierra Nevada began to fill what is now the Sacramento Valley and then lapped onto the Klamath Mountains. In the Sacramento Valley these deposits are assigned to the upper part of the Great Valley sequence. The oldest of these superjacent rocks in the Klamath Mountains contain pelecypods and ammonites of early Cretaceous age. A 300-meter (1,000-foot) thick section of these rocks is exposed at Reading Creek about 16 kilometers (10 miles) south of Weaverville. Thinner remnants of these rocks are exposed elsewhere along the eastern side of the Klamath Mountains, especially in a 50-kilometer (30-mile)-long arc bisected by State Highway 36 east of Beegum in extreme southwestern Shasta County. These early Cretaceous rocks are conglomeratic at their base and are separated by profound unconformity from the rocks below. Conglomerate pebbles at Reading Creek are mostly mica schist much like the rocks of the Central Metamorphic plate on which it rests. At Ono, 24 kilometers (15 miles) southwest of Redding, these early Cretaceous rest on the Shasta Bally batholith and are the main reason the batholith is believed by some to be of Jurassic age, despite the radiometric dates that suggest a Cretaceous age.

Late Cretaceous sedimentary rocks belonging to the Hornbrook Formation crop out along Interstate Highway 5 from 5 kilometers (3 miles) south of Hornbrook north of Yreka, to the Oregon border. Although this exposure is only about 30 kilometers (18 miles) long in California, it extends some distance into Oregon where it continues eastward beneath the lavas of the Cascades. The conglomerate beds of the Hornbrook form prominent ridges and the sandstones and shales tend to form valleys. The fossils present, as well as the clast size, shape, and composition indicate that the Hornbrook was deposited in shallow marine waters quite close to shore. Further, the distribution of outcrops and contact relationships suggest a shoreline with numerous offshore islands.

Interestingly, the basal conglomerate of the Hornbrook contains some gold. Apparently the gold was derived for quartz veins in the subjacent rocks, which were exposed along the Cretaceous shoreline. Erosion of these older lode deposits led to the development of placer concentrations in the streams, river deltas, and offshore bars, which eventually became part of the Hornbrook Formation. These ancient placer deposits were mined during the nineteenth century near the town of Hornbrook in northern Siskiyou County. As current gold prices of about $400 an ounce, some parts of the Hornbrook Formation contain as much as $5 of gold per cubic meter, but extreme variability of these ancient placers makes mining them highly speculative.

The Weaverville Formation is exposed in an area of about $15\frac{1}{2}$ square kilometers (6 square miles) northeast of Weaverville at Reading Creek and around Hayfork; there are smaller patches elsewhere. This formation is composed of nonmarine sandstone, shale, tuff, and conglomerate with beds of lignite. It was deposited on a swampy flood plain covered with lakes. The depositional environment and resulting rocks are quite similar to the Eocene Ione Formation present at the western base of the Sierra Nevada. Plant fossils from the Weaverville Formation have been variously assigned to the Miocene, Oligocene, and Eocene, but today's best estimate is Oligocene, a little younger than the Ione. The lignite was formerly mined at Reading Creek, Hyampom, Hayfork, and Big Bar south of Weaverville. Phosphate comprises as much as 20 percent of the formation at Hyampom, almost enough to support an economic phosphate mine. Stream gravels in the Weaverville Formation have been mined for the gold they contain.

The marine Wimer Formation of late Miocene age is a yellow and red poorly consolidated sandstone, shale, and conglomerate that corps out on some ridges in the extreme northwestern portion of the California Klamaths. It is the youngest marine unit in the province.

GLACIATION AND GLACIAL DEPOSITS

During the Pleistocene the Klamath Mountains supported dozens of glaciers from North Yolla Bolly Mountain in southeastern Trinity County to the Oregon border. The jagged topography left by the alpine glaciers is most distinctive in the middle of the province and on granitic peaks above an elevation of 1,500 meters (5,000 feet). Cirque lakes, aretes, and U-shaped valleys with polished and striated boulders are very much in evidence in the Siskiyou and Salmon mountains, and in the Trinity Alps (Figure 4-8).

The Trinity Alps lie north of Weaverville and northeast of Trinity Reservoir. Tributaries of the Trinity River drain the Alps on the northeast, east, and south sides. Although the region was once heavily glaciated, only two small glacierets remain, one on the north side of Thompson Peak and another on adjacent Sawtooth Ridge. These glaciers have elevations between 2,500 and 2,600 meters (8,200–8,500 feet). The Grizzly glacieret on Thompson Peak and the Thompson glacieret on Sawtooth Ridge each cover about 2 hectares (5 acres).

The extensive glacial deposits of the Trinity Alps were studied in the 1960s by R. P. Sharp, who was able to document four glacial stages. The three most recent are all Wisconsin and the oldest is possibly Illinoian. It is likely that this same glacial chronology applies to other parts of the Klamath province that were glaciated during the Pleistocene (Table 4-1).

Considerable glacial detritus is found in Coffee, Swift, and Canyon creeks (Table 4-2). Less extensive deposits occur in other drainages in the Trinity Alps. Because most of the glaciers headed on exposures of quartz diorite, but terminated on metamorphic rocks, the moraines contain erratic boulders that are quite distinctive in their present settings. Furthermore, the light-colored quartz diorite boulders are easily identifiable in outwash several kilometers downstream from the large moraines in Swift and Canyon creeks. One complication in determining the exact extent of glaciation in the Trinity Alps is the tendency for debris flows to form at the snout of the glaciers. Morainal material, mobilized by melting ice, may form debris flows that are nearly indistinguishable from ordinary till. In addition, when slippery ser-

TABLE 4-1 Principal Episodes of Pleistocene
Glaciation on the Trinity Alps

Age	*Time Designation*	*Local Name*
Wisconsin	Late	Morris Meadow
	Middle	Rush Creek
	Early	Alpine Lake
Illinoian or older	Ancient	Swift Creek

Source: After R. P. Sharp, 1960.

TABLE 4-2 Summary of Lengths and Terminal Elevation of the Largest
Pleistocene Glaciers in the Trinity Alps

Glacial Substage	Length (km	mi)	Valley Occupied	Elevation (m	ft)	Number of Glaciers
Morris Meadow	12.6	7.8	Canyon Creek	1,021	3,350	30
Rush Creek	15.1	9.3	Canyon Creek	850	2,800	23
Alpine Lake	18.3	11.3	Swift Creek	760	2,500	6
Swift Creek	22.2?	13.7?	Swift Creek	750	2,450	1

Source: After R. P. Sharp, 1960.

pentinite is present in moraines, mass movements are greatly facilitated. Both Deer
and Swift creeks head in the serpentinites of the Trinity ophiolite east of Trinity
Reservoir. The glaciers that once occupied these valleys carried large quantities of
serpentinite boulders and the debris flows from the moraines choked the valleys
downstream, sometimes forming deposits 100 meters (300 feet) thick.

Later glacial erosion commonly obscures evidence of earlier glacial episodes, but
in the Trinity Alps the Swift Creek glaciation reached lower elevations than later
advances (to as low as 750 meters or 2,450 feet) and thus its terminal moraine is
preserved. We do not know how much area was under ice during the Swift Creek
advance, but during the Alpine Lake glaciation the glaciers of Swift Creek, Stuart
Fork, and Canyon Creek covered 47, 43, and 24 square kilometers (18, 16.7, and
9.2 square miles), respectively. Although these later glaciations did not reach such
low altitudes as the Swift Creek advance, there appear to have been more glaciers
and the ice coverage may have been similar (Table 4-2).

QUATERNARY GRAVELS

The entire province is characterized by narrow valleys with rushing streams and very
little alluvial fill. Scott Valley, about 32 kilometers (20 miles) southwest of Yreka on
the eastern side of the province, is an exception to this rule. Here as much as 120
meters (400 feet) of alluvium rests on rocks of the Eastern Klamath plate and the
Central Metamorphic plate. Gold in the Scott Valley gravels has led to extensive
dredging along the South Fork of the Scott River (Figure 4-9) and elsewhere (see
Mineral Resources). Numerous terraces occur along the Trinity, Smith, and Klamath
rivers, and since the Pleistocene some rivers such as Weaver Creek near Weaverville,
have downcut their channels as much as 120 meters (400 feet), leaving well-defined
river terraces or benches on the valley sides. The rapid downcutting by rivers, even
in resistant rock, indicates that the province experienced considerable uplift in the
last several million years.

PRE-CRETACEOUS STRUCTURE AND PLATE TECTONICS

The dominant structural features of the Klamath province are related to subduction
and accretion of exotic terranes to the North American continent. Indeed, long before
the geometry or even the existence of crustal plates was discerned, J. S. Diller

FIGURE 4-9 Gold dredge on the South Fork of the Scott River, 1973. Gravels that previously made the valley a wasteland are now partly recycled as dredging for gold proceeds. (Courtesy of California Department of Water Resources)

defined the boundary of the Klamath Mountains mostly on the basis of these subjacent rocks. As we have already noted, these rocks are divisible into at least four separate plates that were joined to the North American continent between the Devonian and the Jurassic periods. Each collision crumpled the rocks of the adjacent plates and produced fold axes parallel to the suture. Quartz diorite and elongate ultramafic bodies mark the trace of many of these sutures.

The oldest suture joins the Eastern Klamath plate to the Central Metamorphic plate. The Eastern Klamath plate was overthrust 380 to 400 million years ago, an event dated by the metamorphism it produced in the rocks that became part of the Central Metamorphic plate. The boundary is most easily recognized east of the Bully Choop Mountain 32 kilometers (20 miles) southeast of Weaverville, northwesterly to the Trinity Reservoir. This boundary is called the Bully Choop thrust fault, and north of the south end of the reservoir this fault also forms the edge of the Trinity ophiolite. The rocks on both sides of the thrust appear to be about the same age, with the westward rocks being the metamorphosed equivalent of the rocks east of the fault. As previously noted, the fossil and paleomagnetic nature of these rocks indicates that they were part of an island arc environment close to the continental boundary of the time.

The Siskiyou fault zone is the western boundary of the Central Metamorphic plate, and the rocks of the Western Paleozoic and Triassic plate were underthrust along this structure between 133 and 158 million years ago. The Siskiyou fault forms a prominent lithologic boundary from Selvester Ranch 26 kilometers (16 miles) southwest of Ono, northwest to Hayfork Summit and northward to near Cecilville. At Cecilville, the fault is obscured by a large ultramafic body, and to the north it is much less prominent. The southern part of the fault is marked by east-dipping shear planes and a crushed rock called phyllonite.

We have been referring to *underthrusts* and *overthrusts*, both dipping to the east. Because motion on faults is usually expressed in relative terms, it is understandable if these sounded like the same thing seen from different perspectives. It is very difficult to be certain which sort of motion actually occurred,but when the structure is referred to as an underthrust, the geologist believed the evidence suggested active thrusting beneath a relatively immobile upper slab. Similarly, an overthrust suggests that the motion was confined chiefly to the upper plate, but these are judgments based on field evidence, and the actual motion may have been different than that which was deduced.

Rocks of the Central Metamorphic plate are tightly folded and overturned to the west. In the Trinity Alps, the crests of some of these folds have been eroded deeply enough to expose what appear to be less metamorphosed rocks of the Western Paleozoic and Triassic plate beneath the Salmon schist.

Some geologists argue that the Permo-Triassic Sonoma orogeny, a mountain-building episode perhaps best displayed in central Nevada, also bought profound changes to the outline of this part of North America. Prior to this orogeny, the continental boundary presumably ran through central Nevada and an island arc, perhaps like modern Japan, lay off the coast. The orogeny joined this arc (the Klamaths and the northern Sierra Nevada) to the continent and produced a new shoreline somewhere to the west of what we now recognize as the Central Metamorphic plate. This explains why some of the older rocks in the northern Sierra seem to match older rocks in the Klamaths. We have not found the subduction zone that marked the western edge of the island arc that joined North America; perhaps it was subducted by the younger Siskiyou thrust fault and today lies buried beneath the Central Metamorphic plate. The blueschists at Fort Jones are Triassic (214–222 million years old), however, and these and the associated remnants of metamorphosed chert, argillite, and mafic igneous rocks may be parts of the lost subduction zone.

The Siskiyou thrust fault of late Jurassic age separates rocks of similar age and lithology just as does the Melones fault in the Sierra Nevada. In the northern Sierra the Shoo Fly complex of Devonian age is equivalent to rocks of the Eastern Klamath plate.West of the Melones fault the Calveras Formation contains metacherts, phyllites, various metavolcanic rocks, and ultramafic plutonic rocks much like those present in the Western Paleozoic and Triassic plate in the Klamath Mountains. Thus, the rocks east of the Siskiyou-Melones fault zone are continental basement rocks joined to what are called *allochthonous* rocks to the west. It will be recalled that allochthonous rocks are those that formed in distant areas and were transported, probably by plate motions, to their present sites. The Western Paleozoic and Triassic plate contains fossils and rocks that seem to have come from the western Pacific.

The western edge of the Western Paleozoic and Triassic plate is a fault that also bounds the western side of the Ironside Mountain pluton and extends from near Hayfork on the south, northward about 130 kilometers (80 miles) to near Happy Camp in Siskiyou County where it veers northeast. The northern 32 kilometers (20 miles) of this boundary lies along ultramafic bodies rather than the granitic rocks of the Ironside Mountain pluton. The northern portion of this structure generally coincides with the Orleans fault along which the Western Jurassic plate was underthrust beneath the Western Paleozoic and Triassic plate. The Orleans fault was originally believed to be a steep reverse fault, but new evidence indicates that instead it is a gently dipping thrust. The fault is well exposed 6 kilometers (4 miles) east of Hoopa in northeastern Humboldt County. At this locality the fault has a gentle east dip and is associated with eastward dipping shear planes in serpentinite. Elsewhere the fault appears steeper, possibly because of later folding.

Faults on both margins of the Western Paleozoic and Triassic plate were active in the late Jurassic. Because they represent old subduction zones it is evident that during the Nevadan orogeny two or more separate crustal plates docked at American shores, or perhaps a plate broken into several discrete parts during transport joined the continent. The history of these plates is very complex and it is not surprising that no consensus has yet been reached about their origin.

Most of the rocks of the Galice and Rogue formations, which represent the majority of the Western Jurassic plate rocks, show slaty cleavage that dips eastward (Figure 4-6). Folds in these rocks are aligned generally with the trends of the major bounding faults.

An outlier of South Fork Mountain schist well within what we normally consider the Coast Range province, forms a band 5 to 10 kilometers (3–6 miles) wide, extending from the coast at Stone Lagoon near Orick, west of Redwood Creek almost as far south as Arcata Bay. Another outlier covering about 10 square kilometers (4 square miles) occurs at Patricks Point just north of Trinidad on the Humboldt County coast. When the South Fork Mountain schist was thought to be a part of the Western Jurassic plate, these outliers were called klippen. Now that we know they are metamorphosed Cretaceous rocks, the exposures are actually fensters (windows) through the thrust sheet where erosion has not removed all the blueschist formed just under the sole of the thrust. Our present view of the geometry of these outliers is exactly the opposite of what we formerly believed; new evidence is always hazardous to ruling theory.

The Klamath Mountain province and the Western Jurassic plate is sharply separated from the Coast Range province by the South Fork Mountain thrust fault. The prominence of this fault as a lithologic and topographic boundary is without rival in the northern part of California. It is along this striking feature that the Klamath Mountains have been thrust westward over the Franciscan rocks of the Coast Ranges (Figure 4-10). This fault also forms the western boundary of the Josephine ophiolite and extends south-southwest for more than 190 kilometers (120 miles) before turning easterly as it rounds the south end of South Fork Mountain. This fault passes just north of North Yolla Bolly Mountain in southeastern Trinity County and reaches the Sacramento Valley west of Red Bluff. The South Fork Mountain thrust joins or passes into the Coast Range thrust from this point. Like the South Fork Mountain thrust, the Coast Range thrust is marked by sheared ultramafic rocks such as serpentinite and is discussed more fully in Chapter 11.

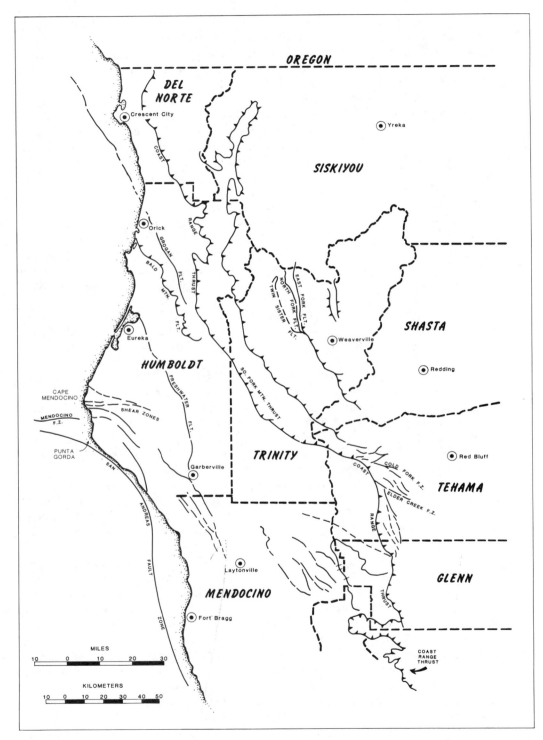

FIGURE 4-10 The South Fork Mountain thrust and related faults. The South Fork Mountain thrust is one of the main segments of the Coast Range thrust zone.

The rocks of the Great Valley sequence have been thrust as much as 80 kilometers (50 miles) westward over the Franciscan rocks along the Coast Range thrust; perhaps the rocks of the Western Jurassic plate have moved a similar distance westward over the Franciscan on the western side of the Klamath Mountain province. Such a history is suggested by the South Fork Mountain schist that crops out almost continuously along the thrust and represents regionally developed blueschist formed as Franciscan rocks were overridden along the western side of the Klamath Mountains. The age of metamorphism and thus the underthrusting is 120 million years. This tells us that the collage of plates that today comprises the Klamath Mountains was moved westward over the Franciscan rocks in early Cretaceous time.

CENOZOIC STRUCTURE

It should be clear that most of the rock types, structural features, and topography of the Klamath province are related to pre-Cenozoic plate tectonic activity. However, structural features of Cenozoic age are also present. For instance, the Oligocene Weaverville Formation is preserved in down-dropped blocks bounded by steeply-dipping faults. These Tertiary faults form two poorly outlined grabens, one from Trinity Reservoir southwest to near Weaverville, about 16 kilometers (10 miles), and another in the vicinity of Hayfork. He most obvious of these faults is the La Grange fault a short distance northwest of Weaverville near Oregon Mountain. Because these are normal faults, they represent an episode of crustal stretching or extension in contrast to the much longer and more prominent compressional history that characterized much of the province during the Paleozoic and Mesozoic.

SPECIAL INTEREST FEATURES

Shasta Lake Recreational Area

The largest reservoir in California and the northernmost element of California's vast Central Valley Water Project is Shasta Reservoir (Figure 4-11). The 182-meter (602-foot)-high dam across the Sacramento River 19 kilometers (12 miles) north of Redding was built in the early 1940s and the reservoir first filled in 1944. The water is ponded on rocks of the Eastern Klamath plate and backs up water of several important tributaries of the Sacramento River. Along the 584 kilometer (365 mile) shoreline of the lake are many fine rock exposures and there is even an opportunity to go inside a formation. At Shasta Caverns the McCloud limestone has been dissolved to the form an extensive series of caverns, and groundwater has reprecipitated the limestone to form many fine travertine structures.

This area also has some economic importance. The McCloud limestone is mined commercially and ores of zinc and copper occur in the metamorphic rocks, mainly of Devonian and Permo-Triassic age. Further, more than 500,000 ounces of gold were mined from the West Shasta district.

Castle Crags State Park

West of the Sacramento River canyon, midway between Redding and Mount Shasta, is Castle Crags State Park. The main attraction is an unusually eroded Jurassic granodiorite pluton, which has intruded the Trinity ophiolite. The granitic rocks are

FIGURE 4-11 Shasta Lake and Shasta Dam. The lake has drowned major segments of the Sacramento, McCloud, and Pit rivers, thus emphasizing the ridge-valley topography of the Klamath province. Mount Shasta dominates the skyline. (Photo courtesy of California Department of Water Resources)

interesting because they contain potash feldspar crystals 2.5 centimeters (1 inch) long or more and quartz dikes up to 15 centimeters (6 inches) wide, which often form resistant ridges that stand out, keel-like above the most easily weathered granitic rock. The granite is more resistant than the serpentinized periodite into which it has intruded, however, so the granitic rocks form a topographic eminence. In addition, nearly vertical jointing formed during the cooling or unroofing of the pluton have provided avenues for more rapid weathering. The net result is a cluster of granitic pinnacles and tower like features that rise prominently above the surrounding area (Figure 4-12).

Marble Mountain Wilderness

It is fortunate that Herbert Hoover enjoyed fishing Wooley Creek east of Somes Bar, because his familiarity with this area was probably the reason Wooley Creek and the region to the north and east were designated the Marble Mountain Primitive Area in 1931, during his administration. In 1953 the designation was changed to the Marble Mountain Wilderness Area. The area is named for Marble Mountain (2,098 meters or 6,880 feet) and lies about 45 kilometers (30 miles) west of Mount Shasta.

The rocks in the area are metachert, slate, limestone, quartzite, greenstone, and peridotite breccia of the Western Paleozoic and Triassic plate, intruded by late Jurassic plutons. Rocks of the Central Metamorphic plate have been thrust over similar rocks in the Western Paleozoic and Triassic plate along the South Fork thrust which is about parallel to and a little over a kilometer east of North Russian Creek, a small tributary of the Salmon River a few kilometers northeast of Sawyers Bar.

FIGURE 4-12 Castle Crags, an intrusive body of Jurassic granodiorite located in the eastern part of the Klamath Mountains, Shasta County. (Photo by R. M. Norris)

Detailed studies in the Western Paleozoic and Triassic plate show that the rocks in this unit have been folded at least three times, producing overturns and isoclinal folds that have the effect of repeating the section over and over. The earlier folding was associated with the subduction of one plate beneath the other, and later folding is attributed to stresses resulting from the intrusion of plutonic rocks during the Nevadan orogeny.

Although the most common rocks are metamorphosed clastic and volcanic rocks, the most prominent rocks are marble and intrusive rocks. A number of small plutons are found in the area, but the two largest are the English Peak and Wooley Creek plutons (Figure 4-8). These plutons are 147 to 157 million years old and were forcefully intruded into the country rock, deforming and truncating these older rocks. The dominant intrusive rock is quartz diorite or granodiorite, but there is an enormous diversity of other intrusives, ranging from gabbro to quartz monzonite. This variation results from the different kinds of country rock assimilated during intrusion. For example, the resulting plutonic rock is more mafic as if a lot of metamorphosed basalts (greenstones) were assimilated during intrusion.

The peridotites and other ultramafic rocks, now mostly altered to serpentinite, are probably late Paleozoic or Triassic in age. Red cherts found just outside the Wilderness area are associated with ophiolitic rocks and contain Triassic radiolaria. The late Jurassic English peak pluton has further metamorphosed the serpentinite at Red Bank campground on the North Fork of the Salmon River, proving that the serpentinite predates the intrusion. As in commonly seen elsewhere in California's higher mountains, the reddish-brown weathering of the ultramafic rocks accounts of the color of Red Mountain, Red Rock Valley, Red Rock, Buckhorn Mountain, and Huckleberry Mountains, all in the north and northeastern part of the Wilderness Area.

Marble (2,098 meters or 6,880 feet) and Black Marble (2,269 meters or 7,442 feet) mountains are found in the north central part of the Wilderness. Rocks forming these peaks are carbonate rocks that were originally coral reefs, later metamorphosed and

FIGURE 4-13 Marble Mountain Wilderness area looking east toward Mount Shasta. Rainy Lake, in the center of the photo, occupies a well-developed glacial cirque. The light-colored elongate ridge in the upper left included Marble Mountain and consists of Duzel limestone, as does the pointed summit of Elk Peak in the lower left corner. Siskiyou County. (Photo courtesy U.S. Geological Survey)

joined to the continent with Western Paleozoic and Triassic plate. Black Marble Mountain has a dark top because of the dark metasedimentary rock that caps the lighter marble below.

Most of the Marble Mountain Wilderness high country was glaciated during the Pleistocene. The ice, which extended down to an elevation of 1,370 meters (4,500 feet), left abundant glacial erosional and depositional features. Most of the lakes are cirque lakes and many are dammed by moraines. Morainal deposits and outwash are particularly prominent along the North Fork of the Salmon River and along its Right Hand Fork. Topography of the area is dramatic because knife like arete ridges divide the valleys and connect the peaks, which are composed chiefly of the plutonic rocks of the English Peak and Wooley Creek plutons (Figure 4-13).

MINERAL RESOURCES

Metallic resources of the Klamath province include gold, chromium, copper, zinc, pyrite, lead, silver, iron, platinum, mercury, and manganese. Nonmetallic resources include limestone, sand and gravel, building stone, clays, and various crushed rocks. However, the most precious mineral may prove to be water. The Klamath River

System and the Smith, Mad, and Eel rivers represent one of the large untapped sources of water for the more arid parts of the state to the south.

Gold

The Klamath province is second only to the Sierra Nevada in historic gold production. About 20 percent of the 4 million ounces of gold recovered in California has been obtained from the Klamath Mountains. In this part of California, the first gold was discovered in 1848 in the Trinity River near Douglas City, by Major Pierson B. Reading. Productive gold placers were soon located elsewhere on the Klamath River drainages. The northern placers were at Hornbrook, Yreka, Scott Bar, Hamburg, Somes Bar, Orleans, Sawyers Bar, Forks of the Salmon, Callahan, and at Cecilville. Placers along the Trinity River were richest at Carrville, Trinity Center, Minersville, Lewiston, Weaverville, Junction City, and Salyer; some of these are now flooded by Clair Engle Lake (Trinity Reservoir).

The placer deposits were mined from both modern stream gravels and old stream terraces. In addition, some gold was mined from the Weaverville Formation and some from the older Hornbrook Formation along Cottonwood Creek, a stream that parallels Interstate Highway 5 near the Oregon border. Although the Weaverville and Hornbrook formations are consolidated rocks, the gold ores are not lodes in a strict sense, but fossil placers, because both these formations were originally stream deposits.

Because gold in the placers were derived from eroded lode deposits in the region, these lodes were quickly located after the placer discoveries. By 1852, gold production was underway at the Washington mine, 5 kilometers (3 miles) north of State Highway 299 just east of French Gulch, or 24 kilometers (15 miles) west of Redding. The French Gulch district proved to be the most productive gold district in the Klamaths. The gold occurs in quartz veins cutting slate, shale, and siltstone of the Mississippian Bragdon formation within the Eastern Klamath plate. The source of these gold-quartz veins is apparently the large Shasta Bally batholith a few kilometers to the south. The ore is associated with numerous porphyrytic quartz diorite and diorite dikes called "bird's-eye porphyry."

Other lode deposits have been mined at Harrison and Whiskeytown in Shasta County and at Deadwood, Dillon Creek, Callahan, Oro Fino, and Sawyers Bar in Siskiyou County as well as at a number of other localities. More lode gold is associated with the three eastern plates; the richest is the Eastern Klamath plate. Although most of these deposits have not been studied in detail, it is likely that the proximity of plutons and birdseye porphyry demonstrates that the ores were formed during the Nevadan orogeny. All the host rocks for the lodes predate the late Jurassic age of the plutons.

In the 1980s, gold was being recovered from an ore body at the Grey Eagle mine near Happy Camp in western Siskiyou County. Many of the old gold districts are being reevaluated as new exploration methods, improved technology, and higher prices have made the tailings left by earlier miners today's economic deposits.

The Shasta district in the vicinity of Shasta Reservoir has long been known as a copper and zinc district, but the area has recently been explored for its gold potential. The massive sulfide deposits of the East Shasta district occur in the Triassic Bully

Hill rhyolite and in the Pit Formation. The West Shasta district, on the other hand, is underlain by the Devonian Balaklala rhyolite.

Chromium

Chromite, an oxide of iron and chromium occurs throughout the Klamath province in the ophiolite bodies or is sometimes concentrated by weathering in the soils or found as placers. Some beach placers near Crescent City in Del Norte County contain as much as 7 percent chromite. The weathered peridotite in the Josephine ophiolite at Gasquet Mountain in Del Norte County is also a potential source of chromium, and attempts have been made to develop this site about 1980. The host rock is a deeply weathered soil containing 2 percent chromium plus about 0.8 percent nickel and 0.1 percent cobalt. Thirty million tons of ore have been identified in the upper 20 feet of soil near the crest of Gasquet Mountain. Although the United States currently imports almost all of its chromium from South Africa and produces no cobalt and very little nickel, mining development has proceeded very slowly because the operations are near the Smith River, an officially designated Wild and Scenic River.

Copper and Zinc

Copper and zinc rank with gold in terms of past production. Almost all of the copper and zinc came from the West Shasta district. The deposits are massive sulfide ores typically composed of 90 to 95 percent sulfide minerals. Of these, pyrite is the most abundant, and lesser amounts of other sulfides such as chalcopyrite and sphalerite are present, ores of copper and zinc, respectively. The massive sulfides are replacement bodies formed as the Balaklala rhyolite was invaded by hot solutions that are believed to have come from nearby plutons. These sulfides were derived from the Copley Greenstone, which underlies much of the area, or were from vents on the sea floor active at the time the Balaklala rhyolite was erupted. The West Shasta district has produced 340,000 tons of copper and 30,000 tons of zinc, but is not currently active. Similar ores occur in the East Shasta district, but to date that district has yielded only about 15 percent as much ore as the West Shasta district.

Other Metals

Platinum weathered from ultramafic bodies is sometimes found in gold placers. Just east of Interstate Highway 5 as it crosses Shasta Reservoir, the Shasta iron mine is visible at the beginning of the McCloud Arm. This is a replacement magnetite deposit in the McCloud limestone and in a quartz diorite dike that carried the ore. Little ore has been produced from this deposit to date. A significant mercury deposit, probably formed by hot-spring activity replacing diorite and peridotite, occurs at the Altoona mine in northeast Trinity County in the drainage of the East Fork of the Trinity River.

REFERENCES

General

Burchfiel, B. C., and G. A. Davis, 1981. Triassic and Jurassic tectonic evolution of the Klamath Mountains-Sierra Nevada geologic terrane. *In* Ernst, W. G., ed., The Geotectonic development of California. Prentice-Hall, Englewood Cliffs, NJ: pp. 51–70.

Irwin, W. P., 1966. Geology of the Klamath Mountains. *In* Bailey, E. H., ed., Geology of Northern California. Calif. Div. Mines and Geology Bull. 190, pp. 19–38.

———, 1981. Tectonic Accretion of the Klamath Mountains. *In* Ernst, W. G., ed., The Geotectonic Development of California. Prentice-Hall, Englewood Cliffs, NJ:, pp. 29–49.

Rapp, J. S., 1985. Gold-bearing conglomerate Beds of the Hornbrook Formation. California Geology, v. 38, pp. 219–224.

Sharp, R. P., 1960. Pleistocene Glaciation in the Trinity Alps of Northern California. Am. Journal of Science, v. 258, pp. 305–340.

Special

Albers, J. P., 1966. Economic Deposits of the Klamath Mountains. *In* Bailey, E. H., ed., Geology of Northern California. Calif. Div. Mines and Geology Bull. 190, pp. 51–63.

Aune, Quentin, 1970. A Trip to Castle Crags. Mineral Information Service (now California Geology), v. 23, pp. 139–144.

Green, D., 1980. Marble Mountain Wilderness, Wilderness Press, Berkeley, 162 p.

five

CASCADE RANGE AND MODOC PLATEAU

Nothing tones up a volcano like blowing off a bit of lava.

—*S. J. Perelman*

CASCADE RANGE

Geographically, the Cascade Range occupies only a very small part of California. Nevertheless, the Californian part of the range has two notable distinctions. It includes the second-highest volcano in the entire range, Mount Shasta (Mount Ranier is slightly higher) and until the eruption of Mount St. Helens in 1980, the most recently active volcano in the coterminous 48 states, Lassen Peak.

The Cascade Range extends from southern British Columbia's Mount Garibaldi (817 meters or 2,678 feet) to Lassen Peak (3,188 meters or 10,453 feet), about 800 kilometers (500 miles) apart. Between lie 12 major and many minor cones built on or near the western edge of the Columbia Plateau. Some Cascade volcanoes straddle the escarpment at the edge of the plateau, others are built from orifices penetrating the Columbia Plateau lavas, and one major cone, Mount St. Helens (2,549 meters or 8,364 feet at present), lies west of the Columbia Plateau.

The boundaries of these provinces, like most others, are somewhat arbitrary. Much of the Modoc Plateau and adjacent parts of the southern Oregon are underlain by thick, mainly Tertiary lava flows characteristic of the larger Columbia Plateau, but also are broken by north-south trending fault blocks reminiscent of the Basin and Range province. Depending upon one's point of view this area could be assigned with equal logic to either the Basin and Range or Columbia Plateau provinces.

In its California section, the Cascade Range grades east into the Modoc Plateau,

152

whose volcanism is dominated by lava flows rather than explosive eruptive products characteristic of the Cascades. Despite this, the Medicine Lake Highlands are normally included in the Cascades, although the style of eruption there has been less explosive.

Geography

The Cascade Range is famous for Mount Shasta and Lassen Peak. Mount Shasta (4,319 meters or 14,162 feet) stands boldly above a generally undulating timbered surface with an average elevation of 1,200 meters (4,000 feet). The mountain is double-crested: The higher and younger cone, called Hotlum Cone, occupies the summit crater of Mount Shasta. The lower, larger and somewhat older peak is Shastina. The pair make an imposing landmark visible more than 150 kilometers (100 miles). Adjacent to Mount Shasta are smaller cones. To the north many cones rise to elevations of 2,100 to 2,600 meters (7,000–8,000 feet), including older cones like 4.8-million-year-old Miller Cone, 3.5-million-year-old Eagle Rock, and younger cones like Whaleback and Goosenest. All of the cones are Quaternary, but Whaleback— 75,000 years old—and Goosenest—50,000 years old—are so young they are scarcely touched by erosion. Black Butte—9,360 years old—is a young plug dome next to Interstate Highway 5 (Figure 5-1). Between Shasta and Lassen are several lesser cones, including Crater Peak (Magee Mountain), Burney Mountain, and Cinder Cone. These break the undulating volcanic platform that is the southwest extension of the Modoc Plateau (Figure 5-2).

Lassen Peak is a very large plug dome that dominates the skyline and is visible from as far away as Sacramento. Snow-capped in most years, it erupted sporadically from 1914 to 1921. Its most recent activity consisted of steam emissions that concluded

FIGURE 5-1 Black Butte, a young hornblende dacite plug dome between Mount Shasta and Weed, Siskiyou County. The symmetrical form of the mountain is the result of talus shed from the central plug and differs from a cinder cone in which eruptive materials accumulate close to the vent. (Photo by G. R. Wheeler)

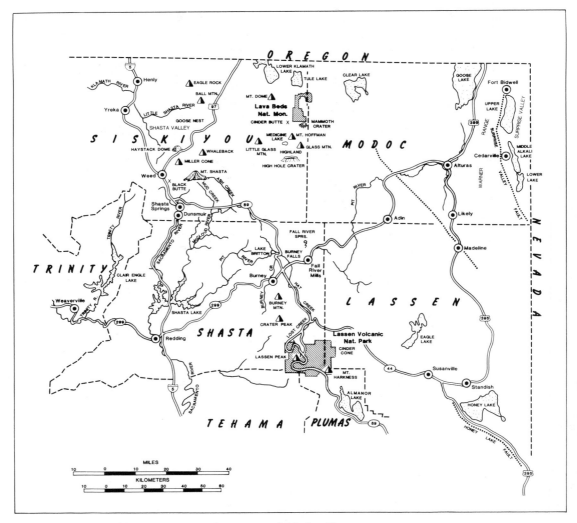

FIGURE 5-2 Place names, Cascade Range and Modoc Plateau.

the episode initiated in 1914. The upper part of the mountain is still barren and is easily reached by road and trail.

Rocks

The exposed rocks of the California Cascades are predominately volcanics of great variety and form. Some lake-bed deposits including freshwater diatomite, some water-laid tuff and ash, and a few glacial morainal deposits occur on Mount Shasta, but they make up less than 10 percent of the province's rocks (Figure 5-3).

The oldest volcanic unit related to the formation of the Cascade Range is the Tuscan (maximum thickness of 450 meters or 1,500 feet), composed primarily of mudflow, ash, and breccias. The Tuscan Formation forms a ramplike transition more

Composite Geologic Columns, Modoc Plateau, and Cascade Range

Modoc Plateau		Cascade Range	
HOLOCENE	Alluvium, dunes, various volcanics, chiefly basalt		Alluvium, Shastina Andesite, Pluto Cave Basalt, other volcanics
PLEISTOCENE	Glacial deposits, lake beds, various volcanic rocks		Lake deposits, Glacial deposits, Butte Valley, and other volcanics
PLIOCENE	Various volcanics, diatomite	Warner Basalt	Cinder cones, basalts, andesites, Tuscan, Tehama (nonmarine)
MIOCENE	Cedarville		
OLIGOCENE			
		CRETACEOUS	Hornbrook (marine)
		JURASSIC	Potem (marine)
			Bagley Andesite
		TRIASSIC	Modin (marine)
		PERMIAN	Dekkas, Nosoni (metavolcanics) McCloud Limestone (marine)
		MISSISSIPPIAN	Baird (marine) Bragdon (marine)
		DEVONIAN	Kennett (marine) Copley Greenstone

FIGURE 5-3 Geologic columns, Cascade Range and Modoc Plateau.

than 95 kilometers (60 miles) long between Lassen Peak and the Sacramento Valley. It is of Pliocene age and is tentatively considered the partial equivalent of the oldest exposed rocks of the Modoc Plateau, the Cedarville series.

Stratigraphically above the Tuscan Formation are thick, extensive andesitic flows extruded from vents throughout the California Cascades. After deep and prolonged erosion, late Pleistocene basaltic lavas were erupted, burying most of the andesites that had escaped erosion. Some basaltic flows cover more than 100 square kilometers (60 square miles), with the total volume of all erupted basalt being approximately 8 cubic kilometers (2 cubic miles). Cones of varying composition, represent several eruptive episodes and culminate in Mount Shasta and Lassen Peak, followed and overlapped the basaltic flow eruptions. Activity probably has not ceased altogether, but the California Cascades have been dormant since 1921.

Many cinder cones occur, usually composed of black or red clinkery fragments, sometimes blown from local vents as hardened chunks of rock. Cinders frequently develop in place as a result of vesicular lavas being fragmented by gas or liquid discharges during eruption. Hills are often domal, pushed upward by trapped gas. Sometimes basaltic scoria, pumice, and obsidian intermingle. On occasion such rocks "float" upward in the rising magma because the porous cindery materials are lighter than the new enclosing magma. Cinders are mined to make concrete blocks for lightweight construction and for road metal. Shades of red, brown, gray, and black can be obtained by using appropriate cinder sources, and color change from one highway jurisdiction to another is a striking feature of northeastern California roads.

Structure

Faulting has been important throughout the development of the California Cascades. In the Tuscan Formation, the lowest beds are folded and eroded but the highest are horizontal, indicating significant local deformation between earlier depositional stages. During andesitic (earlier) volcanism, block faulting occurred; magma emerged along some of the faults, and cones and domes developed. Before and during basaltic (later) flow eruptions, there was vertical faulting along which cinder cones aligned. More-over, important young faults cut the basaltic sequences. These continuing fault move-ments substantiate the importance of north-south and northwest-southwest fault patterns reflected in the geologic history of the Basin and Range and Sierra Nevada provinces. Abundant evidence of recent movement on similar faults is found in the Nevada section of the Columbia Plateau, the Sierra Nevada of eastern California, and the Warner and adjacent ranges of the Modoc Plateau.

The patterns of the Cascade volcanoes are not adequately explained by assuming that faulting has occurred along parallel linear fractures. The volcanoes have a nearly north-south alignment that transgresses the arcuate structure of the basement plat-form of Sierran-Klamath-Cascade rocks. Furthermore, Cascade peaks of Oregon and Washington are built on thick marine and nonmarine interbedded clastic rock and volcanics that include flows as old as Eocene. The chain of volcanoes is primarily Quaternary. Present understanding of the Cascade volcanism relates it to subduction of the Gorda plate beneath California north of Cape Mendocino and subduction of the Juan de Fuca plate beneath Oregon and Washington farther north. It should be remembered that California, Oregon, and Washington are all part of the North American plate and it is only happenstance that the boundary between the two oceanic plates, the Gorda and the Juan de Fuca, reaches shore near the political boundary between California and Oregon.

The block faulting that was more or less continuous during early Tuscan deposition produced enclosed drainage in which water collected. Alluvial fan, delta, and lake deposits, including water-laid tuff, ash, and diatomite accumulated in these lake basins. Sometimes deposition of the lake sediments was interrupted by lava flows.

Many basaltic flows throughout the province show well-developed columnar joint-ing and possess weathered zones and fossil soils. These occasionally contain charred remains of vegetation that grew on the flows between eruptive periods. In rare instance, woody material has petrified.

Shasta Valley

Interstate Highway 5 enters the southern part of Shasta Valley near the town of Weed and leaves the valley near Henley north of Yreka. The valley covers an area of about 1,100 square kilometers (450 square miles) and is underlain mainly by a

FIGURE 5-4 Mount Dome, with Mount Shasta in the distance. (Photo by Burt Amundson)

sequence of Cenozoic volcanic rocks of Eocene to Holocene age. Geologists have subdivided these rocks into two groups, the older Western Cascade group including all rocks from Eocene to latest Miocene, and the High Cascade group including all younger rocks, some as recent as 500 years old.

The Western Cascade group is a thick pile of lava flows, pyroclastic deposits, volcanic necks, and dikes plus some stream and lake deposits. It clearly demonstrates a long volcanic history for the area, but because the younger High Cascade group lies to the east, it also shows that the locus of volcanic activity shifted eastward about the end of the Miocene, 5 million years ago, where it has remained to the present.

The High Cascades are dominated by Mount Shasta, a huge mainly andesitic strato-volcano. North of Mount Shasta, on the eastern side of Shasta Valley, is a group of low, very young shield volcanoes including the Whaleback, Miller Mountain, Goosenest, Ball Mountain and others (Figure 5-4). These are chiefly basaltic and associated with them are some extensive olivine basalt flows. The largest of the flows was erupted from a fissure on the northwestern base of Mount Shasta and covers more than 130 square kilometers (50 square miles). Like many other basalt flows, it has numerous open tunnels, one of which is the locally well-known Pluto's cave. This flow as extruded about 300,000 years ago, but other flows in Shasta Valley are much younger.

Mount Shasta

Mount Shasta is perhaps California's grandest mountain, rising some 3,000 meters (10,000 feet) above its base to an elevation of 4,316 meters (14,162 feet), fully comparable in elevation to the higher peaks of the Sierra. The enormous bulk of Mount Shasta, about 335 cubic kilometers (80 cubic miles), makes it one of the world's largest composite volcanoes (Figure 5-5).

FIGURE 5-5 Mount Shasta (higher peak) and the parasitic cone, Shastina, the lower peak in the foreground. (Photo by Spence Air Photos, courtesy of Department of Geography, University of California, Los Angeles)

The mountain consists of at least four different overlapping cones, all of Quaternary age. The main mass of the mountain shows the effects of Pleistocene glaciation, but the summit cone, called Hotlum Cone, and the large parasitic cone on the west flank, Shastina, are both postglacial and therefore unglaciated. Recent studies show that Hotlum Cone is the most recently active area on Mount Shasta, having been active only 200 years ago. Conversely, the youngest flow so far dated on Shastina has yielded a radiocarbon date of about 9,200 years, or just shortly after the close of the Pleistocene, which accounts for the absence of glacial scour on that feature.

Interestingly, the French explorer La Perouse, sailing along the northern California coast in 1786, noted what appeared to him to be a large volcanic cloud rising from some point well into the interior of California. This report may well be the only recorded eyewitness account of the last eruption of Mount Shasta.

Along the basal circumference of Mount Shasta, especially on Shastina's northwest flank, are domal upwellings of magma and overlapping steeply sloping mudflows and breccia flows of very recent date, almost untouched by erosion. The most prominent extrusive dome is Haystack Dome, so young its formation is documented with unusual topographic clarity. South of Haystack Dome is an avalanche-like protrusion of overlapping lava flows with an expanded bulbous base developed during late stages of magma congealment. Furrows and ridges are visible behind the escarpment at the base of these flows. The ramparts or levees of earlier flows confine the later magmatic effusions (Figure 5-6). Such features are unusual in most eruptions but may develop as eruptions subside.

The recency of volcanism at Mount Shasta reminds us that we should regard both

FIGURE 5-6 Mount Shasta and Shastina. Note the flow ramparts, almost like natural levees on a river flood plain. (Photo by Burt Amundson)

this peak and others in the Cascades as dormant rather than extinct. At least seven of the major Cascade volcanoes have been active in the past 200 years (Mount Shasta and Lassen Peak in California, Mount Hood and The Three Sisters in Oregon, and Mounts St. Helens, Ranier, and Baker in Washington) and two of these have been active in the past 70 years. Recent work, moreover, has revealed a hitherto unappreciated number of eruptive events in the Cascades of California. These events are

TABLE 5-1 Holocene Volcanic Activity in the California Cascades

Date	Location	Eruptive Style
1914–1921	Lassen Peak	Steam and ash eruptions. Extrusion of short lava flow.
1910	Glass Mountain, Medicine Lake Highlands	Minor ash eruption.
1854–1857	Dome near Chaos Crags, Lassen Peak	Steam and gas.
1850–1851	Cinder Cone near Lassen Peak	Basalt lava flow.
1786	Mount Shasta (Hotlum Cone)	Steam and ash. Pyroclastic flows and lahars.
1750	Glass Mountain, Medicine Lake Highland	Lava flow and at least 8 domes. Fissure eruption.
1650(?)–1690	Chaos Jumbles, Lassen Peak	Three volcanic debris avalanches.
1667 ± 200	High Hole Crater, Medicine Lake Highland	Lava flow.
ca 1260	Hotlum Cone, Mount Shasta	Pyroclastic debris flow.
ca 1215	Hotlum Cone, Mount Shasta	Pyroclastic debris flow.
960 ± 300	Chaos Crags, Lassen Peak	Pyroclastic flows, west bank of Lost Creek.
910 ± 90	Little Glass Mountain, Medicine Lake Highlands	Pumice, lava flow, and dome.
ca 890	Hotlum Cone, Mount Shasta	Pyroclastic flow
880 ± 300	Chaos Crags, Lassen Peak	Pyroclastic flow, Manzanita Creek.
800 ± 300	Chaos Crags, Lassen Peak	Pyroclastic flow, Manzanita Creek.
ca 120, 180 or 230	Hotlum Cone, Mount Shasta	Pyroclastic flows.
<2000 years	Cinder Butte, Medicine Lake Highland	

Source: R. T. Kilbourne and Catherine L. Anderson, California Geology, 1981.

summarized in Table 5-1, and even if some of the dates are revised as more work is done, all but the most extreme skeptics will agree that volcanism in this province is quiescent, not dead.

Medicine Lake Highland

Medicine Lake Highland is an area of youthful volcanism east and north of Mount Shasta. With an average elevation of nearly 1,500 meters (5,000 feet), the Highland is astride the Cascade Range-Modoc Plateau boundary and contains features that are built on a broad volcanic platform. This platform was a Mio-Pliocene volcano, Mount Hoffman, about 25 kilometers (16 miles) in diameter, whose original summit area collapsed following a violent eruption, forming an elliptical caldera covering about 65 square kilometers (25 square miles). Ejecta from this eruption was voluminous enough to remove support and allow the summit to collapse. The caldera was then buried by younger lavas extruded from cones that developed along the caldera's rim.

FIGURE 5-7 Steep-sided rhyolitic flows at Glass Mountain resting on older Hoffman dacite. The young flow, about 2,000 years old, is barren but the dacite supports a few trees. Medicine Lake Highland, Siskiyou County. (Photo by G. R. Wheeler)

In the Pleistocene and Holocene, rhyolitic, andesitic, and basaltic eruptions have produced more than 100 cinder and lava cones whose explosive action produced pumice and tuff. The youngest of these rocks are basaltic flows and additional cinder cones, some erupted as recently as 1910. Glaciation produced minor morainal deposits, and some volcanic units are glacially eroded and others postdate glaciation. Faulting occurred during both volcanism and glaciation.

Medicine Lake Highland is bounded by volcanic escarpments that enclose a high central tableland. Medicine Lake itself is a recreational area located near the center of the old caldera, but the lake basin is the result of lava dams not the caldera (Figure 5-7).

Of the more than 100 cinder cones, domes and lava flows, four of the vents in the Medicine Lake Highland have been active in the past 2,000 years (Table 5-1)— Glass Mountain, Little Glass Mountain, High Hole Craters, and Cinder Butte. The volcanic history of the Highland shows that the earlier eruptions were basaltic and the more recent ones mostly rhyolitic. It has been pointed out that if this trend continues, the volcanic hazards in the area will increase because rhyolitic eruptions are more likely to be explosive rather than basaltic eruptions.

Is Eruption Imminent?

Volcanologists are aware that volcanoes, though they may appear dead, are part of the continuous progression of earth history. Recently, concern has been expressed regarding future Cascade volcanism and possible hazards to the population. What are the facts and potentials of Mount Shasta and Lassen Peak, California's two Cascade giants?

The May 1980 eruption of Mount St. Helens was the first volcanic activity in the 48 coterminous states in about 60 years, and was much larger, more dramatic, and devastating than any that occurred at Lassen Peak between 1914 and 1921. The renewal of Cascade volcanism came as no surprise to geologists, even though no one had made any specific predictions that Mount St. Helens was to be the next active Cascade volcano until frequent earthquakes and other manifestations began on 20 March 1980. The past record of Mount St. Helens suggests that the current eruptive cycle may last as long as 50 years before the mountain enters another period of dormancy. As has been noted, approximately seven of the major Cascade volcanoes, including two in California, have been active in the past 200 years, so some sort of new activity every 30 to 50 years can be expected somewhere in the Cascades.

What specific volcanic hazards exist in the California Cascades? Both Mount St. Helens and Lassen Peak give us some guidance. There are at least four different sorts of hazards in active volcanic regions like the Cascades. One of these is the eruption of large amounts of ash and cinders (tephra), which may damage vegetation, machinery, cause respiratory problems in humans and animals, and may in some cases lead to the collapse of buildings. Second, explosive activity may produce nearly horizontal blasts of the sort that devastated a large area (about 600 square kilometers or 288 square miles) north of Mount St. Helens, destroying a mature forest and leaving a barren, desolate landscape. If these blasts are accompanied by a cloud or avalanche of glowing, incandescent rock particles, they are referred to as nuées ardentes.

The third danger comes from lava flows extruded during an eruption. Lava flows cause little or no loss of human life because they move slowly and it is easy to walk to safety. They overwhelm and bury anything in their path, however, and often set fire to trees and other flammable materials with which they come in contact. In the case of andesitic or dacitic volcanoes like Mount Shasta and Lassen Peak, the flows are likely to be short, stiff, and slow-moving and would rarely extend beyond the base of the volcano. Extrusion of basaltic flows, however, would threaten a much larger area because of their greater fluidity.

The fourth hazard is from destructive mud flows or lahars, which have accompanied many—perhaps most—Cascade eruptions and can be expected in the future. These may move swiftly, up to 30 kilometers per hour (18 miles an hour), following stream valleys, sometimes for tens of kilometers from the volcano. These flows destroy bridges, buildings, highways, and indeed virtually anything in their path. They result from various mixtures of the ash erupted, the soil and rock mantle on the slopes of the volcano, stream, rain, and lake water plus melting snow or glacial ice if any is present. These mixtures often take the form of rapidly moving slurries the consistency of very wet concrete. Owing to their considerable density, they are capable of bearing huge boulders or even bulldozers or bridge spans that are transported like corks on water for considerable distances.

In addition to the four hazards just mentioned, the slopes of the volcano may experience large landslides during eruptions, as was the case at Mount St. Helens in 1980. Ordinarily, these would not extend beyond the volcano itself, but such slides may be converted to lahars by moving into lakes or rivers.

Icelanders have named mud flows caused by the sudden melting of glacial ice *jökulhlaups*, and one of Mount Shasta's glaciers, Konwakiton, has a particular propensity for producing *jökulhlaups*, and has done so on a number of occasions at least as far back as about A.D. 800. In recent times, the most destructive of these mud flows occurred in 1924 following a period of very warm summer weather, which promoted unusually rapid melting of the ice. A series of muddy torrents was produced, which swept with a thunderous roar that was audible for many kilometers, down Mud Creek, the outlet stream for the glacier. The ground shook nearby according to eyewitnesses, who also reported blocks of rock weighing several hundred kilograms and chunks of ice "as big as cabins" bobbing along on the surface of the flows.[1] These effects were observed by a team of employees of the McCloud Lumber Company, who were sent to investigate the failure of the McCloud town water system that tapped Mud Creek.

Volcanic mudflows, or lahars, can be expected along any stream valley draining an active volcano regardless of whether glacial ice is present. The necessary water can be provided by snow melted by volcanic heat or even by the torrential rains that sometimes accompany the rising steam cloud during an eruption. These flows are likely to result in more widespread and more severe damage than either the relatively localized blast effects or lava flows, or the more widespread deposition of ash.

In short, geologists recognize active and dormant volcanoes; "extinct" is a dubious usage. This recognition is a value judgment, however, and some apparently dormant volcanoes have erupted quite unexpectedly. Despite sophisticated techniques for recording preeruption symptoms, objective data on which to base reliable prediction are still not available. Cascade volcanism is far from extinct, and even considering Shasta and Lassen dormant (with their considerable hot spring and steam activity) may be intellectually and practically hazardous. Future eruptions must be anticipated, because another could occur within the next 30 or 40 years—or tomorrow.

Stratigraphic Record

Geologic studies of the Cascade Range—Modoc Plateau are still fragmentary and involve problems relating mostly to Mount Shasta and Lassen Peak. Correlating volcanic sequences is usually extremely difficult. Many simultaneous eruptions are localized over small areas and must be correlated not only with each other but also with earlier and later episodes that may be a part of the same, or different (older or younger), volcanic pulsation. Often large areas are involved, sometimes complicated further because the rocks may be deeply weathered and covered by vegetation. Avalanches of gas, water, and comminuted ash are often strewn by explosive eruption simultaneously with mudflows, cinders, and occasional lavas. Recurrence of all these events through time complicates the problem of unraveling geologic history. Yet the mineralogical character of ash showers may be so distinctive that layers only millimeters thick can be traced for hundreds of kilometers from source vents. Solving correlation problems usually depends on availability of complete geologic maps of the province in question. Fewer than 15 percent of the Cascade–Modoc provinces have been mapped on a detailed geologic basis.

---

[1]Mary R. Hill, 1970, Calif. Geo., v. 23.

Special Interest Features

Mount Shasta Recreational Area. Mount Shasta covers about 600 square kilometers (230 square miles) and the area has been developed into a major recreational location. High annual snowfall and upper slopes barren of timber above about 2,500 meter (8,000 feet) have made this a popular skiing area. Many graded county roads encircle Mount Shasta, enabling visitors to see countless volcanic features, some of which are scarcely studied. The Everitt Memorial Highway near the town of Mount Shasta winds nearly 1,200 meters (4,000 feet) up the mountain's southwest flank and is a dramatic route for travelers.

Volcanic regions where rainfall is abundant usually contain springs, some of which are quite large because of the channel ways and openings that occur in volcanic terrains. Volcanic rock is more porous than most rocks because of the vesicles formed when gas is driven off during eruption and because basalts and some andesites have large, open tunnels. Thus surface water is minimal, but large springs may emerge at contacts between flows from a different source or of a different age; from soil zones and gravels that are buried by overlying rock layers (such soil zones often carry water trapped in earlier geologic episodes, representing connate as well as meteoric water); from local drainage lines as new streams develop where volcanic flows have displaced earlier drainages; and from lava tubes and tunnels.

Shasta Springs is a famous old spa located in the canyon of the Sacramento River, between the towns of Dunsmuir and Mount Shasta. Here a series of andesitic lavas with prominent columnar joints overlies a porous, gravelly conglomerate from which huge springs emerge. The lava flow that caps these water-bearing gravels can be traced for more than 60 kilometers (40 miles). Some of the springs are charged with carbon dioxide, and their water was marketed in the San Francisco area for many years until the demand exceeded the natural supply.

Glaciers of Mount Shasta. Mount Shasta has five small glaciers on its upper slopes, and it was these glaciers that were the first reported to exist in American territory. Although J. D. Whitney, the state geologist, climbed to the top of this mountain in 1862, his party evidently did not see any of the glaciers, and it was not until Clarence King visited the mountain in 1870 that glaciers were reported. King named the largest of these glaciers in Whitney's honor. The other glaciers, in order of decreasing size are Bulam (Bolam), Hotlum, Wintun, and Konwakiton (McCloud). The following remarks are excerpted from the description of these glaciers published in 1895 by J. S. Diller whose pioneering work in the Klamath Mountains has already been mentioned.

Whitney glacier lies on the northwest slope of the mountain between Shasta and Shastina as an ice stream about 300 to 600 meters (1,000–2,000 feet) wide and about 4 kilometers (2½ miles) long, ending at about an elevation of 2,500 meters (9,500 feet). This glacier is heavily crevassed near its head and has well-developed lateral and terminal moraines.

Bulam glacier is also on the northwestern side of the mountain and is similar to Whitney glacier, but does not occupy such a well-defined valley.

Hotlum is more of an ice cap on a slope. It has a convex form and is not confined to any well-developed valley. It is about as wide as it is long (2 by 2½ kilometers or

1.2 by 1.6 miles). It lies on the northeast side of the mountain and extends down-slope to an altitude of about 3,300 meters (10,800 feet).

Wintun lies on the eastern side of the mountain and is over 3 kilometers (2 miles) long, ending at an abrupt ice front with little or no moraine.

Konwakiton is on the southeast side of the volcano, and although it is small today, it was much larger in the past. Clear evidence in the form of polished and striated surfaces and prominent moraines show that it was once longer than 8 kilometers (5 miles) and covered about 20 times its present area. Its propensity of generating destructive mudflows has already been mentioned and may be due in part to its location on the sunny side of the mountain.

Lassen Volcanic National Park. Shortly after the volcanic eruptions of 1914 and 1915, Congress established the National Park to preserve the array of volcanic features in the area. It should be noted that President Theodore Roosevelt, recognizing the importance of the area, established the Lassen Peak Forest Preserve in 1905. Logging nevertheless took place on parts of the reserve, causing local residents to petition the president for greater protection from commerical development. This led the president to establish Cinder Cone National Monument and Lassen Peak National Monument as two separate areas. In 1916, the eruptions galvanized Congress into establishing the National Park, which embraced the two monuments and enlarged them. Though the eruptive events of 1914 to 1921 were relatively modest by geologic standards, the park does include a wide range of volcanic features.

Lassen Peak is the most recent volcanic edifice and is built on the wreckage of what was probably a higher, larger mountain known as Mount Tehama (Figure 5-8). The visible remnants of Mount Tehama are preserved today southwest of Lassen

FIGURE 5-8 Lassen Peak from the east. (Photo by Don Norris)

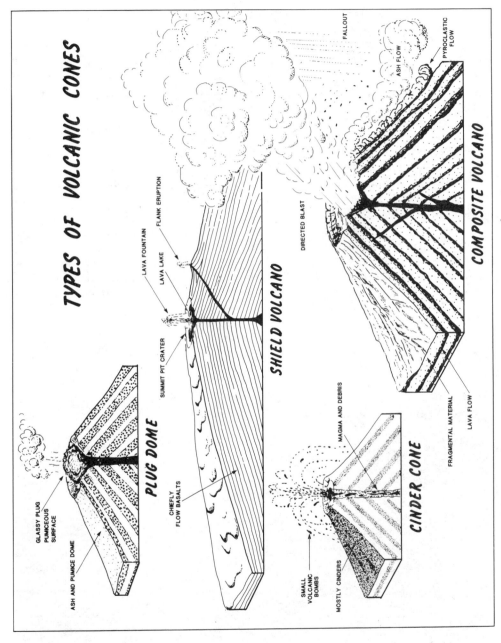

TYPES OF VOLCANIC CONES

GLASSY PLUG
PUMICEOUS
SURFACE

ASH AND PUMICE DOME

PLUG DOME

CHIEFLY
FLOW BASALTS

LAVA FOUNTAIN

LAVA LAKE

FLANK ERUPTION

SUMMIT PIT CRATER

SHIELD VOLCANO

DIRECTED BLAST

FALLOUT

ASH FLOW

PYROCLASTIC
FLOW

COMPOSITE VOLCANO

FRAGMENTAL MATERIAL

LAVA FLOW

MAGMA AND DEBRIS

CINDER CONE

SMALL
VOLCANIC
BOMBS

MOSTLY CINDERS

FIGURE 5-9 Comparison of idealized internal structures of a plug dome (such as Lassen Peak), a cinder cone, and a composite volcano (such as Mount Shasta).

166

Peak as Mount Diller (2,771 meters or 9,087 feet) and Brokeoff Mountain (2,817 meters or 9,235 feet). Mount Tehama was a composite volcano similar to Mount Shasta and had an elevation estimated to be between 3,300 and 3,600 meters (11,000–12,000 feet). This mountain was destroyed by a violent eruption followed by collapse of the summit area much like the process that formed Crater Lake in the Oregon Cascades. There is some reason to believe that these events occurred about 450,000 years ago, accompanied by eruption of a widespread ash deposit. Today's Lassen Peak was formed about 11,000 years ago at the end of the Pleistocene epoch. It has been built upon the wreckage of Mount Tehama in much the same way as the small plug now forming in the caldera at Mount St. Helens. However, Lassen Peak is one of the very largest plug domes known anywhere and is also one of the few plug domes to erupt a second time (Figure 5-9).

The principal events in Lassen Peak's most recent active episode were the blast and accompanying mudflows (lahars) that destroyed considerable forest and some property along Hat and Lost creeks on the northern side of the volcano in mid-May 1915. These events were not observed directly because they occurred on a cloudy night and there were only a few people in the area. A few days later, on May 22, an explosive eruption occurred in which a volcanic cloud composed of steam, ash, and some rocks was propelled high into the air. The cloud was estimated to have been more than 7,500 meters (25,000 feet) high and was seen from as far away as Sacramento and Redding. Subsequent activity consisted mainly of minor steam explosions until 1921, when the final one occurred (Figure 5-10).

Several other eruptive events have occurred in Lassen Peak in the last several hundred years. A steam and gas eruption occurred from a dome near Chaos Crags on the north side of Lassen Peak between 1854 and 1858. Cinder Cone was active in 1850 and 1851 when an unusual basalt that contained quartz xenocrysts was

FIGURE 5-10 Lassen Peak during its most recent eruptive event, 1915. (Photo by B. F. Loomis, courtesy of National Park Service)

TABLE 5-2 Holocene Volcanic Activity in the Lassen Peak Area

Date	Location	Activity Type
1914–1921	Lassen Peak, main dome	Steam and ash explosions, mudflows. Lava extrusion.
1854–1957	Dome near Chaos Crags	Phreatic steam and gas eruption.
1850–1951	Cinder Cone, northeastern Lassen Park	Basalt flow.
1650–1990	Chaos Jumbles	Three separate debris avalanches.
980 ± 300	Chaos Crags	Pyroclastic flows.
880 ± 300	Chaos Crags	Pyroclastic flows.
800 ± 300	Chaos Crags	Pyroclastic flows.

erupted. Quartz rarely occurs in basalts, but in this case the quartz was provided by rhyolitic rocks through which the basaltic magma passed. Table 5-2 lists presently known eruptive events in the Lassen area during the last 2,000 years, but only a few studies of the area have been completed; more evidences of Holocene activity can be expected as new investigations are undertaken.

Chaos Jumbles is a collection of broken and shattered rock from Chaos Crags, spread over an area of a square kilometer or two northwest of the Crags. The Jumbles appear to have formed following a steam blast from the base fo Chaos Crags, which removed support from the rock above, abruptly forming an avalanche that spread the debris comprising the Chaos Jumbles. In 1980, concern that the same type of event might occur again led the National Park Service to close their headquarters' facilities and campground at Manzanita Lake and plan for their relocation in a more protected spot.

The park has an abundance of special volcanic features, including, Geysers, Boiling Lake, and the sulfurous fumarolic areas at Bumpas Hell, probably the best American example outside Yellowstone Park, and Devil's Kitchen. Brokeoff Mountain and Mount Diller, the remnants of Mount Tehama have already been mentioned. The mudflows in the Lost and Hat creek devastation area of 1915 created spectacular scenery. Cinder Cone and Mount Harkness are very youthful-looking features showing almost no erosion. Chaos Crags, Chaos Jumbles, and the plug dome of Lassen Peak are well marked and easily accessible. Many large volcanic bombs and spatter cones are found throughout the park. Lava caves and tunnels, some with large underground streams, have been mapped in the park and in the surrounding area. Subway Caves, near the junction of State Highways 44 and 89 north of the park is a well-known and easily accessible example. Recent faulting has produced significant scarps, especially in the valley of Hat Creek. It is often difficult to separate such fault scarps from the scarps produced by flow fronts, where pasty lava has congealed to form an escarpment.

MODOC PLATEAU

The Modoc Plateau is an undulating platform 1,200 to 1,500 meters (4,000–5,000 feet) high, composed of assorted volcanic materials, principally Miocene to Holocene basaltic lava flows. The region covers about 26,000 square kilometers (10,000 square

FIGURE 5-11 Eastern face of the Warner Range seen from near Eagleville, Modoc County. (Photo by R. M. Norris)

miles) of the southwestern corner of the Columbia Plateau. As such, it is merely a small part of the larger physiographic unit that embraces about 518,000 square kilometers (200,000 square miles) of eastern Oregon, eastern Washington, southern Idaho, northern Nevada, northern Utah, and western Wyoming. Locations and geography are shown in Figure 5-2.

The Modoc Plateau grades west into the Cascade Range and south into the Sierra Nevada. The flows thin to the south, overlapping the older formations of the Sierra Nevada. The Plateau also abuts the Basin and Range province to the east where volcanic ramparts form escarpments with the valleys of the Basin and Range province. Basin and Range faults, some of which are still active, either displace the volcanics or disappear beneath them. Much of the plateau itself is broken by faults that have produced block mountains. Except for the Warner Range, these are generally low with small escarpments, exposing only lava flows and none of the older rocks below. Because of its structure, some workers assign the Warner Range to the Basin and Range province. Whatever its assignment, the Warner Range exposes a nearly complete record of Modoc Plateau volcanism from middle Tertiary to the present (Figure 5-11).

Despite its small size, the Modoc Plateau is important as a transitional area between the Cascades, the Basin and Range, and the Sierra Nevada, because it provides details of geologic history not available elsewhere in the Columbia Plateau. There are impressive records of extensive freshwater lakes that occupied parts of the Modoc Plateau during the middle and late Tertiary. Fragmental and nonbasaltic rocks that are exceptions to the usual Columbia sheetflood basalts also occur in the Modoc Plateau. Nevertheless, basaltic magmas poured from fissures and flowed like water for vast distances. These records are studied best in the Columbia Plateau province proper, where only restricted areas of recent volcanism are present. Much of the Modoc Plateau is veneered with Holocene cones, glass flows, pumice, and fragmental materials interspersed with soil layers, lacustrine sediments, and volcanic plugs and domes.

FIGURE 5-12 Looking across the Modoc Plateau over the city of Alturas and the North Fork of the Pit River. (Photo courtesy of California Department of Water Resources)

Drainage

The Modoc Plateau is a rolling upland drained by a master stream, the Pit River, which crosses the plateau from northeast to southwest and enters Shasta Lake and the Sacramento River to drain into the Pacific (Figure 5-12). For much of its course, the Pit River flows lazily across meadows developed on lava surfaces and around and along escarpments. Finally, as it leaves the plateau, it enters a gorge before reaching Shasta Lake, although much of this gorge has been flooded by lakes Britton and Shasta.

Rocks and Structure

The pioneering work in the Warner Range was done in the late 1920s by Richard J. Russell, then a graduate student at the University of California, Berkeley. In those days, Russell had only very limited paleontologic information with which to date the lava flows and pyroclastic deposits of the range, so it is not surprising that radiometric dating based on potassium-argon methods has revealed that Russell's "Warner Basalt" includes lavas ranging in age from middle Miocene to Pleistocene. Until new names are assigned and accepted for these rocks, however, it is convenient to refer to the older andesitic and rhyolitic pyroclastic rocks as the "Cedarville Series" and to the basalts younger than about 15 million years as the Warner Basalt.

Most of the higher parts of the Warner Range and a sizable portion of the plateau surface are made up of Warner Basalt, but the core of the range as well as some exposures on the southern and western margins of the plateau proper are assigned to the Cedarville series, which is Oligocene to middle Miocene in age.

The Surprise Valley fault zone along the eastern side of the Warner Range seems to have become active near the middle of the Miocene, about 15 million years ago, when the first of the Warner basalts were being erupted. Recently faulted alluvium shows that this fault zone remains active. Because drilling has revealed that the volcanic floor of Surprise Valley lies about 3,600 meters (11,800 feet) below the matching level in the Warner Range, it is possible to calculate a minimum rate of uplift for the range over the past 14 or 15 million years. This works out to be about 26 centimeters per 1,000 years (0.85 foot per 1,000 years), a geologically rapid rate.

The basaltic lavas that occupy so much of the present Modoc surface were erupted into the sediment-filled basins between block-faulted ranges. Extensive lava floods occurred throughout the late Miocene and Pliocene and have continued sporadically. Most Holocene volcanic events have been concentrated on Modoc borderlands. Large

FIGURE 5-13 Lava Beds National Monument: view from Schonchin Butte, looking west toward the Cascade Range. Note the scarp-step topography of the middle distance. (Photo by Mary Hill)

continuous sheets of Warner basalt occupy the Devils Garden Rim and cover about 1,800 square kilometers (800 square miles) there. The total thickness of the Warner basalt is unknown, but estimates are as high as 335 meters (1,100 feet) thick. The gross thickness of materials of the Modoc Plateau is also unknown. It is assumed that the Cedarville series underlies the Warner basalt as far west as the Medicine Lake Highland. The Cedarville's southern counterpart may be the Tuscan Formation of the southern Cascades and northwestern Sierra Nevada. Pleistocene and Holocene volcanic features include obsidian and basalt flows, domes, plugs, and cinder cones, many of which are particularly well displayed in Lava Beds National Monument (Figure 5-13).

Special Interest Features

Fall River Springs. The Modoc Plateau generally has low rainfall and sparse vegetation, except for the conifer forests at higher elevations on the western edge. One of the largest springs in the United States, however, is Fall River Springs near Fall River Mills. With a daily flow of 4,720 million liters (1,250 million gallons), the springs discharge water collected in underground channels from as far away as 80 kilometers (50 miles). The water moves through weathered soil zones buried beneath the Warner basalt flows, through porous fragmental volcanics and along open tunnels. An important source of this water is the Pit River, which is unusual because from Goose Lake to Fall River Mills it loses water to percolation. In fact, the regional water table lies below the Pit's channel through most of the upper half of the river's drainage.

FIGURE 5-14 Valentine Cave, a typical lava tunnel in a basalt flow. Modoc County. (Photo by David Kimbrough)

Ice Caves. Lava tubes and tunnels are features typical of most basaltic volcanic areas. Where magmas that fed flows were highly fluid, lava surged forward after crusting over by initial cooling. Uncongealed magma drained from beneath the hardened crust, leaving open passages and forming extensive systems of lava tunnels. These tunnels may be small with diameters ranging from only a few centimeters to monsters big enough to accommodate a large truck (Figure 5-14). In some cases, as in Craters of the Moon National Monument (Idaho) and Lassen Volcanic National Park, caves or tunnels almost a kilometer long have been mapped. Approximately 300 lava tubes are known in Lava Beds National Monument, and many others elsewhere in the Modoc Plateau remain to be explored or discovered. These tunnels are accessible only where the roof has collapsed or the end of the flow eroded or faulted off. Often these entrances establish air movement through the tubes, allowing rain and snow to enter. If cold, heavy air is trapped, ice accumulates and is preserved in the so-called ice caves. Even in summer little ice melts in these caves, although there is a constant loss of ice by direct evaporation into the usually dry air. Visitors to ice caves or even to the lava tunnels of the Modoc Plateau are well advised to wear warm jackets, because of these places often have temperatures near the freezing mark even in midsummer, when above-ground temperatures may be hot (Figure 5-15).

FIGURE 5-15 Crystal Cave, Modoc County. Because some lava tunnels trap cold, heavy air during the winter, they may remain very cold throughout the year and often accumulate appreciable amounts of ice. (Photo by David Kimbrough)

FIGURE 5-16 Large springs emerging from the eroded edge of Modoc Plateau lava flows which, owing to numerous tunnels and openings, permit copious flows of ground water. Burney Creek is the surface stream that produces the main part of the falls. McArthur-Burney Falls State Park, Shasta County. (Photo by Don Norris)

Burney Falls. McArthur-Burney Falls State Park is a recreational area that preserves an outstanding example of a waterfall and stream fed by ground water trapped between lava flows. Part of the the volume is attributable to surface flow in Burney Creek, which was formerly a small tributary of the Pit River and now flows into man-made Lake Britton. Burney Creek drops 37 meters (120 feet) over the lip of a durable basalt flow. Beneath this flow lie tuffs and other porous materials from which large springs issue to augment Burney Creek. These springs discharge an estimated 4.6 million liters (1.2 million gallons) per day. The gorge of Burney Creek has been developed below the falls by rapid downcutting of the weak rocks that underlie the durable basalt cap rock (Figure 5-16).

Features like Burney Falls are common elsewhere in the Modoc Plateau, but on a smaller scale. Between volcanic eruptions soils are developed and vegetation is established, only to be obliterated by the next episode or eruptive activity. Ground water enters these soil zones, often assisting in the petrifaction of woody plant material, and the zones act as aquifers for water storage.

Lava Beds National Monument. This small, little-known section of the Modoc Plateau contains examples of most volcanic features characteristic of the plateau. The area is adjacent to and north of the Medicine Lake Highland and covers about 190 square kilometers (73 square miles).

From Medicine Lake Highland, the surface slopes gently to the north with local irregularities where cinder cones, explosion craters, chimneys, east-facing scarps, and individual lava flows appear (Figure 5-17). The surface basalt is generally a dark-colored flow rock that produces a jagged topography on which minimal soil or vegetation has developed. Some flows are less than 1,000 years old. The monument has at least 12 prominent cinder cones from 15 to more than 200 meters 50–700 feet)

FIGURE 5-17 Young normal fault scarps at Gillems Bluff, Lava Beds National Monument, Siskiyou County. (Photo by R. M. Norris)

high. Mammoth Crater, the most spectacular of several funnel-shaped explosion craters, lies about 115 meters (375 feet) below the surroundng surface.

Among the most recent features are several groups of spatter cones or chimneys formed by gradual accumulation of semi-molten lumps and clots of lava around vents. These chimneys range from 1 to 15 meters (3–50 feet) high and from 1 to 30 meters (3–100 feet) in diameter. Sinuous ropy ridges of lava, known as *pahoehoe*, extend as far as 100 meters from the base of some chimneys and often have hollow interiors, forming miniature lava tubes and tunnels. Small lava domes, which probably formed as blisterlike gas bubbles, occur at many places on the surface.

Perhaps the monument's most spectacular features are its caverns and lava tubes. Of the 293 known, only 130 have been fully explored. There are usually two or more open caverns per tube, and each shows a rich diversity of dimension, form, wall, and floor detail. Fossil remains of a mastodon and a prehistoric camel have been found, and various stalactitic forms (lavacicles) occur in many of the caves. Pictographs, stone artifacts, obsidian weapons, bones, and human skeletons are evidence of Indian occupation of some of the caves.

The Modoc War and Captain Jack. On the northern boundary of Lava Beds National Monument the last of the Indian battles in California took place between November 1872 and June 1873. Fifty-three Modoc men and perhaps twice as many women and children held off assaults first by 300 Army troops and later by 650 troops with field guns. The Modocs suffered almost no casualties, but the U.S. Army lost about 70 men. Much of the fighting took place in cold, windy weather, sometimes in dense ground fog.

Captain Jack and Hooker Jim led two small bands of Indians that defeated a far greater number of U.S. troops. The Modocs had an accurate and detailed knowledge of the ragged northern edge of the lava plateau on the south side of Tule Lake, which gave them a tremendous advantage over the U.S. troops sent to round them up.

The Modoc's stronghold is an area with numerous collapse pits, 3 to 8 meters (10–25 feet) deep, which communicate with lava tunnels beneath, and which pro-

FIGURE 5-18 Schollendome showing typical crestal cracking. (Photo by David Kimbrough)

vided many refuges for small bands of Indians. In addition, the ends of the lava flows were often blunt, downward-curving walls behind which clefts, many meters long and usually 2 to 3 meters (6–10 feet) deep, served as remarkably effective trenches for Indian snipers.

In some cases, the downward-curving ends of the flows have been rafted away from the main part of the flow during lava emplacement, leaving detached, isolated, tilted blocks of lava called *schollendomes* (Figure 5-18). These are typically elongate in form and many preserve the deep clefts characteristic of the downward-turned ends of the flows. The Modoc's stronghold area included many of these schollendomes, each of which provided an uncommonly effective site for snipers to harrass the Army troops.

At the time of the Modoc War, the southern edge of Tule Lake was close to the stronghold, providing the Indians with a ready and easily protected access to food and water. The modern Tule Lake is much smaller and farther away, owing to drainage and reclamation schemes. Visitors to Captain Jack's stronghold can only imagine how the lake looked 100 years earlier when it played such an important role in the Indian defense.

When the Modocs withdrew, they used their detailed local knowledge to make an easy and efficient escape across the nearly treeless lava plateau to the south under the cover of darkness.

The entire affair was a disaster from the Army's standpoint, and might have dragged on for months longer had it not been for treachery on the part of some Indians, which led to the eventual capture of all of them.

Lieutenant Colonel Wheaton summed up the situation succinctly in commenting on the battle of January 7, 1873. He fought 52 Indians in an area that he described as littered with boulders from the size of matchboxes to churches. He noted that in his 23 years of fighting both civilized peoples and savages, he had never encountered an enemy occupying such a position of strength.   His troops fought all day from 8 A.M. to 10:30 P.M. without ever seeing an Indian, with only puffs of smoke at which to fire. The weather was freezing and foggy, and at the end of the day he said the attack could only be regarded as a complete failure.

Geothermal Prospects. Geothermal activity has been reported from many areas of the Modoc Plateau and the Cascade Range. To date, however, none in the California

Cascades has sparked commercial interest. In Lassen Volcanic National Park, geothermal activity has long been a favorite tourist attraction, but under present regulations the park prohibits commercial ventures. In the Modoc Plateau, notable geothermal activity occurs in Surprise Valley east of the Warner Range. The heat appears to be related to residual volcanism. Drilling for prospective geothermal power was conducted in 1959, but additional exploration is required before the full potential can be properly evaluated. Results thus far do not suggest any large-scale energy availability.

REFERENCES

Alt, David D., and Donald W. Hyndman, 1975. Roadside Geology of Northern California. Mountain Press Publ. Co., Missoula, Montana.

Anderson, Charles A., 1941. Volcanoes of the Medicine Lake Highland. Calif. Univ. Dept. Geol. Sci. Bull., v. 25, pp. 347–422.

Aune, Quentin, 1964. A Trip to Burney Falls. Mineral Information Service (now California Geology), v. 17, pp. 183–191.

―――, 1970. Glaciation in Mount Shasta-Castle Crags. Mineral Informatin Service (now California Geology), v. 23, pp. 145–148.

Chesterman, Charles W., 1971. Volcanism in California. California Geology, v. 24, pp. 139–147.

―――― and George J. Saucedo, 1984. Cenozoic Volcanic Stratigraphy of Shasta Valley. California Geology, v. 37, pp. 67–74.

Evans, James R., 1963. Geology of some Lava Tubes. Mineral Information Service (now California Geology), v. 16, pp. 1–7.

Geologic Map Sheets of California, Alturas (1968), Redding (1962), Weed (1964), and Westwood (1960). Calif. Div. Mines and Geology.

Hill, Mary R., 1970. Mt. Lassen Is in Eruption and There Is No Mistake about That! Mineral Information Service (now California Geology), v. 23, pp. 211–224.

―――, 1976. A California Jökulhlaup. California Geology, v. 29, pp. 154–158.

―――, 1977. Glaciers of Mount Shasta. California Geology, v. 30, pp. 75–80.

Johnston, David A., and Julie Donnelly-Nolan (eds.) (1981). Guides to Some Volcanic Terranes in Washington, Idaho, Oregon and Northern California. U.S. Geol. Survey Circular 838.

Kane, Phillip, 1982. Pleistocene Glaciation in Lassen Volcanic National Park. California Geology, v. 35, pp. 95–104.

Kilbourne, Richard T., and Catherine L. Anderson, 1981. Volcanic History and Active Volcanism in California. California Geology, v. 34, pp. 159–168.

MacDonald, Gordon A., 1966. Geology of the Cascade Range and Modoc Plateau. In Geology of Northern California. Calif. Div. Mines and Geology Bull. 190, pp. 65–95.

―――, and T. E. Gay, Jr., 1968. Geology of the Southern Cascade Range, Modoc Plateau, and Great Basin Areas of Northeastern California. Mineral Information Service (now California Geology), v. 21, pp. 108–111.

Slosson, James E., 1974. Surprise Valley Fault. California Geology, v. 27, pp. 267–270.

Stearns, H. T., 1928. Lava Beds National Monument, California. Geogr. Soc. Philadelphia Bull., v. 26, pp. 239–253.

Williams, Howel, 1932. Geology of Lassen Volcanic National Park, California. Calif. Univ. Dept. Geol. Sci. Bull., v. 21, pp. 195–385.

Woods, Mary C., 1974. Geothermal Activity in Surprise Valley. California Geology, v. 27, pp. 271–273.

six

BASIN AND RANGE

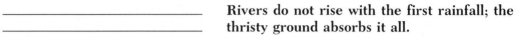

Rivers do not rise with the first rainfall; the thristy ground absorbs it all.

—Seneca

Although the Basin and Range province is more extensive in neighboring Nevada, its most dramatic development is in California between the Sierra Nevada crest and the Nevada state line. The province's western boundary is the Sierran crest itself, and some geologists think that the Sierra should be considered the highest and grandest of the Basin and Range mountains. The Garlock fault separates the province from the Mojave Desert to the south. The California part of the Basin and Range province includes almost all of Inyo and Mono counties, plus northeastern Kern and northern San Bernardino counties (Figures 6-1 and 5-2).

The province also extends into California in the Susanville-Honey Lake area and extreme northeastern California. Though underlain by rocks like those of the Modoc Plateau, the northeastern corner of California is characterized by high north-south trending fault-block mountains like the Warner Range and elongate, narrow depressions like those occupied by Upper, Middle, and Lower Alkali lakes east of the Warner Range in Surprise Valley. These areas and the province generally are characterized by interior drainage. The Honey Lake basin is usually included in the Basin and Range province because it is a triangular wedge dropped down between the granitic rocks of the Sierra Nevada and the volcanic rocks of the Modoc Plateau. Recent faulting is reflected by a 15-centimeter (6-inch) scarplet produced at the base of the Fort Sage Mountains during a 1951 earthquake. A similar earthquake, but without ground rupture, affected the Honey Lake area in 1979.

A certain amount of confusion exists regarding the terms *Basin and Range* and *Great Basin*. Basin and Range, as originally defined, includes a large part of southwestern United States in which elongate mountain ranges are separated by broad, nearly flat valleys. It includes the area lying south of the Columbia River Plateau

178

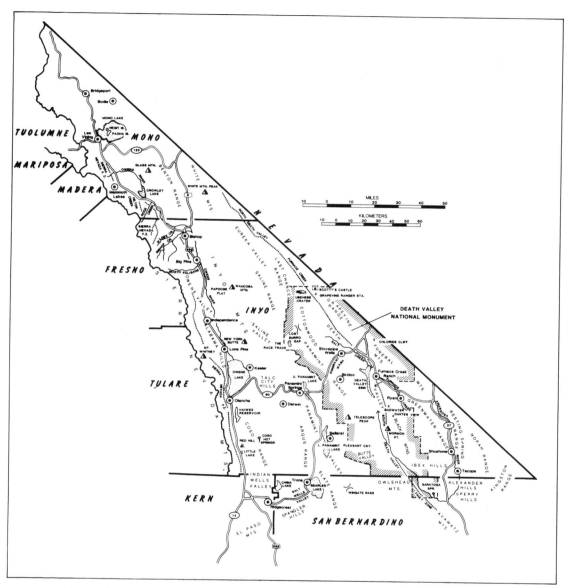

FIGURE 6-1 Place names Basin and Range province.

and between the Sierra Nevada on the west and the Wasatch Range on the east. Included also are the Mojave and Colorado deserts of California, much of southern Arizona and southwestern New Mexico, as well as parts of adjacent northern Mexico.

The Great Basin usually refers to the northwestern part of this area lying between the Sierra and the Wasatch Range and generally north of Las Vegas. The area discussed in this chapter is thus a part of the Great Basin and shares the interior drainage characteristic of the Great Basin subdivision. However, the California Division of Mines and Geology designates the three California desert provinces as (1)

FIGURE 6-2 Eastern escarpment of the Sierra Nevada at the Alabama Hills, Inyo County. Mount Whitney is the highest peak in the sawtooth group left of center. Note the well-developed granitic pediment in the foreground. (Photo by R. M. Norris)

the Basin and Range, (2) the Mojave Desert, and (3) the Colorado Desert. It is this usage we follow, although all three are considered parts of the Basin and Range in the original sense.

The Basin and Range province includes some of the greatest relief in North America. For example, there are only 130 kilometers (80 miles) between Mount Whitney (4,421 meters or 14,495 feet), the highest peak in the 48 coterminous states and the lowest spot on land in the Western Hemisphere near Badwater in Death Valley (−86 meters or −282 feet). In Inyo County alone, at least four spectacularly deep valleys occur between towering mountains. The best known is Owens Valley, lying between the Sierra Nevada and the White-Inyo mountains, each range containing peaks more than 4,250 meters (14,000 feet) high (Figure 6-2). The floor of Owens Valley rises gradually from about 1,050 meters (3,500 feet) at its south end to 1,600 meters (5,300 feet) at the foot of Montogomery Pass in the north.

Immediately east of the Inyo Mountains is deep, almost circular Saline Valley, whose floor is nearly 600 meters (2,000 feet) lower than that of Owens Valley. Although the western side of Saline Valley is a 2,750-meter (9,000-foot) escarpment, perhaps the valley's most interesting aspect is its closure. If water filled Saline Valley to overflowing, a 900-meter-(3,000-foot) deep lake would form, deeper than any lake in the Western Hemisphere.

Southeast of Saline Valley is Panamint Valley, about 90 kilometers (55 miles) long. From the valley floor with an elevation of about 300 meters (1,000 feet), the Panamint Mountains to the east rise to just over 3,350 meters (11,049 feet) at Telescope Peak. From this peak there is an impressive view into Panamint Valley to the west, about 3,000 meters (10,000 feet) below, or Death Valley to the east more than 3,350 meters (11,000 feet) below. The eastern wall of Death Valley is also very steep. About 3.2 kilometers (2 miles) east of Badwater is Dante's View (1,670 meters or 5,475 feet), atop the Black Mountains. The lower portion of the slope from Dante's View to Badwater is so steep that a person standing at Dante's View can see neither Badwater

FIGURE 6-3 Looking northwest across Death Valley from Dante's View in the Black Mountains. Devil's Golf Course is the large flat central area. The extensive alluvial fans at the east base of the Panamint Mountains are characteristic. Note the channel of Salt Creek, the major stream of the valley, leading into Devil's Golf Course from the north. (Photo by Mary Hill)

nor the base of the slope (Figure 6-3). Death Valley possesses superb examples of alluvial fans and some remarkably recent fault features. In addition, the surrounding ranges contain pre-Phanerozoic to Tertiary rocks that reflect nearly all the main subdivisions of geologic time.

DRAINAGE

Although interior drainage characterizes California's Basin and Range province today, parts of the region probably had external drainage as recently as Pleistocene time. The basins were produced by block faulting initially, and interior drainage resulted because rain and snowfall produced less water than was lost by evaporation. Lakes in the Basin and Range province commonly lose 2 to $2\frac{1}{2}$ meters (7–8 feet) to evaporation annually. When this water is not replaced, the lakes become dry lakes or playas. Most of the province's lakes are ephemeral, holding water only after heavy rains. The few permanent lakes such as Mono Lake are fed by streams rising in high, well-watered ranges like the Sierra Nevada. Mono Lake has no outlet and is saline.

FIGURE 6-4 Owens Lake playa from the south at Dirty Socks spring. White saline crust is chiefly sodium chloride, but includes a number of other carbonates, chlorides, and borates. (Photo by R. M. Norris)

Because of construction on the extension of the Los Angeles aqueduct into the Mono Basin in the 1940s, the streams feeding Mono Lake have been partially intercepted and evaporation now exceeds inflow in most years. Similarly, when the Los Angeles aqueduct first went into operation in 1913, much of the Owens River's flow was diverted, and by the 1920s Owens Lake had shrunk so rapidly that it became the vast, salty playa it remains today. Unusually heavy runoff from the Sierra sometimes forms a small lake in the Owens basin, but only temporarily; the lake disappears during dry years (Figure 6-4).

There are few permanent streams in the California portion of the Basin and Range province. The most prominent is the Owens River, which owes its existence to the Sierra Nevada. Actually, few permanent streams entering Owens Valley south of Bishop manage to reach the Owens River as live streams. Huge fans built out onto the valley floor by these Sierran streams during flooding provide permeable gravel cones that soak up the stream flow of normal years. Water moves toward the Owens River beneath the surface, emerging at the toe of the fans near the river, the lowest place in the valley.

Apart from a few short streams confined to mountain canyons and one or two special cases, the only permanent stream in California's southern Basin and Ranges is the Amargosa River, which rises in western Nevada. It flows south in a broad valley between the Greenwater and Resting Spring ranges, across the eroded bed of former Lake Tecopa, through a narrow canyon cut across the Sperry Hills, and finally north into Death Valley (Figure 6-5). The river surfaces where bedrock is near ground level, but disappears where the alluvium is deep. High evaporation has produced a white saline crust along much of the river's course. In wet years the Amargosa may persist all the way to Badwater and form a short-lived lake on the floor of Death Valley.

FIGURE 6-5 Amargosa River south of China Ranch, Sperry Hills, San Bernardino County. Note the white encrustation on the rocks, demonstrating high evaporation and the mineralized nature of the water. Flow in this reach of the Amargosa is permanent because the channel is cut into bedrock. (Photo by R. M. Norris)

ROCKS AND GEOLOGIC HISTORY

Because of the prevailingly dry climate and the resulting sparse vegetation, Basin and Range rocks are unusually well exposed. California's finest Proterozoic and lower Paleozoic sedimentary sequences appear in this province (Figures 6-6 and 6-7).

Pre-Phanerozoic

Oldest Basin and Range rocks are a complex of early to middle Proterozoic schists and gneisses (probably of sedimentary origin) and associated granitic rocks much like those exposed in the inner gorge of the Grand Canyon. Some of these ancient rocks have been dated at 1,800 million years old. Good exposures are found in the western Black Mountains, which form the steep eastern wall of southern Death Valley.

Only slightly metamorphosed sedimentary rock, the younger Proterozoic is a few thousand meters thick and is composed of regularly bedded sandstones, siltstones, shales, limestones, dolomites, and conglomerates. The rocks are well displayed in the Funeral, Nopah, and southern Panamint mountains. Outcrops continue southeast to the Kingston Range and into Nevada. This sequence constitutes the Pahrump Group: from oldest to youngest, the Crystal Spring Formation composed chiefly of sandstone 1,200 meters (4,000 feet) thick, the Beck Spring Dolomite (300 meters or 1,000 feet), and the Kingston Peak Formation, 60 to perhaps 1,000 meters (200–3,300 feet) thick. These sedimentary rocks are probably nearshore marine materials representing the earliest deposits in the Cordilleran geocline. Some of the Kingston

Proterozoic and Cambrian Strata

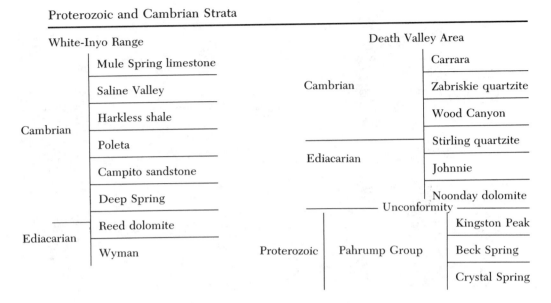

FIGURE 6-6 Proterozoic and early Paleozoic strata, Basin and Range province.

Peak beds appear to be glacial marine deposits containing striated drop-stones rafted offshore by floating ice.

Long presumed to be unfossiliferous, Pahrump Group rocks have in the past quarter century or so yielded numerous fossil Procaryotes: single-celled and filamentous cyanobacteria or blue-green algae, possibly the forerunners of more complex forms (Eucaryotes) found in younger rocks. As a matter of fact, the Beck Spring Formation contains fossils believed to be the earliest eucaryotes in the geologic record. In addition to the microfossils, there are large—1 to 10 centimeters ($\frac{1}{3}$–4 inches)—layered, mound-shaped structures called *stromatolites*. The stromatolites were produced by the sediment-trapping and binding propensity of the cyanobacteria. Microfossils and stromatolites have now been found in all formations of the Pahrump Group. The age of the Pahrump Group remains the subject of considerable debate, however, and ages are estimated from 800 million to 1,200 million years.

Proterozoic-Paleozoic Transition

Unconformably overlying the Pahrump Group is a thick sequence of latest Proterozoic carbonate and clastic rocks consisting, in ascending order, of the Noonday Dolomite (300 meters or 1,000 feet), Johnnie Formation (750 meters or 2,300 feet), and the Stirling Quartzite (600 meters or 2,000 feet). Above these is a very thick sequence of Cambrian rock (5,200 meters or 17,000 feet). This sequence and equivalent units in the White-Inyo Mountains are of considerable interest to paleontologists because somewhere in this thick package of sedimentary rock is the boundary between two major Eons—the Proterozoic and the Phanerozoic. The base of the Cambrian is recognized on the basis of the first occurrence of invertebrates with hard shells or exoskeletons. It probably occurs in the upper part of the Stirling Quartzite.

Composite Geologic Column, White-Inyo Range and Vicinity

HOLOCENE	Alluvium, dunes, lake beds, volcanics
PLEISTOCENE	Bishop Tuff and other volcanic rocks
MIO-PLIOCENE	Various volcanic rocks
JURA-CRETACEOUS	Granitic rocks of Sierra Nevada and adjacent areas
TRIASSIC	Tuff, quartzite, and limestone (marine)
PERMIAN	Reward Conglomerate (marine) Owenyo Limestone (marine)
PENNSYLVANIAN	Keeler Canyon Limestone (marine) Resting Spring (marine)
MISSISSIPPIAN	White Pine (marine) Perdido (marine) Tin Mountain Limestone (marine)
DEVONIAN	Lost Burro (marine)
SILURIAN	Hidden Valley (marine)
ORDOVICIAN	Ely Spring (marine) Eureka Quartzite (marine) Barrel Spring (marine) Pogonip Limestone (marine) Mazourka (marine)
CAMBRIAN	Mule Spring Limestone (marine) Saline Valley (marine) Harkless (marine) Poleta (marine) Campito (marine) Deep Spring (marine)

FIGURE 6-7 Geologic column, White-Inyo Range.

Some geologists would assign the Johnnie and much of the Stirling as well as the equivalent Wyman Formation and Reed Dolomite in the White-Inyo Mountains to a discrete geological system, the Ediacarian, on the basis that much of the time represented by the sequence was the time when the earliest multicellular animals first appeared, but before hard-shelled animals had evolved. This Ediacarian Period and System (roughly equivalent to the term *Vendian* used by Soviet and other geologists) is named after a locality in the Flinders Range of South Australia where these earliest soft-bodied metazoan fossils were first discovered. Although there is still no general agreement on the appropriate name for this period, we will use Ediacarian.

In the Death Valley area, the Ediacarian strata include three conformable units, much of the Noonday, the Johnnie, and most of the Stirling formations. The lowest, the Noonday, in most places rests unconformably on the Proterozoic Pahrump Group. Previously, the Precambrian-Cambrian boundary was placed at this unconformity, although the oldest definitely Cambrian fossils were found as many as 3,000 meters (10,000 feet) higher in the section (Figure 6-6).

Although it is reasonable to set the base of the Cambrian at the unconformity, it is equally reasonable to place it at the base or within the Wood Canyon Formation, because that unit contains the earliest Cambrian fossils in the Death Valley area. However, in the White-Inyo range, *Wyattia*, an enigmatic shelly fossil, probably a primitive mollusk and presumably of earliest Cambrian age, has been found in the upper Reed Dolomite. The Reed Dolomite is regarded as the time equivalent of the middle Stirling Quartzite farther southeast, but the demonstrably earliest Cambrian fossils in the White-Inyo Mountains occur in the Deep Spring Formation, which appears to be equivalent to the upper Stirling and lower Wood Canyon. To further confuse the picture, what appear to be Cambrian fossils have been found in the middle Stirling in the Funeral Mountains.

Where does that leave us today? Most geologists would draw the boundary between the Proterozoic and the succeeding Phanerozoic near the base of the Noonday, and some would put the lower Stirling, Johnnie, and much of the Noonday formations in the Ediacarian, the earliest of the Phanerozoic periods. The Cambrian then would begin with the middle Stirling. This boundary question is one of the more hotly debated topics in historical geology today, and its final resolution will be greatly influenced by the California record. Clearly, the last word on the subject is not yet in.

Paleozoic

Cambrian strata dominate the Paleozoic in the Death Valley area, but most later Paleozoic periods are also represented as extensive but thin limestone and dolomite deposited in a miogeoclinal setting. The Paleozoic section in the Inyo Mountains (Figure 6-7) represents nearly all the periods and is extremely thick, implying more or less continuous deposition in a tectonically stable setting. Aggregate thickness is about 7,000 meters (23,000 feet), and nearly half of this is Cambrian.

Post-Cambrian Paleozoic rocks reach a maximum thickness in the Inyo Mountains, thinning toward the east. Most are carbonate rocks (limestones and dolomites), indicating shallow, warm waters with little incoming detritus. This, in fact, is the thickest section of Paleozoic carbonate anywhere in North America. Although the extensive development of carbonates indicates a generally tranquil depositional environment, some interruptions are evident. The Ordovician Eureka Quartzite represents an influx of abundant, clean-washed sand from adjacent land, and upper Permian rocks that include conglomerates resting by well-defined angular unconformity on Pennsylvanian deposits. Both the angular unconformity and the conglomerates imply initiation of orogenic activity along the continental margin from which the Sierra Nevada was eventually raised.

These same post-Cambrian Paleozoic rocks are well displayed in the Cottonwood, Funeral, Nopah, and northern Panamint mountains near Ubehebe Crater. Richly fossiliferous Devonian rocks of the Lost Burro Formation are exposed in Lost Burro Gap, about 32 kilometers (20 miles) southwest of the crater.

Mesozoic

Marine conditions persisted through much of the Triassic. Some of the rocks deposited then are now well exposed in the Inyo Mountains, the Panamint Mountains near Butte Valley, and the Alexander Hills, where a xenolith of Triassic volcanic rock is surrounded by a younger intrusive porphyry. Lower Triassic rocks are primarily sedimentary, but by the close of the period huge amounts of volcanic material were being erupted in the area now represented by the Sierra Nevada, Owens Valley, and Inyo Mountains.

Numerous granitic intrusives of Mesozoic age are found in California's Basin and Range province. These developed during a period of subduction that was particularly active in middle and late Cretaceous time. The largest of these intrusions occurs in the Owlshead Mountains; smaller exposures are present in the Greenwater, Black, Argus, Inyo, White, Last Chance, and southern Panamint ranges.

Cenozoic

Any rocks deposited between the middle Mesozoic and early Oligocene time have not been preserved in the California Basin and Range province. The oldest Cenozoic rocks yet recognized are the Oligocene Titus Canyon beds northeast of Stovepipe Wells, and their probable correlative, the Artists Drive Formation near Furnace Creek. Titanothere remains were discovered in the Titus Canyon beds in the 1930s. Related to the horse, the titanothere was a gigantic herbivore that lived in association with such animals as the rhinoceros, tapir, and camel. The fossil remains of these animals suggest an Oligocene environment of savanna type, with moist climate and succulent vegetation. Later Tertiary rocks are abundant in the Basin and Range province, but few are fossiliferous.

The region became increasingly arid during Tertiary time, and by early Pliocene rainfall was probably about 580 millimeters (15 inches) annually. Grasslands were widespread and hot summers and mild winters the rule. The climatic pattern suggests to some geologists that erosion had reduced the ancestral Sierra Nevada to a range of low hills, allowing the moderating effects of the Pacific to extend farther inland than is now possible. The late Pliocene Coso Mountains Formation has yielded impressive mammalian fossils, including hyaenid dogs, horses, short-jawed mastodons, camelids, and peccaries. Other Pliocene rocks found at Furnace Creek have been dated by fossil leaves, diatoms, and animal footprints.

As more radiometric dates of plutonic rocks become available, some intrusive rocks formerly thought to be of Mesozoic age are proving to be Miocene. Examples are now known from the Panamints, the Black Mountains, from the Kingston Range, and the Alexander Hills south of Death Valley.

Beginning at about this time, the Basin and Range province was subjected to crustal extension (stretching) as increasingly oblique subduction near the continental margin was replaced by transform movement on the San Andreas fault system. This change in plate motion as well as the extension that is pulling the Basin and Range province northwestward away from the continental interior are probably both driven by poorly understood motions in the upper mantle. The crustal extension has produced the horst and graben fault block pattern, which is so evident in the province today. This same extension has produced the detachment faults, generally best displayed in the eastern Mojave Desert, but also present in the Basin and Range. It should be remembered, however, that the high-angle normal faults we see at the surface in the Basin and Range province may in fact flatten at depth, forming *listric* faults that may merge downward into detachment faults.

In the Greenwater and northern Black Mountains, along the Furnace Creek and Artists Drive fault zones, is a down-dropped wedge of rock partly occupied by Furnace Creek Wash. This wedge contains a Tertiary section totaling almost 4,000 meters (13,000 feet) thick. The Artist's Drive Formation is nearly 1,600 meters (1 mile) thick and has both sedimentary and volcanic members. This is the sequence's oldest unit and is, as noted, at least partly contemporary with the Titus Canyon beds to the north. The overlying Furnace Creek Formation of Pliocene age is composed of volcanic rocks, fanglomerates, and playa lake sediments that include borates and gypsum. These sediments indicate not only an arid or semiarid climate, but also the presence of undrained topographic depressions in which waters saturated with sulfates and borates could accumulate. The fanglomerates indicate an environment of vigorous erosion, flash floods, steep mountain slopes, and deep basins (Figure 6-8).

Late Pliocene to Holocene volcanic rocks are abundant in the Basin and Range province. Prominent Pliocene basaltic flows are exposed near Ryan in the Greenwater Mountains. Pleistocene and Holocene cinder cones and flows are well displayed throughout the southern Owens and Rose valleys, particularly near Little Lake where basaltic flows and several red cinder cones dominate the landscape. Three ages of volcanic activity are evident at Little Lake. The oldest of these, about 400,000 years old, is a very prominent lava flow on the east side of the valley south of Little Lake. This flow extends for several kilometers and was erupted from a cinder cone just a little south of Little Lake at the base of the Coso Range. Although the straight edge

FIGURE 6-8 Badland topography developed of tilted and eroded nonmarine shale and siltstone of the Furnace Creek Formation, Zabriskie Point, Death Valley National Monument, Inyo County. (Photo by R. M. Norris)

FIGURE 6-9 Erosional escarpment developed on the Lower Little Lake Ranch lava flow 440,000 years old. Although this straight escarpment appears to be a fault, it was cut by the ancestral Owens River that flowed through the area in late Pleistocene time. The lava flow forms only the resistant cap; most of the escarpment is cut in granitic alluvium and bedrock, which is concealed by talus from the lava cap. Near Little Lake, Inyo County. (Photo by R. M. Norris)

of this prominent flow appears faulted, it is an erosional feature produced by the Pleistocene river, which flowed south through the valley (Figure 6-9).

The next-youngest flow forms a low cliff on the east side of the lake itself and has an age of about 130,000 years. The youngest flow, a little over 10,000 years old, is related to Red Hill, the prominent cinder cone just north of Little Lake. This flow is the one cut by glacial Owens River at Fossil Falls, a well-displayed dry fall about 30 meters (100 feet) high that shows old plunge pools and excellent potholes formed during the latest pluvial episode of the Pleistocene. Other youthful cinder cones and flows are scattered northward in Owens Valley and on the slopes of the Sierra Nevada, Inyo, and Coso mountains, especially near Big Pine. Just south of Big Pine is a group of bold volcanic peaks, the highest of which—Crater Mountain—rises about 600 meters (2,000 feet) above the valley floor.

A center of recent activity is in the southern Coso Range, where numerous basaltic flows and cinder cones have developed. Some eruptions discharged dark flows that poured over low divides and down modern canyons, appearing almost as fresh today as if they had just congealed. Obsidian and pumice cones and domes like those in the Mono Lake area are also present (Figure 6-10), associated with boiling springs and steam vents.

The presence of Plio-Pleistocene saline beds and fanglomerates in the Basin and Range province implies that basins resembling today's were taking shape during these epochs. Saline deposits associated with clay-rich playa beds in Death Valley, Saline Valley, Searles Lake, and numerous valleys in the Mojave Desert province are chiefly of Pleistocene age. The saline layers in these basins are often separated

FIGURE 6-10 Obsidian and pumice cones and domes of the southern Coso Range, Inyo County. Compare with Mono Craters, Figure 3-27. (Photo by James Babcock)

by layers of silt and clay deposited during wetter periods that may have corresponded to glacial epochs in the Sierra Nevada.

The origin of these saline deposits is worth considering. Weathering of igneous rock like granite or sedimentary rocks like sandstone will not yield much ordinary salt. Although crystalline igneous rocks have plenty of available sodium, they contain virtually no chlorides. Some marine sedimentary rocks, especially shales, contain traces of salt from the seas in which they were deposited, and these rocks may supply some of the salt found in playa lake sediments. Volcanic activity, particularly hot springs and steam vents, can and does supply both chlorides and borates, but perhaps the most important if unheralded source of ordinary salt in the Basin and Range deposits is airborne salt from the sea. This *cyclic salt* is delivered in minute but continuing amounts by westerly winds from the Pacific Ocean.

The chemistry of these basins differs. Owens Lake, for example, accumulated deposits rich in sodium carbonates and borates that have been mined intermittently since 1904. Searles Lake has yielded the greatest array of saline minerals of any Basin and Range valley; end products include potassium salts, borax, boric acid, salt, soda ash, bromine, and lithium carbonate The lake has two main saline deposits. The upper, porous one is 21 to 25 meters (70–80 feet) thick, with many voids filled with saturated brines. It covers about 78 square kilometers (30 square miles). Below this deposit is about $3\frac{1}{2}$ meters (12 feet) of mud, and below the mud is another saline body with brine about 11 meters (35 feet) thick. The intervening mud layer contains organic material dated at 16,000 years before the present or late Pleistocene.

Death Valley was an early historic source of borate minerals, which had accumulated in the valley's muds during the Pleistocene. The deposits were first worked in 1882; the ore was hauled to the railhead at Mojave, 266 kilometers (165 miles) to the southwest by the famed 20-mule teams. Borate mining continued in Death Valley until recently, chiefly in the Furnace Creek area, where commercial deposits of three different boron-bearing minerals (colemanite, ulexite, and probertite) occur in lenses and veins in the Furnace Creek Formation. A much larger borate deposit at Kramer near Mojave (see Chapter 7), produces most of the borates mined in the United States today. This deposit was crossed by the 20-mule teams from Death Valley, although the deposit was not discovered until 1913.

STRUCTURE

The high mountain blocks of the Basin and Range province have a north-northwest trend and complex internal structure. Complex folds, flat thrust faults, high- and low-angle normal faults and some strike slip faults occur in or along these ranges.

Garlock Fault System

The Garlock fault, which demarcates the Basin and Range from the Mojave Desert provinces, is a near vertical fault with about 64 kilometers (40 miles) of left lateral separation (Figure 6-11). It is no older than early Tertiary and may be much younger. Although the Garlock is extensive, only its eastern half is considered here: from Red Rock Canyon in the El Paso Mountains to its junction with the Death Valley fault system at the eastern base of the Avawatz Mountains.

The Garlock system is an interlacing series of shorter faults enclosing narrow, elongate slices and wedges of rock. On the south face of the El Paso Mountains, for example, the El Paso fault is more prominent than the Garlock proper, which tends to be south of the range and buried beneath desert gravels. The El Paso fault is clearly exposed at the south entrance to Red Rock Canyon, where the Miocene Ricardo beds are in contact with the Mesozoic granite (Figure 6-12). Here, the two faults are separated by about $1\frac{1}{2}$ kilometers (1 mile). They continue east for 16 kilometers (10 miles) until the El Paso fault joins the Garlock. In this section, the El Paso fault shows a strong component of vertical offset, whereas the Garlock shows equally strong left slip displacement. The slice between the two faults contains several small depressed blocks, springs, and offset stream channels. About 4.8 kilometers (3 miles) east of Garlock Station, the faults enclose a well-developed graben or sag, 15 meters (50 feet) deep and about $1\frac{1}{2}$ kilometers (1 mile) long. This feature is just out of sight from, but parallel to the Garlock road. For 100 kilometers (60 miles) from its juncture with the El Paso fault to its intersection with the Death Valley fault zone, the Garlock is marked by a fault-line valley until it becomes the northern base of the Avawatz Mountains.

Evidence of left-slip separation on the Garlock is abundant, but the amount of movement is not entirely clear. Recent movements up to about 0.8 kilometer ($\frac{1}{2}$ mile) are demonstrated by offset stream courses and other topographic disruptions. In 1925, left-lateral slip up to $9\frac{1}{2}$ kilometers (6 miles) was first suggested, an estimate

FIGURE 6-11 View to the east showing the trace of the Garlock fault. Town of Ridgecrest is in the lower left. Searles Lake is in the left center, with the Slate Range almost enclosing the lake on the east. The El Paso Mountains are in the center foreground. The intersection of the Death Valley, Garlock and Furnace Creek fault zones is evident in the upper center. (Photo courtesy of U.S. Air Force and U.S. Geological Survey)

based on inferred displacement of Paleozoic formations in the Randsburg area. Subsequently, displacement up to 64 kilometers (40 miles) was suggested, based on the presumed equivalence of a swarm of basic dikes on each side of the fault. These volcanic dikes belong to what is known as the Independence Dike Swarm and extend, as a group, southward about 250 kilometers (155 miles) from the Sierra Nevada near Independence. They are present in the Inyo and Argus ranges, in the Alabama Hills near Lone Pine, and in the Spangler Hills near Trona where they are offset from counterparts farther east on the south side of the Garlock fault. These counterparts extend about another 100 kilometers (62 miles) southward into the ranges of the Mojave Desert (Figure 6-13).

Death Valley-Furnace Creek Fault System

Another predominately strike-slip fault zone is the Death Valley-Furnace Creek system. North of Furnace Creek, this fault is known as the North Death Valley fault

FIGURE 6-12a El Paso fault, eastern side of the south entrance to Red Rock Canyon, Kern County. Compare with Figure 6-12b. (Photo by R. M. Norris)

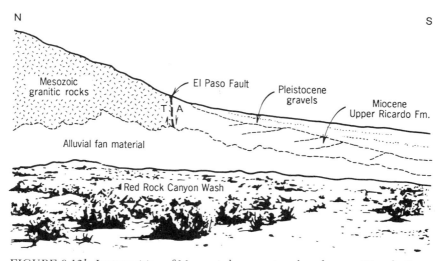

FIGURE 6-12b Juxtaposition of Mesozoic basement rock and upper Ricardo Formation along the El Paso fault. Note the angular unconformity between the north-dipping Ricardo beds and the gently inclined Pleistocene alluvial gravel. The symbols T/A on the fault refer to "toward" and "away," the direction of slip with respect to the viewer. (Courtesy C. A. Nelson, Department of Earth and Space Sciences, University of California, Los Angeles.

zone. The Furnace Creek portion extends from the vicinity of Furnace Creek Ranch eastward across the Amargosa River Valley to the Resting Spring Range. The Furnace Creek portion is often concealed by alluvial gravels, but the North Death Valley portion displaces alluvial gravels south and east of Stovepipe Wells, and northward,

FIGURE 6-13 Columnar jointing in a thin vertical basalt dike, Argus Range, near Trona, San Bernardino County. This dike is an example of the widespread Independence dike swarm, which extends southeast from near Independence, Inyo County, to the central Mojave Desert. (Photo by R. M. Norris)

more or less along the base of the Grapevine range. This portion of the fault system is marked by scarps in alluvial fans as much as 15 meters (50 feet) high, as well as by prominent lines of springs between the Scotty's Castle road and Ubehebe Crater (Figure 6-14).

The fault zone continues many kilometers farther northwest, bounding the Last Chance Range, then Fish Lake Valley, and the east side of the White Mountains.

The Furnace Creek portion is marked by sharp boundaries between fan deposits and the older rocks of the Funeral Mountains between Furnace Creek Ranch and the Amargosa River Valley, but as previously noted the Furnace Creek fault cannot be traced across the Amargosa River Valley, though it is believed to do so and merge with the frontal fault of the Resting Spring Range.

The well-exposed frontal faults along the western base of the Black Mountains are collectively known as the Central Death Valley fault zone. Some segments of this portion have been given local names, especially where large vertical movement has occurred, producing the precipitous west-facing escarpment near the deepest part of the valley.

The Southern Death Valley fault zone is best exposed at Shoreline Butte, in the Confidence Hills, and along the northeastern Avawatz Mountains where the system joins the Garlock fault zone. In one place, a short distance north of Shoreline Butte, a small cinder cone called Cinder Hill is offset right-laterally by this fault.

Many geologists now believe that crustal extension manifested by right slip on these various portions of the Death Valley-Furnace Creek fault system is responsible for the central part of the Death Valley depression, a feature known as a *pull-apart basin* (Figure 6-14).

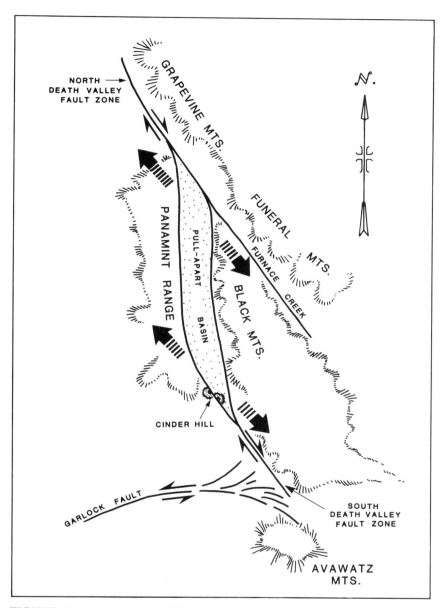

FIGURE 6-14 Main structural features in the Death Valley area. Note the effect of the southeasterly motion of the Black Mountains and the northwesterly motion of the Panamint Range in producing the deep pull-apart Death Valley depression. (After Wright and others, 1974)

The junction of the left-slip Garlock and the right-slip Death Valley fault systems poses some interesting problems. Evidence suggests about 64 kilometers (40 miles) of left-lateral separation occurs along the eastern Garlock fault only 13 kilometers (8 miles) west of its junction with the Death Valley fault zone. At least three interpretations of this unusual situation have been offered.

1. Some believe that the Garlock terminates at the Death Valley fault zone and that the 64 kilometers (40 miles) of displacement is accommodated by crustal extension or stretching in a roughly triangular area whose apex is about 10 to 12 (6 to 7½ miles) kilometers west of the Death Valley fault zone and whose base is along that fault zone.

2. Others suggest that the Garlock has been offset by a younger Death Valley fault and that an eastward continuation of the Garlock exists to the south. Thus far, no evidence for such a continuation has been found.

3. Another view is that the Garlock crosses the Death Valley fault and continues directly east. The differing structures of the Basin and Range north of this extension and in the Mojave Desert to the south may substantiate this, but again, no conclusive evidence has been forthcoming.

Although it is true that most geologists favor something like the first explanation, no generally accepted version is yet available.

Special Fault Features

Wineglass or goblet valleys are canyons whose lower walls are nearly vertical in contrast to their upper walls, which are more open and broadly V-shaped. Such valleys indicate rapid and recent uplift, which causes valley deepening to outpace valley widening and subsequently form the stem of the goblet. In the Basin and

FIGURE 6-15 Western face of Black Mountains, Inyo County, showing gullied Badwater Turtleback cut by a single deep wineglass valley on the right. Note the coarse alluvial material that has been swept out of the canyons onto the floor of Death Valley. Light-colored rocks on the left are parts of the Tertiary upper plate and the darker rocks in the Turtleback are Proterozoic gneisses of the lower plate. (Photo by R. M. Norris)

Range province, the volume of material visible in alluvial fans typically associated with these valleys is insufficient to account for the total material excavated from the drainage basins mentioned earlier. The explanation is that rapid, relatively downward movement of blocks like Death Valley depress old fan deposits and force streams emerging from the steep valleys to construct new alluvial cones on top of older, buried fans (Figure 6-15).

Like the Garlock, the Death Valley and Furnace Creek fault zones are composed of en echelon faults that incorporate elongate, narrow slices of crust. Where right-slip faults step to the right, the crust is stretched between the fault strands and a pull-apart basin is formed. Where right-slip faults step left, the opposite occurs and compression produces ridges more or less at right angles to the faults. Such geometries produce varied topography and, in the relatively down-dropped blocks, preserve younger rocks that elsewhere are largely destroyed by erosion. Colorful examples occur at Zabriskie Point and Artists Drive, where thick sections of mostly yellow, late Tertiary nonmarine Furnace Creek Formation are well exposed between strands of the Death Valley and Furnace Creek fault zones.

In the northeastern Avawatz Mountains are the colorful, predominately red, Salt Basin beds. These are thought by some geologists to correlate with the Furnace Creek Formation, although the characteristic colors of each formation differ and the age of the Salt Basin beds is not well established. The Salt Basin beds have been preserved because they are part of a block compressed between the merging Garlock and Death Valley fault zones.

Turtleback Faults

For many years there has been debate regarding the complex structure of the Black Mountains. The frontal fault system is not at issue; it is considered a set of high-angle normal faults separating the uplifted range from the depressed Death Valley block. Within the range, however, are features formerly thought to be folded, broken, and disordered thrust sheets resting on a core of Proterozoic crystalline rocks. In several places the younger rocks of these "thrust" sheets have been stripped away, exposing smooth, but folded fault surfaces. Some of these folds form *antiforms* (upwardly convex folds in which the stratigraphic order of rocks is unknown; anticlines are similar, but older rocks occur in the core). These antiforms are as much as 20 kilometers (13 miles) long and are called *turtlebacks* (Figures 6-15 and 6-16). Prominent turtlebacks occur at Badwater, Mormon Point, and at Copper Canyon.

Levi F. Noble spent many years studying the Black Mountains, where rock exposures are almost too good and too complex. (The broad picture often can elude the geologist presented with an abundance of well-exposed detail.) It was Noble, however, who established that the range's core is overlain in several places by disordered and exceedingly complex rock masses, presumed to be thrust sheets, but now generally regarded as the result of fracturing or tectonic mixing along extensional or detachment faults. He termed these disordered sheets *chaos*. The lowermost is the Virgin Spring chaos, which consists chiefly of Proterozoic to middle Cambrian brecciated and shattered blocks 60 to 760 meters (200–2,500 feet) long. It records four major periods of deformation. The earliest is about 1,700 million years old, and was followed by a period of metamorphism. The second event occurred near the end of the Proterozoic and opening of the Phanerozoic. The third deformation was

FIGURE 6-16 Turtleback south of Badwater at Mormon Point. Note the small alluvial fans that indicate rapid uplift of the mountains along the Central Death Valley fault. (Photo by Spence Air Photos, courtesy of the Department of Geography, University of California, Los Angeles)

near the end of the Mesozoic and the most recent in the middle Cenozoic. Atop the Virgin Spring is the Calico chaos, composed of colorful Tertiary volcanic rocks. Highest is the Jubilee chaos, which consists mainly of sedimentary breccia containing granitic and volcanic rocks. The Jubilee also contains a megabreccia that includes many of the rock types found in the Black Mountains.

Geomorphic Evidence of Faulting

Evidence of recent faulting in Death Valley and Panamint Valley is provided by well-preserved scarps in unconsolidated alluvial gravels. The very existence of a well-preserved scarp in such weak Holocene deposits corroborates their youth, because in arid climates stream erosion can quickly modify or eliminate such features. Many of the scarps are probably only a few hundred to a few thousand years old. Apart from features known to be connected with the 1872 Owens Valley earthquake, none of the young scarps in the province's basins is associated with a recorded earthquake.

The best known of these youthful fault features is the Wildrose graben near the mouth of Wildrose Canyon on the fan near the western base of the Panamint Mountains (Figure 6-17). Here, a block of fan gravel about $6\frac{1}{2}$ kilometers (4 miles) long and nearly $1\frac{1}{2}$ kilometers (1 mile) wide has dropped down about 60 meters (200 feet). This has diverted some water courses, forcing them to follow along the base of the western scarp instead of flowing directly across it. The old courses of these desert streams form beheaded valleys notching the uplifted western wall of the graben.

FIGURE 6-17 View to the north of the graben at the mouth of Wildrose Canyon. Note the streams beheaded by recent downfaulting and the subsequent readjustment. Northern Death Valley is in the far distance, with the Grapevine Mountains as backdrop. (Photo by John S. Shelton)

Other excellent examples of faulted, very young alluvial gravels can be seen just a few meters from the road at the base of the Black Mountains for a kilometer or so south of Furnace Creek. About 12 kilometers ($7\frac{1}{2}$ miles) south of the Grapevine Ranger Station near Scotty's Castle, the road follows the base of an east-facing scarp about 10 to 15 meters (30–50 feet) high along the North Death Valley fault zone.

It is curious that so few earthquakes have originated in the California Basin and Range east of the White-Inyo mountains in historic times, because evidence of geologically young faulting is so pervasive. However, it should be noted that a recurrence interval of 200 to 300 years is geologically very frequent, but the documented history of the California Basin and Range is only about 150 years at most. We should therefore derive very little comfort from the current seismically quiet interval; it could end tomorrow.

The Argus Range is capped with Quaternary basalt flows, which disappear beneath alluvial gravels at the range's base. The basalt reappears $1\frac{1}{2}$ kilometers (1 mile) east as an east-dipping cap on Ash Hill. The lava cap shows that both Ash Hill and the

Argus Range have been faulted and tilted. This is an example of *louderback caps*, named by William Morris Davis after George Davis Louderback, who first cited such features as evidence for tilting of fault blocks in the Basin and Range of Nevada.

Many of the ranges in the province tend to be asymmetrical, elongate blocks with bounding faults on their flanks. Conversely, some ranges such as the Panamint, Black, and other ranges are tilted fault blocks. Each has a steep western face bounded by young normal faults and a more gentle eastern slope that passes gradually from foothills into pediment and fan.

SPECIAL INTEREST FEATURES

Pleistocene Lakes and Streams

The Basin and Range events of the past 2 million years are reflected everywhere in today's landscape. Because of this, it is considerably easier to assess the Quaternary environment than Tertiary or earlier environmental settings.

Geologic time distinctions within the Basin and Range province are continuously in dispute. As mentioned previously, for instance, the Proterozoic-Phanerozoic boundary in the area is still unconfirmed. Similarly, there has been much debate about the Tertiary-Quaternary boundary. Although recognizing that this matter is not yet settled, the authors arbitrarily equate the opening of the Quaternary with the earliest recorded glaciation in the Sierra Nevada—about 1.8 million years ago.

The glacial epochs produced little permanent ice in the Basin and Range province. Only the highest ranges like the Sierra Nevada had any glaciers at all. Small valley glaciers probably existed near the summit of the White Mountains and on Charleston Peak in Nevada's Spring Mountains.

The glacial epochs nevertheless profoundly affected Basin and Range climates, which were cooler, perhaps quite dry, but characterized by less evaporation than today. In addition, climatic and associated vegetative zones shifted to lower altitudes. Thus the boundary between desert scrub vegetation and pinon-juniper woodland, which today occurs high on Basin and Range slopes, lay at much lower elevation. Evidence from the desert southwest confirms this lowering of vegetative zones and seems to suggest that the climatic contrast between upland and valley was stronger than it is today. The deep basins like Death Valley seem to have received low rainfall during the glacial epochs, although they were probably much cooler and experienced appreciably less evaporation than at present.

Subsequently permanent drainages and lakes developed in basins that today contain only salt flats, playas, and ephemeral streams. The seeming paradox of permanent lakes and streams existing in cool but arid valleys perhaps can be accepted more easily if modern parallels are considered. For example, alkaline Mono Lake lies in a cool, arid valley about 1,950 meters (6,400 feet) high. Rainfall is low, but the lake and entering streams are fed by the much better watered Sierra to the west. The recent marked shrinkage of Mono Lake shows how delicate is the balance between an arid climate and snow-fed supplying streams. Shrinkage is not due primarily to climatic change, but to withdrawal of water from entering streams to supply the Los Angeles aqueduct system. Another example is Owens Lake. Until construction of the Los Angeles aqueduct and subsequent interception of the Owens River, Owens Lake was a large, rather shallow body of saline water occupying the extremely arid southern end of Owens Valley. Except perhaps for a brief period about 2,000 years ago, both these lakes have lacked outlets since the end of the Pleistocene, and both have become mineralized, further confirming the aridity of their surroundings.

Extensive lakes and connecting streams probably occupied many of the province's shallower basins in each of the glacial epochs, but most of the evidence was erased during the intervening dry periods. For the most part, only the record of the latest chain of lakes and connecting streams can be seen today (Figure 6-18).

Alkaline Mono Lake is appreciably smaller than its freshwater Pleistocene ancestor, Lake Russell, which probably was about 225 meters (750 feet) deeper than the modern lake. Lake Russell spilled east into the valley of Adobe Lake and from there into the upper Owens Valley.

Water from Lake Russell and the greatly augmented streams draining the heavily glaciated Sierra undoubtedly made the ancient Owens River a formidable stream. Recent studies of the Pleistocene lake system suggest that the maximum discharge of the Owens River during the Pleistocene was approximately 10 times its present discharge. Owens Lake, now only a few shallow saline ponds in a vast shimmering salt flat, was then at least 80 meters (250 feet) deep and extended south to the vicinity of Haiwee. The lake stayed fresh at its higher levels because it maintained an outlet channel across Rose Valley, and a well-defined gorge was cut in the young volcanic rocks near Little Lake. This gorge, its falls, and the associated potholes developed during the final glacial stage because the lava flows involved are no older than last interglacial, as mentioned earlier.

The now-vanished river that produced the falls continued south to Indian Wells and Salt Wells valleys, where a broad shallow lake formed. During this period, lime-secreting cyanobacteria promoted development of tufa pinnacles around the lake's edge. The shallow lake occupying Indian Wells Valley spilled east across the southern Argus Range, cutting the winding gorge of Salt Wells Canyon through which the highway to Trona now passes. During the wettest stages, the level of Lake Searles rose high enough so that that lake and China Lake in Indian Wells Valley formed a single body of water, submerging Salt Wells Canyon and accounting for some of the patchy, light-colored lake deposits that can be seen in protected places on the sides of the canyon.

The stream that cut the Salt Wells Canyon fed glacial lake Searles, which was once about 195 meters (640 feet) deep. As noted earlier, the floor of this basin contains two thick beds of saline minerals separated by a clay layer that accumulated during the final wet period, when the lake probably spilled east into Panamint Valley. These saline beds show that the Lake Searles basin was often too deep and too arid for the entering streams to maintain a freshwater lake for any extended period. The Lake Searles depression was thus the final basin in the Owens-Death Valley chain of lakes for much of the Quaternary. Conversely, the lakes of Indian Wells Valley accumulated little saline material and were so shallow, only about 12 meters or 40 feet, that they overflowed with even minimal inflow.

Both Mono and Searles basins are rimmed with many kilometers of well-preserved shorelines cut into the surrounding ranges. Where the shorelines cross desert and washes and alluvial fans, they form fossil beaches and sweeping curved sand and gravel bars. The better-developed lower shorelines in the Searles Lake basin are further proof that the high-level, freshwater Lake Searles was more transitory than its lower saline counterparts.

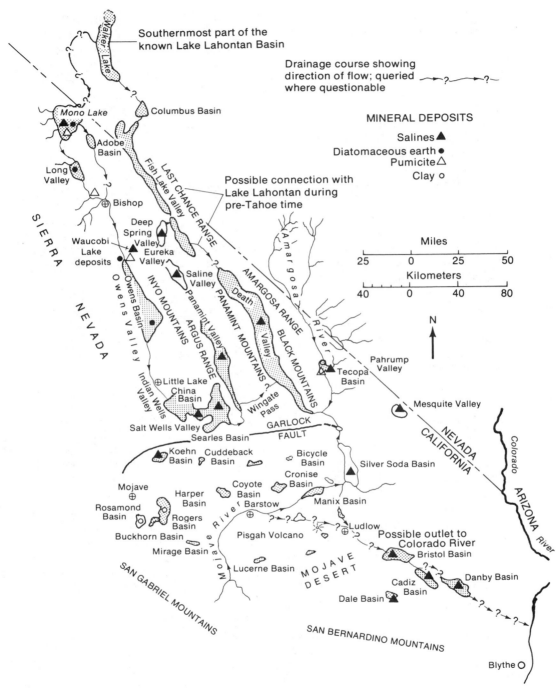

FIGURE 6-18 Inferred Pleistocene drainage of the Basin Ranges. (Source: California Division of Mines and Geology)

FIGURE 6-19 (top) Tufa pinnacles at the southwestern shore of Searles Lake. Note old shorelines in lower left. (Photo by Fairchild Aerial Surveys, courtesy of Department of Geography, University of California, Los Angeles) (bottom) Detail of tufa pinnacles. (Photo by Robert M. Norris)

California's finest tufa domes and pinnacles occur in the southern Lake Searles basin, where there are more than 500 pinnacles (Figures 19a and 19b). Some are as much as 43 meters (140 feet) high and more than 160 meters (500 feet) in diameter, but most are 3 to 12 meters (10–40 feet) high and 6 to 11 meters (20–30 feet) in diameter. Like those of the Indian Wells Valley, these pinnacles were formed by a combination of spontaneous precipitation of calcite near lime-rich springs and the influence of lime-secreting cyanobacteria (blue-green algae) living in the lake. A modern analog can be seen at Mono Lake where pinnacles of this sort are forming along the southern shore (see Figure 3-39).

During the wettest part of the Pleistocene, Lake Searles drained east around the southern end of the Slate Range into Panamint Valley. Evidently Lake Panamint received limited local runoff and was quickly isolated from Lake Searles whenever the discharge of the Owens River system faltered. Until recently, salt deposits were thought to be absent or nearly absent in the Panamint basin, but core drilling has revealed several buried salt beds, the thickest of which is about 40 meters (130 feet) and appears to have formed between 24,000 and 34,000 years ago. At its maximum, Lake Panamint was 290 meters (950 feet) deep, about 90 kilometers (55 miles) long, and 16 kilometers (10 miles) wide. When the lake was nearly full, several large streams entered it from the Panamint Mountains to the east. One of these streams deposited a large delta at the mouth of Pleasant Canyon, which can be seen just south of Ballarat (Figure 6-20). As the lake shrank, recessional shorelines were cut into this delta and elsewhere along the steep western face of the Panamint Mountains.

Because a single evaporation of 290 meters (950 feet) of water could scarcely account for the 40 meters (130 feet) of salt that occurs beneath present valley floor clays, either repeated, closely spaced fillings of the basin or long-continued intermittent local runoff is required to account for the salt.

At its highest level, Lake Panamint's depth was limited by the level of Wingate Pass, which leads into the southern end of Death Valley. Death Valley was occupied by Lake Manly, about 145 kilometers (90 miles) long and 275 meters (900 feet) deep. Even during wet cycles, Lake Manly did not drain, but was the sink for minerals

FIGURE 6-20 Deltaic deposit formed in Pleistocene Lake Panamint at the mouth of Pleasant Canyon near Ballarat, Inyo County. Thickness of the delta is about 120 meters (400 feet). (Photo by R. M. Norris)

from Lake Panamint and from other streams entering from the south and east. Substantiating this are the thick accumulations of saline minerals in the valley bottom. Shorelines left by Lake Manly are preserved at several places in Death Valley, most notably at Shoreline Butte, a hill of young volcanic rock and Quaternary nonmarine beds located at the foot of Jubilee Pass. Other prominent shorelines occur along the west face of the Black Mountains near Badwater. A fine fossil sandy beach bar occurs between Beatty Junction and Hell's Gate on the Daylight Pass road to Beatty, Nevada, in central Death Valley.

It seems likely that streams connected Lake Manly southward with Silurian and Soda Lakes in the Mojave Desert (Figure 6-18) and from there to the Colorado River. Another possible connection existed with lakes in Fish Lake Valley and Walker Lake to the north, in Nevada. These intricate links have been proposed because of some unusual fish that live in Salt Creek and Saratoga Springs in Death Valley, in stretches of the Amargosa River, and in Devil's Hole springs northeast of Death Valley Junction. The small fish are known as Desert Pup Fish, and belong to three species of the genus *Cyprinodon*, all related to fish in the Colorado River. These species of Desert Pup Fish are believed to have had free movement through the drainage system just described, but as the climate became drier the fish were isolated in permanent springs and streams.

The most recent flooding of modern Basin and Range valleys probably occurred about 2,000 to 5,000 years ago, when small shallow lakes briefly occupied many basins. For instance, the shallow lake in Death Valley was about 10 meters (30 feet) deep and its water persisted long enough to form faint shorelines. As this pond evaporated, a surface salt crust accumulated. In addition, young faulting has deformed these shorelines, making the eastern shore about 6 meters (20 feet) lower than the western. This reflects the pattern of Quaternary fault movement also indicated by goblet valleys, steep frontal faults, and the eastward tilt of the valley floor.

The Basin and Range province extends into California northwest of Lake Tahoe in the vicinity of Susanville in Lassen County. In this area are two large lakes, Eagle and Honey. Both are parts of ancient glacial Lake Lahontan, a large Pleistocene lake that at its maximum covered about 22,400 square kilometers (8,700 square miles) chiefly in northwestern Nevada, but including parts of eastern California as well. Today Honey Lake is dry most of the time, but Eagle Lake is permanent, though subject to some fluctuation. Both lakes lack outlets, and the water of Eagle Lake is quite alkaline (pH 9.5). The high alkalinity has prevented the survival of introduced fish species, and only the five native species have managed to survive. Like other surviving remnants of ancient Lake Lahontan, Eagle Lake persists only because local conditions provide more water than is lost annually to evaporation (Figure 6-21).

Late Cenozoic Volcanism

Near Little Lake and Coso Hot Springs are many volcanic features, including about 15 late Quaternary basaltic cinder cones and about 30 rhyolitic tuff cones and domes (Figure 6-10). Most of these rest on the typical Sierran granite that composes the core of the Coso Range. Some of the lava flows are so young that they rest on Quaternary lake beds or form congealed lava streams in the canyons cut into the range. The canyons are almost certainly Quaternary, so the flows are quite recent.

The Coso region contains a variety of igneous rocks. Apparently, dark, rather fluid

FIGURE 6-21 Eagle Lake near Susanville, Lassen County. This is one of the larger Basin and Range lakes in California and is a surviving remnant of Pleistocene Lake Lahontan. (Photo by R. M. Norris)

basalts were the first volcanic erupted. These were followed by explosive rhyolitic eruptions that produced tuff cones of light-colored, pumiceous frothy rocks often underlain by glassy obsidian. The basaltic eruptions continued, and some may be contemporary with the rhyolitic eruptions. The most recent eruptive activity has been basaltic, producing rather fluid lava flows and red and black cinder cones. Field evidence suggests that the basaltic magma rose from the upper mantle through the crust and the granitic rocks, melting some of the granites to produce the rhyolitic magmas. The chemical composition of the rhyolites resembles that of the underlying granite.

The Coso-Little Lake volcanic episode began during the Pleistocene, and there is nothing to suggest that it has yet concluded, although the most recent activity is more than 10,000 years old. The basaltic flows that poured down essentially modern canyons are one proof of continued activity. Another is the fresh, scarcely eroded shape of the cones themselves. Even eroded but still intact cones reflect youth, for there are few recognizable cones whose last eruption predates late Tertiary.

Probably contemporary with the Coso field is the volcanic center in the northern Argus and Coso ranges on either side of State Highway 190, from Lone Pine and Olancha to Panamint and Death valleys. South of Keeler are several distinct cones and a thick pile of sheetlike basaltic flows cut by north-south trending faults. Most of these volcanics are late Pleistocene, although a few are Pliocene. Additional flows occur east of the junction with the Darwin road. Beyond the crest, a spectacular cross-sectional view is provided by Rainbow Canyon, which cuts the Argus Range north of the highway. The highway then slices across three cinder cones as it descends into Panamint Valley, giving the traveler an excellent worm's-eye view of the inside

FIGURE 6-22 Ubehebe and Little Hebe craters, northern end of Tin Mountain on west side of northern Death Valley. Note that the material in the walls of Ubehebe is bedded sediment. (Photo by Spence Air Photos, courtesy of Department of Geography, University of California, Los Angeles).

structure of a cinder cone. Toward Panamint Springs, another roadcut reveals a thin basaltic dike cutting across extremely young, unconsolidated gravels. This little feature is not unusual in itself, but it does demonstrate that volcanic activity has continued into essentially present time.

Ubehebe Crater, north of Tin Mountain near Death Valley, displays an unusual kind of volcanic activity (Figure 6-22). Some geologists consider the crater distinctive enough to represent special features designated *ubehebes*. Most investigators, however, call such a wide, flat-floored, low-rimmed vent a *maar* or *tuff ring*. Some geologists think that magma coming into contact with ground water generates steam, with subsequent ejection of fragmental material and escaping steam, ash, cinders, and rock fragments torn from the throat of the vent. The structure is believed to be the product of *base surge* during the eruption. A base surge is a ring of gas and suspended solids that moves radially outward at high velocity from the base of a vertical eruption column. At Ubehebe Crater, the erupted material is basaltic, but no lava is present.

Sand Dunes and Wind Erosion

Despite the popular association of dunes and deserts, dunes are not widespread in any of California's deserts. In the Basin and Range, for example, rugged topography has precluded development of large tracts of dune sand. Certainly nothing in the

American Basin and Range province compares with the sand seas that cover about a third of the Sahara or the seemingly endless linear dune ridges of the flat Australian desert. Basin and Range dune systems are typically localized sand accumulations. They have developed where wind patterns allow the sand to move to and fro, but not to move in any one direction for long.

Many of the desert basins have more than one group of dunes. Owens Valley has only a few, but they include several small areas southwest of Owens Lake that are blanketed with low, partially vegetated dunes probably derived from old lake beaches. Saline Valley also has a cluster of low dunes, none much higher than 15 meters (50 feet). They are northwest of the salt lake and indicate a wind pattern different from that prevailing in the Owens Valley. Panamint Valley has a dune field at its far northern end, suggesting that winds transport sand north until they are slowed by being forced up the steep southern face of Hunter Mountain. These dunes are about 90 meters (300 feet) high.

Although they are not more than 25 meters (80 feet) above the valley floor, the dunes in Death Valley are the best known in the province. They have been seen by countless visitors and have provided the desert settings for many motion pictures. Associated with these dunes is the Devil's Cornfield, where deflation of sand and silt has left clumps of arrow-weed atop hummocks of sand and silt anchored by the plant roots. North of Stovepipe Wells and near Midway Well are some low dunes anchored by mesquite. The mesquite thrives in sandy areas where water is available at shallow depth, and its tangled branches trap sand until mounds about 6 meters (20 feet) high develop. Many dunes of this sort are found in northern Death Valley.

Both the Panamint and Death Valley dunes show little persistent orientation of slope or slip face, although at any given moment the pattern of slip faces will be

FIGURE 6-23 Eureka Valley dunes. These dunes are the highest in California and rival those at Great Sand Dunes National Monument in Colorado, which are generally regarded as the highest in North America. Bedded rocks of the Last Chance Range behind the dunes are chiefly lower Paleozoic marine limestones and dolomites. Inyo County. (Photo by R. M. Norris)

internally consistent. (A slip face is a steep active slope at the angle of repose for dry sand, about 33 degrees.) In addition, because both sets of dunes have relatively little vegetation, they can respond rapidly to wind changes.

The dunes of southern Eureka Valley are the highest and most striking dunes in the Basin and Range province, rising approximately 210 meters (700 feet) above the valley floor. These dunes are so high, in fact, that some geologists think they may cover a rocky outcrop. Like the Panamint dunes, the Eureka dunes apparently formed where sand-laden winds were forced up against the mountain front (Figure 6-23).

The Dumont dunes have formed on a flat gravelly surface trenched by the Amargosa River near the southernmost part of its course. The center dune rises 128 meters (420 feet) above the desert floor and resembles the dunes of Death, Eureka, and Panamint valleys. The margins however, include many small barchans, star dunes, and seif ridges that extend away from the main dune.

Wind activity has not only produced dunes in many Basin and Range valleys, but also generates occasional severe dust storms. Strong winds are common in the area, particularly in the spring months, and they often whip up clouds of blinding and irritating dust from saline playa surfaces. The alkaline crusts that form on Owens Lake after winter runoff has evaporated readily give rise to such events. Depending upon wind direction, appreciable amounts of Owens Lake dust occasionally reach the cities south of the Transverse Ranges or is blown over the high Inyo Mountains into the desert valleys to the east.

Alluvial Fans

Although alluvial fans occur throughout the Mojave and Colorado deserts and in the semiarid Transverse and southern Coast Ranges, their most spectacular California development is in the Basin and Range province. Perhaps the main reason for this is the extreme relief that characterizes the province, locally aided by vigorous or very steep streams. For example, along the steep eastern front of the Sierra is a series of large, coalescing alluvial fans (Figure 6-2). These are partly the result of more than 3,000 meters (10,000 feet) of relief, but they are also due to the large streams fed by snow at the Sierran crest. In contrast, the White-Inyo mountains have few major streams and far fewer conspicuous fans than the Sierra.

The largest fans are those built out from the eastern Panamint Mountains into Death Valley. Of these, the Emigrant Wash and Hanaupah fans are the best examples (Figure 6-24). Starting at the fan's base near sea level at Stovepipe Wells, State Highway 190 climbs Emigrant Wash for 19 kilometers (12 miles), reaching an elevation of 1,300 meters (4,300 feet) at the top of the fan. Hanaupah fan is more than 700 meters (2,300 feet) high.

Alluvial fans are quite sensitive to subtle changes in climate or to tectonic events. Careful analysis of fans has proved useful in reconstructing both the recent climatic and tectonic history of the area in which the fan occurs. For example, fans along the west base of the Grapevine Mountains and east base of the Panamints in northern Death Valley clearly show several stages of fan development; the older portions are darkest, heavily coated with desert varnish, whereas the younger are the lightest. Between 1978 and 1983, a number of intense rainstorms affected the California Basin and Range. Floods swept across the fans leaving broad new channels filled with fresh, lighter-colored rocks lacking desert varnish and contrasting sharply with the older, darker parts of the fans.

FIGURE 6-24 Hanaupah alluvial fan in the Panamint Mountains. The giant fans encroaching high into the Panamint Mountains are in contrast to the small alluvial fans and cones at the base of the Black Mountains on the east side of Death Valley. Telescope Peak, often snow-capped until midsummer, overlooks Death Valley. (Photo by Spence Air Photos, courtesy of Department of Geography, University of California, Los Angeles)

Desert Pavements and Desert Varnish

Basin and Range desert pavements usually occur on the flat-topped divide surfaces between active wash channels. There are at least two plausible explanations for these mosaics of shiny pebbles typically encrusted with desert varnish. They may be produced by selective erosional removal of the finer fractions of a sand-clay-gravel mixture, leaving the coarsest materials behind as a surface armor protecting the undisturbed mixed materials below. Similar armors may also develop on surfaces of expansive clay layers that rest on gravel. As the clays shrink and swell (owing to alternate wetting and drying), loose pebbles from the gravel move upward and eventually accumulate on the surface. Especially fine examples of desert pavements are found in the southern Greenwater Valley.

Desert varnish is a glossy dark-brown to black coating that develops on the exposed surfaces of many rocks in arid climates. For many years it was thought to be produced by weathering processes that released iron and manganese oxides from the rock's interior despite the fact that it was known to develop on rocks with little or no iron and manganese. It was also known to be a surface process because the varnish never covered the undersides of rocks. Recent studies have demonstrated that the coatings are about 70 percent wind-blown clay and about 30 percent various metallic oxides. Maganese-oxidizing bacteria cement the clays and other minerals to rock surfaces and seem to thrive in environments where little organic material is present and alternate wetting and drying occurs, perhaps from dew. Varnish takes an appreciable time to develop—perhaps hundreds to thousands of years—although under controlled and very favorable conditions in the laboratory, varnish has developed in less than six months.

Springs

The Basin and Range of California contains a surprising number of large springs for such an arid region. Several distinct types are known, but the precise origins of many of the springs are still undetermined.

Most of the streams draining into the Owens Valley from the Sierra carry substantial flows initially, but dwindle away as they cross the alluvial fans. Some stream water is lost by evaporation, but most of it sinks into the porous fan gravels. The flows that emerge at the base of the fans are called cienaga springs. Los Angeles now taps much of this underflow for its water supply, so the cienaga springs are less productive than formerly.

Most of the large springs are controlled by faults, which are often concealed by superficial gravels. The faults provide extensive fracture systems along which water may collect, move, or emerge. Springs of probable fault origin include Saratoga Springs in southern Death Valley, Tecopa and Resting Springs near Shoshone, and springs at the east base of the Argus Range among others.

Sometimes water moves to the surface along the frontal faults of ranges and supports conspicuous linear patches of bright green vegetation that contrast strongly with the prevailingly olive drab desert vegetation. Such linear springs can be seen along the west base of the Grapevine Mountains near Scotty's Castle, at the base of the Panamint Mountains at Indian Ranch, and along the Garlock fault at the base of the El Paso Mountains.

Hot and warm springs occur throughout the Basin and Range. The distinction among cold, warm, and hot springs is somewhat arbitrary, but warm springs are those with temperatures ranging from about 25° to 45° C (80–110°F). True hot springs are the least common and are usually associated with recent volcanic activity. Two notable occurrences are Coso Hot Springs just east of Rose Valley in the Coso Range and the boiling springs at Hot Creek and Casa Diablo in the Crowley Lake area. These and other hot springs have provoked considerable recent interest as possible sources of geothermal power.

Minerals and Mining

Many mineral species have been found in the Basin and Range province, and some of the metal mines have been worked since the middle of the nineteenth century. Most mines are no longer active, however, except for small operations that produce chiefly nonmetallic minerals.

Among the more famous metal mines is the Cerro Gordo, near the crest of the southern Inyo Mountains. First operated about 1860, considerable lead and silver was mined initially, but zinc became the main product in the early twentieth century. Mineralization was localized in Paleozoic carbonate rocks as replacements and fissure fillings adjacent to granite. The same pattern of rock type and mineralization is typical of many mining districts in the region, although few have yielded the array of unusual minerals mined at Cerro Gordo.

One of the famous silver camps was at Panamint City, in the upper part of Surprise Canyon in the Panamint Mountains. Rich ore was produced from 1873 to 1876, and the town population grew to more than 1,500. Theft of silver bullion from wagons creaking slowly down the steep canyon became a severe problem, so the mine operators stopped casting silver in the standard bars of 9 to 13 kilograms (20–30 pounds); instead they cast it into cannonballs weighing 220 to 320 kilograms (500–700 pounds). Thieves were utterly frustrated and the cannonballs reached San Francisco without so much as a guard present en route. Panamint City ended about as spectacularly as it had begun. In 1876, when the ore was nearly exhausted and only the optimists were still in town, a cloudburst and flash flood descended on the canyon. Much of the town and the smelter were swept down to Panamint Valley in a torrent of rocks and water.

Other famous lead-zinc-silver mines are the Minnietta and Modoc in the northern Argus Range, the Ubehebe in the northern Panamints, and the War Eagle, Columbia, and Noonday in the southern Nopah Range. Gold was mined at Skidoo, Chloride Cliff, and Ballarat about 1900, and at Bodie in the Mono Lake area in the 1870s and 1880s.

Apart from saline minerals, mining today is predominately for talc, though gold is enjoying a resurgence. Talc mines operate intermittently in the Cottonwood and Ibex mountains, the Talc City Hills, the Argus Range, and particularly around the southern end of Death Valley in the Panamint and Black mountains. The talc is concentrated in contact metamorphic deposits, chiefly in magnesian limestones near intrusive igneous bodies. Most talc deposits occur in the Proterozoic Crystal Spring Formation.

Titus Canyon

Death Valley National Monument maintains a passable but fairly rough one-way road across the central Grapevine Mountains from the Nevada side near the town of Rhyolite into Death Valley via Titus Canyon. This road provides means to see a wine glass or goblet valley at close range as well as an opportunity to see the rocks and structures in a representative Basin and Range mountain. The scenery is spectacular and the trip is well worth the three or four hours it requires.

The trip starts on the broad alluvial fans of the upper Amargosa Desert in Nevada, enters the Grapevine Mountains from the east where the rocks are mainly colorful volcanic deposits including lava flows, plugs, and multicolored tuffaceous sedimentary rocks. Patches of dark, pebbly Titus Canyon Formation are exposed near the summit of Red Pass. The Oligocene titanothere fauna comes from the Titus Canyon Formation.

Beyond Red Pass, the road descends steeply into the headwaters of Titus Canyon where most rocks belong to a thick sequence of marine Cambrian limestones and dolomites. The canyon gets steadily narrower and more slotlike below the remains of the old ghost town of Leadfield, and virtually all the rocks from Leadfield to the mouth of the canyon are marine Cambrian carbonates (Figure 6-25).

The last few kilometers of the canyon are in the stem of the goblet. Sheer polished rock walls rise hundreds of meters nearly vertically before widening. At places, the

FIGURE 6-25 Deeply undercut wall of a wineglass or goblet valley, lower Titus Canyon, Grapevine Range, Inyo County. The rocks are marine Cambrian limestones. (Photo by R. M. Norris)

sinuous canyon is wide enough for only a single vehicle. Folds and faults are well displayed in the Cambrian bedrock, and the canyon walls show many dramatic evidences of recent flash floods and mud flows. Visitors are cautioned to avoid the canyon if heavy rains threaten.

Moving Stones of Racetrack Playa

North of Panamint Valley are several small upland valleys lying between the Cottonwood Mountains and Saline Valley. These valleys are undrained, and each contains a small playa. One of these playas is known as The Racetrack and lies about 1,130 meters (3,700 feet) above sea level. The Racetrack is so named because of its oval shape, smooth surface, and the presence of a syenite island that resembles and thus is called The Grandstand.

The Racetrack is unusual not only for its smooth surface, but also for its peculiar "moving stones" (Figure 6-26). Many years ago, visitors noticed that curving grooves $2\frac{1}{2}$ centimeters (1 inch) or so deep and up to 30 centimeters (1 foot) wide were etched into the playa surface. When followed, the grooves often led from the shore toward the center of the playa where they ended abruptly. At the ends of the grooves were rocks that weighed as much as 45 kilograms (100 pounds). The shapes of the grooves and rocks showed that the grooves had been formed as the rocks were dragged and shoved across the surface when it was wet and soft. Although there was agreement that the rocks had indeed formed the grooves, much difference of opinion existed concerning the method.

It was suggested that, following winter rains, strong winds had blown the rocks over the slick surface when the clays were moist and plastic. No one had ever observed this, though it may occur on some occasions. Another view was that the stones were swept onto the lake by mudflows following rainstorms, but there was little or no associated debris as confirmation. Still another interpretation was that Indians had been involved, but no footprints could be found.

Perhaps the most satisfactory explanation offered thus far was proposed by George Stanley of Fresno State University. He envisioned that The Racetrack occasionally would be covered by 30 centimeters (1 foot) or so of water following heavy winter rains. This currently happens every few years. Furthermore, temperatures may drop well below freezing in the clear and windy weather that typically follows a winter storm, and the lake could easily freeze to a depth of 15 centimeters (6 inches) or more. According to Stanley, the ice along the shore of The Racetrack would freeze around some of the pebbles and boulders littering the playa margin. Several stones might be included in a slab of ice that strong winds might blow away from shore. If large enough, an ice slab could certainly support the incorporated rocks. In shallow water, the lower parts of the rocks would scribe the soft clay bottom. When the ice melted and the water evaporated, the stones would be left at the ends of the grooves they had cut. Stanley found some examples in which three or four rocks had cut parallel tracks in the clay until the supporting ice had been rotated by the wind,

FIGURE 6-26 Racetrack playa, showing moving stones and tracks. (Photo by Derek Rust)

causing the trails to make loops and turns in unison. This showed that the rocks had remained in the same position relative to one another, even though their supporting ice slab had twisted and turned.

In recent years people have driven on the playa when the surface was damp, leaving wheel tracks that may persist for several years, leading the National Park Service to close the playa to automobiles. They did so, however, by cutting a trench a meter or so deep around the western side of the playa, arguably altering the natural environment more seriously than the wheel tracks.

Moving stones, like those of The Racetrack, have been found on a few other playas in the Basin and Range province, for example Bonnie Clare playa (in Nevada) east of Scotty's Castle. It is clear that their occurrence requires conditions that seldom arise together on most playas in the Basin and Range province.

Redrock Canyon

Located near the western end of the El Paso Mountains, Redrock Canyon is a short, narrow gorge noted for rock outcrops of striking color and shape. The canyon has been the setting for many films and television productions (Figure 6-27).

Ricardo Creek and its tributaries cut the gorge of Redrock Canyon in response to uplift along the El Paso fault. Uplift may have been slow enough for Ricardo Creek to maintain its course and cut the channel, making the stream antecedent. On the other hand, the creek may have established its channel across an alluvial cover on Ricardo beds and was then able to maintain its course as uplift accelerated. In this interpretation Ricardo Creek is superimposed.

FIGURE 6-27 Miocene Ricardo Formation, Red Rock Canyon, Kern County. The Ricardo beds are chiefly soft nonmarine sandstones and siltstones, but include several lava flows and ash beds. The vertical fluting results from rainwash acting on the softer parts of the rock. Note the black lava flow caprock on the left. (Photo by R. M. Norris)

In the Redrock area, the Paleozoic and Mesozoic basement of the El Paso Mountains is overlain unconformably by Ricardo Group nonmarine sandstones and water-laid ash and tuff beds. Interbedded lava flows near the base of the Dove Springs Formation of the Ricardo Group often contain zeolites and opal. The color of the canyon's red and pink rocks results from oxidation of iron-bearing minerals and from baking of underlying rocks where lava flows spread across older rocks. Blacks and browns are from basaltic flows. Grays and whites are sandstone layers that are rich in quartz and light-colored feldspars.

Vertebrate fossils have been found in the younger layers of the canyon. Included are camel, primitive horses, mastodons, rhinoceros, antelope, saber-toothed cats, and many rodents. Some petrified wood has been found as well. In the Miocene, when these beds were deposited, savanna grasslands and moderate rainfall prevailed.

Northwest of the canyon is a broad surface that slopes gently upward to the base of the Sierra Nevada. This surface is a rock-floored plain or pediment that has been cut across the inclined Ricardo beds. The feature has long been known as the Ricardo surface, and can be seen by looking southwest from the highway as it leaves the head of Redrock Canyon.

Papoose Flat

A number of small flats or upland valleys occur in the White-Inyo mountains, some with undrained depressions and small playas. All are interesting and scenically attractive features. Papoose Flat is one of these and lies almost directly east of the Big Pine volcanic field in the Owens Valley. It provides a commanding view of the Sierran crest and the Owens Valley from its western rim. The flat lies at an elevation of about 2,600 meters (8,500 feet), more than 1,200 meters (4,000 feet) above the Owens Valley (Figure 6-28).

FIGURE 6-28 Papoose Flat, White-Inyo Range, Inyo County. Knobs rising from the valley floor are parts of the Papoose Flat pluton. Cambrian marine sedimentary rocks intruded by this pluton form the distant rim as well as the rocks in the foreground. (Photo by R. M. Norris)

Topographically, Papoose Flat is an oval, nearly flat-floored valley dotted with granitic hills or inselbergs 30 to 60 meters (100–200 feet) high and rimmed on three sides by higher peaks, the most notable of which is Waucoba Mountain on the east with an elevation of 3,393 meters (11,123 feet). Because the western side of the valley is not walled in, but is more or less open to Owens Valley 1,200 meters (4,000 feet) below, headward erosion by streams draining the steep western face of the Inyo Mountains can be expected to destroy the flat in the not too distant geological future.

As interesting as the flat is geomorphically, its underlying geology is what makes it special. Much of the flat and the mountains immediately to the east are underlain by the Papoose Flat pluton, a body of quartz monzonite about 75 to 81 million years old (late Cretaceous). Quartz monzonite plutons of this age are quite common in the Basin and Range, but this one is interesting because during its intrusion, it cut across

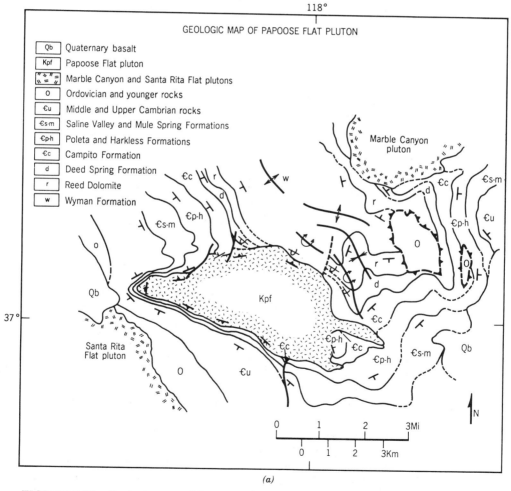

(a)

FIGURE 6-29*a* Geologic map of Papoose Flat area. Note how the outcrop thickness of the Cambrian units decreases as they wrap around the southwestern margin of the pluton. (Courtesy, C. A. Nelson, Department of Earth and Space Sciences, University of California, Los Angeles)

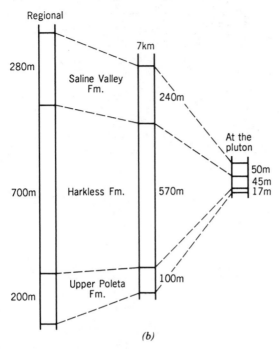

ATTENUATION OF STRATIGRAPHIC UNITS
PAPOOSE FLAT PLUTON

FIGURE 6-29*b* Comparison of measured thicknesses of three Cambrian formations in the Papoose Flat area. The regional thickness, the thickness 7 kilometers north of the pluton, and the thickness at the pluton are compared. (Courtesy, C. A. Nelson, Department of Earth and Space Sciences, University of California, Los Angeles)

the covering Cambrian sedimentary rocks and spread some of them apart, stretching a few units to an astonishing degree and forming a sort of granitic blister in Cambrian sedimentary rocks. On the western margin of the pluton, Cambrian rocks have been metamorphosed, stretched, and thinned dramatically so that they are no more than a tenth as thick as they are elsewhere in the range. Stratigraphic details that are well displayed in the unmetamorphosed Cambrian Poleta, Harkless, and Saline Valley formations of normal thickness can be faithfully traced into the stretched, thinned, and metamorphosed equivalents that wrap around the western side of the pluton, demonstrating the extreme degree to which normally brittle rocks may be deformed under metamorphic conditions, while at the same time retaining many characteristics present in their unmetamorphosed equivalents (Figure 6-29).

The metamorphosed rocks on the west margin contain a calc-silicate mineral assemblage that contains no mineral contributions from the metamorphosing pluton. On the east side, however, the calc-silicate mineral assemblage contains important contributions from the pluton. Perhaps the hotter eastern pluton margin cut through the rocks while the cooler west side merely distended the Cambrian rocks into the blister we see today.

REFERENCES

Anonymous, 1968. Geology of the Basin Ranges. Mineral Information Service (now California Geology), v. 21, pp. 131–133, 137.

———, 1978. Origin of Desert Varnish. California Geology, v. 31, p. 58.

Berkstresser, C. F., Jr., 1974. Tallest (?) Sand Dune in California. California Geology, v. 27, p. 187.

Blackwelder, E., and others, 1948. The Great Basin. Bull. Univ. Utah, v. 38, no. 20, Biol. Series, v. x.

Cole, O. N., 1975. Panamint, City of Silver. California Geology, v. 28, pp. 278–279.

Cooke, Ronald, 1965. Desert Pavement. Mineral Information Service (now California Geology), v. 18, pp. 197–200.

Cooper, J. D., and others, eds., Geology of Selected Areas in the San Bernardino Mountains, Western Mojave Desert, and Southern Great Basin, California (Guidebook). Shoshone, Calif., Death Valley Publ. Co., 202 p.

Duffield, Wendell and George I. Smith, 1978. Pleistocene River Erosion and Intercanyon Lava Flows Near Little, Lake, Inyo County, Calif., California Geology, v. 31, pp. 81–89.

Hazzard, John, 1937. Paleozoic Section in the Nopah and Resting Springs Mountains, Inyo County, Calif. Calif. Jour. Mines and Geology, v. 33, pp. 273–339.

Hill, Mary R., 1972. A Centennial . . . the Great Owens Valley Earthquake of 1872. California Geology, v. 25, pp. 51–54.

Knopf, Adolph, 1918. A Geologic Reconnaissance of the Inyo Range and the Eastern Slope of the Southern Sierra Nevada, California, U.S. Geological Survey Prof. Paper 110.

Noble, Levi F., and L. A. Wright, 1954. Geology of Central and Southern Death Valley Region, California. In Geology of Southern California. Calif. Div. Mines and Geology Bull. 170, pp. 143–160.

Norris, Robert M., 1986. A Geologic Guide to Titus Canyon, Death Valley National Monument, Inyo County. California Geology, v. 38, pp. 195–202.

Oakeshott, Gordon B., and others, 1972. One Hundred Years Later. California Geology, v. 25, pp. 55–61.

Slosson, James E., 1974. Surprise Valley Fault. California Geology, v. 27, pp. 267–270.

Stewart, J. H., 1970. Upper Precambrian and Lower Cambrian Strata in the Southern Great Basin, California and Nevada. U.S. Geological Survey Prof. Paper 620, 206 p.

Sylvester, A. G., and others, 1978. Papoose Flat Pluton: A Granitic Blister in the Inyo Mountains, California. Geol. Soc. America Bull., v. 89, pp. 1205–1219.

Troxel, Bennie W., 1963. Mineral Resources and Geologic Features of the Trona Sheet, Geological Map of California. Mineral Information Service (now California Geology), v. 16, pp. 1–7.

———, 1974. Man-made Diversion of Furnace Creek Wash, Zabriskie Point, Death Valley, California. California Geology, v. 27, pp. 219–223.

———, ed., 1974. Guidebook. Shoshone, Calif. Death Valley Publ. Co.

———, and Lauren E. Wright, eds., 1976. Geologic Features of Death Valley, California. Calif. Div. Mines and Geology Spec. Report 106, 72 p.

———, 1987. Tertiary Extensional Features. Death Valley Region, Eastern California. In Centennial Field Guide Volume 1, M. L. Hill, ed., Geological Society of America, pp. 121–132.

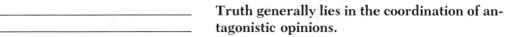

seven

MOJAVE DESERT

Truth generally lies in the coordination of antagonistic opinions.

—*Will Durant*

The Mojave and Colorado deserts are so closely related geologically that considering them separately is almost more a matter of convenience than geological unity. Conversely, sufficient difference in late Cenozoic events warrants treating each area as a discrete unit, and tends to result in a clearer overall picture.

The Mojave Desert occupies about 65,000 square kilometers (25,000 square miles) of southeastern California (Figure 7-1). It is landlocked, enclosed on the southwest by the San Andreas fault and the Transverse Ranges and on the north and northeast by the Garlock fault, the Tehachapi Mountains, and the Basin and Range. The Nevada state line and the Colorado River form the arbitrary eastern boundary, although the province actually extends into southern Nevada and western Arizona. The San Bernardino-Riverside county line is designated as the southern boundary.

The Mojave area contains Proterozoic, Paleozoic and lower Mesozoic rocks, although Triassic and Jurassic marine sediments are scarce. The marine sediments of that age may have been eroded away, or parts of the Mojave may have been an early Mesozoic upland on which no such sediments were deposited. Jurassic and Cretaceous granitic rocks of the Nevadan orogeny are widespread throughout the region's mountain blocks (Figure 7-2).

The desert itself is a Cenozoic feature, perhaps formed as early as the Oligocene, presumably from movements related to the San Andreas and Garlock faults and their predecessors. Prior to the development of the Garlock, the Mojave was part of the Basin and Range and shares Basin and Range history possibly through the first part of the Miocene.

Today the region is dominated by broad alluviated basins that are mostly aggrading surfaces receiving nonmarine continental deposits from adjacent uplands. The de-

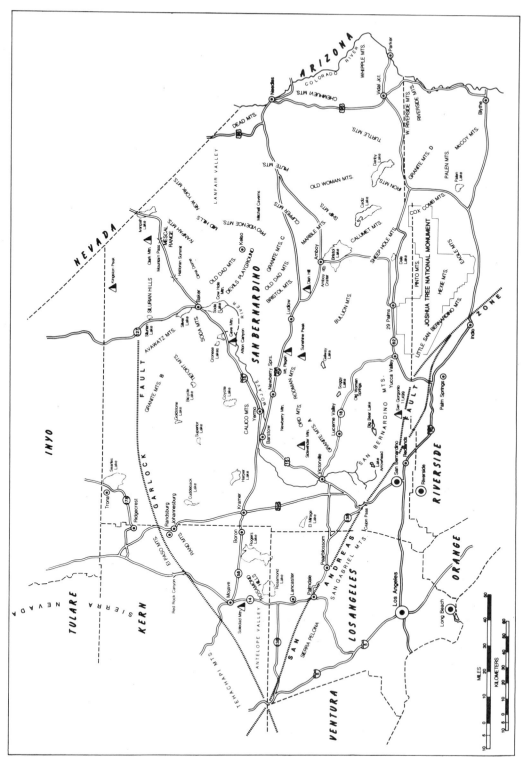

FIGURE 7-1 Place names, Mojave Desert Province.

221

Composite Geologic Columns

EASTERN MOJAVE DESERT

QUATERNARY	Alluvium, stream and lake deposits. Cima dome volcanics
TERTIARY	Fountain Peak Rhyolite
CRETACEOUS	Teutonia Quartz Monzonite
JURASSIC	Aztec (nonmarine)
TRIASSIC	Moenkopi
PERMIAN	Kaibab Limestone Bird Spring Limestone
MISSISSIPPIAN	Anchor Limestone Monte Cristo Limestone
DEVONIAN	Sultan Limestone
CAMBRIAN	Bonanza King Cadiz Chambless Limestone Latham Shale Zabriskie (Prospect Mountain) Quartzite Wood Canyon
EDIACARIAN	Stirling Johnnie
PROTEROZOIC	Noonday Fenner Gneiss & Essex Series

WESTERN MOJAVE DESERT

QUATERNARY	Black Mountain Basalt, Manix Lake Beds, Alluvium
PLIOCENE	Horned Toad, Ricardo (nonmarine), Bedrock Spring
MIO-PLIOCENE	Almond Mountain volcanics, Lane Mountain volcanics, Gem Hill volcanics, Red Mountain Andesite, Tropico Group (nonmarine), Pickhandle, Bopesta
MIOCENE	Barstow (nonmarine), Kinnick
LOWER TERTIARY	Jackhammer (nonmarine), Witnet
PALEOCENE	Goler (nonmarine), San Francisquito (marine)
CRETACEOUS	Atolia Quartz Monzonite, Rand and Pelona schists.

FIGURE 7-2 Geologic columns, Mojave Desert.

Composite Geologic Columns (*Continued*)

UPPER PALEOZOIC and LOWER MESOZOIC	Sidewinder volcanics
PERMIAN	Garlock, Fairview Valley, Hinkley Valley
ORDOVICIAN	Mesquite Schist
PROTEROZOIC	Johannesburg Gneiss, Oro Grande Metamorphics Waterman Gneiss

Source: California Division of Mines & Geology.

FIGURE 7-2 (*Continued*)

posits are burying the old topography, which was previously more mountainous. In the late Tertiary, these mountains shed debris to the Pacific, but with the northward slip of the west side of the San Andreas and accompanying elevation of coastal ranges, drainage began entering interior basins.

The highest general elevation of the Mojave Desert approaches 1,200 meters (4,000 feet), but most valleys lie between 600 and 1,200 meters (2,000–4,000 feet). A double chain of lower valleys extends northwesterly across the eastern Mojave, one from Soda Lake northwest to Death Valley including Silver and Silurian lake playas and lying at elevations slightly below 300 meters (1,000 feet). The other is a broad depression extending southeast from Bristol Lake near Amboy (180 meters or 600 feet) through Cadiz and Danby lakes to the Colorado River between Parker and Blythe. Rock-floored pediments are somewhat more extensive in the northeastern Mojave than elsewhere, and the thickness of valley fill is greatest in the Antelope Valley lying between the Garlock and San Andreas faults in the westernmost Mojave.

North of Barstow, extending to the Garlock fault, are several middle Tertiary depositional basins containing thick sections of Miocene continental rocks. Eastward, along the Mojave River valley and at progressively lower elevations, are several Quaternary depressions. Northeast from Baker, the Mojave surface is largely an erosional one, and rises rapidly to almost 1,200 meters (4,000 feet) at Halloran Summit. The most prominent erosional feature in this area is Cima Dome, an extensive granitic pediment, partly capped by Miocene to Holocene lava flows and cinder cones. Quaternary erosion on the western side of Cima Dome has removed much of the volcanic cover and lowered the land surface as much as 180 meters (600 feet).

Finally, Clark Mountain, almost 2,440 meters (8,000 feet) high, forms the northeastern corner of the province in California.

Throughout the Mojave, small hills rise above the alluvial valley fill, islandlike in seas on gravel. These are remnants of the mountainous topography that is partly erased by erosion or buried by debris. Other prominent features of today's surface are the many playas, including Rosamond, Rogers (Muroc), Mirage, Bristol, Cadiz, and Danby. Every local internal drainage basin contains at least one playa, and the linear valleys of the eastern Mojave often have several, because drainage is blocked by almost imperceptible rises between playas.

In contrast with the Basin and Range, valleys in the Mojave are proportionally broader, mountains are correspondingly more widely spaced, and generally do not stand as high above their surroundings. But, like desert mountains everywhere, the ranges of the Mojave rise abruptly from bajadas or pediments surrrounding them with few or no foothills.

Mountain ranges in the Mojave show much less consistency in orientation than those of the Basin and Range. The mountains of the eastern and northern Mojave have a rough north-south orientation, but mountains elsewhere have little consistency in this regard. The Bristol, Old Dad A and B, Marble, Avawatz, and Bullion mountains trend east-west, and the Turtle and Whipple mountains in the southeastern part of the Mojave have no discernible trend whatsoever.

GEOGRAPHY

Like most deserts, the Mojave has highly variable rainfall—from about 50 to 165 millimeters (2–6½ inches) annually. A single stream, the Mojave River, drains most of the region. This river rises in the northern San Bernardino Mountains and flows northeast about 160 kilometers (100 meters). Flow is primarily underground, except where bedrock at shallow depth forces the water to the surface. Exceptional flow occurs only during periods of unusually high rainfall. The Mojave terminates either in the two small Cronese Lakes or in the much larger Soda Lake south of Baker (Figure 7-1). Some water reaches these lakes every few years, but most of the time they are dry playas.

Most Mojave Desert mountains have elevations reaching between 1,050 and 1,500 meters (3,500–5,000 feet), but the ranges in the northeastern corner of the province are higher. One chain of mountains extending northeasterly from the Granite Mountains (one of four ranges with that name in the Mojave) through the Providence and Mid Hills to the New York mountains and Nevada's McCullough Range has a number of peaks near 2,100 meters (7,000 feet). The highest mountain group in the Mojave is the Ivanpah-Mescal-Clark range extending north and south from Mountain Pass on Interstate Highway 15. Clark Mountain at 2,418 meters (7,929 feet) is the highest peak in the Mojave Desert of California.

Some confusion is bound to exist with the four "Granite Mountains" and the two "Old Dad" mountains in the Mojave, all in separate areas. These are designated Granite mountains A, B, C, and D and Old Dad mountains A and B; their locations are given in Figure 7-1.

The Mojave province displays an unusually large variety of landforms produced by intricate erosional, depositional, and structural processes. Volcanic features like the late Cenozoic basaltic flows and cinder cones near Pisgah, Amboy, and on Cima Dome, and the plugs, domes, sheets, and dikes of older rhyolitic volcanic episodes (Rosamond Hills and Soledad Mountain) are prominent.

Fault-related landforms are present along the Garlock fault on the northern edge of the province as well as along the San Andreas fault that forms the province boundary on the southwest (Figure 7-3). A number of other smaller, but active strike-slip faults subparallel to the San Andreas lie southwest of a line between Newberry Springs and Twentynine Palms. One prominent cross-fault, the Manix, parallels the Mojave

FIGURE 7-3 View of the western Mojave Desert, looking northwest from Palmdale. The general monotony of nearly zero relief is apparent. On the left is the San Andreas fault. The Tehachapi Mountains form the skyline. (Photo by Spence Air Photos, courtesy of Department of Geography, University of California, Los Angeles)

River northeast of Barstow. Recent movement on all these faults has offset fans, lava flows, drainages, produced sags, squeeze-ups, shutter ridges, and other landforms associated with strike-slip faulting. In the ranges of the southeast Mojave such as the Dead, Whipple, Chemehuevi, and Riverside mountains, some landforms produced by detachment faulting can be seen. This interesting style of faulting will be considered more fully later in this chapter.

Despite the fact that faulting has played an important role in the history of most Mojave Desert ranges, recent activity has been slight in the northeastern part of the province with the result that tectonic landforms are fewer there than elsewhere.

The common landforms of pediment, bajada, bolson, alluvial fan, badlands, and mudflow were first defined or are particularly well demonstrated in the Mojave. There are abundant examples of such weathering features as desert varnish, exfoliation, caliche horizons, and efflorescenes. Desert pavements, sheetwash, and sheetflood forms, and many features produced by wind erosion occur repeatedly. Most of these features result from the complex interactions of climate, rock type, and geologic history.

ROCKS

Proterozoic

It will be recalled from the chapter on the Basin and Range that the term *Precambrian* is becoming less appropriate as a transitional period (Ediacarian or Vendian) becomes more generally accepted as the first period of the Paleozoic, the time when metazoan organisms had appeared, but before any of them had developed hard parts. Readers encountering references to the Precambrian of California will, in most instances, be correct in equating that time period with the Proterozoic.

The Proterozoic is well represented in the Mojave Desert. The age is established in some instances because igneous and metamorphic complexes lie unconformably below sedimentary beds carrying lowermost Cambrian or Ediacarian fossils. In other cases, radiometric dates from basement crystalline rocks are Proterozoic. Oldest Proterozoic rocks are highly schistose and gneissic and often are dynamically metamorphosed. Such rock units are exposed in the Ord Mountains (granitic gneiss and marble with intruding porphyry), the Old Woman Mountains (gneiss, marble, quartzite, schist), the Marble Mountains (granitic basement), and near Kelso (granitic gneiss). Some age assignments are still unconfirmed, and their Proterozoic designation has been made primarily on apparent position in the rock column and on degree of metamorphism, both unreliable criteria.

The younger Proterozoic, equivalent to the Pahrump group of the Basin and Ranges, is now known to occur at a number of places in the Mojave, though Mojave examples are generally strongly metamorphosed and occur as more or less isolated blocks surrounded by Mesozoic granitic rocks, in other words as roof pendants.

These rocks, being relatively unfossiliferous even where unmetamorphosed, have been recognized in the Mojave mainly by their stratigraphic position and by matching similar lithologic sequences. Examples with varying degrees of alteration occur in the Kingston, Ivanpah, Providence, Marble, Soda, Avawatz, and Silurian mountains. It is probable that additional study will demonstrate an even wider occurrence as metamorphosed roof rocks in other mountains are more thoroughly investigated.

Paleozoic

The Paleozoic system is not completely represented in the Mojave block (Figure 7-2). Partial sections occur in some of the Mojave's eastern ranges, but the main Mojave contains few Paleozoic rocks. In the Marble Mountains, early Cambrian quartzite and fossiliferous shale lie unconformably on Proterozoic granites. The province's lower Paleozoic is less than 1,500 meters (5,000 feet) thick, even when all fragments from the various exposed units are pieced together. Nowhere is there a continuous section more than 750 to 900 meters (2,500–3,000 feet) thick. Upper Paleozoic is recognized in the Ord Mountains and near Victorville (Oro Grande series, probably Mississippian and in the Soda Mountains [Permian]).

One of the thickest marine Paleozoic sections found in the Mojave is exposed in the Providence Mountains where Ediacarian, Cambrian, Devonian, Mississippian, Pennsylvanian, and Permian rocks are present, but no Ordovician or Silurian is yet recognized. The total thickness of these rocks is about 3,000 meters (10,000 feet), a maximum for the Mojave province but much less than the thickness in the Basin and Ranges to the north, where as much as 11,000 meters (36,000 feet) of equivalent

FIGURE 7-4 West face of the Providence Mountains. The light-colored, cliff-forming rocks are chiefly limestones of Devonian, Mississippian, and Permian age. The dark-colored peak on the right is Fountain Peak (1,829 meters or 5,996 feet), composed of Tertiary rhyolite. The higher peak to the left is Mount Edgar (2,187 meters or 7,171 feet), composed of the Permian Bird Spring limestone. The darker colored rocks in the lower foothills are Cambrian and Ediacarian sedimentary rocks resting on Proterozoic crystalline rocks. (Photo by R. M. Norris)

strata is present (Figure 7-4). As noted earlier, the thick Paleozoic strata in the Basin and Range province represent a miogeoclinal environment (subsiding outer shelf), but the much thinner equivalent section in the Mojave represents deposition on a relatively stable continental platform (the craton). Changes in relative sea level on the craton account for the presence of at least one nonmarine unit in the eastern Mojave—the Jurassic Aztec sandstone.

In recent years, in many ranges in the western Mojave, metamorphosed sedimentary rocks intruded by Mesozoic plutonic rocks have been correlated with the Proterozoic and Paleozoic sequences displayed in the eastern ranges. This demonstrates that the pre-Mesozoic history of much of the Mojave province had strong affinities with the history of the Basin and Range.

Mesozoic

Mesozoic bedded rocks have been recognized at a number of places in the Mojave, but the best examples occur in the eastern ranges where they have escaped severe metamorphism. Examples can be seen in the Soda, Old Dad B, Mescal Range, Providence, Cave, and Cow Hole mountains. In the Barstow area, equivalent rocks, more strongly metamorphosed, are present in the Rodman and Sidewinder ranges.

Two rock units, best known from the Colorado Plateau to the east, are representative of this group of rocks. The marine, early Triassic Moenkopi is present in the Providence Mountains and the Jurassic nonmarine Aztec sandstone has been mapped in the Old Dad B, Soda, Cow Hole, and Mescal mountains, and tentatively in the Rodman Mountains, where it is called the Fairview Valley Formation. During deposition of the Moenkopi, a vast shallow sea advanced and retreated several times across broad mud flats. The shoreline lay somewhere to the northwest of the eastern Mojave. By the early Jurassic, when the Aztec sandstone was laid down, the sea had retreated well to the northwest and extremely arid conditions affected much of southwestern United States, a condition that persisted into the Jurassic. Both the Aztec Formation and its equivalents to the east include deep-red wind-blown sands. Conditions must have been much like those that prevail in the Arabian Desert today.

Volcanic rocks near Barstow have been designated Triassic because of their stratigraphic position. Sediments carrying fossil wood no older than Cretaceous have been found in the McCoy and Palen mountains on the border of the Colorado Desert. Rocks of similar age probably occur in the New York Mountains and Mescal Range.

Prominent exposures of similar-looking schists occur in the Rand Mountains and across the Garlock fault in the El Paso Mountains as well as in the Transverse Ranges (Pelona and Orocopia schists). In all these areas, the schists have similar appearance and field relations and are widely believed to be a single lithologic unit, now dismembered by large-scale strike-slip faulting. These schists are unfossiliferous and must be dated by their field relationships. They appear to have been metamorphosed near the end of the Cretaceous or early in the Paleocene, perhaps by pressures associated with thrust faulting. Metamorphism as the result of granitic intrusion is thought unlikely because no clear-cut contacts with intrusive rocks have been found. The majority view is that these rocks originally were oceanic sediments of Cretaceous age.

The Mojave's mountain blocks and hundreds of square kilometers of pediment are floored primarily by Nevadan granitic intrusives. Many plutons with composition similar to Sierra Nevada plutons have been described. Eastern Mojave plutons are younger and more silicic (quartz monzonites to granites) than western Mojave plutons, which include the silicic types plus gabbros and quartz diorites. Radiometric dates have established a middle Jurassic to late Cretaceous age for the silicic plutons. The gabbros and diorites are older, but presumably do not reflect a separate orogenic episode predating the Nevadan. Instead they probably accompanied major intrusions identified in the Sierra Nevada.

Cenozoic

Cenozoic rocks appear throughout the Mojave Desert. Except for thin, restricted, lower Miocene marine sediments in western Antelope Valley, deposition is all nonmarine, with extensive thicknesses of Quaternary alluvium. Tuff, ash, and other volcaniclastics interbedded with lake-bed sediments and evaporites are widespread. Volcanic flows and flow breccias are common, with andesites and rhyolites in older extrusives and basaltic flows and cinder cones characterizing younger episodes. Accumulations were in localized basins. Correlation between these basins is difficult, because drainage integration between depositional areas was temporary and the evidence therefore unsubstantial. Some basins accrued 3,000 meters (10,000 feet) of deposits (practically all post-Eocene) from pre-Cretaceous basement. Dozens of these isolated Cenozoic basins once existed, and some are evident today as playas. Most have been drilled for water, potash, nitrate, other evaporite deposits, and even petroleum. A 1967 publication lists well logs for over 200 exploratory holes drilled in 15 different basins in the western Mojave alone.

Miocene rocks are particularly prominent among Cenozoic basin deposits. Some of these are of notable economic interest such as the borate deposits at Kramer, which are discussed later in this chapter.

Perhaps the best known and best displayed sequence of Miocene rocks is found north of Barstow in what is called the Barstow Basin or Barstow syncline. This is an area of about 20 by 100 kilometers (12 by 60 miles) containing more than 1,000 meters (3,300 feet) of lake, stream and fan sediments, some of which are fossiliferous and many of which are very colorful, and a few of which have yielded commercial deposits of borate and strontium minerals (Figure 7-5).

Fossils recovered from Barstow beds give considerable insight into environmental conditions that prevailed during the Miocene. Among the land animal remains that have been recovered are several kinds of horses (both grazing and browsing types), two or more kinds of camels, chalicotheres (related to the rhinos), pronghorns, peccaries, dogs, cats, rabbits, saber-toothed cats, as well as many smaller mammals such as mice and chipmunks. Several types of tortoises were present as well. Beau-

FIGURE 7-5 The nonmarine Barstow Formation of Miocene age, showing an unusually well-exposed synclinal fold. Rainbow Basin, San Bernardino County. (Photo by R. M. Norris)

tifully preserved silicified aquatic insects occur in some concretionary beds in the Barstow Formation.

These organisms as well as fossil pollen suggest an environment like northern Mexico today with summer rains, but cool enough to permit the development of semipermanent alkaline lakes.

The oldest Cenozoic unit yet described is a thin (45-meter or 115-foot maximum) mudstone, with limestone and red sandstone. It is found in limited exposure in only one locality on the pre-Tertiary basement and is all unfossiliferous and nonmarine. Sometimes conformably overlying this unit, but usually lying unconformably directly on the older basement, is a sequence of pyroclastics up to 1,130 meters (3,700 feet) thick. Included are flow breccias, tuff, and conglomerate, with some pumice, perlite and opalite. This sequence is the Pickhandle Formation and is present in all the basins studied in the western Mojave.

So far, radiometric dating has revealed no Cenozoic rocks older than early Miocene in the Mojave Desert.

FAULTING

The Mojave block is approximately bounded by the San Andreas and Garlock faults. Both are at least Cenozoic, and a few geologists think the San Andreas may have originated in the middle or early Mesozoic. The western Mojave Desert is broken by major faults that roughly parallel the San Andreas and seem to be truncated by the Garlock. Many faults undoubtedly occur in the eastern Mojave also, but since most of this area is underlain by rather uniform granitic rocks, the faults are notoriously difficult to map. Some faults are known positively, but many can only be inferred. Basin and Range types of faults have been mapped in the eastern Mojave ranges. In the Clark Mountains, thrust faults of considerable magnitude occur and reflect several episodes of movement. These faults are believed to establish pre-Cenozoic faulting related to the Cordilleran orogeny. Our interest centers on post-Mesozoic faulting, however, since these structures are revealed most clearly in today's landscape. Principal faults in the Mojave are shown in Figure 7-6.

In recent years, numerous listric and *detachment faults* have been recognized, particularly in the ranges of the eastern Mojave and in adjacent Arizona and Nevada. Listric faults are a special type of normal fault whose fault planes curve and flatten at depth. Because of this flattening at depth, it is believed that adjacent listric faults merge at depth along a detachment plane or surface (Figure 7-7). These faults seem characteristic of some crustal regions that have been subjected to tension. If, as some geologists believe, the steep normal faults so common in the Basin and Range also flatten at depth, this would indicate that the eastern Mojave ranges showing detachment faulting have been more deeply eroded than Basin and Range terrains.

It is surprising that these prominent faults were not recognized until recently, because as shown in Figure 7-8, they are obvious features unlikely to be overlooked by field geologists. They weren't overlooked, but because of their low angle of dip they were assumed to be thrust faults, an understandable conclusion until the regional fault mechanics and rock sequences were better understood.

The lesson bears repeating. Readers are cautioned to retain a healthy skepticism of all theory and hypothesis and to remember that just because an explanation is printed does not confer upon it any infallibility.

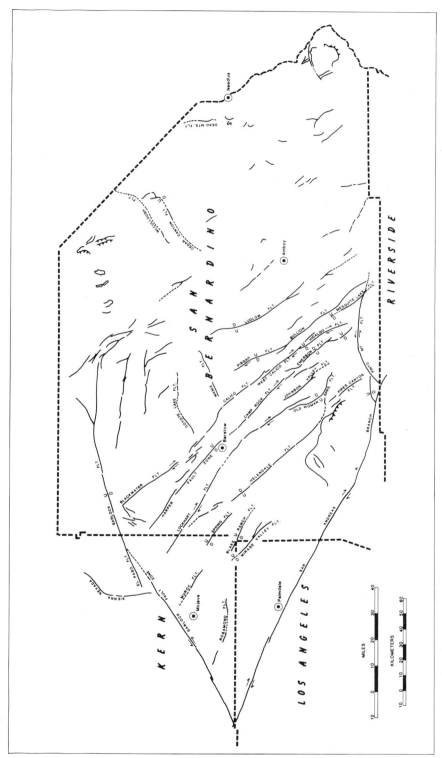

FIGURE 7-6 Principal faults of the Mojave Desert province.

231

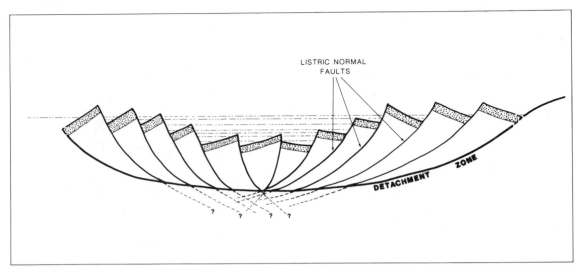

FIGURE 7-7 One possible relationship of listric normal faults to detachment zones. Structures of this sort result from crustal tension or stretching, and in California are best displayed in the eastern and southeastern desert provinces. Where relatively little erosion has affected the structure, only the steeper upper parts of the listric faults are exposed. Where deeper erosion has occurred, the flatter portions of the listric faults or even the detachment zone beneath may be exposed as in the eastern Mojave and Colorado deserts.

FIGURE 7-8 Detachment fault, Whipple Mountains, looking north toward the Chemhuevi Mountains in the distance. The dark-colored rocks above the detachment surface are Miocene volcanic rocks, which rest on thin slices of Proterozoic rock. Below the detachment are the light-colored crystalline rocks of Proterozoic and Mesozoic age. (Photo by J. C. Crowell)

Garlock Fault

The Garlock fault extends from Frazier Mountain easterly for more than 240 kilometers (150 miles). It is characterized by left-slip displacement that is variously estimated from 10 to 60 kilometers (6–40 miles). Displacement during the Quaternary appears to be no less than 18 kilometers (11 miles), and as noted earlier, Holocene slip is about 80 meters (260 feet). Thus despite the absence of any historical earthquakes clearly attributable to this fault, it must be regarded as active. The Garlock shows the characteristic features of high-angle faults with major strike-slip comments. The fault zone is narrowly confined through most of its length and is seldom more than a kilometer and a half wide (1 mile) (Figure 7-9).

East from its junction with the San Andreas, the Garlock forms two parallel segments that extend nearly 32 kilometers (20 miles) until they merge into a single structure. This single unit then extends another 32 kilometers (20 miles) to Red Rock Canyon and the El Paso Mountains. As noted in Chapter 6, near the junction of the Garlock and the southwest-curving Sierra Nevada fault, the El Paso fault begins, parallel to the Garlock and within the Garlock zone of faulting. The El Paso fault is well exposed at the mouth of Red Rock Canyon (see Figure 6-12).

The Garlock displaces the Sierra Nevada and Basin and Range faults and is therefore younger. The Garlock may have been a boundary of the Mojave block as early as Eocene, but it certainly was a boundary of major movement by late Miocene, because by then the Mojave had developed internal drainage, reversing the previous drainage pattern.

San Andreas Fault

The San Andreas fault appears to form as sharp a boundary on the southwest of the Mojave Desert as the Garlock fault does on the north. In fact, however, the San Andreas lies about $1\frac{1}{2}$ kilometers (1 mile) inside the foothill belt terminating the desert floor, and minor subparallel faults seem to be the usual boundary between the Mojave and Transverse Ranges provinces. The San Andreas fault is considered more fully in Chapter 13, but it is important to note here that the Garlock appears possibly to have been offset by the San Andreas where the two intersect at the western end of the Mojave block.

Internal Faults

The internal faults of the southwestern Mojave block show remarkable parallelism with the strike of the San Andreas fault and equally impressive divergence from the strike of the Garlock, but none of these faults quite reaches the Garlock, so their relationship to that important fault is not yet clear. This group of faults is regarded as a part of the greater San Andreas system and all are active, right-slip faults. Both the Galway Lake and Homestead faults have generated earthquakes and produced ground breakage in 1975 and 1979, respectively. The internal faults are more numerous than is indicated on the generalized map of Figure 7-6. Furthermore, it is likely that many individually named faults are actually parts of single fault trends that are continuous in the basement beneath the deep alluvium.

Important vertical displacement occurs on the Muroc, Blake Ranch, and Mirage

FIGURE 7-9 A small sag-pond on the Garlock fault near Garlock, Kern County. This left-slip fault steps left just behind the viewer, resulting in a small pull-apart basin or sag. View northeast. (Photo by R. M. Norris)

Valley faults in the western Mojave and on the Johnson Valley, Pipes Canyon, and Spring faults in the central Mojave. Right-slip displacement of uncertain amount is inferred for the Cottonwood, Lockhart, and Blackwater faults in the west and for the Lenwood, Hidalgo, West Calico, and Pisgah faults in central and south-central Mojave. The extensive work of T. W. Dibblee has permitted the determination of strike-slip movement on the numerous internal faults. Strike-slip seems to follow vertical displacement in almost every case.

Vertical displacements are continuing, as shown by radiometrically dated Pleistocene basaltic flows that have been vertically displaced as much as 100 meters. Thick alluvial deposits are cut by some of the internal faults to form present-day scarps, further demonstrating continuing vertical movement. Internal faulting also is reflected by uplifted bedrock surfaces of low relief. Such a surface is the Ivanpah upland of the eastern Mojave, which might well be extended to the less continuous but nonetheless impressive surface that bevels the pre-Cenozoic rocks of the western Mojave.

East-west trending faults are present in the northeastern Mojave more or less parallel to the Garlock fault. These faults appear to be high-angle left-slip faults like the Garlock. The Manix fault extends eastward from near Yermo about 40 kilometers (25 miles) to Afton Canyon on the Mojave River. The canyon, in part, has developed along this structure. This fault is believed to have generated an earthquake in 1947, and geophysical evidence suggests a total of about 3.5 kilometers (2 miles) of left slip.

Several prominent east-west faults mark the boundary between the Mojave and eastern Transverse Ranges. One of these is the Pinto Mountain fault, which joins the north branch of the San Andreas (Mission Creek fault) southeast of San Gorgonio Peak. This fault extends eastward through Morongo Valley to Joshua Tree and

Twentynine Palms, where it terminates against the Mesquite fault, one of the group of northwest-trending strike slip faults already mentioned. The Pinto Mountain fault shows a maximum of about 18 kilometers (11 miles) left slip in its central part.

GEOLOGIC HISTORY

The Mojave block in some respects shares the pre-Cenozoic geologic history with the Basin and Range and Sierra Nevada provinces, but in the eastern Mojave there is a major unconformity between the Cambrian and Devonian; in the Basin and Range this interval was marked by more or less continuous deposition. Degradation after the Nevadan orogeny reduced the Mojave and the Sierra to low relief by the close of the Mesozoic. Subsequently uplift was initiated, and by the opening of the Miocene the Mojave block had been lifted between 3,000 and 4,500 meters (10–15,000 feet). This estimate is deduced from the amount of detritus shed west and south to form lower Tertiary units in the Transverse and southern Coast ranges. By early Miocene or late Oligocene, the Ivanpah erosional surface existed, rising gently eastward from the Garlock-San Andreas convergence. Depression then began, probably because of faulting, transforming the Mojave block into an area of internal drainage. Thick Miocene, Pliocene, and Pleistocene nonmarine sections accumulated in local basins, a condition that persists today. Pleistocene lake sequences developed extensively during periods of cool temperatures and low evaporation.

Volcanic activity in the Mojave began as early as late Oligocene and reached a climax during the Miocene, tapering off in the Pliocene and Quaternary. Most of the volcanics are pyroclastic rocks, but flows of rhyolitic and basaltic composition are widespread; some of these are considered in more detail later in this chapter.

SUBORDINATE FEATURES

Cima Dome and the Ivanpah Upland

The Ivanpah Upland slopes from eastern elevations of 1,370 to 1,500 meters (4,500–5,000 feet) west to the central Mojave at 450 to 600 meters (1,500–2,000 feet). Its topography is irregular (probably the result of old mountain block roots and resistant Miocene to Holocene volcanic covers), but the surface is widely underlain by uniformly grained quartz monzonite of Nevadan age. The basement has low relief wherever younger rocks have preserved it, and apparently much of the Mojave block was planed before uplift to its present position.

Following this most recent uplift, considerable erosion has taken place, lowering the surface below the old Ivanpah upland as much as 200 meters (600 feet) in places. Lava flows and cinder cones have protected the underlying granitic rocks and their old soils from erosion and now form lava-capped mesas and volcanic hills perched atop granitic pedestals (Figure 7-10). As a rule, the younger the volcanic feature, the less the difference in elevation between the modern desert surface and the base of the volcanic rock. The maximum relief, about 200 meters (600 feet), occurs below lava flows dated at about 8 million years, which rest on remnants of the Ivanpah upland.

One tiny peak, Squaw Tit, in the northern part of the Cima volcanic field, has a lava cap about 10 meters (30 feet) across and about 3 meters (10 feet) thick resting

FIGURE 7-10 Cima Dome volcanic field, San Bernardino County, showing Pleistocene cinder cones and associated flows about 630,000 years old. The thickness of the flow is shown by the low cliff at its edge. The difference between the bottom of the flow and the light-colored pediment is a measure of the amount of stripping or downwasting that has occurred in the last 630,000 years. The flow originally occupied a shallow valley, so erosion has reversed the topography. Compare with Figure 3-19. (Photo by John Dohrenwend, U.S. Geological Survey)

on a granite pedestal (Figure 7-11). This peak illustrates very clearly the important protective effect of lava caps on remnants of the Ivanpah upland.

On the eastern margin of the Ivanpah upland is Cima Dome, an upland segment of about 260 square kilometers (100 square miles). Cima Dome has received considerable attention because it has been studied in attempts to understand the mechanisms of desert erosion, especially slope retreat of block-faulted rock units under desert regimens. In 1915, A. C. Lawson studied slope retreat and concluded that as erosion progressed, a fault scarp would retreat by being degraded essentially parallel to the original dip of the scarp slope, with gradual shift of the scarp declivity. Thus rock floors (pediments) formed at the base of the scarp, progressively widening as the erosional cycle advanced and panfans developed. Cima Dome was thought to be a prime example of the panfan concept. It was later found, however, that the dome is wholly bedrock and not covered with alluvial material as Lawson had supposed (Figure 7-12).

In 1933, William Morris Davis tried to explain Cima Dome as a special case of mass wasting of a desert fault block, unique because the rock was of uniform texture and composition, producing a convex surface. This concept too, was not supported by the field evidence.

R. P. Sharp has demonstrated that the fundamental premises of both Lawson's and Davis's ideas do not apply to Cima Dome. Instead, like adjacent domes, Cima appears to result from slight upwarping of the Ivanpah upland during or after a thin cover of Miocene and younger volcanics buried the upland. Subsequent stripping of the cover exposed the domal surface of the granite. The stripped material appears

FIGURE 7-11 Squaw Tit, an unusually symmetrical hill of Mesozoic granitic rock capped by a very small remnant of a Miocene·(?) lava flow. A small volcanic neck can be seen on the extreme right. Northwestern Cima Dome, San Bernardino County. (Photo by R. M. Norris)

to have consisted chiefly of weathered soils developed on the land surface before the late Miocene volcanic activity was initiated. Stripping has not only removed most of these old soils, but perhaps any later ones as well, exposing the bedrock of the pediment that makes up Cima Dome.

FIGURE 7-12 Cima Dome in profile, viewed from the north near Valley Wells, San Bernardino County. The knob on the left is Teutonia peak (1,572 meters or 5,155 feet). The dome is largely a pediment or bedrock surface developed on the Mesozoic Teutonia quartz monzonite with only a thin, discontinuous mantle of soil. (Photo by R. M. Norris)

FIGURE 7-13 The youngest lava flow in the Cima Dome volcanic field. This flow is approximately 13,000 years old. The flow rests on an essentially modern surface, which shows negligible downwasting since the flow was formed. (Photo by R. M. Norris)

Cima Volcanic Field

Marginal to Cima Dome and concentrated on the west and southwest are about 40 volcanic cinder cones with associated basaltic flows up to 90 meters (300 feet) thick. These late Tertiary and Quaternary flows overlie much of the southwestern edge of Cima Dome and the surrounding alluvial fans, and once may have covered as many as 650 square kilometers (250 square miles). The younger volcanic cones to the south, on the higher levels of Cima Dome, are built on basaltic flows. W. S. Wise has noted compositional variations in the flows and cones that suggest the origin of their magmas and associated magmatic contaminations. Commercial quarries produce such items as cinder block and road metal from the cinder cones.

The older northern flows are at least 7.6 million years old, but one flow at the southwestern corner of the field, easily visible from the county road, is about 13,000 years old (Figure 7-13).

Volcanic Cones: Barstow-Amboy Axis

From Barstow to Amboy, there is a conspicuous line of volcanic cones including Malpais Crater, Pisgah Crater and lava field, Dish Hill (also known as Siberia Crater), Sunshine Peak, and the Amboy cone and lava field. Pisgah and Amboy are the best known, because of their prominence as desert landmarks. Their degree of dissection suggests that they are the youngest members of the chain.

Malpais Crater is about 24 kilometers (15 miles) southeast of Barstow. It is built on a base of older rocks and emerges from the flank of Newberry Mountain. The crater is a large volcano with extensive lava flows extending toward the valley through which Interstate Highway 40 passes.

About 8 kilometers (5 miles) southeast of Malpais Crater is Pisgah Crater, an extremely young cinder cone. Prior to the crater's eruption, basaltic lava emerged from fissures and flowed south and northwest in thin sheets extending up to 9.6 kilometers (6 miles) from the vent source. The Pisgah flows were highly fluid with typically vesicular tops. This suggests a relatively high gas content that probably accounts for the viscosity. The vesicles have no significant filling, indicating a lack of mineralizing solutions in the magma. The flows cover thick lake-bed clays. The montmorillonite clay called *hectorite* is mined as are the cinders in the volcanic cone. Superimposed on the Pisgah flows is the cone itself, which rises about 90 meters (300 feet) above the volcanic platform. The cone is almost circular, with a distinct crater, and is composed of red and black cinders, some lava, and clinkery breccias. Because it is being actively quarried, its appearance has been greatly changed.

Dish Hill (Siberia Crater) is the largest of a group of flows and cones between Ludlow and Amboy. Although not so spectacular as Pisgah and Amboy, the area has long been famous for volcanic bombs that have olivine or granite cores. Mineral collections throughout the country contain olivine bombs from Dish Hill, although good specimens are increasingly more difficult to find.

Easternmost of the Barstow-Amboy chain is Amboy Crater (Figure 7-14). This almost circular hill is probably less than 2,000 years old and forms a prominent landmark on Bristol playa. The crater is breached on its western side and rises about 90 meters (300 feet) above the playa on basaltic flows from 3 to 10 meters (10–30 feet) thick. These flows cover about 13 square kilometers (5 square miles) of the playa's northern side. The variety of volcanic features—lava tubes and tunnels,

FIGURE 7-14 Amboy Crater, from the east. Sand dunes are forming because the prevailing northwesterly winds are slowed by the cone's elevation. A small playa is in the center of the crater. (Photo by Spence Air Photos, courtesy of Department of Geography, University of California, Los Angeles)

blisters, collapsed domes, schollendomes, and ropy lavas—in Amboy and Pisgah volcanic fields is as great as any volcanic field south of the Cascade Range.

Mojave River

The Mojave River, the only major stream crossing the Mojave block, is intermittent through most of its course from its head in the San Bernardino Mountains to its present terminus in Soda Lake (Figure 7-15). In earlier postglacial time, the river continued north and joined the Amargosa River flowing into Death Valley. The unusual variations in the river's channel pattern are at least partially due to the complex local history of segments of the Mojave block.

The Mojave River has three widely separated areas of constriction where surface flow usually occurs. The Victorville water gap, on which the river is superimposed,

FIGURE 7-15 Mojave River during the great flood of March 1938. The large lake in the upper left is Silver Lake, normally a dry playa. The lake in the lower left is Soda Lake, most of which is usually dry also. (Photo by Spence Air Photos, courtesy of Department of Geography, University of California, Los Angeles)

FIGURE 7-16 Mojave River water gap at Victorville (center), during unusual flood conditions. (Photo by Spence Air Photos, courtesy of Department of Geography, University of California, Los Angeles)

is a mass of bedrock formerly buried by alluvium (Figure 7-16). The bedrock is now being exhumed because of local shift in base level. The volcanic barrier at Barstow has deflected and impeded flow as the more durable volcanics along the river channel are exhumed, but in this area surface flow is rare. At the Afton Canyon water gap, poorly consolidated bedded sediments have been cut through to crystalline rock. This segment of the channel is thus superimposed. Alternatively, the Afton Canyon water gap may have formed by downcutting of the outlet of glacial Lake Manix, probably along the Manix fault.

Since the Mojave province shifted from external to internal drainage 15 or 20 million years ago, the river has responded to a complex geomorphic history involving structural activity and repeated alternation of regional erosion and deposition.

Like most desert streams, the Mojave River is characterized by wide swings in water volume. Its average runoff is 101 million cubic meters (82,300 acre feet), but it has varied from as few as 5.3 million cubic meters (4,340 acre feet) in 1951 to as many as 425 million cubic meters (345,000 acre feet) in 1922, a volume greater than observed in any other southern California stream apart from the San Gabriel River.

Pleistocene Lake Manix

The modern Mojave River crosses the floor of the basin occupied by Lake Manix during glacial times. Covering about 500 square kilometers (200 square miles), the lake was a major feature of the drainage connecting Manix, Cronese, Soda, and Silver lakes. These lakes may once have been connected with Bristol, Cadiz, and Danby playas, subsequently draining into the Colorado River. The basin of Lake Manix has special relevance because there is some indication that prehistoric man lived on its shores. In addition, early Pleistocene vertebrate fossils (horses, jackrabbits, camels, deer, pronghorns, and tapirs) have been recovered. These findings establish climatic characteristics of the Mojave during part of the Pleistocene.

The lake beds may be easily seen south of Interstate Highway 15, beginning a few kilometers northeast of Yermo and extending as far as Afton Canyon.

Kelso Dunes and the Devils Playground

The Kelso dunes are the highest and most prominent portion of the Mojave's largest dune field, an area known by the forbidding name of the Devil's Playground. This dune field extends about 56 kilometers (35 miles) east from the lower end of Afton Canyon, where the Mojave River enters the Soda Lake basin. Though some sand may have been added to the Devil's Playground from old beaches around Soda Lake, most of it was derived from material carried by the Mojave River. Apart from the Kelso dunes proper, the Devil's Playground does not contain particularly distinctive dunes. The area consists primarily of low dune groups and featureless sandy areas anchored by vegetation, although a few good examples of seif dunes and barchans are included (Figure 7-17).

The Kelso dunes are approximately in the center of a broad alluviated valley that imposes no obvious barriers to further eastward movement of sand. Although the valley floor slopes upward east of the dunes, in traveling from the mouth of Afton Canyon to the Kelso area the sand already moved uphill at least 450 meters (1,500 feet), so a further climb would not seem to present a problem. Studies on sand

FIGURE 7-17 Symmetrical barchan dune at the Devil's Playground, San Bernardino County. Dunes of this kind advance in the direction the horns point, in this case to the left. (Photo by R. M. Norris)

FIGURE 7-18 Highest ridge, Kelso dunes. The west face of the Providence Mountains is in the distance. (Photo by R. M. Norris)

behavior in the Kelso dunes show that the sand moves to and fro rather than in a consistent direction. Apparently present conditions differ from those prevailing when the dunes accumulated. (This may have occurred as recently as a few thousand years to as many as 20,000 years ago.) Consequently, the Kelso dunes and much of the Devil's Playground are probably relics of past conditions now greatly altered. Furthermore, the dune sand has a light tan color rather than the white or pale gray that characterizes most river sand. This also suggests an older age for the dunes, because older sands often become brownish or even red as iron-bearing minerals stain surrounding grains with red and brown iron oxides. The four approximately parallel ridges of the dunes show little relation to existing wind patterns and may be another indication of origin under previous conditions. The southern ridge is the highest, rising nearly 168 meters (550 feet) above the desert floor, and is nearly 6.4 kilometers (4 miles) long (Figure 7-18).

Two stream channels cross the eastern portion of the Kelso dunes. The larger is Cottonwood Wash, which lies about 30 meters (100 feet) below the sand bordering the channel. Neither Cottonwood Wash nor Winston Wash to the east reveals any bedrock under the higher parts of the dunes. The channels are cut primarily in alluvial deposits, so it is unlikely that the height of the dunes results from a buried bedrock core.

The Kelso dunes have several unusual features. Among the California dunes they are the best known barking dunes. When certain dune sand is disturbed by the activity of people sliding down slip faces or rapidly shuffling their feet in the sand, low-pitched barking noises can be heard. There are also reports of sudden, apparently spontaneous initiations of low booming noises from the dunes, sometimes as loud as the noise from passing aircraft. The noise appears to result from sand shear or layers of grains moving over one another. In order to produce the best sounds, the sand must be well-rounded, dry, well-sorted, with frosted surfaces—mostly quartz—and on steep slopes. Other California dunes known to bark include the Eureka Valley, Panamint Valley, and Olancha dune groups.

Most Kelso dune sand is composed of quartz and feldspar, like dune sand in many places. The Kelso dunes also contain an appreciable quantity of dark heavy minerals often concentrated by the wind into streaks and patches. The dark minerals are mostly magnetite, probably derived from the iron ores of Cave Mountain in Afton Canyon. Minor amounts of such minerals as zircon, ilmenite, monazite, rutile, garnet, and cassiterite are also present. In recent years, the presence of these heavy minerals has prompted efforts to mine the sand, but as yet no successful production has occurred.

SPECIAL INTEREST FEATURES

Providence Mountains State Recreation Area (formerly Mitchell's Caverns)

Addition of Mitchell's Caverns to the California park system has assured preservation of a long-known but obscure group of caverns in the Providence Mountains. Mitchell's Caverns are limestone solution features in the thick Permian Bird Spring Formation. The caverns are quite dry and small, but they are varied and show most features seen in larger caves. The caves show quite clearly that the climate of the area has changed markedly since they were formed. The last significant development of dripstone in the caves probably occurred during the wetter episodes of the Pleistocene, the so-called pluvial stages (see Figure 1-29).

Trilobites of Marble Mountains

The Marble Mountains east of Amboy include a famous collecting locality, the lower Cambrian Latham shale of red, green, and gray fissile mudstones. The unit is not confined to the Marble Mountains, but other exposures are less accessible. In the sourthern Marble Mountains, the Latham shale is 15 to 24 meters (50–75 feet) thick and is underlain by the Cambrian Prospect Mountain Quartzite. The shale is highly fossiliferous, with brachiopods and trilobites the principal forms. A prominent trilobite at the Marble Mountains locality is *Fremontia fremonti*. Specimens with bodies up to 20 centimeters (8 inches) long have been collected, though these are quite rare.

An unusual feature of the Marble Mountains locality is that the underlying Prospect Mountain Quartzite is exceedingly thin, from 10 to 15 meters (30–50 feet) thick. Elsewhere, this quartzite is up to 350 meters (1,100 feet) thick. Consequently, the unconformity between the underlying Proterozoic granites and the Cambrian is quite distinct. The quartzite layers contain small, whitish quartz pebbles, particularly in the lower few meters. This lowermost basal conglomerate grades upward into crossbedded sandstone (quartzite) and then into the Latham trilobite-bearing shale.

Vulcan Mine

Iron ore reserves of commercial grade have been known in California since at least 1914. The largest deposits are in the Eagle Mountains of the Colorado Desert, but significant deposits occur in the middle Cambrian Bonanza King limestone of the southern Providence Mountains. Magnetite and hematite were processed from these ores at the Vulcan mine from 1942 to 1947, when the property was abandoned. Besides piles of waste and low-grade ore, the mine has an enormous glory hole— an open pit several hundred meters deep where visitors can see the relations of the ore body to the limestone country rock. Today the pit is partially filled with water, but many minerals are present in the dumps, including magnetite, hematite, pyrite, chalcopyrite, calcite, dolomite, serpentine, limonite, and epidote. The associated intrusive rock from which the mineralizing solutions probably came is a dark green syenite. Syenite is a quartz-poor granitic rock, often with lots of pink orthoclase and dark minerals like augite and hornblende; it is a somewhat uncommon plutonic rock.

Bristol Playa

Near the town of Amboy is Bristol Lake, a large playa partially covered by flows from the Amboy volcanic center. Just below its surface, the playa has a saline water body from which evaporite minerals are obtained. One of the mining operations consists of trenching the playa as much as 6 meters (20 feet) deep, stacking the playa clays in rows along the trenches, and allowing the trenches to fill with the brines from below. Evaporation precipitates the salts like table salt and calcium chloride, which can be harvested. Operations similar to those at Bristol Lake are carried on sporadically in other Mojave playas such as Koehn and Danby.

Mountain Pass Rare Earth Deposit

Mountain Pass is the highest point on Interstate Highway 15 between Barstow and Las Vegas, where the road crosses the southern part of Clark Mountain. The area possesses one of the most unusual mines in North America. It annually produces about 25 percent of the rare earths in the world and is the largest single known deposit of such minerals. The deposit was first discovered in 1949.

The ore mineral bastnaesite is made up of seven rare earth carbonates in order of elemental abundance: cerium, lanthanum, neodymium, praseodymium, samarium, gadolinium, and europium. These occur in an unusual intrusive igneous rock called a carbonatite, composed largely of carbonate minerals. At Mountain Pass, the carbonatite body has the form of a sill intruded into Proterozoic gneiss. The ore body is about 60 meters (200 feet) thick and about 700 meters (2,300 feet) long, dipping about 50 degrees into the gneiss. The rare earth district covers an area of about 7 by 11 kilometers (3 by 7 miles).

Interestingly, the area is an old gold-producing property in which the rare earth minerals were overlooked until prospectors with geiger counters began to check out rumors of a uranium deposit there. The gold property was first worked in 1936.

Randsburg Mining District

The Rand Mountains adjoin the Garlock fault immediately south and east of the El Paso Mountains. Their rocks include several Proterozoic formations, with some Paleozoic, Mesozoic, and Tertiary intrusive rocks and surficial volcanics. This complex has yielded gold, silver, and tungsten, each producing a mining boom in the district.

Gold

The Yellow Aster. Several famous gold mines produced the Rand's first boom. Towns that sprang up were Randsburg and Johannesburg, obviously reflecting the hope of rivaling the great South African mining areas. About the only common factor, however, is the great age of some of the rock units, which in South Africa contain gold but in the Randsburg district generally do not. The Rand ore was in mineralized veins, presumably derived from ore fluids generated by granitic bodies that invaded the Rand schist. Gold occurs in the granites and along the contacts with the schists.

Gold production from the Yellow Aster is estimated to total about 580,000 troy ounces. The Yellow Aster's glory hole, about 100 meters deep, dominates the landscape of hillside Randsburg and was mined by open-pit methods. Major production in the Yellow Aster was between 1917 and 1918. Some placer mining was undertaken in the 1930 when squatters eked out a bare livelihood by mining gravels of the alluvial fans derived from the outcrops of the gold areas. No significant production occurred, but the high price of gold has made the Yellow Aster an active mine in the late 1980s.

Silver

The Kelly Rand. A chance discovery of high-grade silver ore in the eastern Rand district in 1919 led to the development of the Kelly Rand silver mine. Production was from rich ores that occurred in irregular veins and blocks, permitting stoping of large bodies of rock from 80 to 90 meters (250–300 feet) below ground. Large underground "caverns" (stopes) were created, sometimes resulting in surface cave-ins. The primary mineral was miargyrite, an unusual sulfantimonide of silver related to the more common pyrargyrite. Associated minerals were a few sulfides such as stibnite and marcasite. The ore apparently was formed from magmatic waters accompanying shallow rhyolitic intrusives. Though figures vary, when production ceased in 1928, ore valued at about $13,580,000 had been processed. Gold ores were discovered at the 450-meter (1,500-foot) level in the Kelly mine, but production was only a by-product of silver mining.

Tungsten

The Atolia Mine. Scheelite, a principal ore of tungsten, was recognized in 1903 in a Nevadan granitic pluton on the southwestern slopes of the Rand Mountains. The demand for tungsten for steel manufacture during World War I promoted both lode and placer production. Production subsequently languished but was revived during World War II. Small production by leaseholders has been almost continuous, but except for the peak periods, annual production has been minimal. The placer ground is characterized by rows of conically shaped mounds that resemble the tips or waste piles of the tin mines in Cornwall, England. The Atolia tips are the residues from placer mining in the gravels derived from the bedrock carrying the scheelite veins. Scheelite has a specific gravity of 6 and therefore is readily separated from associated minerals by either wet or dry placer mining. Scarcity of water in the Rand district necessitated only dry placer methods.

FIGURE 7-19 Open pit mine at Kramer, near Boron. (Photo by Spence Air Photos, courtesy of Department of Geography, University of California, Los Angeles)

Kramer Borate Deposits

One of the more important Cenozoic basin sequences in the Mojave Desert is the early middle Miocene Kramer deposit, now one of the world's larger sources of borate minerals. It lies about 56 kilometers (35 miles) east of Mojave and 1½ kilometers (1 mile) north of the hamlet of Boron (Figure 7-19). The borates evidently were deposited in a permanent but saline lake fed in part by borate-bearing thermal springs associated with volcanic activity. The borate-bearing lake beds are underlain by basaltic lavas.

The deposit is buried beneath younger gravels and was discovered as a result of a homesteader's water well drilled in 1913. Production of the borates did not begin until 1926, however. Many new and several unusual borate minerals have been described from this deposit. The major ore body is composed of borax and kernite, a hydrous borate of soda with 7 water molecules instead of the 10 of borax. Until 1926, California's commercial production of borax had been from the Death Valley colemanite deposits. Colemanite, a hydrous borate of lime, requires processing in which the calcium is replaced by sodium to make borax; however, this is a relatively expensive process compared with that required to convert kernite to borax. By 1928, borax mining operations were concentrated in the Kramer district, where the operation continues today. The present mine is open pit, with a huge annual production of raw ore from ancient lake sediments.

Calico Mountains

The restored ghost town of Calico in the Calico Mountains was one of southern California's early silver camps. Silver was discovered in 1881, and production of high-grade ore continued until 1896, with negligible activity until recently. A large-scale secondary recovery operation is now in the final planning stages and new silver production is expected whenever the price of silver rises sufficiently. The productive ores (argentite and cerargyrite) come from mineralized sandstones and siltstones of the Barstow Formation. Ore-bearing solutions evidently were provided during Pliocene time when shallow andesite and dacite porphyries were emplaced. The primary mining areas occurred in a mineralized belt extending about 8 kilometers (5 miles) along the south face of the range. Approximately 20 million troy ounces of silver have been recovered.

Prior to 1914, colemanite was mined on the northeastern side of the Calico Mountains from folded lake beds, and today's visitors are impressed with the enormous waste piles that hug the slopes. The colemanite was found in narrow seams, veins, and geodes in the clay shale of the Barstow Formation and constituted only a small percentage of the volume of material excavated. Hand sorting of much of the ore was required.

REFERENCES

General

Bassett, A. M., and D. H. Kupfer, 1964. A Geological Reconnaissance of the Southwestern Mojave Desert, California. Calif. Div. Mines and Geology, Spec. Rept. 83, 43 p.

Dibblee, Thomas W., Jr., 1967. Areal Geology of the Western Mojave Desert, California. U.S. Geological Survey Prof. Paper 522, 152 p.

_____ and D. F. Hewett, 1970. Geology of the Mojave Desert. Mineral Information Service (now California Geology), v. 23, pp. 180–185.

Geologic Map Sheets of California. Kingman, 1961, Trona, 1963, Needles, 1964, Bakersfield, 1965, Los Angeles, 1969, and San Bernardino, 1969. California Div. Mines and Geology.

Hewett, D. F., 1954a. General Geology of the Mojave Desert Region, California. In Geology of Southern California. Calif. Div. Mines and Geology Bull. 170, pp. 5–20.

_____, 1954b. Fault Map of the Mojave Desert Region. In Geology of Southern California. Calif. Div. Mines and Geology Bull. 170, pp. 15–18.

Special

Beeby, David J., and Robert L. Hill, 1975. Galway Lake Fault. Calif. Geology, v. 28, pp. 219–221.

Clark, William B., 1985. Gold in the California Desert. Calif. Geology, v. 38, pp. 178–185.

Evans, James R., 1966. California's Mountain Pass Mine now producing Europium Oxide. Mineral Information Service (now California Geology), v. 19, pp. 23–32.

_____, 1974. Relationship of Mineralization to Major Structural Features in the Mountain Pass Area, San Bernardino County, California. California Geology, v. 27, pp. 147–157.

Gale, Hoyt S., 1951. Geology of the Saline Deposits, Bristol Dry Lake, San Bernardino County, California. Calif. Div. Mines and Geology Spec. Rpt. 13.

Gardner, Dion L., 1954. Gold and Silver Mining Districts in the Mojave Desert Region of Southern California. *In* Geology of Southern California. Calif. Div. Mines and Geology Bull. 170, pp. 51–58.

Harthrong, Deborah S., 1983. Renewed Mining Activity in the Calico Mountains. Calif. Geology, v. 36. pp. 216–225.

Hulin, C. D., 1925. Geology and Ore Deposits of the Randsburg Quadrangle, California. Calif. Div. Mines and Geology Bull. 95, 152 p.

Lamey, C. A., 1949. Vulcan Iron-ore Deposit, San Bernardino County, California. Calif. Div. Mines and Geology Bull. 129, pp. 87–95.

MacDonald, Angus A., 1970. The Northern Mojave Desert's Little Sahara. Mineral Information Service (now California Geology), v. 23, pp. 3–6.

Merriam, Charles W., 1954. Rocks of Paleozoic Age in Southern California. *In* Geology of Southern California. Calif. Div. Mines and Geology Bull. 170, pp. 9–14.

Morgan, Vincent, and Richard C. Erd, 1969. Minerals of the Kramer Borate District, California. California Geology, Part 1, v. 22, pp. 143–153; Part 2, v. 22, pp. 165–172.

Olson, J. C., and L. C. Pray, 1954. The Mountain Pass Rare-earth Deposits. *In* Geology of Southern California. Calif. Div. Mines and Geology Bull. 170, pp. 23–29.

Parker, Ronald B., 1963. Recent Volcanism at Amboy Crater, California. Calif. Div. Mines and Geology Spec. Rept. 76.

Schaller, Waldemar T., 1929. Borate Minerals of the Kramer District, Mojave Desert, California. *In* Shorter Contributions to General Geology. U.S. Geological Survey Prof. Paper 158-I, pp. 137–170.

Seiple, Eric, 1983. Miocene Insects and Arthopods in California. Calif. Geology, v. 36, pp. 246–248.

Sharp, Robert P., 1954. The Nature of Cima Dome. *In* Geology of Southern California. Calif. Div. Mines and Geology Bull. 170, pp. 49–52.

————, 1966. Kelso Dunes, Mojave Desert, California. Geol. Soc. Am. Bull., v. 77, pp. 1045–1074.

Thompson, D. G., 1929. The Mohave Desert Region, California: A Geographic, Geologic and Hydrologic Reconnaissance. U.S. Geol. Surv. Water Suppl. Paper 570, 759 p.

Trexler, Dennis T., and W. N. Melhorn, 1986. Singing and Booming Sand Dunes of California and Nevada. Calif. Geology, v. 39, pp. 147–152.

Weber, F. Harold, Jr., 1966 and 1967. Silver Mining in Old Calico. Mineral Information Service (now California Geology), v. 19, pp. 71–80; v. 20, pp. 3–8.

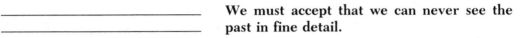

eight

COLORADO DESERT

We must accept that we can never see the past in fine detail.

—Konrad Krauskopf

As indicated in Chapter 7, there is a certain arbitrariness in distinguishing between the Colorado and Mojave deserts. Nevertheless, the Colorado corresponds approximately to the "low desert" and the Mojave to the "high desert" of southern California weather reports. Although this distinction is based chiefly on general altitude, it also reflects climate and indigenous vegetation. Figure 8-1 shows the locations of some of the important features of the Colorado Desert.

For purposes of this book, the Colorado Desert province is bounded on the east by the Colorado River, on the south by the Mexican border, and on the west by the Peninsular Ranges. The northern border lies along the southern edge of the eastern Transverse Ranges, approximating the San Bernardino-Riverside county line. In the northeastern corner of the province is the Coachella Valley, tucked in between the highest parts of the Peninsular and Transverse ranges at San Gorgonio Pass. We have assigned the Orocopia Mountains to the Colorado Desert province, even though some geologists consider them the eastern unit of the Transverse Ranges. The ranges of eastern Riverside County are included in the Colorado Desert, though they could be placed in the Mojave province with equal logic. In short, the boundaries of the Colorado Desert, like those of some other provinces, are somewhat artificial. It is well to remember that classifications are artificial conveniences; nature cares little for our pigeonholes but seems to prefer gradations instead.

GEOGRAPHY

Much of the Colorado Desert lies at low elevation. The Colorado River Valley at the San Bernardino-Riverside county line is 107 meters (350 feet) above sea level, and

250

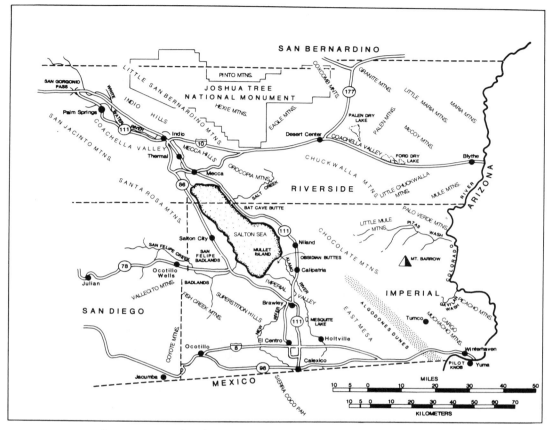

FIGURE 8-1 Place names, Colorado Desert Province.

southeast, at Winterhaven, the elevation is only 40 meters (130 feet). The province's largest low area is the Salton Basin, divided into the Imperial Valley in the south and the Coachella Valley in the north. The Salton Sea, an inadvertently artificial lake with many natural but long-vanished predecessors, occupies the central part of the basin. It has a surface elevation of about 72 meters (235 feet) below sea level. The total portion of the basin lying below sea level extends from near Indio south for about 145 kilometers (90 miles) and is about 40 kilometers (25 miles) wide. This nearly flat depression includes practically all the towns of Imperial County and has developed into one of California's most productive agricultural areas (Figure 8-2).

Apart from a narrow band along the Colorado River and the northeastern quarter of Imperial County, drainage in the Colorado Desert is internal. In eastern Riverside County, much of the drainage ends in the broad Chuckwalla Valley, which contains playas separated from one another by sand dunes. The largest playa is Palen Dry Lake southwest of the Palen Mountains, but Ford Dry Lake, southwest of the McCoy Mountains, is almost as large.

The Colorado Desert's major drainage is into the Salton Sea. For example, the northeastern side of the Peninsular Ranges and the southeastern face of the Trans-

FIGURE 8-2 Salton Sea, Imperial and Riverside counties. Colorado River is in the extreme upper right. The cultivated area to the left of the Salton Sea is the Coachella Valley and the larger area to the right is the Imperial Valley. Swirls in the Salton Sea show water circulation and are due to sediment introduced from the north (left) by the Whitewater River, and from the south (right) by the New and Alamo rivers. (Photo courtesy National Aeronautics and Space Administration)

verse Ranges are drained by the Whitewater River and its tributaries, which reach the northern end of the Salton Sea not far from Mecca. Salt Creek drains the southern slope of the Orocopia Mountains and the northern end of the Chocolate Mountains. The most important western drainage is San Felipe Creek, with headwaters in the Peninsular Ranges near Julian about 80 kilometers (50 miles) west of the Salton Sea. The Imperial Valley is drained by the New and Alamo rivers. Together these rivers account for most of the flow into the Salton Sea. Both cross extensively irrigated farmland and carry water from seepage or flooding during irrigation. Their channels were greatly widened and deepened between 1904 and 1907, when most of the Colorado River entered the valley through man-made openings in the levee south of the international border.

Salton Trough

The dominant feature of the Colorado Desert province is the Salton Trough, a large structural depression that extends from near Palm Springs about 290 kilometers (180 miles) south to the head of the Gulf of California. The term *Salton Trough* (or Salton Sink) refers to the entire basin, from San Gorgonio Pass to the Gulf of California, whereas the term *Salton Basin* applies only to the region draining directly into the Salton Sea. The lower portion of the trough, entirely in Mexico, is occupied by the delta of the Colorado River. In addition, much of the sediment fill in the Imperial Valley is deltaic, because on occasion the river has flowed into the California portion of the basin.

When William Phipps Blake, the first geologist known to enter the Salton Basin, crossed the area in 1853, he correctly surmised that the basin's floor lay below sea level. He further suggested that construction of the deltaic barrier by the Colorado River had isolated the northern extremity of the Gulf of California, allowing it to evaporate and form the extensive salt flats he observed. (There was no Salton Sea at that time.) Later investigations confirmed the existence of elevations below sea level, but they also showed that the basin's history was considerably more complicated than Blake had realized.

STRATIGRAPHY

The oldest rocks exposed in the Colorado Desert are Proterozoic crystalline gneisses, anorthosites, and schists. These are known from the Chocolate, Cargo Muchacho, Orocopia, Chuckwalla, and Little Chuckwalla mountains and from Pilot Knob, an isolated hill near the international border. These ancient rocks have been intruded by several younger plutonic bodies, ranging from late Paleozoic to middle Cenozoic. In fact, this district contains some of the youngest plutonic rocks in California. These young plutonic rocks in the Chocolate Mountains have yielded radiometric ages of 23 to 31 million years (early Miocene to late Oligocene). The Orocopia Schist was thought to be Precambrian when it was originally named by W. J. Miller in 1944. However, based on correlations with the Pelona Schist in the Transverse Ranges, and equivalent schists in the Chocolate Mountains, recent workers have assigned a latest Mesozoic age. All these schists are lithologically similar, and all have the same age relationships; they are collectively referred to as the Pelona-Orocopia Schist. Detailed study has shown that these schists were derived from oceanic basin or trench deposits consisting chiefly of graywacke with lesser amounts of mafic lavas, cherts, and limestone. They were probably deposited on the eastern side of a magmatic arc, remnants of which survive today as the gneisses and metamorphosed granites of the upper plate along the thrust fault exposed in the Orocopia and Chocolate mountains. This thrusting occurred in the late Mesozoic and resulted in the metamorphism of both the upper and lower plates.

In addition, some of these recent investigations, notably those of J. Dillon, G. Haxel, and others, suggest that the Garlock and Rand schists in the northwestern

Mojave fit the same lithologic description, and can be correlated. If this proves to be true, it will require considerable displacement along strike-slip faults.

Most sedimentary and volcanic rocks in the Colorado Desert are Cenozoic and generally rest on the eroded surfaces of the older rocks. The oldest Cenozoic deposits are marine Eocene beds in the Orocopia Mountains and localized nonmarine Eocene deposits found in the Palo Verde Mountains. The Orocopia Mountains also contain the only Oligocene rocks found thus far in the province. These are continental deposits resembling the Vasquez Formation in the Soledad Basin of the central Transverse Ranges. Although early Cenozoic sedimentary rocks appear only sporadically, post-Oligocene sequences are thick, widespread, and well displayed. These young rocks often provide spectacular scenery, as in Anza-Borrego Desert State Park. In southwestern Imperial Valley, 5,000 meters (16,500) of late Cenozoic sedimentary rocks are exposed, in northwestern Imperial Valley 5,700 meters (18,700 feet), and in Coachella Valley 2,650 meters (8,600 feet). Beneath the valley floor, just south of the international border, geophysical evidence has indicated that these same sediments reach a thickness of more than 6,400 meters (21,000 feet). Most of these beds are nonmarine, but the Mio-Pliocene Imperial Formation is a widespread marine unit sometimes nearly 1,200 meters (4,000 feet) thick. Figure 8-3 summarizes several of these sequences.

Pliocene rocks west of the Salton Sea include the Canebrake Conglomerate, which was deposited near the mountains at the edge of the basin. Toward the basin's center, the Canebrake grades into its finer-grained counterpart, the Palm Spring Formation. Much of the desert floor west of the Salton Sea is underlain by this soft, nonmarine

Composite Geologic Column, Colorado Desert

HOLOCENE	Dune sand, Lake Cahuilla Beds, Alluvium, Rhyolites
PLEISTOCENE	Brawley, Ocotillo Conglomerate, Rhyolites, and basalts
PLIO-PLEISTOCENE	Mecca, Palm Spring, Borrego, Canebrake Conglomerate (all nonmarine)
PLIOCENE	Imperial (marine)
MIOCENE	Split Mountain (mostly marine) Anza (nonmarine), Alverson Andesite
OLIGOCENE	Diligencia (nonmarine)
EOCENE	Maniobra (marine)
CRETACEOUS	Orocopia schist, granitic intrusive rocks
PRE-CRETACEOUS	McCoy Mountain, Tumco
PROTEROZOIC	Pinto Gneiss, Vitrefrax Metamorphics, Chuckwalla Complex

FIGURE 8-3 Geologic column, Colorado Desert.

FIGURE 8-4 The folded Palm Spring Formation makes intricate patterns in the barren rocky desert west of the Salton Sea. Tule Wash barchan dune is in the right foreground. (Photo by John S. Shelton)

formation, which has been tightly folded and beveled by erosion (Figure 8-4). The Palm Spring Formation also contains abundant petrified wood, plus numerous zones of sandstone concretions. In some zones the concretions are only a few centimeters long and usually dumbbell-shaped; in other zones they are 20 centimeters or more across and usually spherical.

Sediments deposited in Lake Cahuilla (forerunner of the present Salton Sea) during the late Pleistocene and possibly the early Holocene form a nearly horizontal cover on many exposures of the older rocks. This cover is normally only a few meters thick, but in places it is as much as 90 meters (300 feet) thick. Like the Palm Spring Formation, the Lake Cahuilla beds are soft, weakly consolidated siltstones and clays readily cut by streams whenever there is any runoff. Sufficient runoff occurs only two or three times every five years or so. When it does, stream channels are cleared of wind-deposited sand, and steep banks are swiftly undercut, releasing blocks of the Palm Spring Formation or the overlying lake beds. These blocks are rolled along by the water, becoming rounded in the process, and commonly acquire a coating of gravel and small rocks. When the water subsides, the channels often retain hundreds of such armored mud-balls (Figure 8-5).

Although the Mio-Pliocene Imperial Formation represents the final, large-scale marine invasion into the basin, Pleistocene marine mollusks occur at several localities. These may reflect a brief, relatively recent marine invasion. All the shells are found

FIGURE 8-5 Large Armored Mud Ball, Tule Wash, Imperial County. (Photo by R. M. Norris)

on the surface, not in the lake beds, at 45 to 80 meters (150–250 feet) below sea level. Their significance is strongly debated. The wide distribution, occurrence of articulated pelecypod shells, and lack of attached matrix indicate that the animals lived in one of the later stages of Lake Cahuilla. Moreover, although they are marine species and may have been introduced into water with salinity similar to the ocean's, such a chance introduction would not require invasion by the ocean. Furthermore, these marine species are mixed with freshwater forms. During World War II, marine barnacles were introduced into the Salton Sea on the floats of seaplanes. The barnacles could live in the lake because its salinity was then about two-thirds that of the ocean, although the salt mix is somewhat different (Table 8-1). By 1980, the salinity of the Salton Sea was about 4.4 percent compared with the ocean's 3.5 percent.

The barnacle has developed into a plague in the Salton Sea because thousands of dead specimens are washed up on the beaches. Their sharp shells are a menace to bathers, and the aroma from the decay of these organisms is extremely unpleasant.

The various lakes that occupied the basin have left recognizable deposits, some of which are quite thick. For example, the Pliocene Borrego Formation in the Borrego badlands west of the Salton Sea is 1,800 meters (6,000 feet) thick.

TABLE 8-1 Comparative Composition of the Salton Sea and the Ocean, 1980

Component	Salton Sea		Ocean
	(in parts per thousand)		
Chloride (Cl)	19.060		18.971
Bromide (Br)	—		0.065
Sulfate (SO_4)	8.733		2.649
Carbonate (CO_3)	—		0.071
Bicarbonate (HCO_3)	0.490		0.140
Fluoride (F)	0.004		0.001
Calcium (Ca)	1.066		0.400
Strontium (Sr)	—		0.013
Potassium (K)	0.236		0.380
Boron (B)	0.010	H_2BO_3	0.026
Sodium (Na)	13.185		10.556
Magnesium (Mg)	1.126		1.272
Lithium (Li)	0.004		—
Silica (SiO_2)	0.044		—
TOTAL	44.058 = 4.4058%		34.482 = 3.4482%

STRUCTURE

The Colorado Desert shows the northwesterly structural trends characteristic of most geologic provinces of California (Figure 8-6). The faults of the eastern Peninsular Ranges facing the Salton Basin have a more northerly trend than the main faults entering the basin, however. Some, like the San Jacinto fault, follow well-defined courses in the Peninsular Ranges, cutting obliquely across the mountains to either dissipate completely or vanish beneath the basin deposits. The San Jacinto and its branches enter Borrego Valley and cut across the Ocotillo badlands near Ocotillo Wells before disappearing beneath lake beds. During the 1968 Borrego earthquake, low scarps were formed along this segment of the San Jacinto. A possible buried extension of this fault, known as the Superstition Hills fault, produced surface ruptures during a 1987 earthquake. Similarly, the more complicated Elsinore fault system disappears as it enters the valley near the Coyote Mountains.

The San Andreas fault system is prominent in the Coachella Valley and along the northeast side of the Salton Sea as far south as Bat Cave Butte (Figure 8-7). Most geologists now believe that the San Andreas ends at the northernmost segment of a series of spreading centers that can be traced southward all the way to the mouth of the Gulf of California, where the spreading center is known as the East Pacific Rise. Nevertheless, there is some geophysical evidence for an inactive extension of the San Andreas, the Sand Hills fault, from Bat Cave Butte southeasterly toward Yuma.

The Brawley and Imperial faults lie southwest of the San Andreas and both show right-slip displacement. Recent movement on these two faults has formed a small sag pond, Mesquite Lake, about halfway between Brawley and El Centro.

On a larger scale, the entire Salton Basin can be considered a complex pull-apart

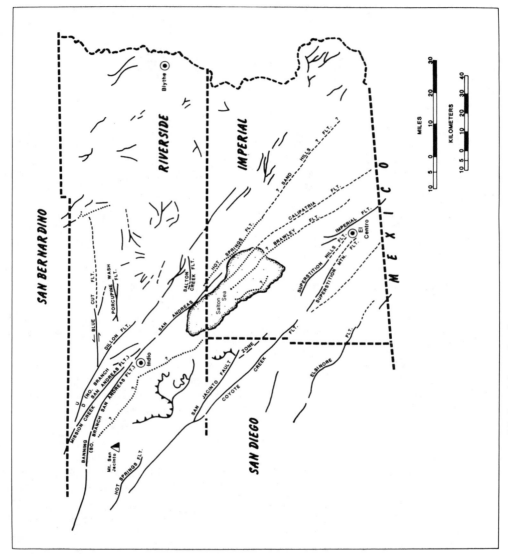

FIGURE 8-6 Principal faults of the Colorado Desert Province.

FIGURE 8-7 View to the northwest along the San Andreas fault, near Indio. Intensely folded Palm Spring Formation is on the right. (Photo by Spence Air Photos, courtesy of Department of Geography, University of California, Los Angeles)

structure resulting from the northwesterly drift of the Peninsular Ranges away from the North American continent, accompanied by crustal spreading beneath the valley. The northwesterly movement along the San Andreas fault zone seems to be about 210 kilometers (130 miles) in the past 8 to 10 million years, including several kilometers in Holocene time (the past 10–12,000 years). The apparently younger Elsinore and San Jacinto faults each show displacements of about 40 kilometers (25 miles) since late Pliocene time.

During the Holocene epoch, the Salton Trough has been seismically very active and remains so today. It is perhaps the most active area in California at the present time. Between 1900 and 1987, the area has been shaken by 32 or more moderate to large earthquakes. Surprisingly, none of the active faults in the basin were known to have showed any surface displacement until the magnitude 7.1 earthquake of May 18, 1940, which was accompanied by ground displacement of as much as 6.3 meters (21 feet). This right slip occurred on the previously unrecognized Imperial fault, offsetting trees, roads, and presumably the nearby international border east of Calexico (Figure 8-8). The trace of this fault is still evident despite more than 45 years of road repair and cultivation that have obliterated many of its features. With the

FIGURE 8-8 Offset trees near Calexico. (Photo by Spence Air Photos, courtesy of Department of Geography, University of California, Los Angeles)

advantage of hindsight, it is now evident that some of the scarps associated with the Imperial were in existence prior to 1940, but their significance was not appreciated until the earthquake renewed them. The Imperial fault parallels other major faults in the basin.

Since 1900, three earthquakes on the San Jacinto fault have been accompanied by surface offset—one in 1918, another in 1968, and one in 1987 on the Superstition Hills segment. Despite the paucity of documented historical examples of surface movement, there are numerous examples of displaced geomorphic features of Quaternary age, showing that surface expression of fault movement is in no sense unusual.

Investigations in the Orocopia, Chocolate, and Cargo Muchacho mountains just east of the San Andreas fault have begun to unravel the history of an important thrust fault system and to reveal the significance of mid-Tertiary detachment faulting mentioned previously in the chapter on the Mojave Desert.

The Chocolate Mountain thrust separates Proterozoic and younger igneous and metamorphic rocks of the upper plate from the Orocopia Schist in the lower plate. In its best exposures it can be located exactly, and is a knife-sharp surface visible even in thin-section. Regionally, it is a gently-dipping surface that is discontinuously exposed from the Orocopia Mountains southeastward into Arizona. Extensive work by John Dillon on the ages of the granitic rocks cut by the thrust and the ages of the sedimentary and volcanic rocks younger than the thrust (not disturbed by it), indicates that movement took place between 60 and 80 million years ago. This also proves to be the age of the related Vincent thrust in the Transverse Ranges. Evidence from the southern Chocolate Mountains indicates that the upper plate of this thrust moved at least 48 kilometers (30 miles) northeastward.

Mid-Tertiary detachment faults are exposed in several ranges of the Colorado Desert of southeastern California and adjacent Arizona. As is the case in the Mojave, these faults result from major extension or stretching of a thick surficial slab of crust above a regional detachment surface at depth. Below the detachment surface is a zone of mylonitic rocks, which many workers think is genetically related to the detachment faulting. This is complicated, however, by the presence of late Mesozoic and early Tertiary mylonites.

Folding is prominent in the Colorado Desert, particularly within young rocks close to major faults. The Indio and Mecca hills, for example, are young anticlinal structures that contain numerous tightly buckled small folds, especially near faults. Low hills in western Imperial Valley, such as the San Felipe and Superstition chains, are similarly anticlinal and show intense deformation adjacent to faults.

Pliocene sedimentary rocks exposed in the flat, barren country west and southwest of Salton City display a wide variety of small-scale folds that are outlined by thin sandstone beds or beds containing abundant sandstone concretions. The smaller folds were produced chiefly during movement along faults. Most of the faults have affected young sediments and provide good evidence of the area's continuing tectonism (Figure 8-4).

The Salton Basin is underlain primarily by land-laid deposits, which are mainly or entirely Cenozoic. The great thickness of these deposits demonstrates that considerable sinking of the basin floor has occurred as the sediments accumulated. Continuing tectonism is also reflected by the eroded crumpled beds exposed around the basin's margin. The general form of the basin's sedimentary fill is synclinal, with the upturned edges of many units exposed in the surrounding hills.

As Baja California and the Peninsular Ranges drift northwest alongside mainland North America, local pull-apart basins, floored by thin, stretched continental crust, will continue to trap sediments delivered by the Colorado River and other streams. In addition, thick prisms of sediment tend to subside as water is squeezed from them, as they compact under their own weight, and as the crust sags beneath the accumulating load.

SUBORDINATE FEATURES

Most mountains in the Colorado Desert province lie between the Salton Basin and the Colorado River, though the northern tip of Sierra Cocopah extends from Mexico into California west of Calexico and several low hills occur in the western Salton Basin. As far as bedrock geology is concerned, the Colorado River and the international border are irrelevant boundaries; the ranges of western Arizona and northern Baja California are similar in many ways to the ranges considered here. Topographically, all the Colorado Desert ranges resemble those of the Mojave Desert, although the Colorado's are generally slightly lower and drier and usually are separated by broader valleys.

Orocopia Mountains

East of the town of Mecca and just north of the Salton Sea are the Orocopia Mountains. This range is comparable topographically to the neighboring Chuckwalla and Chocolate mountains, but in addition it has characteristics whose interpretations have contributed substantially to our understanding of California's geologic history. The range exhibits a variety of probably Proterozoic rocks, including gneisses along its eastern edge and intrusives like gabbro, diorite, anorthosite, and syenite. Many small bodies of titaniferous magnetite and ilmenite are also evident. These Proterozoic rocks form a belt lying between two major faults, the Clemens Well on the northeast and the Orocopia thrust system on the southwest. Along this belt that crosses the range from northwest to southeast, the gneisses are cut by younger intrusives. Interestingly, this entire suite of Proterozoic crystalline rocks is almost perfectly duplicated in the San Gabriel Mountains, on the opposite side of the San Andreas fault to the northwest. Lithology, degree of metamorphism, and field relationships are so alike that these two terranes are regarded as parts of a single mountain mass that has now been separated by 210 kilometers (130 miles) of right-slip displacement on the San Andreas fault. J. C. Crowell has been primarily responsible for establishing these conclusions.

This Proterozoic complex is cut by younger intrusive rocks, mostly of Mesozoic age but with some of possible Paleozoic age. Again, the relationships are strikingly similar to those in the San Gabriel Mountains.

West of the Orocopia thrust fault is the lower plate of the Orocopia Schist; the upper, overriding plate is the complex of Proterozoic and Mesozoic rocks just described. Thus far the Orocopia Schist has not yielded any direct evidence of its age. As indicated previously, however, it lithologically resembles the Pelona Schist of the San Gabriel Mountains, and the Rand and Garlock schists found further north along the Garlock fault. Most geologists believe these units were originally part of the same depositional and metamorphic terrane — probably eugeoclinal rocks consisting mainly of Cretaceous graywackes. Perhaps even more notable than the lithologic similarity is the matching structural pattern seen in the San Gabriel, Orocopia, and Chocolate mountains. In all cases, a complex of Proterozoic gneisses and intrusives forms the upper plate of a thrust. In the presumed position of all three ranges prior to offset, the Orocopia and Pelona schist terrane would be in close proximity to one another. It seems likely that present rock distribution reflects a single structural feature disrupted and separated by long-term right slip on the San Andreas fault system.

The Orocopia Mountains is the only Colorado Desert range east of the San Andreas fault known to contain marine Eocene and nonmarine Oligocene sedimentary rocks. The Eocene rocks cover about 67 square kilometers (26 square miles) in the northeastern part of the range. The section is about 1,460 meters (4,800 feet) thick and consists of a coarse conglomerate with clasts up to 10 meters (30 feet) across in areas where the formation rests on underlying granitic rocks. The upper part is generally finer grained but includes many boulder beds.

Above the Eocene is a thick unit of nonmarine beds and associated basaltic and andesitic flows. Not only do these beds and the Vasquez Formation of the Soledad Basin look alike, but also radiometric dating of their upper, volcanic members has yielded quite similar ages: 24 and 25 million years for the Vasquez and $22\frac{1}{2}$ million years for the Diligencia Formation of the Orocopia Mountains. The lower units may be Oligocene, but age determination is not available. Both formations apparently were deposited in relatively arid interior basins, probably close to one another. So far, however, evidence does not indicate that deposition was in the same basin.

Youngest deposits in the Orocopia Mountains are Pleistocene sandstones and gravels found on the range's eastern slopes and around its western flank. As is typical of desert regions, recent alluvial deposits surround the range and blanket the adjacent valley floor.

Chocolate and Cargo Muchacho Mountains

The Chocolate Mountains extend from Salt Creek, which separates them from the Orocopia Mountains, 96 kilometers (60 miles) southeast at least as far as Gavilan Wash. Some authorities extend them another 32 kilometers (20 miles) to the Colorado River and include the Picacho Mountain area, which constitutes the southeastern end of what is actually a continuous range. Highest elevation is at Mount Barrow, 755 meters (2,475 feet) in the southeastern end.

The Cargo Muchacho Mountains are a small, somewhat isolated group of hills that rise from the flat desert floor in the southeastern corner of the Colorado Desert province. Most of the surrounding desert is less than 180 meters (600 feet) above sea level, so the range is prominent although its highest peak is only 677 meters (2,221 feet). The range is about 13 by 9½ kilometers (8 by 6 miles).

Recent mapping in the Chocolate Mountains from Salt Creek south to the Colorado River has led to a better understanding of the history of the late Mesozoic Chocolate Mountain thrust, the Orocopia Schist, mid-Tertiary plutonic activity, and the late Cenozoic San Andreas fault system. As previously noted, the Chocolate Mountains, like the Orocopia Mountains, have a complex Proterozoic crystalline basement thrust over Orocopia Schist. These Proterozoic gneisses and associated rocks are intruded by at least five different granitic plutons. The oldest pluton so far dated is about 235 million years (early Triassic) but most are Mesozoic; one from the Cargo Muchacho Mountains has yielded a radiometric date of 140 million years (early Cretaceous). The Proterozoic gneisses and associated rocks above and the Orocopia Schist below the thrust fault, plus the thrust fault itself, are all cut by quartz monzonite stocks dated radiometrically at 23 million years—among the youngest granitic intrusives dated in California. Volcanic rocks of similar age and possibly derived from the same magma body as the quartz monzonite intrusive rocks are widely distributed in the Chocolate Mountains.

Miocene fanglomerates with interbedded basaltic flows overlie these older rocks by unconformity and are, in turn, overlain unconformably in a few localities by the marine, lagoonal, and nonmarine Bouse Formation of Mio-Pliocene age. Finally, late Pliocene and Quaternary alluvial deposits lap onto all the older formations. The Miocene rocks are of special interest because some geologists believe they were derived from highlands to the southwest and were deposited by streams flowing toward the northeast, generally opposite to the modern drainage of the Colorado River.

Palen and McCoy Mountains

These two ranges in eastern Riverside County have been studied in some detail and have proved to be less complex than the ranges along the eastern side of the Salton

Basin. The Palen Mountains extend almost due north from Chuckwalla Valley for about 19 kilometers (12 miles) and have a maximum elevation of 900 meters (3,000 feet). The McCoy Mountains, about the same length and only slightly lower, lie southeast of the Palen Mountains and form the northeastern wall of the Chuckwalla Valley.

Oldest rocks present in either range are the intensely faulted and folded marble, quartzite, and gypsum of the late Paleozoic Maria Formation. The only exposure is in the northern Palen Mountains at Palen Pass. Maria Formation rocks are intruded by a quartz porphyry that seems to underlie most of both ranges, though it is not widely exposed. The porphyritic character of the rock suggests it was a shallow-depth intrusion first exposed and eroded during the late Mesozoic.

Resting unconformably on the quartz porphyry are about 7,000 meters (23,000 feet) of sandstone, mudstone, and conglomerate. All are assigned to the McCoy Mountains Formation, which forms the bulk of both ranges. A few rather poorly preserved plant fossils have been recovered from the upper part of the formation, but they indicate only that the rocks are Cretaceous or younger. At the southern ends of both ranges, the intrusive quartz porphyry has been thrust over the younger McCoy Mountains Formation; some believe this thrust is an extension of the Orocopia-Chocolate Mountain thrust system.

Palo Verde and Little Mule Mountains

The Palo Verde Mountains form the southwest wall of the Palo Verde Valley south of Blythe. This arcuate mountain group reaches a maximum elevation of 547 meters (1,795 feet). The Little Mule Mountains are a group of low hills about halfway between the southern Chuckwalla and central Chocolate Mountains. This group of hills is drained by eastward-flowing Milpitas Wash, a generally dry tributary of the Colorado River.

The Palo Verde and Little Mule mountains are parts of an extensive volcanic field whose stratigraphy is broadly similar to that seen in the Chocolate Mountains. The lower part of the volcanic sequence, of latest Oligocene age, rests unconformably on pre-Cenozoic basement rocks. It consists of andesitic to rhyolitic flows, plugs, domes, and extensive volcaniclastic deposits. These are intruded by middle Cenozoic granitic rocks and capped by the eruptive equivalents of the intrusive rocks, which range in composition from rhyolite to basalt with flows, dikes, and pyroclastic deposits. Some of these rocks are interbedded with fanglomerate.

What does this tell us about the structural history of the area? Some geologists believe it indicates a period of crustal deformation in the Oligocene, culminating in the eruption of extensive volcanic materials. By about 21 million years ago, crustal extension initiating Basin and Range normal faulting occurred, followed by a second period of volcanic and intrusive activity between 12 and 17 million years ago, when the San Andreas fault south of the Transverse Ranges appears to have first developed.

Fish Creek Mountains

The Fish Creek Mountains are on the west side of the Imperial Valley, south of San Felipe Creek. They extend about 16 kilometers (10 miles) with a maximum elevation of 712 meters (2,334 feet) near the southeastern end. They are separated from the larger Vallecito Mountains to the west by the gorge of Fish Creek, which is known locally as Split Mountain Canyon.

The basement of the Fish Creek Mountains is a mixture of gneisses, marbles, and granitic rocks. These rocks are of uncertain age, but probably are partially equivalent to the Cretaceous crystalline rocks that form much of the Peninsular Range to the west. It is logical, in fact, to interpret both the Fish Creek and Vallecito mountains as merely extensions or outliers of the Peninsular Ranges, although the former are true desert ranges and are surrounded on three sides by the Colorado Desert.

During at least the first half of the Cenozoic, the crystalline basement was unroofed and deeply eroded. By Miocene time, erosion was replaced by a depositional environment in which nonmarine sediments accumulated. This presumably occurred under arid or semiarid conditions, because coarse-grained fanglomerate was laid down on the exposed basement at the base of mountains that rose steeply to the west.

The geologic section exposed in the walls of Split Mountain Canyon ranges from the nonmarine, probably Miocene, Anza Formation that is beautifully displayed in the 80-meter-high cliffs in lower Split Mountain Canyon (Figure 8-9), upward through

FIGURE 8-9 An 80-meter-high cliff cut in the coarse fanglomerate of the Anza Formation of probable Miocene age. Lower Split Mountain Canyon (Fish Creek), San Diego County. (Photo by R. M. Norris)

the marine and nonmarine Split Mountain Formation to the overlying Mio-Pliocene Imperial Formation present in the badlands area at the head of the canyon.

The Anza Formation is the oldest known sedimentary rock on the west side of the Salton Trough. Most geologists consider it to be an alluvial fan deposit—a fanglomerate—on the basis of its lenticular bedding, large fresh clasts of granitic rock, and its coarse pebbly conglomerate and sandstone beds.

Capping the Anza Formation and equally well exposed in Split Mountain Canyon is the varied Split Mountain Formation. The base of this unit is a spectacular dark gray breccia containing huge quartz diorite clasts. A second very similar breccia occurs higher in this formation, but contains schist clasts and some even larger diorite blocks. The origin of these interesting members has been debated for years. Some geologists claim they are mudflow deposits, but others point to the immense size of some of the blocks—one is estimated to weigh 15,000 tons! Such enormous blocks, they say, are much too large to be transported by mudflows. They assert that these two massive breccias are not mudflows at all but are airborne landslide sheets similar to the Blackhawk landslide deposit, which is described in Chapter 10 on the Transverse Ranges. In addition, the large Martinez Mountain landslide on the east side of the Santa Rosa Mountains, about 15 kilometers (10 miles) southwest of Indio, contains huge blocks not unlike those seen in the two breccias.

The middle part of the Split Mountain Formation is a sequence of well-bedded marine sandstone and mudstone very much like the overlying Imperial Formation. The beds lack good diagnostic fossils but seem to be late Miocene or Pliocene in age. In some places, the middle part of this formation contains as much as 30 meters (100 feet) of gypsum, which for many years has supported a large mining operation just east of Split Mountain Canyon. These gypsum beds are capped locally by a thin bed of celestite (strontium sulfate), which was also mined for a time. The gypsum mine has sometimes yielded large plates of the nearly transparent variety of gypsum called selenite.

In a few places the Anza Formation is overlain by the Alverson Andesite rather than by the Split Mountain Formation, but in others the Alverson rests directly on the basement, and in still others it interfingers with the Anza beds. It is Miocene in age, about 13 to 16 million years old.

Much of the ancestral Salton Basin, including the Fish Creek area, was invaded by a shallow sea near the end of Miocene time or early in the Pliocene. During this time, the clays, silts, and sands and the oyster beds of the middle Split Mountain and Imperial formations were widely deposited. Probably the thickest section of the formation is on the southern side of the Fish Creek Mountains, where about 1,130 meters (3,700 feet) have been eroded into low, barren hills often capped with dark brown oyster beds (Figure 8-10).

Before the Pliocene epoch closed, the sea was excluded from the Salton Basin, possibly by construction of the Colorado River delta across the head of the presently-developing Gulf of California. Nonmarine conditions again prevailed. The Palm Spring Formation was deposited in the center of the basin, grading west into its coarse-grained equivalent, the Canebrake Conglomerate. As mentioned earlier, the Palm Spring Formation contains abundant petrified wood thought to be ironwood, a desert tree still extant throughout the Colorado Desert. Where the Palm Spring Formation is exposed, the resistant gray petrified wood often weathers out of the soft enclosing siltstone and claystone.

FIGURE 8-10 The marine Mio-Pliocene Imperial Formation, upper Fish Creek, San Diego County. These soft beds display badland gullying and are capped with a thin, resistant oyster-shell bed. (Photo by A. G. Sylvester)

Sand Dunes

Although comparatively little of the state's arid regions contains sand dunes, sand deposits are more extensive and varied in the Colorado than in other California deserts. The largest Colorado dune tracts are the Algodones dunes or Imperial sand hills, which are about 72 kilometers (45 miles) long and extend about 8 kilometers (5 miles) into Mexico, to the edge of the Colorado River flood plain. The dunes are sometimes more than 8 kilometers (5 miles) wide, but the average width is about 5 kilometers (3 miles). Their highest peaks rise nearly 80 meters (250 feet) above the desert floor. The dune chain is made up of overlapping, slightly arcuate ridges lying at right angles to the length of the dune field and alternating with sandy-floored depressions (Figure 8-11). Toward the southern end of the chain, the depressions become larger, up to 1.6 by 0.8 kilometers (1 by $\frac{1}{2}$ miles) in area. Some are free of sand, and faint traces of old drainage channels may be visible on the gravel floors. Presumably these channels predated the dunes and existed when drainages developed from the uplands in the eastern Salton Basin. In a few of these flat-floored and sand-starved hollows, swarms of small barchans have formed. The entire dune chain is migrating southeast in response to strong northwesterly winds that rake the area, especially in late winter and spring.

Some geologists believe these dunes represent a long plume of sand blown inshore from the beaches of former Lake Cahuilla. Others think that ground water moving up along the buried Sand Hills fault, which may pass beneath the dunes, either trapped blowing sand directly or promoted a greater growth of vegetation, which then trapped the sand. In addition, some recent studies seem to indicate that the dune sand has closer affinities with sand transported by the Colorado River than

FIGURE 8-11 Algodones dunes, from the south. Note the areas of sand starvation inside the dune clusters. (Photo by Spence Air Photos, courtesy of Department of Geography, University of California, Los Angeles)

with sand delivered to Lake Cahuilla from the Coachella Valley area. However the dunes formed, they represent the largest tract of desert dunes in North America and for many years imposed a serious barrier to travel between the Imperial Valley and Arizona. In 1914, a narrow plank road was built across the dunes to link El Centro and Yuma, but sand was continually being blown from beneath or on top of it. In the late 1920s, the first paved road was built across the dunes. Sand had to be scraped from the road surface frequently and dunes sloping down from the road had to be oiled to prevent sand being removed from under the road's surface. The same problems persist today, although a four-lane divided highway now crosses the area.

Until the middle 1930s, the canal system servicing the Imperial Valley collected water from near the international boundary, transported the water in canals around the southern end of the Algodones dunes in Mexico for some 64 kilometers (40 miles) before turning north into the United States. Early canal builders were unable to construct the canal entirely on the American side of the border across the dunes. The modern canal does cross the dunes but requires constant dredging to remove dune sand and silt contributed by the sediment-laden Colorado River.

Other dunes occur in the Imperial and Coachella valleys, though none compares with the large Algodones dune chain. Well-formed barchan dunes are found west of the Salton Sea, north of San Felipe Creek (Figure 8-4). Some are moving 30 meters (100 feet) per year toward the Salton Sea, whereas others move more slowly. Small dunes tend to move more rapidly than large ones, causing the configuration of dune fields to change as small dunes overtake and merge with larger ones.

Extensive tracts of dune sand exist in the Coachella Valley east of Palm Springs. Perhaps because of the great volume of new sand brought into the valley by the Whitewater River and its tributaries, these dunes generally lack the well-defined geometry of those seen in the Imperial Valley. Nevertheless, the windiness of the upper Coachella Valley, coupled with the availability of large volumes of sand, provides many good examples of wind abrasion, deposition, and erosion. Urbanization of the valley plus the establishment of numerous shelter belts of trees have reduced the area of active dunes, and have made the remainder less obvious to the traveler.

Low rocky ridges extend from the base of San Jacinto Peak into the flat floor of San Gorgonio Pass at right angles to the strong winds that regularly sweep through this gap into the valley below. On the windward side of each of these spurs, wind-blown sand and gravel have etched the hard crystalline rocks. On the lee sides, plants often grow, but their tops are perfectly even with the ridge. New shoots are trimmed off by the sharp windblown sand as soon as they venture above the protecting rock barrier. William Phipps Blake observed these features during his visit in 1853 and correctly described the general cause of the persistent winds. His words are instructive.

> . . . They both (the San Francisco Golden Gate and San Gorgonio Pass) appear to be great draught channels from the ocean to the interior, through which the air flows with peculiar uniformity and persistence, thus supplying the partial vacuum caused by the ascent of heated air from the surface of the parched plains and deserts. . .

In connection with the effects of sandblasting in San Gorgonio Pass, he says:

> . . . I had before me remarkable and interesting proofs of the persistence and direction of this air-current, not only in the fact that the deep sand-drift was on the east side of the spur, but in the record which the grains of sand engrave on the rocks in their transit from one side to the other.[1]

The highway traveler coming into Indio from the west often encounters windblown sand and gravel despite the many shelter belts of trees and the sand fences. Fence posts, highway signs, telephone poles, bottles, and just about anything that is exposed is subjected to frequent sand blasting. Many automobiles have undergone severe damage during a single sandstorm. Windbreaks alleviate the problem but do not eliminate it. The stong winds still manage to seek out loose sand, and processes of erosion, abrasion, and deposition continue.

---

[1]William Phipps Blake, *in* R. S. Williamson, Report of Explorations in California for Railroad Routes (1853–54). Thirty-third Congress, Second Session; Senate Executive Document n. 78, v. 5, part 2, Chap. 8, pp. 91–92.

SPECIAL INTEREST FEATURES

Lake Cahuilla and Its Fossil Shorelines

On the west side of the Salton Sea near the base of the Santa Rosa Mountains is a feature resembling a bathtub ring. Myriads of tiny gastropod and pelecypod shells litter the desert below the ancient water line. Travertine Rock on the Riverside-Imperial county line is a prominent granitic knob. Its lower portion is heavily encrusted with a pale brown spongelike coating of travertine. This limy deposit was secreted by cyanobacteria (blue-green algae) that once inhabited a freshwater lake—ancient Lake Cahuilla. Above the highest lake level, the rocks are darker colored, owing to a coat of desert varnish. The old, highest lake level is thus plainly marked for about 20 kilometers (12 miles) along the base of the Santa Rosa Mountains (Figure 8-12).

Most of the high shoreline does not reach the bases of the other mountains that rim the Salton Basin, because the shoreline's average elevation is only 9 to 12 meters (30–40 feet) above sea level. Its existence is manifested in other ways, however. Old beaches, sand spits, and bay-mouth bars can be seen along the base of the Santa Rosa Mountains, but the finest examples occur along the Coachella canal between Niland and Mecca. Smooth, curved beaches link former headlands that once projected into the lake (Figure 8-13). Where flat pebbles of Orocopia Schist were carried to the lake by streams draining the Orocopia and northern Chocolate mountains, shingle beaches of neatly stacked, smooth, flat schist pebbles now stretch for several

FIGURE 8-12 Lake Cahuilla high shoreline, Santa Rosa Mountains south of Indio, Riverside County. The darker rocks below the shoreline are encrusted with travertine. Above are light-colored rocks eroded by wave action. Still higher, above the reach of waves, the rocks retain a coating of darker desert varnish. (Photo by R. M. Norris)

FIGURE 8-13 Ancient shorelines of Lake Cahuilla, near Niland. Salton Sea is on the left (west). The zigzag pattern along the railroad is made by flood revetments. (Photo by Spence Air Photos, courtesy of Department of Geography, University of California, Los Angeles)

kilometers (Figure 8-14). Elsewhere, waves in Lake Cahuilla cut low cliffs into soft sedimentary rocks such as the Palm Spring Formation; good examples can be seen near Niland and in the San Felipe badlands west of the Salton Sea. And still elsewhere, broad flats represent wave-washed beach flats where older, highly folded lake beds are beveled and exposed like a geologic map (Figure 8-4).

The complete history of Lake Cahuilla and its predecessors is yet to be resolved. Nevertheless, it is likely that the lake filled on several occasions when the distributaries of the Colorado River changed their courses across the delta. Sometimes these distributaries entered the closed Salton Basin to the north, at other times they flowed into the Gulf of California, and sometimes they did both simultaneously. Between fillings, evaporation quickly reduced the lake level, often leaving a salty crust on the central part of the basin floor.

It seems certain that Lake Cahuilla did not follow the pattern of the ancient lakes of the Mojave Desert and Basin Ranges to the north; these seem to correlate with wetter periods associated with glaciation. Lake Cahuilla may well have responded to wetter conditions in the high ranges nearby, but the lake was almost certainly as much a consequence of the Colorado River floods that are independent of glacial cycles. The last filling, possibly to the highest shoreline described earlier, is thought to have occurred between about A.D. 900 and 1400.

FIGURE 8-14 Beach bar composed of wave-rounded Orocopia Schist shingle. Orocopia Mountains in the distance. Riverside County. (Photo by R. M. Norris)

Salton Sea

The present Salton Sea is a product of the twentieth century and the result of real estate promotions and farming development in the last decade of the nineteenth century. By 1900, more than 1,000 people had settled in the Imperial Valley, even though the available water supply was insufficient for farming and development. Beginning that year, a canal system was constructed from a point on the Colorado River just north of the Mexican border. By 1902, 640 kilometers (400 miles) of canals and lateral branches were in service, and nearly 40,500 hectares (100,000 acres) had been prepared for cultivation. By the end of 1904, the canals had silted up considerably, the river was low, and settlers were clamoring for more water. To increase the flow into the canal system, the control gates on the river were bypassed by making cuts in the riverbank south of the border. Wing dams were built from shore to deflect the water into the canal system. Although several of these openings promptly silted up, they did provide some badly needed water. Plans for new control gates were completed in November 1904, but were not approved by the Mexican government before the next major flood.

The first important flood to occur during this perilous situation was in January 1905, followed by three large flash floods in February. By then the river was out of control, and considerable water was entering the canal system. The engineers in charge hoped to close the gaps in the west bank during the normal period of low water before the late spring floods. This period failed to materialize, and during the late summer attempts to plug the gaps were curtailed by continued high water.

Early in the summer of 1905, about 16 percent of the river was entering the canal system, and by October virtually the entire Colorado River was flowing into the valley. The Southern Pacific Railroad tracks had been inundated, so the company built a new barrier on the river. Unfortunately, a violent flash flood in November again allowed much of the river to enter the valley. In early 1906, new control gates were completed and the river was contained again on November 4, 1906. On December 10, still another violent flash flood swept down the river, and the entire flow once again poured into the canal system and from there into the channels of the Alamo and New rivers and into the valley bottom. On February 11, 1907, the breach was sealed for the last time. The level of the Salton Sea then stood at 60 meters (198 feet) below sea level, its highest point. The lake was then about 26 meters (80 feet) deep.

Ironically, many canal branches were left totally dry because the flooding had widened and deepened the channels of the New and Alamo rivers, which had previously been tiny gullies across a nearly featureless desert floor. These two rivers now became prominent channels cut into the valley's soft alluvium, sometimes measuring 0.4 kilometer (¼ mile) wide and 15 meters (50 feet) deep. They, rather than the canals, then carried much of the water across the Imperial Valley and into the Salton Sea. Subsequently, these rivers also became valuable drainage channels for saline waste irrigation water.

Spectacular examples of headward erosion developed as the river channels were widened, deepened, and lengthened. During 1905 and 1906, the process began at the shore of the Salton Sea and rapidly extended about 80 kilometers (50 miles) up the New River to Mexicali. The erosion was so rapid near Mexicali that banks were undercut in a few hours, destroying many buildings. The process was less dramatic in the Alamo River, but even there headward and lateral erosion eventually extended upstream about 50 kilometers (30 miles).

By 1925, evaporation had reduced the lake level to almost 80 meters (250 feet) below sea level; the lake was then about 9 meters (30 feet) deep. Beginning in the late 1930s as agricultural development was augmented and more water was brought in from the Colorado River, the net inflow into the Salton Sea reversed the lake's shrinkage pattern. The present level is about 72 meters (235 feet) below sea level. This is probably near the stable level because as the lake rises and its area increases, annual evaporative loss approaches a balance with annual inflow, which is itself limited by the amount of water that California may legally withdraw from the Colorado River.

Recent Volcanic Features

Near Niland, at the southeastern end of the Salton Sea, are some volcanic features that include several low pumice and obsidian domes. Five low volcanic hills have been formed, rising 35 to 45 meters (100–150 feet) above the basin floor. These hills extend along a 6½ kilometer (4 mile) axis from southwest of Mullet Island to Obsidian Butte, and are young (presumably Quaternary) rhyolitic domes of pumice and obsidian. Several still produce warm gases, and at Mullet Island these gases were acidic enough to decompose the rocks. Mud volcanoes, mud pots, and boiling springs lie along a northwest-southeast line almost at right angles to the trend of the volcanic domes. Principal gases produced are steam, carbon dioxide, and hydrogen sulfide (Figure 8-15).

Beginning in 1927, attempts were made to recover the carbon dioxide from wells. In the 1930s, the operation produced gas from depths of 60 to 210 meters (200–700 feet), but the field was abandoned in 1954 because of the rise of the Salton Sea, scaling problems, and for other economic reasons. In 1957, geothermal exploration was initiated, and since then at least 60 production and hundreds of exploration wells

FIGURE 8-15 A small mud volcano about a meter high, south of Niland, Imperial County. Note the resemblance to volcanic features such as plugs, craters, flows, domes, subsidence basins, and shrinkage cracks. (Photo by R. M. Norris)

have been drilled. The hot brines from these wells sometimes are extremely high in sodium and calcium chlorides and several base metals. The content of such metals as copper and zinc may someday be economically important.

At present, some six geothermal fields have been identified in the Salton Basin with reservoir temperatures greater than 150°C. The U.S. Geological Survey has estimated that these could eventually yield 6,800 megawatts over a 30-year period. As of 1989, twelve geothermal power plants were operating in the Imperial Valley of California. Others are operating at Cerro Prieto, a short distance south of the border.

Because of the high dissolved mineral content of these hot brines (10–39 percent dissolved solids), the power plants have adopted a largely closed system in which heat exchangers are used to draw off energy and the cooler brines are returned to the underground reservoirs. This method reduces scaling and corrosion problems as well as severe surface disposal problems.

Mining in the Cargo Muchacho Mountains and Picacho District

Legend has it that in 1781, two prospectors' sons came out of these hills with their shirts loaded with gold, and the Cargo Muchacho "Loaded Boy" Mountains were thus named. The rocks within this group of hills are generally similar to those of the Chocolate Mountains and include Proterozoic and Jurassic gneisses and Mesozoic granitic intrusive rocks. For some reason, however, mineralization was more intense in the Cargo Muchacho district, and gold has been mined there intermittently since 1781. One particularly well-known gold-producing area was at Tumco, which is now a ghost town. The town's name is an acronym for The United Mining Company, which operated the local mines around the turn of the century. In the early 1900s, Tumco had a population of 2,000; there were four saloons on the main street—known as Stingaree Gulch—but only a few foundations remain today.

Perhaps as many as 400,000 troy ounces of gold have been mined from the Golden Queen, Golden Crown, and Golden Cross mines of the Tumco district. The ore occurs in dark gray siliceous zones in the Tumco Formation, a gneiss of probable Jurassic age, which is associated with some schist in the vicinity of the ore deposits. The origin of the ore is still vigorously debated. Some geologists have attributed the ore to mineral-rich brines seeping from the sea floor at the time the sedimentary predecessors of the Tumco Formation were deposited. Other workers assert that ore deposition resulted from fluids introduced much later than the metamorphism, and followed Basin and Range type faults. Still others suggest the ores resulted from fluids associated with the metamorphism of the Tumco Formation, possibly derived from the intruding plutons that caused metamorphism to take place.

Whatever the reason, the high price of gold in the 1980s rekindled interest in the Tumco and other areas in the Cargo Muchacho Mountains.

The Picacho mine is located at the extreme southeastern end of the Chocolate Mountains near Picacho Peak. The mine currently produces about 22,000 troy ounces of gold per year from the metamorphic rocks exposed below the Chocolate Mountains thrust. This occurrence is in a structurally complex basin overlain by alluvium. Some investigators have argued that the mineralization is related to the detachment faults that occur here. The mineralization and the enclosing rocks were once thought to be of Proterozoic age, but it now seems more likely that they are Jurassic.

The second largest gold mine in California is the Mesquite mine, which began production in 1987. Production of 180,000 troy ounces per year comes from an ore body estimated to contain about 2.6 million ounces of gold. As in the Picacho district southeast of the Mesquite mine, the ore occurs in metamorphic rocks of probable Jurassic age.

In the southern part of the Cargo Muchacho Mountains, the blue mineral kyanite has been mined. Like other aluminum silicates, kyanite is used in making high-temperature ceramic materials. Annual production was about 80 metric tons in the early 1940s, but the mines have been inactive in recent years.

REFERENCES

Abbott, Patrick L., ed., 1979. Geological Excursions in the Southern California Area. San Diego State University, Dept. of Geol. Science, Geol. Soc. Amer. Guidebook.

Biehler, S., and R. W. Rex, 1971. Structural Geology and Tectonics of the Salton Trough, Southern California. *In* Geological Excursions in Southern California. Univ. Calif. Riverside Museum Contrib. no. 1, pp. 30–42.

Crowell, John C., ed., 1975. San Andreas Fault in Southern California. Calif. Div. Mines and Geology Spec. Rpt. 118.

———, and A. G. Sylvester, eds., 1979. Tectonics of the Juncture between the San Andreas Fault System and the Salton Trough, Southeastern California. University of California, Santa Barbara, Dept. Geol. Sci, Geol. Soc. Amer. Guidebook.

Dibblee, T. W., Jr., 1954. Geology of the Imperial Valley Region, Calif. *In* Geology of Southern California. Calif. Div. Mines and Geology Bull. 170, pp. 21–28.

Hamilton, Warren B., 1969. Geology of the Colorado Desert. California Geology, v. 22, pp. 96–98.

Herber, Lawrence J., 1985. Guidebook Geology and Geothermal Energy of the Salton Trough. Calif. Poly. University Pomona. National Assoc. Geol. Teachers Far Western Section Guidebook.

Miller, William J., 1944. Geology of the Palm Springs-Blythe Strip, Riverside County, California. Calif. Jour. Mines and Geology, v. 40, pp. 11–72.

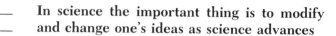

nine

PENINSULAR RANGES

In science the important thing is to modify
and change one's ideas as science advances

—*Claude Bernard*

The Peninsular Ranges are one of the largest geologic units in western North America. They extend 200 kilometers (125 miles) from the Transverse Ranges and the Los Angeles Basin south to the Mexican border and beyond another 1,250 kilometers (775 miles) to the tip of Baja California. The total province varies in width from 48 to 160 kilometers (30–100 miles), includes—according to some interpretations—the offshore area, and is bounded on the east by the Colorado Desert and the Gulf of California. The locations of important features are shown in Figure 9-1.

The Peninsular Ranges contain minor Jurassic and extensive Cretaceous igneous rocks associated with Nevadan plutonism and a few remnants of roof rock that establish continuity with the better-known pre-Nevadan history of the Sierra Nevada. Marine Cretaceous sedimentary rocks are well represented. Post-Cretaceous rocks form a restricted veneer of volcanic, marine and nonmarine sediments, although adjacent to the Peninsular Ranges in the Los Angeles Basin, post-Cretaceous marine sections up to 12,200 meters (40,000 feet) are found (Figure 9-2).

GEOGRAPHY

The Peninsular Ranges are a northwest-southeast oriented complex of blocks separated by similarly trending faults. Highest elevations are found in the San Jacinto-Santa Rosa Mountains of the easternmost block: San Jacinto Peak is 3,296 meters (10,805 feet), and summits in the Santa Rosa Mountains average about 1,800 meters (6,000 feet). Toward the Pacific are the Agua Tibia, Laguna (Cuyamaca), and Santa Ana mountains, which are not as high as the eastern ranges. The escarpment between the San Jacinto Mountains and the adjacent Coachella Valley is one of the boldest

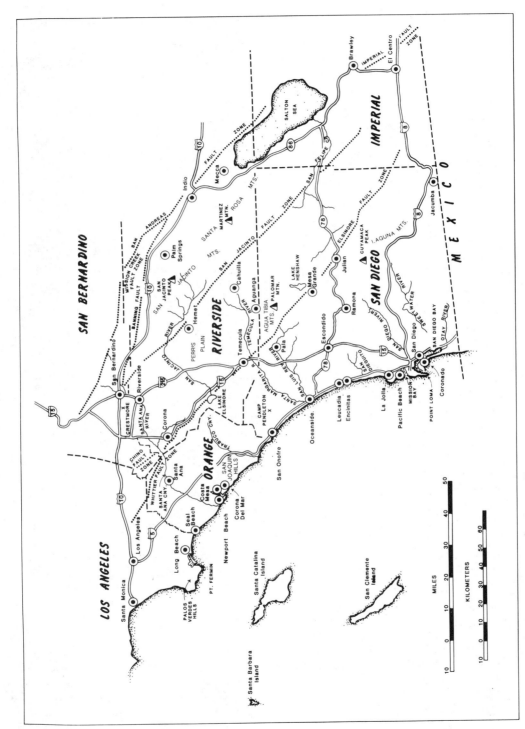

FIGURE 9-1 Place names, Peninsular Ranges Province.

278

Columnar Section, Northern Peninsular Ranges

QUATERNARY	Alluvium Temecula Arkose and Bautista Beds Marine Terraces
PLIOCENE	San Timoteo and Mt. Eden (nonmarine) Repetto (San Diego Fm. to south) Santa Rosa Basalt
MIOCENE	Puente San Onofre Breccia Topanga Vaqueros
OLIGOCENE	Sespe (nonmarine)
EOCENE	Santiago
PALEOCENE	Silverado
CRETACEOUS	Williams and Ladd Formations Trabuco (nonmarine) Granitic rocks of Southern California batholith
JURASSIC	Santiago Peak volcanics Bedford Canyon French Valley (position uncertain)
PALEOZOIC (?)	Roof pendants in granitic rocks

Source: California Division of Mines and Geology.

FIGURE 9-2 Geologic column, northern Peninsular Range.

in North America, especially near Palm Springs where the San Jacinto summit rises abruptly 3,000 meters (10,000 feet) above the valley (Figure 9-3). The boundary between the Santa Rosa Mountains and the Imperial Valley is more irregular and less striking, and if this boundary is a fault, it has been inactive for some time.

The general cross section of the Peninsular Ranges resembles that of the Sierra Nevada, since each range has a gentle westerly slope and, normally, a steep eastern face. The western side of the Peninsular Ranges is composed of discrete blocks that slope progressively lower to the west and are produced by the breaks of major fault zones. In addition, the plutonic rocks are similar to those in the Sierra.

Drainage of California's Peninsular Ranges is primarily by the San Diego, San Dieguito, San Luis Rey, and Santa Margarita rivers. The Santa Ana River drains only the extreme northwesterly part of the province, and short streams drain the rest of the Pacific slope of the Peninsular Ranges. The main stream flowing east is San Felipe Creek, which enters the Salton Sea. Rainfall varies from about 250 millimeters to almost 900 millimeters (9–35 inches), so most streams are intermittent.

FIGURE 9-3 Bold eastern face of Mount San Jacinto, with Palm Springs in the center and San Gorgonio Pass at the far right. Mount San Jacinto has an elevation of 3,295 meters or 10,804 feet and Palm Springs has an elevation of only 144 meters or 475 feet. Photo was taken in 1956. (Photo courtesy U.S. Geological Survey)

In the Baja California section of the Peninsular Ranges, essential geographic units from north to south are: (1) the Sierra San Pedro Martir, the Sierra Juarez, and some minor ranges, all of which are direct southeasterly extensions of the California mountain blocks; (2) the Vizcaino Desert; (3) the Sierra de la Giganta; and (4) the southern tip, the Sierra Victoria. Elevations are generally low, but reach 3,088 meters (10,126 feet) at La Providencia Peak in the northern Sierra San Pedro Martir.

ROCKS

Cretaceous and Pre-Cretaceous

Metasedimentary and metavolcanic rocks occupy restricted areas in the Peninsular Ranges. Near Riverside there are restricted outcrops of late Paleozoic limestone (from which cement is processed). Some older rocks are exposed in the San Jacinto and Santa Rosa mountains, where altered schist and gneiss may reach total thicknesses of 6,700 meters (22,000 feet). Adjacent to the Los Angeles Basin, rocks up to 6,100 meters (20,000 feet) have been mapped as the Bedford Canyon series in the Santa Ana Mountains and southward. Fragmentary fossils from these rocks have been dated as Jurassic. Thick volcanics up to 700 meters (2,300 feet) are assigned to the Santiago Formation and overlie the Bedford Canyon beds. Similar volcanics inland from San Diego may be equivalent. These volcanics are also considered Jurassic (Figure 9-2).

Igneous rocks are age-equivalents of Nevadan plutonism. In declining age, typical Peninsular Range intrusives are gabbro, quartz diorite or tonalite (the commonest), and granodiorite. The entire plutonic complex of the Peninsular Ranges is known as the Southern California batholith and includes plutons like the San Marcos Gabbro, the Bonsall Tonalite, and the Woodson Mountain Granodiorite. Typically all units have widespread dark inclusions of older rocks. Radiometric dating suggests Nevadan plutonic emplacements from 90 to 120 million years ago. Younger dates, to 70 million years, have been determined in some instances, but the intrusives are definitely part of the batholithic episode associated with the Nevadan orogeny.

Peninsular Range rocks are distinctly less silicic and more calcic than typical Sierran intrusives. Nevertheless, the abundance of Nevadan-age intrusives may imply a relationship between Sierran and Peninsular histories—albeit a tenuous one, because the thick Paleozoic sections in the Sierra are not repeated in the Peninsular Ranges. Possibly an extensive Paleozoic sequence has been eroded away, but no evidence substantiates this. Pre-Phanerozoic history is not recorded in Peninsular Range rocks or structure, whereas the Transverse Ranges immediately to the north include Proterozoic igneous intrusives and metamorphic gneisses. Similarly, latest Paleozoic intrusives of the Transverse Ranges have not yet been identified in the Peninsular Ranges.

Although the plutonic rocks of the Peninsular Ranges are roughly similar in age and composition to those of the Sierra Nevada and the Salinian block of the central and southern Coast Ranges, they appear to have developed far to the south of these counterparts and to have reached their present position only as a result of the long and complex tectonic history of California. In one view, for example, it is suggested that the northern end of the Salinian block, now north of San Francisco, originally matched the southern end of the Sierra Nevada, and the southern end of the Salinian block, now perhaps as far south as Ventura County, once joined the northern end of the Peninsular Ranges near Corona. Prolonged right slip on the San Andreas fault and displacement on other faults of the system is thought to be the cause of present rock distribution.

Late Cretaceous sedimentary rocks (including the marine Williams, Ladd, and Rosario formations and the largely nonmarine Trabuco) are extensively exposed in the western flank of the Santa Ana Mountains from Santa Ana Canyon southward into the Camp Pendleton area near San Onofre. To the south there are other more limited exposures of these sedimentary rocks, inland from Leucadia and Encinitas, around the La Jolla headland and at Point Loma near San Diego. The basal portion of these sedimentary rocks is composed of conglomerate, especially well developed in the Trabuco Canyon area of the Santa Ana Mountains where the Trabuco Conglomerate is very thick and coarse. The upper part of the Cretaceous section is composed chiefly of sandstone and shale, which has yielded a few enormous ammonite fossils. One specimen of *Pachydiscus catarinae* displayed in the San Diego Museum of Natural History is about a meter (3 feet) across. Late Cretaceous (Rosario Formation) rocks at Point Loma and farther south in Baja California have also yielded more examples of this unusually large extinct cephalopod.

The sandstone exposed at La Jolla's Cove in which the famous sea caves are developed, are examples of these late Cretaceous sedimentary rocks.

Post-Cretaceous

All post-Cretaceous rocks lie unconformably on either the Cretaceous sedimentary rocks or on basement. Lengthy, widespread erosion of crystalline units contributed substantially to Cenozoic deposits in the Los Angeles, Imperial, and offshore basins and to thick sediments in localized interior basins like the Elsinore trough and Perris Plain.

Early Tertiary rocks are confined to coastal margins, reaching a maximum thickness of about 1,400 meters (4,500 feet) in the Santa Ana Mountains. The Silverado Formation is about 425 meters (1,400 feet) of Paleocene nonmarine rocks that occur in the northern Santa Anas. The basal, nonmarine portion of the Silverado contains minor amounts of coal as well as considerable glass sand and high-alumina clays. The sands and clays have sustained a mining operation for more than 80 years. Only about 180 meters (600 feet) of marine sedimentary rock occur in the San Diego area, overlain by about 300 meters (1,000 feet) of mostly nonmarine sediments of the Poway Formation. These are small thicknesses as compared with those of adjacent basins like the Los Angeles, Ventura, and some offshore areas. Nonmarine Oligocene up to 900 meters (3,000 feet) is found in the Santa Ana Mountains, but the Peninsular Range Province was rejuvenated by significant uplift near the close of the Oligocene, further restricting sedimentation.

One of the more interesting early Tertiary rock units is the Eocene Poway Formation mentioned earlier, which was first described from a locality near San Diego. This rock, which contains both marine and nonmarine beds, is believed to represent a number of different environments from stream channel to submarine fan. The Poway is characterized by its hard, well-rounded silicified metavolcanic pebbles from an unknown source area. No suitable parent rock is known from anywhere in southern California, which has led some geologists to look farther afield. The most likely source identified to date is in northwestern Sonora in Mexico. If this is indeed the correct one, it provides yet another demonstration of how strikingly different today's geography is than that which existed in early Tertiary time. These great geographic changes are also shown by the present distribution of Poway rocks. In addition to the type locality east of San Diego, exposures occur on San Nicolas, Santa Cruz, Santa Rosa, and San Miguel islands. To reconstruct the Eocene geography of southern California, we not only need to bring the various exposures of the Poway group back together again to form a coherent pattern, but we must reassemble the fragments so as to account for the current and channel directions recorded by these rocks. Since the Eocene the islands must have moved northwesterly away from the mouth of the ancient "Poway River" near San Diego. However, neither the exact path followed nor the time of movement is yet clear.

Late Tertiary beds are also limited in extent. Little Tertiary volcanism occurred in the Peninsular Ranges (in contrast to the widespread Tertiary volcanism in other southern California provinces). A few late Tertiary marine and continental deposits are found in intra-montane valleys. A well-known formation from this period is the San Onofre Breccia, which is composed of irregular blocks (some very large) of Franciscan Catalina schist. The San Onofre is distributed from the Santa Monica Mountains and offshore islands of the Transverse Ranges to Oceanside and the Los Coronados Islands (Baja California) in the Peninsular Ranges.

The San Onofre Breccia raises some of the same kinds of questions concerning the puzzling distribution of the Poway conglomerates. This rock is a very coarse, largely marine breccia containing large blocks of glaucophane schist (blueschist) derived from nearby exposures of Catalina schist. It has an oddly scattered distribution, but unlike the Poway conglomerates, not all the exposures are thought to have come from a single stream drainage. It seems more likely that the San Onofre occurrences were derived from a now submerged bedrock ridge or ridges off the coast. Exploratory oil drilling and seismic profiling suggest that there were six or seven basement ridges (some possessing Catalina schist) in the borderland off coastal southern California in middle Miocene time, when the San Onofre was deposited. Only three of these ridges are exposed today, and one, Santa Cruz Island, seems to lack the necessary glaucophane schist. The other two, Palos Verdes Peninsula and Santa Catalina Island, have exposures of blueschist, but do not appear to have been important sources of San Onofre Breccia because no nearby outcrops of San Onofre occur. The principal source area seems to have been a now submerged, long basement ridge extending from near Oceanside to Santa Monica, more or less parallel to the present shoreline.

Any reconstruction of the tectonic history of the Peninsular Ranges as well as coastal and offshore southern California, must account for these scattered exposures of the breccia and provide nearby source areas in order to explain the large and very angular schist clasts found in the San Onofre Breccia.

Pliocene nonmarine rocks, dated by vertebrate fossils, are thick and widespread in the northern Peninsular Ranges. For example, the Mount Eden and San Timoteo Canyon beds of siltstone, sandstone, and conglomerate reach thicknesses of 2,150 meters (7,000 feet). Although much thinner and more localized than equivalent strata to the north in the Los Angeles and Ventura basins, the marine Pliocene San Diego Formation is well exposed along the coast, especially near Pacific Beach north of Mission Bay. It is very fossiliferous. Because the marine Pliocene is limited to localities close to the present coast in the San Diego area, however, it is evident that Pliocene geography of the San Diego region was more like that seen today than is true in either the Los Angeles or Ventura areas, where marine Pliocene was much more extensive.

Quaternary deposits include fluviatile and lacustrine sediments in the interior and restricted volcanic and marine terrace deposits along the coast. Apparently none of the province was high enough to sustain Pleistocene glaciers, because no evidence to corroborate their existence has yet been discovered.

STRUCTURE

Faults dominate the structure of the Peninsular Ranges (Figure 9-4). Moreover, recent faulting has facilitated mapping of the chief through-going faults. Topographic expressions are clear, and major escarpments often can be traced for tens of kilometers.

Major faults are the San Jacinto and related branches within the San Jacinto zone and the Elsinore and associated faults within the Elsinore zone. The San Jacinto originates in the Transverse Ranges, bifurcating from the San Andreas and crossing province boundaries. The fault continues southeast through the Peninsular Range province and into the Colorado Desert. It appears that as they leave the Peninsular Range province southward, the San Jacinto and Elsinore zones converge and eventually join. The Elsinore may become the eastern fault of the Sierra Juarez scarp in Baja California. The San Jacinto fault may extend south into the head of the Gulf of California where it merges with other faults of the greater San Andreas system.

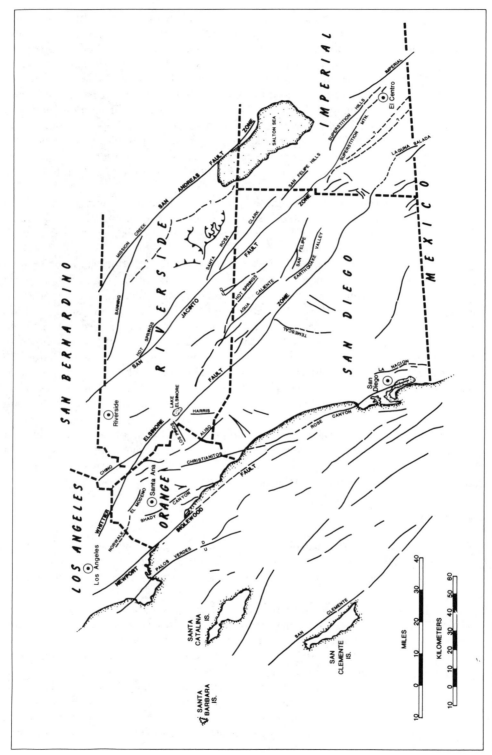

FIGURE 9-4 Principal faults, Peninsular Ranges.

284

San Jacinto Fault Zone

Within the Peninsular Ranges, the primary named faults of the San Jacinto zone are shown on Figure 9-4. Many smaller faults, some unnamed, anastomose between the major strands.

Movement on the San Jacinto fault system is post-Nevadan, probably mostly post-middle Tertiary, since all southern California batholithic rocks of Nevadan age are displaced. Evidence of movement includes crushed zones in dominately plutonic rocks, fault-line and eroded fault scarps, offset roof pendants, and linear topographic trends paralleling known faults (Figure 9-5). Evidence for most recent movement includes epicenter swarms, at least seven earthquakes of magnitude 6 or greater since 1890, and alignment of hot springs. Sags with small playas and sag ponds with permanent springs exist along some faults, and several fault-line valleys are being lengthened and deepened by modern streams.

As a rule, amount and type of movement have not been definitely established. Nevertheless, all offsets appear to be right slip and accumulated movement, since middle Tertiary time is probably about 24 kilometers (15 miles). Since the Pleistocene, slip totals about 5 kilometers (3.2 miles).

Evidence for vertical movement is also compelling. Recent movements have had strong dip-slip components that produced scarps and scarplets. Known faults of the San Jacinto zone are relatively up on the east and down on the west. The Perris Plain, with a general elevation from 450 to 600 meters (1,500–2,000 feet), is a major topographic feature between the San Jacinto on the northeast and the Elsinore fault on the southwest. Similarly, the Borrego Valley, near sea level, lies about 2,150 meters (7,000 feet) below the Santa Rosa Mountains. These features clearly reflect prominent vertical displacement on the San Jacinto fault zone. Vertical displacement

FIGURE 9-5 San Jacinto fault scarp near San Jacinto, Riverside County. Triangular facets are well developed at the base of the mountains, Mount San Jacinto is on the right. (Photo by R. M. Norris)

of similar magnitude presumably is expressed along the Banning (or South Banning) fault on the northeast side of the San Jacinto Mountains.

It has been suggested recently that marked crustal uplift is likely when two strike slip faults converge, forcing the block between to rise in the direction of convergence. This may provide the explanation for the high San Jacinto Mountain block, which lies between the convergence of the San Jacinto and San Andreas fault. This concept is discussed more fully in Chapter 13.

More detailed study is needed before movement patterns for the San Jacinto zone can be appraised realistically. Furthermore, even partial reliance on uncorroborated but tempting interpretations can only hinder our eventual understanding. For instance, geometric relationships with the San Andreas that seem obvious from patterns on a structural map of California may sometimes exert strong psychological pressure on geologists to infer major strike-slip movement, even though supporting evidence is actually weak or nonexistent. Broad-brush generalizations may have merit, but they also deserve healthy skepticism.

Elsinore Fault Zone

The major named faults of the Elsinore zone are the Elsinore, Wildomar, Aguanga, Agua Tibia, Earthquake Valley, and Hot Springs.[1] Many interconnecting faults are also present. Evidence for faulting within the Elsinore zone is similar to that for the San Jacinto zone except that more demarcations between distinctive rock types are mappable and topographic expression is less marked. In addition, more horst and graben features are apparent. (These structures are small topographically but impressive geologically.)

Parallel to the San Jacinto and subparallel to the San Andreas, the Elsinore fault zone predictably has been assigned right-slip movement. On a small scale, such movement may be demonstrable, but apparently dip slip also has persisted throughout the zone's history. In the Elsinore graben proper, vertical movements of several hundred meters are known (Figure 9-6). Buried parallel faults are separated by at least 600 meters (2,000 feet) of nonmarine Quaternary sediment (shown by well records) (Figure 9-7). On the Agua Tibia fault, landslides and rubble from adjacent uplands bury the fault's trace for several kilometers.

The faults of the Elsinore zone disappear northwest beneath the sediments of the Los Angeles Basin, where the Elsinore apparently bifurcates into the Chino fault on the northeast and the Whittier fault on the southwest.

The nature of the junction (if any) of these two northern branches of the Elsinore fault with the Sierra Madre-Cucamonga frontal fault at the base of the San Gabriel Mountains is not at all clear, because it is deeply buried beneath alluvial fans from the mountains.

At present, the Elsinore fault is believed to be about the same age as the San Jacinto, and seems to have accumulated about 30 kilometers (18 miles) of right slip

---

[1]Unfortunately, the latter is not the same as the San Jacinto zone fault of the same name; "Hot Springs" faults are almost as common in California as "Rattlesnake Canyons."

FIGURE 9-6 Elsinore graben: view to the northwest from Temecula. San Gabriel Mountains with snow-capped Mount San Antonio (Baldy) are in the right distance, and the Santa Ana Mountains are on the left. (Photo by Spence Air Photos, courtesy of Department of Geography, University of California, Los Angeles)

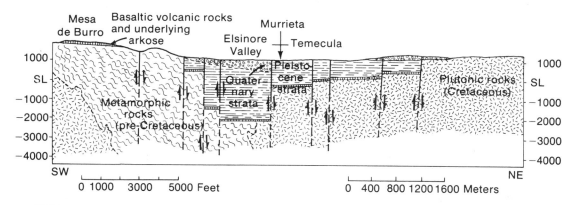

FIGURE 9-7 Cross section of the Elsinore fault zone. Depth shown in feet. (Source: California Division of Mines and Geology)

based on matching facies-change boundaries in Paleocene rocks on both sides of the fault. The Lake Elsinore Basin is regarded as a pull-apart structure of the sort previously described in the Basin and Range and Colorado Desert provinces. The basin itself seems to require about 6 kilometers (3.6 miles) of right slip.

GEOMORPHOLOGY

The Peninsular Ranges show a number of distinctive erosional surfaces the origin and history of which are only partially understood. For convenience, these surfaces may be divided into three groups. First is the Perris Plain (mentioned earlier), a broad, nearly flat surface dotted with bedrock hills, extending from near Corona, southeasterly to Hemet. This plain has an average elevation of about 520 meters (1,700 feet) (Figure 9-8). The numerous bedrock hills that interrupt its surface have been described as residual knobs of resistant rock, which survived prolonged erosion (monadnocks). It has been suggested that a surface of low relief was developed on the crystalline bedrock, leaving behind the scattered monadnocks. Drainage was then to the east via the San Jacinto River, deduced from buried channels located geophysically and by well records. Development of this channeled surface was followed by a period of nonmarine sedimentation, which produced today's Perris surface. Drainage shifted toward the Pacific and the Perris surface was eroded, leaving scraps of the older, higher Perris surface at elevations of about 620 meters (2,100 feet). The lower present-day surface corresponds generally to what we now refer to as the Perris Plain.

Very late Quaternary uplift of the Santa Ana Mountains no doubt played an important role in this story and is indicated by the antecedent nature of the Santa Ana River, which crosses the gently sloping plain northeast of the Santa Ana Mountains only to enter a narrow gorge cut across the northern tip of the range. Apparently it would have been simpler and more logical for the Santa Ana River to have followed

FIGURE 9-8 Perris Plain near Lakeview, Riverside County. The plain is an erosional surface dotted with residual hills composed mostly of Mesozoic intrusive rocks of the Southern California batholith. (Photo by R. M. Norris)

FIGURE 9-9 Broad upland valley near Campo, San Diego County. Valleys of this sort are thought to be remnants of an older subdued topography now being destroyed as erosion attacks the margins of the uplifted Peninsular Ranges. (Photo by R. M. Norris)

the low ground around the northern end of the mountains. It did not do so, however, which indicates that the river established its course before the mountains were formed and that uplift of these young mountains was sufficiently slow that the river was able to maintain its course rather than being diverted to the north. The Santa Margarita River crosses the southern end of the Santa Ana Mountains and is probably antecedent as well.

The second group of erosional surfaces in the Peninsular Ranges of California are those developed in inland San Diego County, almost entirely on crystalline plutonic rocks. These surfaces form a series of broad benches or steps, each tread successively higher and narrower to the east, culminating in the Laguna Mountains at Cuyamaca Peak (1,986 meters or 6,512 feet). These benches are believed to be parts of the same erosional surface, later faulted. Dating of these erosional and faulting events is not well controlled (Figure 9-9).

Streams on the benches typically occupy broad, aggraded valleys, but where they cross the risers between the benches, they occupy deep narrow canyons. Occasionally, elevated blocks rise above the benches, and in such cases streams have cut defile canyons through these higher blocks, suggesting that the streams are antecedent or superimposed. The San Luis Rey is one such river.

The large-scale stream pattern suggests some control by faults, master joints, or perhaps fabric patterns in the plutonic rocks. These structurally controlled patterns are most obvious on the risers between the broad benches. Drainage on the benches is usually dendritic, reflecting little or no structural or lithologic control.

The third set of erosional surfaces is the prominent flights of marine terraces that occur from the Orange–San Diego county line near San Onofre southward to the Mexican border and beyond. Multiple terraces are most obvious in northern San Diego County north of Oceanside, but are wider in the area near San Diego Bay, where the city and its suburbs have spread across these wide, relatively flat platforms.

Between 13 and 22 separate terraces have been recognized, but three or four are most prominent.

Few, if any, marine terraces in California predate the Pleistocene. The San Onofre-Oceanside terraces are cut by stream channels younger than the terraces themselves, but even the channels reflect important geologic changes since their formation. Furthermore, in a geologically active area like California, so far as terrace origin is concerned, it has been difficult to distinguish clearly between the effects of a fluctuating Pleistocene sea level and the effects of vertical tectonic movement. This problem is beginning to be resolved as we get better dates on the terraces themselves as well as on the high and low sea levels of the Pleistocene.

During relatively stable intervals in the Quaternary, marine erosion quickly cut terraces into the coastal margin of the Peninsular Ranges. The width of these terraces was a function of the lengths of the stable intervals, but it was also affected by such considerations as degree of exposure to waves, slope of the sea floor, and rock resistance. Generally, the longer the stable interval the wider the resulting terrace— although there are many exceptions. During a rapid *relative* uplift of the land, terrace cutting gave way to emergence of the sea bottom. During the stable period that followed, the exposed sea bottom was leveled into a new terrace with a cliff at its landward side. Once the cliffs were elevated beyond the reach of the sea, land erosion and mass movement processes began to attack them.

During the final Pleistocene glacial stage, sea level was lowered about 90 meters (300 feet). This forced streams crossing the terraces to cut deep gullies and extend themselves across the newly exposed sea floor. This modification kept pace with sea level changes, because streams adjust quickly to new circumstances.

As the postglacial sea level rose, beginning about 11,000 years ago, stream courses were shortened and the lower ends of the channels were flooded by the rising ocean, forming estuaries. Streams seeking the sea plus tidal action in lagoons oppose the final closing and filling of estuaries, however. Quiet estuaries still are being filled by river sediment, while coastal currents straighten the shoreline by building bay-mouth bars across estuary entrances. The effects of these competing processes are evident along the San Diego County coast. Some former lagoons are now filled completely; others are nearly filled but persist as salt marshes; still others retain open-water lagoons of varying sizes. Most have bay-mouth bars separating them from the open ocean. During the past 50 years or so, most of these features have been subjected to some degree of human management. In few cases has nature been allowed to proceed unchecked (see Figure 1-17).

SUBORDINATE FEATURES

San Jacinto-Santa Rosa Mountains

The highest block of the Peninsular Ranges is the easternmost San Jacinto Mountains, which include the second highest peak in southern California, San Jacinto Peak (3,296 meters or 10,805 feet). This peak is the highest in the entire Peninsular Range, all the way to the southern tip of Baja California. The San Jacinto Mountains and their southern continuation, the Santa Rosa Mountains, contain a thick section of banded gneiss and widespread quartz diorite plutons along the saddle between them. Both ranges are giant horsts caught between the San Jacinto fault on the west and the Banning and other buried frontal faults on the north and east.

FIGURE 9-10 Martinez Mountain landslide, Santa Rosa Mountains south of Indio, Riverside County. (Photo by G. L. Meyer)

The San Jacinto Mountains are relatively well watered with some permanent streams and a snowcap that often persists into summer, but the Santa Rosa Mountains are much drier with few springs and no permanent streams.

The eastern side of the Santa Rosa Mountains is the site of one of California's great fossil landslides—the Martinez Mountain slide, which probably occurred between 15,000 and 20,000 years ago. The slide began just below the summit of Martinez Mountain at an elevation of about 1,900 meters (6,200 feet) and ended at about 70 meters (230 feet). The total drop was thus about 1,830 meters (6,000 feet) in a distance of 7.5 kilometers (4.5 miles). The slide consists of coarse granitic rubble from the east face of Martinez Mountain. The slide is located about 15 kilometers (9 miles) due south of the city of Indio (Figures 9-10 and 9-11).

Elsinore-Temecula Trough

The Elsinore-Temecula trough is the linear, low-lying block northeast of the Santa Ana Mountains and southwest of the Perris Plain. It extends from Corona, on the northwest about 48 kilometers (30 miles) southeast and has a maximum width of 4.8 kilometers (3 miles). The Lake Elsinore Basin is a good example of a pull-apart structure or sag developed along right slip faults where the main faults step to the left. The Glen Ivy fault on the northeast forms the main strand of the Elsinore zone and continues to the northwest. The Wildomar fault represents the main strand continuing to the southeast (Figure 9-12).

Lake Elsinore occupies a shallow closed basin (sag pond). Because of the generally low rainfall in the area, the lake would be intermittent under natural conditions, filling only when the San Jacinto River was in flood, and spilling north into the Santa Ana River drainage only under very rare conditions, such as it did in 1916 to 1917. In order to forestall the total disappearance of Lake Elsinore, water from the Colorado River aqueduct has been used to maintain the lake level in recent years.

The trough contains lower Tertiary strata that originated in brackish water from

FIGURE 9-11 Toe of Martinez Mountain landslide. Compare block size with the person standing in the lower right. (Photo by R. M. Norris)

which lignite, ceramic-quality clays, and glass sand have been mined (Paleocene Silverado Formation). The principal clay and glass sand deposits are at Alberhill, where there are extensive pits and some processing plants. Lignite is no longer mined.

The Elsinore-Temecula trough includes one of the most representative sections of post-Nevadan rocks within the Peninsular Ranges. Its geology is shown by the cross section in Figure 9-7.

SPECIAL INTEREST FEATURES

Cuyamaca Rancho State Park

In eastern San Diego County, on the summits of the Laguna Mountains (elevations to 1,800 meters or 6,000 feet), recreational facilities incorporating the area's flat-topped topography have been developed as Cuyamaca Rancho State Park. The Laguna Mountains are bounded on the east by the Elsinore fault, and a bold, eroded scarp of granitic crystalline rocks extends for 24 to 30 kilometers (15–20 miles). The rocks are Cretaceous igneous plutonics, except for a few remnants of older metamorphic rocks into which the plutons intruded. Occasionally, pegmatitic dikes some-

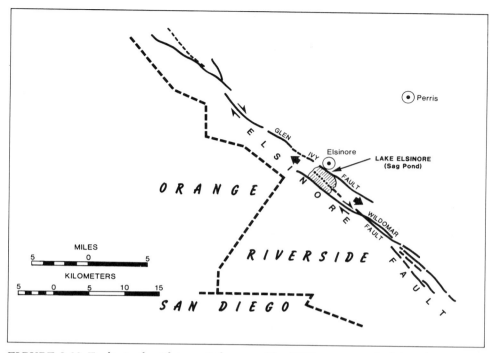

FIGURE 9-12 Faults in the Elsinore Lake area. The Wildomar strand of the Elsinore fault dies out on the southwestern side of the lake and is replaced by the Glen Ivy strand on the northeast (to the right). Right steps of this sort on right lateral faults produce pull-apart basins between the strands.

what like those of the Pala and Mesa Grande gem areas occur. Laguna Mountains pegmatites have yielded few minerals of special interest, however.

Julian Mining Area

Julian is the center of a mining district noted during the nineteenth century for its gold and nickel deposits. The Banner Queen was one of the famous gold mines. Today, however, Julian is probably better known for its apple orchards.

Along the Elsinore fault where igneous intrusives cut the metamorphic sequences of the Jurassic (?) Julian Schist, nickel mineralization occurred in veins and pockets. Since it is in usually short supply, and with the U.S. demand depending upon foreign sources, nickel has always aroused interest. Principal ore was the nickel-bearing iron sulfide pyrrhotite, which occurred abundantly. Percentage of nickel was low and inconsistent, however, and although several mining revivals have occurred, Julian is maintained only to serve tourists and local ranchers. Old mines abound, but few may actually be seen by visitors.

Gem Areas

Pegmatitic dikes occur in two adjacent areas in the upper drainage of the San Luis Rey River and constitute the world-famous Pala and Mesa Grande, Rincon, and

Ramona gem localities. The network of related dikes was formed from hot, watery solutions associated with granitic intrusions of the Nevadan orogeny. The solutions invaded joints and zones of weakness in wall rocks, crosscutting older formations and in some cases the granitic parent rocks. Radiometric dating suggests 100 to 110 million years ago (Cretaceous) as the time of emplacement.

The dikes average about 3 meters (10 feet) thick and are often more than $1\frac{1}{2}$ kilometers (1 mile) long. Podlike thicker zones in the dikes contain some cavities the size of a small room, where giant crystals of pink, green, and yellow transparent tourmaline grew. Individual crystals from 2 to 10 centimeters (5–20 inches) in diameter and up to a meter (3 feet) long have been collected. Other minerals extracted include beryl, kunzite (a gem variety of spodumene first discovered at Pala), gem garnet, and occasional topaz. Such concentrations of rare minerals seldom occur, and the reason for this unique and prolific development in the Peninsular Range is unknown, but many of the minerals are due to the presence of unusual amounts of boron and lithium in the mineralizing solutions.

Mining heyday was from 1900 to 1910, when gem tourmaline was highly prized, particularly in the Orient. Decline in interest in tourmaline forced closure of most mines, but production of specimen materials continues. Many museums throughout the world feature San Diego county gem minerals; some of the best collections are at Harvard University in Massachusetts and at Balboa Park in San Diego. Pegmatites are widespread in the Peninsular Ranges, and in addition to the two most famous areas, the localities of Aguanga, Cahuilla, and the Santa Rosa Mountains have yielded gem material. Useful nongem minerals from the region are commercial quantities of feldspar, lithium products, and quartz.

Crestmore Quarries

The Crestmore limestone quarries in western Riverside County have provided one of the world's largest suites of contact metamorphic minerals. By 1971, 140 minerals had been described, including several new species, and new compounds are still being discovered. The minerals are generally unspectacular in appearance, but they are scientifically important and knowledge of their chemistry has enhanced understanding of many ore- and mineral-forming processes. Occurrence is in a limestone-marble sequence of presumably late Paleozoic age that was intruded by quartz diorite and quartz monzonite of the Nevadan plutonic episode. Silicification of the limestone accompanied by introduction of exotic trace elements have combined to yield the unusual minerals. Suites of Crestmore minerals are displayed in mineral collections worldwide.

Hot Springs of the San Jacinto and Elsinore Fault Zones

Hot and mineral springs have been known in the Peninsular Ranges for some time. Both Indians and Spanish used the hot springs, and Americans developed them as mineral spas. This use has waned somewhat because the medicinal value of most waters is now discounted.

Along the San Jacinto fault zone at the west base of the San Jacinto Mountains are Gilman, Eden, Saboba, and San Jacinto hot springs. Other less well known but nonetheless sizable springs rise along the zone's southern extensions into Borrego Valley. Water volume and temperature vary considerably in such springs.

The Elsinore zone contains springs also. The town of Elsinore was known originally for its sulfur water and spa. Warner Springs is one of the province's larger springs, and Murietta Hot Springs near Temecula is one of the oldest spas.

Although these springs may contain some juvenile water derived directly from igneous processes at depth, virtually all of the hot water represents normal meteoric waters from rain and snow warmed by contact with hot rock at depth along the fault zones. Both fault zones have many smaller springs along traces of their branch faults.

Newport Bay

Newport Bay is located at the southeastern end of the Los Angeles coastal plain, with the San Joaquin Hills a short distance to the east. The bay itself consists of two main parts. The lower portion parallels the coast and is sheltered from the open ocean by a sand spit on which the towns of Newport and Balboa are situated. The upper portion is a winding estuary extending about 8 kilometers (5 miles) inland from the lower bay, widening toward its upper end (Figure 9-13).

It is likely that the upper portion is the drowned channel of the Santa Ana River, which previously emptied into the bay. In 1915, because of severe silting that resulted

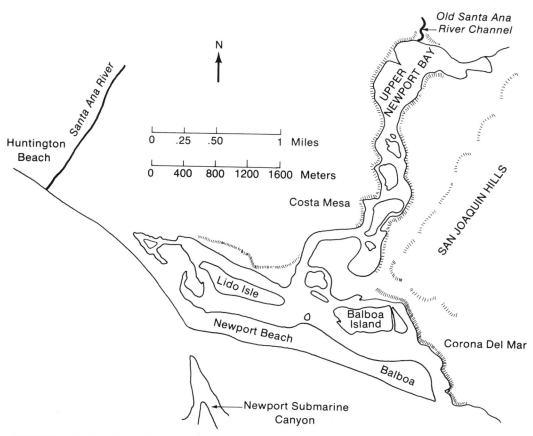

FIGURE 9-13 Sketch of Newport Bay.

FIGURE 9-14 Newport Bay, Orange County, in 1948. (Fairchild Aerial Surveys, courtesy Department of Geography, University of California, Los Angeles)

from flooding in the Santa Ana, the river was diverted into a man-made outlet about 2.4 kilometers (1½ miles) north of Newport harbor. Like most estuaries and salt marshes along the California coast, this bay is a temporary feature that, barring human interference or sea-level change, would be rapidly converted to dry land (Figure 9-14).

Newport Bay differs from some coastal estuaries because it is rimmed with steep cliffs 27 to 30 meters (90–100 feet) high, particularly around the lower bay and on the east side of the upper bay. Cliffs along the bay's west side, and at the edge of flat-topped Costa Mesa, gradually lose height inland and merge with the Santa Ana River plain near the head of the bay. As a result, the bay is actually a steep-sided water-filled valley set down into slightly uplifted mesa lands that extend west from the San Joaquin Hills. The river-cut estuary extends across a block that has been elevated as much as 30 meters (100 feet) above the general level of the main river plain to the north and west. If the present topographic pattern had existed when the Santa Ana River first established its course, the river would have avoided elevated Costa Mesa and entered the sea north of Newport Bay. It is therefore likely that Costa Mesa was elevated after the river had developed a well-established channel to the sea. During uplift, the river maintained its channel even though it had to cut a gorge across a slowly rising upland. If this interpretation is correct, the river channel that now forms Newport Bay is antecedent.

Newport Bay is dotted with islands of various sizes and shapes; many of these islands are so extensively urbanized that examining their composition and structure is difficult. With few exceptions, all the islands result from dredging. Bay floor materials dredged to provide boating channels have been dumped on shallow mud banks to make islands.

The perceptible bulge in the bay's sand spit is probably the only remaining natural irregularity on the coast between Seal Beach and Corona del Mar. This bulge is a direct result of the presence of Newport submarine canyon, which approaches shore at the point of the bulge. In the late nineteenth century, a railroad pier was built into the head of the canyon. The deeper water was an advantage to shipping, and it had been observed that waves were usually lower over the canyon than elsewhere. The presence of submarine canyons close to shore always reduces the height of waves over the canyons and in the adjacent surf. This reduction in wave energy also favors

FIGURE 9-15a Point Loma headland and the entrance to San Diego Bay. Point Loma is an elevated ridge composed of marine Cretaceous sandstones and shales with a thin cap of Pleistocene gravel. (Photo U.S. Department of Agriculture, courtesy Map and Imagery Laboratory, Library, University of California, Santa Barbara)

FIGURE 9-15b Southern tip of Point Loma showing an exposure of the Rosario Group of late Cretaceous age. The Rosario Group is divided into the Point Loma Formation (lower part), and the Cabrillo Formation (upper part). Rocks include sandstones, conglomerates, and shales. (Photo courtesy Dale Frost, Port of San Diego).

deposition of sand transported along the beach by longshore currents, contributing to formation of seaward bulges in the beach. Similar bulges on low sandy coasts occur elsewhere in California and may be caused by reefs, islands, or breakwaters a short distance offshore. Like submarine canyons, if they are close enough to the beach, these features also reduce wave height and encourage sand deposition.

Coronado Island and Silver (Coronado) Strand

San Diego Bay is the only sizable natural harbor on the California coast south of San Francisco. It owes its existence to Point Loma, a mountainous ridge of marine Cretaceous sedimentary rock—the Point Loma and Cabrillo formations—rising almost 120 meters (400 feet) above sea level and forming a prominent barrier west of the city of San Diego and the northern part of the bay (Figure 9-15). Because Point Loma shelters much of the bay from westerly and northwesterly winds and waves, a long, low, curving sand spit has gradually built northward from the mouth of the Tijuana River. At the north end of the spit are two low islands that are probably

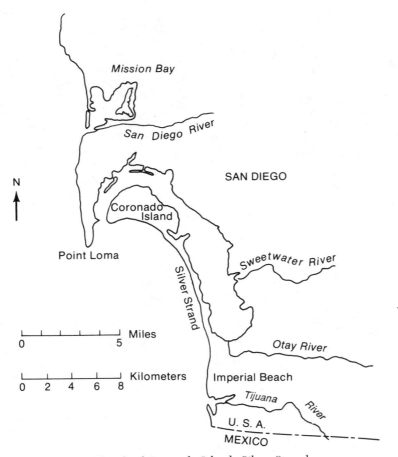

FIGURE 9-16 Sketch of Coronado Island–Silver Strand.

remnants of the Nestor terrace, which is exposed around the city of San Diego at an elevation of about $7\frac{1}{2}$ meters (25 feet). These two islands, once known as North Island and South Island, have been joined by a sand spit, which also connects them to the mainland on the south. North Island is the site of a Naval Air Station and South Island is the site of the city of Coronado. The inlet that once separated the two islands was called Spanish Bight, but it has since been filled (Figure 9-16).

Without human interference in the area, San Diego Bay would have filled with sediment delivered by the Otay, Sweetwater, and San Diego rivers. The San Diego River has been diverted into a channel along the south side of Mission Bay. In addition, it is likely that the northward drift of beach sand that connected Coronado (South) Island with the mainland and Coronado and North Islands together eventually would have blocked or nearly blocked the harbor entrance. Breakwaters, channel maintenance, and tidal action prevent this from occurring.

REFERENCES

General

Anonymous, 1976. History of Coal in California. California Geology, v. 29, pp. 203–203.

Hertlein, L. G., and U. S. Grant, IV, 1954. Geology of the Oceanside-San Diego Coastal Area, Southern California. *In* Geology of Southern California. Calif. Div. Mines and Geology Bull. 170, pp. 53–64.

Jahns, R. H., 1954. Geology of the Peninsular Range Province, Southern California and Baja California. *In* Geology of Southern California. Calif. Div. Mines and Geology Bull. 170, pp. 29–52.

Larsen, E. W., and others, 1951. Crystalline Rocks of South-western California. Calif. Div. Mines and Geology Bull. 159.

Morton, D. M., and C. H. Gray, 1971. Geology of the Northern Peninsular Ranges, Southern California. *In* Geological Excursions in Southern California, W. Elders ed., Univ. Calif. Riverside Museum Contrib. 1, pp. 60–93.

Peterson, G. L., and others, 1970. Geology of the Peninsular Ranges. Mineral Information Service (now California Geology), v. 23, pp. 124–127.

Sharp, R. V., 1967. San Jacinto Fault Zone in the Peninsular Ranges of Southern California. Geol. Soc. Amer. Bull., v. 78, pp. 707–730.

Woodford, A. O., and others, 1954. Geology of the Los Angeles Basin. *In* Geology of Southern California. Calif. Div. Mines and Geology Bull. 170, pp. 65–82.

Special

Burnham, C. W., 1954. Contact Metamorphism at Crestmore, California. *In* Geology of Southern California. Calif. Div. Mines and Geology Bull. 170, pp. 61–70.

Clark, W. B., 1970. Gold Districts of California. Calif. Div. Mines and Geology Bull. 193.

Creasey, S. C., 1946. Geology and Nickel Mineralization of the Julian-Cuyamaca Area, San Diego County, California. Calif. Jour. Mines and Geology, v. 42, pp. 15–29.

Devito, F., and others, 1971. Contact Metamorphic Minerals at Crestmore Quarry, Riverside, California. *In* Geological Excursions in Southern California, W. Elders Ed., Univ. Calif. Riverside Museum Cont. 1, pp. 94–125.

Donnelly, M. G., 1935. Geology and Mineral Deposits of the Julian District, San Diego County, California. Calif. Jour. Mines and Geology, v. 30, pp. 331–370.

Jahns, R. H., 1954. Pegmatites of Southern California. *In* Geology of Southern California. Calif. Div. Mines and Geology Bull. 170, pp. 37–50.

———— and L. A. Wright, 1951. Gem and Lithium-Bearing Pegmatites of the Pala District, San Diego County, California. Calif. Div. Mines and Geology Spec. Rept. 7A.

Kennedy, Michael P., 1973. Sea-Cliff Erosion at Sunset Cliffs, San Diego. California Geology, v. 26, pp. 27–31.

Murdoch, Joseph, 1961. Crestmore—Past and Present. Amer. Mineralogist, v. 46, pp. 245–257.

Smith, Jean DeM., 1977. Gem Pegmatites in San Diego County. California Geology, v. 30, pp. 43–44.

Sutherland, J. C., 1935. Geological Investigation of the Clays of Riverside and Orange Counties, Southern California. Calif. Jour. Mines and Geology, v. 31, pp. 51–87.

Wehlage, E. F., 1984. Black Ghost of the Silverado. California Geology, v. 37, pp. 29–35.

ten

TRANSVERSE RANGES

Let us leave a few problems for our children
to solve; otherwise they might be so bored.

—*Tom F. W. Barth*

Unlike California's other geologic provinces, the Transverse Ranges form a conspicuously east-west trending unit. A few short ranges with east-west trend do occur in
the other provinces, but they are all exceptions to the state's general structural
alignment. It is interesting to note that in North America as a whole, apart from
Alaska, there are few prominent east-west mountain belts. Most of the continent is
dominated by structures that trend roughly north-south. This is no accident and
when we have a better understanding of the cause of these anomalous east-west
ranges, we will also have a more complete understanding of the tectonic history of
the continent.

The Transverse Ranges include California's highest peaks south of the central
Sierra Nevada, the only pre-Phanerozoic rocks in the coastal mountains of the United
States and probably North America, and four of the eight islands off the southern
California coast. The province extends about 520 kilometers (320 miles) from Point
Arguello and San Miguel Island on the west to the mountains of Joshua Tree National
Monument on the east where the province merges with the Mojave and Colorado
deserts. Along the Ventura-Los Angeles County line, the province reaches a maxi-
mum width of 96 kilometers (60 miles); it narrows to about 64 kilometers (40 miles)
at its western end. Figure 10-1 shows the locations of important Transverse Range
features.

From northwestern Ventura County east to Cajon Pass, the San Andreas fault
system forms the northern boundary of the province. The fault deviates markedly
from its usual northwest-southeast trend, suggesting that it too was influenced by
the forces that produced this oddly aligned province. The "Big Bend" as this deviation
is known, indeed results in an east-west zone of compression where the northward

301

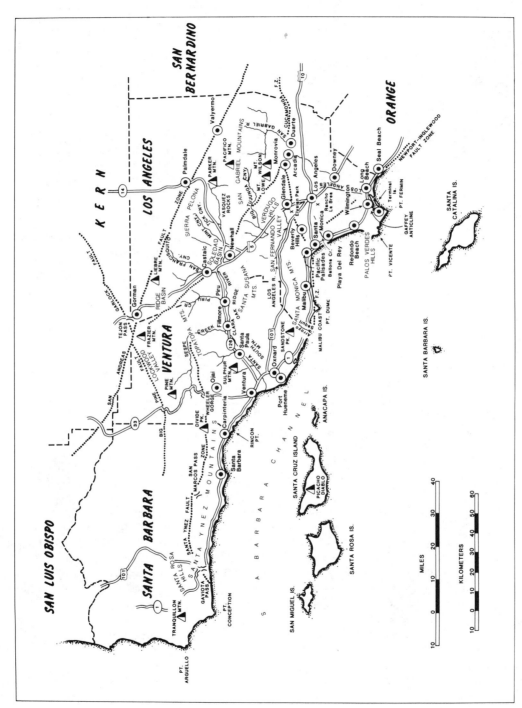

FIGURE 10-1a Place names, western Transverse Range Province.

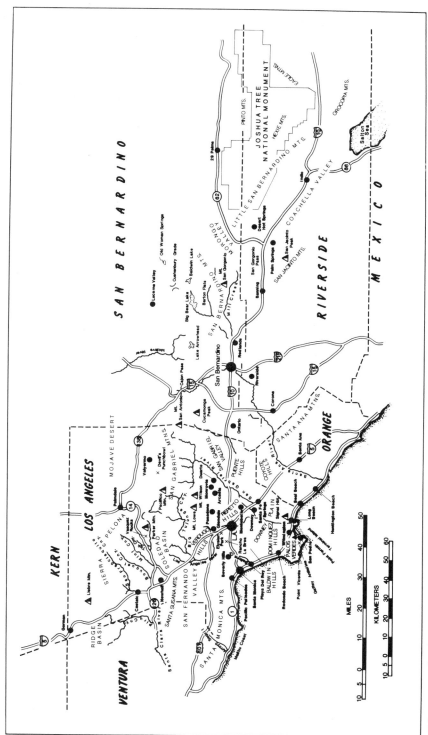

FIGURE 10-1b Place names, eastern Transverse Range Province.

303

movement of the slice of California lying west of the fault pushes against the westward bulge on the opposite side of the fault in the Big Bend area, compressing the rocks on both sides of the fault. This subject is discussed more fully later in this chapter.

The province subdivides into individual ranges with intervening valleys. Several of these units are topographically distinctive and easily distinguished from one another, but some complex groups of ridges and valleys are not readily distinguishable. The ranges of the province are separated by alluviated, broadly synclinal valleys, narrow stream canyons, prominent faults, and sometimes by downwarps of phenomenal magnitude. Although the geologic column for this province is incomplete, it does include rocks of all ages except Archean and possibly some parts of the early Paleozoic (Figure 10-2). Despite their comparatively small area, the Transverse Ranges seem to incorporate a greater spectrum of rock types and structure than any other province in the state.

Composite Geologic Column, Western Transverse Ranges

HOLOCENE	Alluvium, marine, and nonmarine terrace deposits
PLEISTOCENE	Santa Barbara (marine) Casitas (nonmarine)
PLIOCENE	Fernando, Repetto (marine)
MIOCENE	Sisquoc (marine) Monterey (marine) Tranquillon and Conejo volcanics Rincon Mudstone (marine) Blanca (marine) San Onofre Breccia (marine) Vaqueros (marine)
OLIGOCENE	Gaviota (marine) Sespe (nonmarine), Alegria (marine)
EOCENE	Coldwater (marine) Sacate (marine) Cozy Dell (marine) Matilija, Jolla Vieja (marine) Anita (marine) Juncal, Canada, Sierra Blanca (marine)
PALEOCENE	Pozo (marine)
CRETACEOUS	Jalama (marine) Espada (marine) Franciscan (marine)
JURASSIC	Willows Diorite, Alamos Tonalite
PRE-JURASSIC	Santa Cruz Island Schist

FIGURE 10-2 Geologic columns, Transverse Ranges.

Composite Geologic Column, San Gabriel Mountains, Santa Monica Mountains and San Fernando Valley

HOLOCENE	Alluvium
PLEISTOCENE	
PLIOCENE	Saugus (nonmarine) Pico, Repetto, Fernando (marine), Crowder (nonmarine)
MIOCENE	Monterey (Modelo) (marine), Mint Canyon (nonmarine) Conejo volcanics, Tertiary intrusive rocks Topanga (marine), Punchbowl (nonmarine), Tick Canyon (nonmarine) Vaqueros (marine)
OLIGOCENE	Sespe (nonmarine), Vasquez (nonmarine and volcanic)
EOCENE	Llajas (marine)
PALEOCENE	Coal Canyon, San Francisquito (marine), Simi Conglomerate (marine)
CRETACEOUS	Tuna Canyon, Chatsworth (marine) Trabuco Conglomerate (nonmarine) Pelona Schist, Wilson Diorite
JURASSIC	Santa Monica Slate (marine)
PERMO-TRIASSIC	Lowe Granodiorite
PALEOZOIC	Miscellaneous roof pendants
PROTEROZOIC	Syenite, anorthosite Echo Granite Gabbro Mendenhall Gneiss

Composite Geologic Column, San Bernardino and Little San Bernardino Mountains

HOLOCENE	Alluvium
PLEISTOCENE	Alluvium, glacial deposits, landslides
PLIOCENE	Old Woman Sandstone (nonmarine), Crowder (nonmarine)
MIOCENE	Santa Ana Sandstone (nonmarine), Potato (nonmarine), Mill Creek (nonmarine) Pioneertown Basalt

FIGURE 10-2 (*Continued*)

Composite Geologic Column, San Bernardino and Little San Bernardino Mountains (*Continued*)

CRETACEOUS	Pelona Schist
JURASSIC and CRETACEOUS	Pegmatites, Holcomb Quartz Monzonite, Palms Granite, Cactus Granite, White Tank Quartz Monzonite, Woodson Mountain Granodiorite, San Marcos Gabbro, and other intrusive rocks
PERMO-TRIASSIC	Fawnskin Monzonite Lowe Granodiorite
MISSISSIPPIAN to CAMBRIAN	Furnace Limestone (marine)
EDIACARIAN ?	Chicopee Canyon, Saragossa Quartzite
PROTEROZOIC	San Gabriel, San Gorgonio, and Cucamonga complexes, Pinto Gneiss, Baldwin Gneiss, Waterman Gneiss

FIGURE 10-2 (*Continued*)

SANTA YNEZ MOUNTAINS

The Santa Ynez Mountains form a continuous, south-facing rampart along the Santa Barbara coast from Point Arguello east for nearly 110 kilometers (70 miles) to near Ojai, where they merge with the mountains of northern Ventura County. For much of this distance the range crowds the shoreline, leaving scant room for a coastal plain. Figure 10-3 is a view of the western Transverse Ranges.

From Point Arguello to Gaviota Pass, the range is generally less than 600 meters high (2,000 feet), although Tranquillon Mountain and several peaks near Point Arguello are slightly higher. The mountains gain height rapidly to the east of Gaviota Pass, reaching 1,311 meters (4,298 feet) at Santa Ynez Peak. San Marcos Pass, near Santa Barbara, occupies a low saddle formed by a synclinal fold that obliquely crosses the axis of the range. East of San Marcos Pass, the mountains rise again, reaching a maximum elevation at Divide Peak (1,430 meters or 4,690 feet) close to the Santa Barbara-Ventura County line.

The Santa Ynez range is an anticline, with a major fault along its axis. Because the south limb of this fold is so much more prominent than the north, many geologists prefer to describe the range as a south-dipping homocline with beds strongly overturned east of Carpinteria.

Rocks and Geologic History

The oldest rocks exposed in the Santa Ynez Mountains are parts of the Franciscan formation. Here, as in the Coast Ranges to the north, the Franciscan consists of a highly mixed assemblage of deep-water marine sedimentary rocks such as radiolarian

FIGURE 10-3 View to the southeast from the western Santa Ynez Mountains. The steeply-dipping beds of the south flank of the range form strike ridges (durable beds) and intervening strike valleys (weaker beds) that parallel the coastline. The narrow coastal plain widens to the east and borders the Santa Barbara Channel. In spite of protection afforded by the offshore islands (background), the channel waters are usually rough because the channel is open to the west and receives strong prevailing and storm winds. (Photo by Mark Hurd Aerial Surveys)

chert and graywacke, plus altered basalt, serpentinite, and ultrabasic crystalline rocks thought to be derived from oceanic crust. These Franciscan rocks probably make up the basement in this part of the Transverse Ranges and are assigned a late Jurassic to late Cretaceous age.

The processes that formed the Franciscan assemblage evidently terminated during the late Cretaceous in this part of California, although the region remained beneath the sea and sandstones and shales were deposited. Prevailing conditions of the early Tertiary are unknown because no Paleocene rocks are exposed in the Santa Ynez Mountains and none have turned up in well records, although such rocks are known from areas to the east and the islands to the south.

Middle and upper Eocene rocks are all marine and are well represented in the range, attaining a thickness of more than 3,000 meters (10,000 feet). Most of the modern range has been carved from these rocks, and the more resistant sandstones form nearly all the higher peaks. Toward the top of the Eocene section, within the Coldwater sandstone, near-shore conditions are revealed by oyster beds and occasional streaks of red shale, presumably washed into the shallow waters from nearby land.

By Oligocene time, the sea had withdrawn from the eastern region of the Santa Ynez Range, leaving a broad coastal plain on which were deposited the sands, gravels, and silts of the Sespe Formation. Although green, tan, and buff beds are evident, its prominent red color makes the Sespe especially distinctive. It can be traced from west of San Marcos Pass into central Ventura County. A few vertebrate remains such as horse teeth have been found, providing dates for the formation.

In the west, marine conditions continued during the Oligocene, and initially sands resembling the Coldwater Formation were deposited in the shallow seas. Even though the shoreline changed position, as shown by interfingering of the land-deposited Sespe on the east and the marine beds to the west, the western Santa Ynez Mountains record continuous marine conditions throughout the Oligocene.

Before the Oligocene closed, the sea again spread over the Santa Ynez area. The first unit from this marine invasion is the Vaqueros Formation. This sequence contains much shell material, including some heavy-shelled scallops (pectens), which endured the harsh and abrasive conditions on a gravelly or pebbly shore.

As Miocene time began, the sea advanced north and northeast across what is now the Santa Barbara-Ventura area, bringing deeper-water conditions to the site of the Santa Ynez Mountains. The sands and gravels of the Vaqueros were replaced by the fine silts and soft clays of the Rincon Formation. Exposures of the Rincon frequently weather to form expansive and unstable clay soils that have subsequently created problems for buildings placed on this formation.

By middle Miocene time, somewhat unusual conditions developed over much of what is now coastal California between San Francisco Bay and Orange County. These conditions involved development of probably elongate marine basins, some perhaps more than $1\frac{1}{2}$ kilometers (1 mile) deep, that were similar to the deep basins of the Gulf of California, and the widespread Monterey Formation was laid down. Evidently the shoreline was sufficiently distant from the Santa Ynez and Santa Monica mountains of today that much of the material accumulating on the sea bottom was of organic rather then detrital origin. Occasionally there were influxes of terrestrial matter, including thin layers of volcanic ash, but most of the material was ooze derived from tiny diatoms living at or near the sea's surface. Their remains were mixed with fine clays transported from the shore, eventually making the upper part of the Monterey Formation. In addition, the upper beds of the Monterey are sometimes nearly saturated with bituminous material that permeates the rock and seeps from every fracture, crack, and bedding plane. Fresh samples have a strong and unmistakable petroliferous odor.

During the earlier Monterey deposition, widespread volcanic activity broke out in California, related presumably, to a time of active crustal fragmentation in connection with the birth of the San Andreas fault system. In the extreme western Santa Ynez Mountains, rhyolitic and basaltic eruptions produced the volcanic rocks near Tranquillon Mountain and in the Santa Rosa Hills to the north. Apart from these exposures, volcanism in the Santa Ynez Mountains is evident only from the thin ash beds preserved in the Monterey and overlying Sisquoc formations. Elsewhere in the Transverse Ranges, particularly in the western Santa Monica Mountains, Miocene volcanic activity was extremely important, and the lava flows are thick and prominent.

The depositional environment evidently changed little as Sisquoc Formation deposits succeeded those of the Monterey. The Pliocene may have begun before the last Sisquoc beds were laid down, but placement of the Mio-Pliocene boundary is highly disputed. The Pliocene is preeminently a time of change in the entire paleogeography of southern California.

By Pleistocene time, the Coast Range orogeny had elevated many previously marine areas, in particular most if not all of the Santa Ynez and Santa Monica mountains and the offshore islands. During this time of marked uplift, the Santa Ynez Mountains started to rise, perhaps as an anticlinal arching or as a tilted block along the Santa Ynez fault to the north. In addition to the vertical uplift, there is considerable evidence for left slip on the fault. Uplift of the Santa Ynez Mountains produced little coastal plain initially. Marine conditions prevailed to the base of the mountains through the Pliocene and into the early Pleistocene, when the richly fossiliferous Santa Barbara Formation was laid down in shallow near-shore waters. Not until the middle Pleistocene time, near the climax of mountain-building activity, did the coastal plain fully emerge.

Evidence is compelling that the orogeny, or at least its associated faulting, continues today, because earthquakes are common in this part of the state, and near Santa Barbara several Oligocene Sespe beds are in fault contact with late Pleistocene gravels. Moreover, recent surveys across some faults in the city of Santa Barbara indicate that continuing creep is occurring.

MOUNTAINS OF CENTRAL VENTURA COUNTY

In Santa Barbara County, the Transverse Ranges can be distinguished fairly easily from the southern Coast Ranges along the Santa Ynez fault zone. South of the fault, the Santa Ynez Mountains, the Santa Barbara Channel, and the Channel Islands all trend clearly east-west, whereas to the north the fold axes, faults, and topographic grain of the San Rafael Mountains show the distinctive northwest trend characteristic of the Coast Ranges as a whole. In Ventura County, these distinctions are much less clear. Here the rocks and structures of the southern Coast Ranges have changed their trend to merge with the east-west alignment of the Transverse Ranges.

The mountainous part of Ventura County, north of the Ojai and Santa Clara River valleys, is dominated by four prominent east-west faults (Figure 10-4). Two of these, the Big Pine and the Santa Ynez, are considered primarily faults of left-lateral separation along which there has been appreciable vertical slip. The Pine Mountain fault, on the other hand, is a thrust or reverse fault. The southernmost of these faults, the San Cayetano, is a young north-dipping thrust where it faces the Santa Clara River Valley.

Rocks and Geologic History

The Topatopa, Pine, and Santa Ynez mountains are carved chiefly from Eocene rocks, although several wedges of late Cretaceous rocks are exposed along the Santa Ynez fault in the Topatopa Mountains. Along the prominent east-west valley occupied by upper Sespe Creek, an elongate block of rocks has been folded and downfaulted along the Pine Mountain fault so that Oligocene and Miocene rocks are preserved in a topographic depression (Figure 10-5). Other younger rocks, including extensive exposures of the Sespe Formation occur along the southern Topatopa Mountains and around Ojai Valley. North of Ojai Valley, rocks and structural features resemble those of the Santa Ynez Mountains. For example, the overturned or vertical beds first encountered near Santa Barbara continue eastward as prominent features beyond Ojai Valley and are well exhibited in the Matilija overturn (Figure 10-6). This steeply-dipping overturn is clearly displayed in cross section along Wheeler Gorge on State Highway 33.

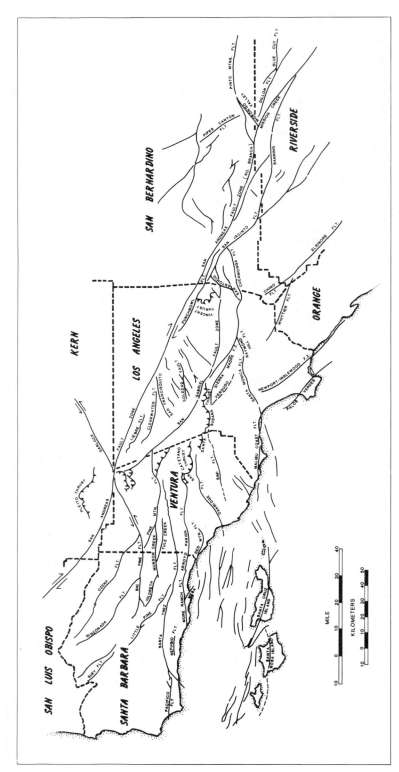

FIGURE 10-4 Principal faults, Transverse Range Province.

310

FIGURE 10-5 Looking east across an anticline in Eocene and Oligocene sedimentary rocks in the Sespe Gorge area north of Ojai, Ventura County. Eocene rocks are exposed in the lower left corner of the photo and Oligocene rocks in the upper part. State Highway 33 crosses the lower left corner of the photo. (Photo by Spence Air Photos, courtesy Department of Geography, University of California, Los Angeles)

Pliocene marine rocks appear infrequently in the Santa Ynez Mountains, but they constitute an important part of the coastal hills from Rincon Point east along the south side of Sulphur Mountain. These same rocks occur sporadically on either side of the Santa Clara River Valley as far east as the San Fernando Valley. Below the valley floor between Santa Paula and Fillmore, the Pliocene beds are 3,650 to 4,250 meters (12,000–14,000 feet) thick (Figure 10-7). These Pliocene beds are famous geologically for their small-scale structural features, which are interpreted as deep-water turbidity current deposits. Included are load casts, graded beds, slump structures, rip-ups, current features, and convoluted bedding. All are well displayed in a classic locality along Santa Paula Creek between Santa Paula and Ojai.

Marine deposition continued into Pleistocene time, with some beds nearly 1,200 meters (4,000 feet) thick. These young rocks are exposed in the lower foothills of Topatopa and Sulphur mountains and in the hills behind Ventura. Some are no older than middle Pleistocene, yet they have been tilted as much as 70 degrees from the horizontal, indicating that intense tectonic activity did not begin here until middle or late Pleistocene time.

Evidence of this tectonic activity is also provided by the appreciably warped and

FIGURE 10-6 Matilija overturn looking west along the north side of the Ojai Valley, Ventura County. Prominent ridges on the north side of the valley are exposures of the Sespe Formation. North-dipping rocks in the mountains are older, overturned Eocene beds. The fact that the older Eocene rocks rest on the younger Oligocene strata demonstrates overturning. The structural relationship can be seen on the far wall of the Ventura River in the top center. (Photo by Spence Air Photos, courtesy Department of Geography, University of of California, Los Angeles)

uplifted late Pleistocene marine terraces. The highest of these is more than 300 meters (1,000 feet) above sea level. Studies by W. C. Putnam were the first to show that some are arched so that they slope down to the north away from the ocean, a direction opposite to the one existing when they were cut. More recent studies using soil-dating techniques show that these terraces were deformed as a result of faulting along bedding planes. Higher beds in synclinal folds have slid over lower beds as folding progressed, breaking and tilting the terraces that had been formed prior to the folding.

The folds, faults, and deformed terraces exposed along the Ventura River prove the canyon is antecedent, one of a very few instances in California where antecedence is unequivocal.

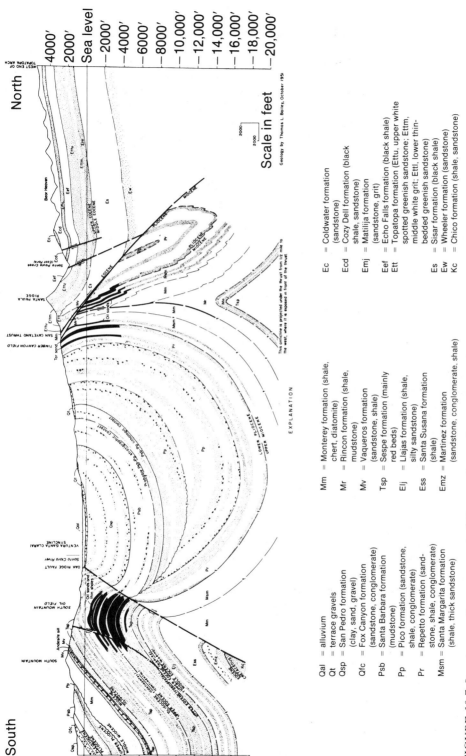

EXPLANATION

Qal = alluvium
Qt = terrace gravels
Qsp = San Pedro formation (clay, sand, gravel)
Qfc = Fox Canyon formation (sandstone, conglomerate)
Psb = Santa Barbara formation (mudstone)
Pp = Pico formation (sandstone, shale, conglomerate)
Pr = Repetto formation (sandstone, shale, conglomerate)
Msm = Santa Margarita formation (shale, thick sandstone)

Mm = Monterey formation (shale, chert, diatomite)
Mr = Rincon formation (shale, mudstone)
Mv = Vaqueros formation (sandstone, shale)
Tsp = Sespe formation (mainly red beds)
Elj = Llajas formation (shale, silty sandstone)
Ess = Santa Susana formation (shale)
Emz = Martinez formation (sandstone, conglomerate, shale)

Ec = Coldwater formation (sandstone)
Ecd = Cozy Dell formation (black shale, sandstone)
Emj = Matilija formation (sandstone, grit)
Eef = Echo Falls formation (black shale)
Ett = Topatopa formation (Ettu, upper white spotted greenish sandstone; Ettm, middle white grit; Ettl, lower thin-bedded greenish sandstone)
Es = Sisar formation (black shale)
Ew = Wheeler formation (sandstone)
Kc = Chico formation (shale, sandstone)

FIGURE 10-7 Structure section across part of Ventura Basin, showing extraordinary thicknesses of the strata, particularly the Pliocene. Rock types and structures are typical of much of the western Transverse Range sedimentary pattern. (Source: California Division of Mines and Geology.)

313

SANTA MONICA MOUNTAINS AND THE CHANNEL ISLANDS

The Santa Monica Mountains and their offshore extensions are the most geologically varied part of the entire Transverse Range province. The area involved extends about 200 kilometers (120 miles) west from Elysian Park in Los Angeles to San Miguel Island. The mountains are not as high as the Santa Ynez and the ranges of central Ventura County. They reach only 949 meters (3,111 feet) at Sandstone Peak on the mainland and 747 meters (2,450 feet) at Picacho Diablo on Santa Cruz Island.

Rocks and Geologic History

The Santa Monica Mountains have a granitic and metamorphic basement more akin to that of the Sierra Nevada than to the Franciscan basement of the Santa Ynez Mountains and the Coast Ranges. The oldest rock present is the Santa Monica Slate of the central Santa Monica Mountains. With its regular bedding, the slate is thought to be of marine origin. It has yielded a few Jurassic fossils and resembles rocks in the Santa Ana Mountains that also contain Jurassic marine fossils. Perhaps correlative with the Santa Monica Slate is the unfossiliferous Santa Cruz Island Schist, which forms an elongate ridge south of the Santa Cruz Island fault. But unlike the Santa Monica Slate, this rock appears to have been derived chiefly from volcanic rocks and to a lesser extent from sandstones and siltstones. Both formations have been intruded by granitic rocks of similar age, but the plutonic rocks on Santa Cruz Island are thought to be more closely related to the Coast Range ophiolite sequences than to the granitic basement of the Santa Monica Mountains, which resembles that of the Sierra Nevada and the Peninsular Ranges.

Upper Cretaceous rocks are exposed in the eastern Santa Monica Mountains and on San Miguel Island. About 1,980 meters (6,500 feet) of marine sandstone and shale are found on San Miguel and about 900 meters (3,000 feet) of primarily marine conglomerate in the eastern Santa Monica Mountains. The lower unit of this conglomerate sequence is a reddish land-laid deposit, which shows that the late Cretaceous shoreline lay in this area. Cretaceous rocks have been encountered in wells on Santa Cruz Island and presumably are concealed beneath younger rocks on Santa Rosa Island.

Paleocene rocks are entirely marine. Their exposures in the Santa Monica chain are limited to shales and limestones in the eastern Santa Monica Mountains and sandstones on Santa Cruz and San Miguel Islands.

Although dominant in the Santa Ynez Mountains, Eocene deposits are widely scattered in the Santa Monica Mountains, but not extensively or thickly. They occur on southwestern Santa Cruz Island and sporadically on Santa Rosa Island and reach a thickness of nearly 600 meters (2,000 feet) on San Miguel Island.

A distinctive unit of the mainland Transverse Ranges is the previously discussed nonmarine Sespe Formation, a late Eocene through early Miocene flood-plain deposit distinctive for its prevailing red color and lenticularity. The Sespe is widespread, and its occurrence indicates either that much of the western Transverse Range province was elevated or that a shallow marine area was simply blanketed by sediments until it became a low coastal plain. In either case, a short-lived land mass was formed, interrupting the marine conditions that had prevailed since the Cretaceous. Before the Oligocene ended, the sea again invaded land, covering most of the site of the western Transverse Ranges by the opening of the Miocene. Sespe rocks occur in the central Santa Monica Mountains and on Santa Rosa Island. They are reported also from a well at the eastern end of Santa Cruz Island, where they include Franciscan schist fragments.

Miocene rocks are the most extensively exposed of any Cenozoic rocks in the Santa Monica chain. They are abundant on Anacapa, Santa Cruz, Santa Rosa, and San Miguel islands. During the Miocene, the Santa Monica-Channel Islands area was the site of a deep marine trough into which 4,550 meters (15,000 feet) of sedimentary rocks were deposited. These vary from the coarse San Onofre Breccia with its large angular blocks of Franciscan schist to the deep-water diatomaceous, dolomitic, and cherty shales of the Monterey Formation. The basin seems to have deepened during the Miocene: earlier units are often coarse, shallow-water marine conglomerates and breccias, and later deposits are fine-grained, deeper-water shales.

These Miocene beds are particularly interesting because many were organically-rich muds when deposited and consequently are the most likely source for much of the petroleum and natural gas found in the Transverse Range province. As in the Santa Ynez Mountains, the Monterey Formation of the Santa Monica Mountains is rich in marine diatom remains and often contains pockets of bituminous material. Where it is buried and fractured, the Monterey is often an important reservoir for gas and oil.

During Miocene time, extensive volcanic activity broke out in this part of the Transverse Range province. Andesitic, diabasic, and basaltic flows, sills, and dikes plus a few silicic rocks were extruded. Many are of submarine origin, as demonstrated by the common occurrence of pillow structure in lava flows. The volcanic rocks total nearly 3,000 meters (10,000 feet) in thickness in the western Santa Monica Mountains, 2,440 meters (8,000 feet) on Santa Cruz Island, and 730 meters (2,400 feet) on San Miguel Island. Lesser amounts occur on Anacapa and Santa Rosa islands.

The middle Miocene San Onofre Breccia is an interesting rock that crops out in the Santa Monica Mountains. It consists of a coarse-grained breccia and conglomerate with prominent clasts of blue glaucophane schist, green schist, gabbro, and limestone. It is extremely poorly sorted and, although marine, apparently was poured into its depositional site from nearby highlands with little opportunity for rounding or sorting of the blocks. As noted in the previous chapter, this rock is so unusual and so distinctive that its occurrences are important for understanding the Miocene paleogeography of southern California.

Post-Miocene rocks in the Santa Monica Mountains are mainly Pleistocene and younger marine and nonmarine terrace deposits. Locally thick marine terrace deposits appear on San Miguel and Santa Cruz islands, thick-bedded, coarse, nonmarine gravels probably deposited as broad alluvial fans occur along the Malibu coast. These make up the high prominent cliffs that have generated destructive landslides at Pacific Palisades and Santa Monica.

A question of long standing is whether or not the Channel Islands have been connected with the mainland at the western end of the Santa Monica Mountains at any time since modern geography developed, probably in the late Pliocene. Because nonmarine beds on Santa Cruz and Santa Rosa islands have yielded fossils of pygmy elephants, related to, but distinct from Pleistocene mainland species, it was assumed that a land connection was required, although geologists had been unable to find any independent evidence for such a land connection at any time after the Oligocene.

But because some geologists claimed there had to be a land connection to account for the elephants on the islands, that became the standard dogma. Biologists had been puzzled about the incomplete island vertebrate fauna. There are no rabbits, rattlesnakes, deer, coyotes, or large skunks on the islands, yet all are common on the mainland.

In an almost classic example of the need for scientists to constantly reexamine their assumptions, a biologist and a geologist questioned the unstated assumption that the elephants reached the island by dry land. They found that the open-water gap between the mainland and the islands was probably as little as 6½ kilometers (4 miles) during at least the last low Pleistocene sea level. Further, all investigators agreed that the present four islands were united into a single large island during these low sea levels of the Pleistocene. But most surprising was the result of a careful search of the literature on elephant behavior. A number of well-documented instances of elephants swimming in the ocean a kilometer or more from shore were found, including one case in which African elephants repeatedly swam across a channel several kilometers wide to a small island off the east coast of Africa.

As a result, it now seems likely that the elephants reached the islands by swimming during one or more of the Pleistocene low sea levels. They subsequently evolved somewhat to produce the smaller forms. The peculiar absence of many mainland land animals on the modern islands is also much more satisfactorily explained.

VENTURA AND SOLEDAD BASINS

A large syncline forms the structure that is known as the Ventura Basin in the west and the Soledad Basin in the east. This structure is about 190 kilometers (120 miles) long and includes the Santa Barbara Channel between the Channel Islands and the Santa Ynez Mountains (Figure 10-7) as well as the hilly inland area between the Santa Monica and Topatopa mountains. The primarily marine Ventura Basin joins the primarily nonmarine Soledad Basin near the San Gabriel fault. The Soledad Basin extends almost 48 kilometers (30 miles) farther east, north of the San Gabriel Mountains to the San Andreas fault.

Within the Ventura Basin are several prominent anticlinal hills, some higher than the Santa Monica Mountains to the south. The highest, the Santa Susana Mountains, enclose the west and northwest San Fernando Valley. Other ridges are Sulphur Mountain, between Ojai and the Santa Clara River valleys, and the South Mountain-Oak Ridge complex, which joins the Santa Susana Mountains to the east.

The Ventura Basin is famous for its remarkably thick section of mostly marine sedimentary rocks, which totals more than 17,700 meters (58,000 feet). The lower 2,440 meters (8,000 feet) are Cretaceous, but the remaining 15,000 meters (50,000 feet) are Cenozoic rock and include perhaps the thickest accumulation of Pliocene deposits in the world.

Pleistocene	1,800–2,150 meters	(6,000–7,000 feet)
Pliocene	3,650–4,250 meters	(12,000–14,000 feet)
Miocene	2,440–3,000 meters	(8,000–10,000 feet)
Oligocene	760 meters	(2,500 feet, nonmarine)
Eocene and Paleocene	5,200 meters	(17,000 feet)
Cretaceous	2,400 meters	(8,000 feet)

These sedimentary rocks reflect a complex tectonic history whose details are becoming increasingly clear. Subduction appears to have prevailed during the late Cretaceous and early Tertiary, and sediments of those ages were deposited in a forearc basin. About 29 million years ago in Oligocene time, subduction had carried the East Pacific Rise (a spreading center) beneath the western edge of the North American continent, and by about 22 million years ago (early Miocene), subduction was replaced by crustal stretching, which began to form the San Joaquin, Ventura, and Los Angeles basins, in which much Miocene sediment was trapped. Near the close of the Miocene, about 6 million years ago, renewed deepening of the basins occurred, providing sites for the accumulation of large thicknesses of Pliocene sediment. Subsidence was so rapid at times that as much as 1,500 meters (5,000 feet) of water may have occupied the basin. Some of the hills south of the Santa Clara River Valley were folded and uplifted in the Miocene, only to be submerged beneath the Pliocene sea. Apart from the Santa Barbara Channel, the entire area was subjected to strong uplift, folding, and faulting during the middle Pleistocene. This produced today's topography and created the structures in which the region's prolific oil fields have developed. The Santa Barbara Channel floor was also folded and faulted, but was not uplifted above sea level.

Most of the valleys in the Ventura Basin are synclines. The Santa Clara River Valley, for example, is structurally a large, thick, synclinal fold. The San Fernando Valley is topographically more striking, but is a broader, flatter syncline incorporating fewer Cenozoic sediments. The other valleys are counterparts of these larger features.

The Soledad Basin east of the San Gabriel fault is dominated by rough, hilly country drained mainly by the Santa Clara River and its tributaries and represents the landward extension of the Ventura Basin. The Soledad Basin contains mainly middle and late Cenozoic nonmarine sedimentary rocks that rest on the crystalline basement of the San Gabriel Mountains to the south and the Sierra Pelona to the north. This nonmarine basin was studied extensively by Richard H. Jahns in the 1940s, and his work provided the framework on which more recent work has been based.

During Oligocene time, the Soledad Basin was cut off for much of the time from direct access to the sea and developed closed basins in which saline lakes briefly formed. Coarse sands and gravels eventually filled the basin and by the end of the Miocene, drainage to the sea was presumably reestablished, because these later deposits show no evidence of deposition in closed interior basins. The Soledad Basin deposits include the coarse clastics that make up the prominent hogback ridges at Vasquez Rocks and the lacustrine borax deposits at Tick Canyon (Figure 10-8).

When the borax lakes were developing, volcanic rocks were being extruded into the Soledad Basin, forming dikes, sills, flows, and breccias. This volcanism very likely provided the boron now found in the lacustrine sediments. By late Miocene time, the basin had filled sufficiently so that lakes were no longer an important aspect of the landscape. During the Miocene subsidence, the sea reached into the western Soledad Basin and persisted there until the early Pliocene. By the end of the Pliocene, the sea had withdrawn, and the nonmarine Saugus beds mark the onset of land conditions that have prevailed since.

Although movement on some of the area's major faults may have started early in the Cenozoic or even sooner, the most intense deformation occurred in the early part of the Miocene. However, tectonic activity continues to the present day as

FIGURE 10-8 Hogbacks in steeply-dipping, nonmarine Vasquez Formation sandstone and conglomerate of Oligocene age, Vasquez Rocks, Los Angeles County. (Photo by R. M. Norris)

FIGURE 10-9 Fault scarp in Lopez Canyon, showing 3-foot displacement from the 1971 San Fernando earthquake. (Photo by V. A. Frizzell, courtesy of U.S. Geological Survey)

demonstrated by the 1-meter (3-foot) vertical movement on the faults in northeast San Fernando Valley during the 1971 earthquake (Figure 10-9).

SIERRA PELONA

The Sierra Pelona forms a block terminated on the west by San Francisquito Canyon and on the north by the San Francisquito and San Andreas faults. The southern boundary lies more or less along the old Mint Canyon (Sierra) highway. Adjacent Liebre Mountain to the northwest, in contrast, is composed of gneisses and granitic rocks.

The Sierra Pelona is composed almost entirely of Pelona schist. This formation, still uncertainly dated, was long regarded as Precambrian (Proterozoic). As indicated previously, however, a strong case is made for a Cretaceous age. This is based on studies of the Pelona and its close correlative, the Orocopia schist of the Colorado Desert. In a pattern resembling that in the Orocopia Mountains, in the San Gabriel Mountains the Vincent thrust fault has moved Proterozoic and Cretaceous crystalline rocks over Pelona schist, causing the metamorphism of the schist. Some geologists have suggested that the Vincent thrust was an ancient subduction zone down which the unmetamorphosed Pelona (its *protolith*) was carried, perhaps to a depth of about 20 to 25 kilometers (12–15 miles) and there metamorphosed, presumably in Paleocene time. The dating of the metamorphism is based on the fact that neither the Pelona nor the Vincent thrust is cut by the Cretaceous plutonic rocks associated with them, but are cut by middle Tertiary plutons. Further, some radiometric dates suggest that the metamorphism of the Pelona occurred about 60 million years ago and that the protolith may be as old as Jurassic.

RIDGE BASIN

This narrow depositional (not topographic) basin lies between the San Gabriel and San Andreas faults near their juncture at Frazier Mountain. It is a down-dropped wedge of sedimentary rocks about 12,000 meters (40,000 feet) thick, of which Miocene and Pliocene beds account for about 8,850 meters (29,000 feet). Only the lower 600 meters (2,000 feet) of Miocene are marine; the other beds are all nonmarine. Remarkably, the vertical thickness of the column is much greater than the width of the basin in which it accumulated. A particularly striking unit is the Violin breccia, about 8,250 meters (27,000 feet) thick, but less than 1½ kilometers (1 mile) across. This and other units of the Ridge Basin accumulated at the base of an active fault scarp along the San Gabriel fault zone. Deposition continued as vertical, and horizontal displacement occurred on the fault. Because of the continuing tectonic activity, nowhere in the Ridge Basin is there more than about 4,000 meters (13,000 feet) of beds in one place, because the site of maximum deposition kept shifting as motion on the San Gabriel fault occurred; the 12,000-meter total is a maximum aggregate thickness. These rocks provide a complete record of the area's Cenozoic structural history and can be seen along Interstate Highway 5 between Castaic and Gorman.

SAN GABRIEL MOUNTAINS

The San Gabriel Mountains are a high, rugged block located between the Los Angeles Basin and the Mojave Desert. They form a continuous feature 96 kilometers (60 miles) long and up to 39 kilometers (24 miles) wide, roughly along a north-south line that passes through Mount Pacifico and Mount Wilson. The Sierra Madre fault zone forms the range's southern boundary. The eastern boundary is the San Andreas fault zone, which crosses through Cajon Pass and separates the similar but higher San Bernardino Mountains (Figure 10-10). The San Gabriel Mountains face the Soledad Basin on the northwest and the San Fernando Valley on the west. Separated by a depressed block from the San Gabriel Mountains are the Verdugo Hills, which form the eastern boundary of the San Fernando Valley. The geology of the Verdugo Hills is essentially similar to that of the San Gabriel Mountains.

Because the San Gabriel Mountains have experienced considerable uplift in recent geologic time, the range has become a deeply dissected, rugged horst. Stream canyons are steep-sided and up to 900 meters (3,000 feet) deep. Many peaks exceed 2,100 meters (7,000 feet) in elevation, the highest being Mount San Antonio (Old Baldy) at 3,074 meters (10,080 feet). The southern and western flanks are very steep and abrupt where they face the lowlands of the Los Angeles Basin and the San Fernando Valley. The north face is less dramatic although almost as steep. The

FIGURE 10-10 Cajon Pass (center foreground), between the San Gabriel and San Bernardino Mountains. Note the linear valley (center) carved along the San Andreas fault. Beheaded alluvial fans now being exhumed are shown in the upper right corner. (Photo by Fairchild Aerial Surveys, courtesy of Department of Geography, University of California, Los Angeles)

difference results from the greater elevation of the Mojave Desert, with its floor nearly 1,200 meters (4,000 feet) high at the base of the range.

Faults

The strong influence of recent uplift along faults is shown by the deep and narrow gorges of streams in the San Gabriel Mountains. To the north, the Santa Clara River has cut Soledad Canyon roughly along the trace of the Soledad fault. In the east, Lytle Creek maintains a straight course along the San Jacinto fault, a major branch of the San Andreas. Especially striking is the strong control on the east and west forks of the San Gabriel River, which follows the San Gabriel fault for nearly 40 kilometers (25 miles) (Figure 10-11).

Right slip on the San Andreas has produced prominent topographic features along the north and northeastern sides of the San Gabriel Mountains. Large alluvial fans built by streams flowing off the north slope have been displaced, resulting in a series of in-facing bluffs from Cajon Pass west to Valyermo. Beheaded stream channels are frequently exposed on the tops of these displaced fans; sometimes the streams that adjusted to separation on the fault show prominent right-angle bends that clearly display the direction of fault movement.

Prominent alluvial fans also occur on the range's southern flank, forming a nearly

FIGURE 10-11 The San Gabriel River is the large stream that crosses the photo from top to bottom; the East Fork is the branch on the right and the West Fork the branch on the left. These two forks are controlled by and follow the San Gabriel fault zone, though the main trace of the fault occupies the linear valley north of and parallel to the two forks. (U.S. Department of Agriculture photo)

continuous coalescing alluvial apron (bajada) from Pasadena to Cajon Pass. Movement on the Sierra Madre-Cucamonga fault zone has been primarily vertical (it appears to be a reverse or thrust fault), so the fans and their source streams are still closely associated. Topography has been somewhat obscured by intensive urban development, however.

Although most evidence suggests that late Cenozoic slip along the San Andreas and San Jacinto fault zones has been right lateral, only substantial vertical slip can account for the current elevation of the San Gabriel Mountains 600 to 1,200 meters (2,000–4,000 feet) above the Mojave Desert. As recently as the 1971 San Fernando earthquake, more than a meter (3 feet) of elevation occurred along several high-angle reverse faults in Lopez Canyon, presumably increasing the difference in elevation between the San Gabriel Mountains and the San Fernando Valley (Figure 10-9). Similar abrupt slopes along the Soledad and Sierra Madre faults also show recent vertical uplift.

The San Andreas and San Jacinto fault systems merge between Palmdale and Valyermo, but toward the southeast they diverge until they are about 9½ kilometers (6 miles) apart. The right slip San Gabriel fault disappears near Cucamonga Peak, where it appears to be offset by the younger San Antonio fault. It may continue east of this fault to merge with the San Jacinto fault zone. The Vincent thrust is well exposed north of Mount San Antonio, but dips south under the highest part of the range. The Vincent thrust is much older than the previously mentioned faults because it is cut by middle Tertiary granitic rocks.

Rocks

A feature of the San Gabriel Mountains that is critical to understanding the geologic history of North America is the presence of ancient crystalline rocks, particularly in the northwest part of the range. Included are extensive exposures of anorthosite, which is known almost exclusively from pre-Phanerozoic terrains (and from lunar sites). The only other California exposure is in the Orocopia Mountains. The San Gabriel anorthosites have been dated as 1,022 million years old. Other ancient rocks are the Mendenhall gneiss (1,045 million years old) and the augen gneisses (1,700 million years old). These rocks are not as old as the Archean rocks of the Lake Superior region, but they are earliest Proterozoic nonetheless.

Other large exposures of metamorphic rocks also exist in the northeast San Gabriel Mountains. These rocks are generally assigned to the Pelona schist, which is thought to have either a Jurassic or Cretaceous protolith and to have undergone metamorphism in either the late Cretaceous or Paleocene. In addition, another thick sequence of metamorphic rocks is located in the eastern San Gabriel Mountains, probably originally Paleozoic sedimentary rock.

Mesozoic granitic rocks dominate the San Gabriel Mountains and constitute perhaps 70 percent of the exposed rocks. The few radiometric dates that exist for these rocks range from 70 to 84 million years (late Cretaceous), but at least one is only 61 million years old (Paleocene). Dates from the Mount Lowe and Parker Mountain granitic rocks yield ages of 220 million years (Permo-Triassic), and 245 million years (Permian). Some small stocks of intrusive granite porphyry occur in the eastern San Gabriels near Mount San Antonio and cut across the Pelona Schist and Vincent thrust as well as all other older rocks in the area. These have been dated as Miocene, between 14 and 16 million years old. Apparently these granitic rocks were emplaced in at least four, possibly five episodes. They range from granites, through grano-diorites and quartz monzonites to syenites and diorites.

Cenozoic beds are located only along the range's margins. The oldest of these is the marine Paleocene assigned to the San Francisquito (formerly Martinez) Formation, which occurs near Devils Punchbowl and near Cajon Pass. At the Punchbowl, the beds lie between the San Jacinto and San Andreas faults, here about 3.2 kilometers (2 miles) apart. The Cajon Pass exposures form little patches northeast of the San Andreas fault. Paleocene marine beds do not recur northeast of the San Andreas fault until the San Joaquin Valley. Apart from a Miocene exposure at the western tip of Antelope Valley, the San Francisquito and some Oligocene Vaqueros near Cajon Pass are the youngest marine beds known from north of the San Gabriel Mountains or in the Mojave Desert proper. Above the San Francisquito and Vaqueros beds are the massive conglomeratic sandstones exposed at Cajon Pass (Figure 10-12) and the Punchbowl as the middle to late Miocene Punchbowl Formation. The Punchbowl Formation contains fossils of land vertebrates and is overlain by the Pliocene Crowder Formation and Pleistocene gravels.

Interestingly, it was the peculiar distribution of the Paleocene San Francisquito on opposite sides of the San Andreas fault, but separated by 39 kilometers (24 miles) from one another, that led Levi Noble in 1926 to first advance the then revolutionary idea that large horizontal displacement can take place in the earth's crust. Noble's perceptive observation was an important early step in building the body of evidence that eventually led to Plate Tectonic theory.

FIGURE 10-12 Massive conglomerates of the late Miocene nonmarine Punchbowl Formation at Mormon Rocks, Cajon Pass, San Bernardino County. (Photo by R. M. Norris)

LOS ANGELES BASIN

The term *Los Angeles Basin* means different things to different people. To those concerned with air pollution, the term refers to the atmosphere of the lowland areas surrounding the Transverse and Peninsular Ranges in the Los Angeles region, where temperature inversions and photochemical smog are recurrent phenomena. Some geologists limit the Los Angeles Basin to the coastal plain between the Santa Monica Mountains on the north, the Puente Hills and Whittier fault on the east, the Santa Ana Mountains and the San Joaquin Hills on the south, and the Palos Verdes Peninsula and the shoreline on the west. The U.S. Geological Survey, conversely, takes a broader view, dividing the basin into four blocks that contain both uplifted portions and synclinal depressions (Figure 10-13).

Southwestern Block

The southwestern block is the seaward part of the Los Angeles Basin. It is bounded on the east by the Newport-Inglewood zone of deformation, which can be traced from Beverly Hills to Newport Bay where it strikes offshore. This structural trend, a combination of folds and faults, is expressed as a chain of low, en echelon (overlapping staggered) anticlinal hills. The distinguishing feature of the southwestern block is its basement. Although actually exposed only in the Palos Verdes Hills, it has been encountered in numerous oil wells at depths of 1,500 to 4,250 meters (5,000–14,000 feet) below sea level. These basement rocks belong to the Catalina schist facies of the Franciscan Formation and are chiefly green chlorite and blue glaucophane schists. They have no known base, are always in fault contact with other basement rocks, and are of undetermined age. The oldest rocks in depositional contact with them are Miocene. Based on lithologic affinities with dated Franciscan in the Coast Ranges, a late Jurassic to late Cretaceous or even younger age is probable.

The main structural features of the southwestern block are the anticlinal Palos Verdes Hills that have been raised along a steep reverse fault, several anticlinal ridges in the basement rocks over which younger sediments have been draped, and intervening broad syclines. The anticlinal structures of the younger rocks have formed important traps for petroleum and natural gas. For example, the Wilmington field is the most productive field in California and the second most productive in United States. The sedimentary blanket in this block is quite thick, up to 6,250 meters (20,500 feet). It is all post-Oligocene and almost entirely marine.

Displacements on the Newport-Inglewood fault have both vertical (to 1,200 meters or 4,000 feet) and horizontal (at least 1,500 meters or 5,000 feet) components. The upper surfaces of the basement rocks typically have about 1,200 meters (4,000 feet) of separation across the fault zone, but the overlying sediments show less vertical separation the younger they are. Movement seems to have begun in the Miocene and is still progressing, as indicated by arching of late Pleistocene and younger strata and by recent seismicity—the Newport-Inglewood fault zone caused the 1933 Long Beach earthquake. The zone displays mainly right slip like that observed on the San Andreas, although of smaller magnitude. This movement probably produced the en echelon wrinkles that are reflected on the surface as the Baldwin, Dominguez, and Signal Hills (Figure 10-14). The surface expression of folding is more prominent than the expression of faulting. Several of these anticlinal hills overlie up-faulted blocks of basement rocks, however, so their origin may be more complex than first believed.

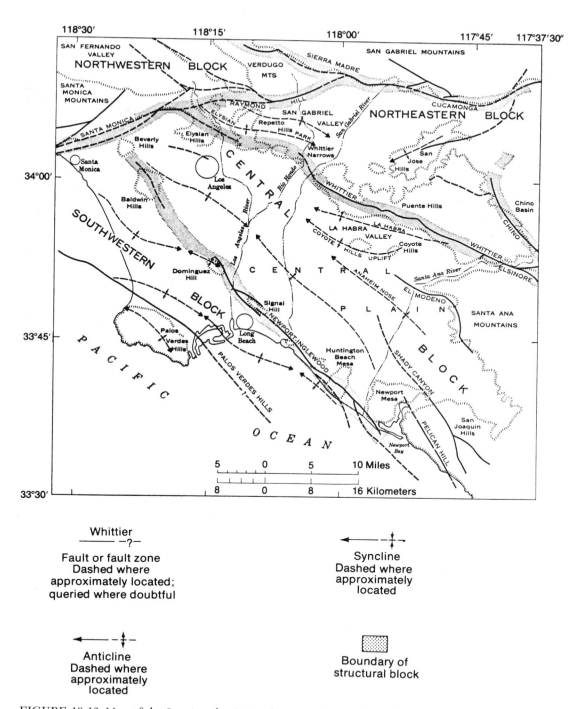

FIGURE 10-13 Map of the Los Angeles Basin. (Source: U.S. Geological Survey)

FIGURE 10-14 Signal Hill in 1941. (Photo by Spence Air Photos, courtesy of Department of Geography, University of California, Los Angeles)

Central Block

The central block of the Los Angeles Basin includes the low portions of the Los Angeles coastal plain from Beverly Hills southeast to central Orange County (the Downey Plain), the Coyote Hills uplift, the La Habra Valley, the San Joaquin Hills, and the Newport and Huntington Beach mesas. The Santa Ana Mountains may be included also, but they are more conventionally placed in the Peninsular Ranges. The block's main portion is occupied by the Downey Plain, a broad synclinal sag about 16 to 22 kilometers (10–14 miles) wide.

There are several folds within the coastal plain. One that lacks surface expression is the anticlinal Anaheim Nose. Another is represented by the Coyote Hills uplift, extending from the low hills near Sante Fe Springs southeast to the Coyote Hills proper, which stand nearly 160 meters (500 feet) above the adjacent lowland. The synclinal trough of La Habra Valley lies northeast of the Coyote Hills uplift and is bounded on the northeast by the Whittier fault zone. The Whittier fault zone also forms the eastern side of the central block from Montebello to near Corona, where it merges with the Elsinore fault. Northwest from Montebello, the presence of the Whittier fault is uncertain. Here the block's eastern boundary is marked by the Elysian and Repetto hills.

Basement rocks are not known from the deep part of the central basin, although they are encountered in some oil wells around the block's margin and are exposed in the Santa Ana Mountains. They are probably equivalent to those seen in the eastern Santa Monica Mountains. They consist of slightly metamorphosed sedimentary Jurassic rocks that have been intruded by late Cretaceous granitic rocks of the southern California batholith. No transition is discernible between the basement rocks of the central and southwestern blocks. This suggests that basement rocks of quite different origin have been brought into contact with one another, probably by appreciable right slip on the fault zones, as well as by about 90 degrees of rotation of the adjacent mountain blocks suggested by paleomagnetic studies.

Younger rocks resting on the basement are best exposed on the western slopes of the Santa Ana Mountains. At least 9,700 meters (32,000 feet) of marine and nonmarine Cretaceous to Pleistocene sedimentary rocks occur here, plus some Miocene volcanic rocks. The older rocks in this sequence are missing from the central part of the block, although Miocene, Pliocene, and Pleistocene beds alone total more than 6,700 meters (22,000 feet) thick. Where older sediments occur, near the junction of the Rio Hondo and Los Angeles River, total thickness is 9,700 to 10,700 meters (32,000–35,000 feet). The basement surface upon which the younger rocks lie is bowed downward from the edges of the central block. Sedimentary rocks in the deepest part of the basin may lie on some sort of volcanic or oceanic crust. In any event, the basement lies about 4,000 meters (13,000 feet) below sea level at the ends and 4,250 meters (14,000 feet) along the Newport-Inglewood fault zone. In the deepest part of the Los Angeles Basin, the probable crystalline floor lies more than 9,150 meters (30,000 feet) below sea level and is known only from geophysical data.

Northeastern Block

The northeastern block is situated between the Whittier fault zone and the base of the San Gabriel Mountains and is separated from the northwestern block by the Raymond Hill fault. This block is a deep synclinal basin that contains mostly marine Cenozoic sedimentary rocks, but includes some thick Miocene volcanic rocks in the east. Basement lies as much as 3,650 meters (12,000 feet) below the surface in the central part of the San Gabriel Valley, and in the eastern Puente Hills more than 6,700 meters (22,000 feet) of Cenozoic sedimentary rock covers the basement.

Northwestern Block

The northwestern block embraces the eastern Santa Monica Mountains and the San Fernando Valley. It is bounded on the south side by the Santa Monica and Raymond Hill faults, on the east and northeast by the San Gabriel Mountains, and on the west and north by ranges usually included in the Ventura Basin portion of the Transverse Ranges. The San Fernando Valley is a broad syncline with the eastern Santa Monica Mountains an adjacent anticline. No faulting of consequence separates the Santa Monica Mountains from the San Fernando Valley, but the Santa Monica block has been appreciably uplifted with respect to the other blocks of the Los Angeles Basin.

Geologic History

Unraveling the geologic history of the Los Angeles Basin is a complicated process. It involves not only vertical movements of great magnitude (such as more than 6,100

meters [20,000 feet] of subsidence in the central basin since the middle Miocene),
but also includes substantial strike-slip movement on the Newport-Inglewood zone
and boundary faults like the Malibu Coast-Santa Monica. This latter fault zone may
have undergone as much as 80 kilometers (50 miles) of left-slip since the Eocene.
Further, paleomagnetic evidence indicates that some of the blocks adjacent to the
Los Angeles Basin may have been rotated as much as 90 degrees in the last 30 million
years or so. Our understanding remains incomplete.

At the beginning of the late Cretaceous, an extensive erosional surface developed
across the older rocks and was subsequently covered by later Cretaceous marine
sediments from an advancing sea. Rocks of this age are known in the Santa Ana and
eastern Santa Monica mountains. A similar pattern probably applies to Paleocene
and Eocene rocks, although they are buried so deeply that they have not been
encountered in wells drilled in the central basin. Late Eocene, Oligocene, and early
Miocene nonmarine sedimentary beds occur in both the Santa Ana and Santa Monica
mountains, and the presumption is that they also extend across the basin.

The great relief of the present basin floor began to evolve in early Miocene time,
when sizable vertical movements began to exert control on the pattern of later
deposition. About this time, approximately 22 million years ago, the area was stretched
and broken as the result of the earlier subduction of the East Pacific Rise beneath
this part of the continent. Between about 15 and 13 million years ago, volcanic
activity was widespread in the basin and rotation of the crustal blocks is believed to
have begun. During the late Miocene, the sea advanced over the Los Angeles Basin
from south to north, eventually covering basement highs at Palos Verdes and other
parts of the southwestern block. By the close of the Miocene, the sea had reached
the base of the San Gabriel Mountains and flooded most of the Los Angeles Basin,
although a shoal existed at the site of today's Anaheim Nose. There is evidence that
the Los Angeles and Ventura basins were connected at this time and that marine
conditions prevailed over the intervening area. It is important to remember that
modern ranges and valleys did not exist until near the end of the Pliocene. Furthermore, although the great thickening of marine Miocene in the Central Los
Angeles Basin is reflected in well records, drills have not yet penetrated to the
probable crystalline rocks below the sedimentary column, which is often more than
6,100 meters (20,000 feet) below the surface.

During the Pliocene, the rate of sinking accelerated in the central basin, and some
sediments were deposited in as much as 1,800 meters (6,000 feet) of water. Concurrently, the basin's margins were undergoing marked uplift, and rocks from the
surface and from oil wells show many unconformities that record continuing tectonism. The central basin continued to receive large volumes of sediment from the
Northeast, and the basin's sea became steadily shallower. By the close of the epoch,
more than 3,000 meters (10,000 feet) of deposits had been laid down. An unconformity
at the top of the Pliocene section shows that deposition was interrupted by deformation about this time. As the Pliocene closed, land areas included an island formed
by the Palos Verdes Hills, much of the Santa Ana and Santa Monica mountains, the
Puente Hills, and probably some small islands along the Inglewood fault zone.

In early Pleistocene time, the Palos Verdes Hills sank below sea level, the central
basin continued to receive marine deposits as did the San Joaquin Hills and the San
Gabriel Valley, and the San Gabriel Mountains and the Puente-Repetto Hills were
rising. The Santa Monica Mountains seem to have persisted as a lowland.

Deposition began to outpace subsidence, and by middle Pleistocene the shoreline probably lay along the southern margin of the Santa Monica chain and along the Whittier fault zone. Exceptions were two embayments: one at Whittier and one at the base of the Santa Ana Mountains and north of the San Joaquin Hills. Near the center of the basin are more than 900 meters (3,000 feet) of Pleistocene land-laid beds, some deposited in near-shore lagoonal environments along a low-lying coast. To accommodate such a thickness with today's low elevations, there must have been substantial subsidence until the present. The region experienced its last deformational episode, the Coast Range orogeny, by the middle Pleistocene. This was expressed in the central basin by more subsidence, but in the surrounding areas by considerable uplift.

By late Pleistocene, the Palos Verdes Hills began to rise along the Palos Verdes fault, producing marine terraces that will be described more fully later. Lowering of sea level, which was partially caused by continental glaciation, made the entire coastal plain emerge and allowed streams to cut channels as much as 76 meters (250 feet) deep at the present shore. (The land area at that time was thus more extensive than today's.) The hills along the Newport-Inglewood uplift were developed, and some of the hills in the eastern basin were uplifted. For example, the San Joaquin Hills rose as much as 300 meters (1,000 feet) during the late Pleistocene and Holocene. The Whittier fault zone is thought to have about 4.8 kilometers (3 miles) of right separation, 1.6 kilometers (1 mile) of which appears to be late Pleistocene and Holocene. This last orogenic episode is apparently continuing, as indicated by historical earthquakes and folding of young deposits in such areas as Signal Hill, the central basin, and the Coyote Hills. There seems to be good evidence too that the Gaffey anticline, north of San Pedro, is still rising.

The thick marine sediments of the Los Angeles Basin are richly fossiliferous, especially within Pleistocene sections. The San Pedro and Timms Point beds from Palos Verdes Hills are particularly rich. In one study, a single exposure of a 30-centimeter (1-foot) layer of Palos Verdes sand near Playa del Rey yielded more than a million shells. Nonmarine deposits are less fossiliferous, but do contain vertebrate remains. Particularly remarkable is the array of animal remains recovered from the tar pits at Rancho La Brea.

The Los Angeles Basin is one of the world's most prolific petroleum-producing areas. Fields like Signal Hill, Huntington Beach, Santa Fe Springs, Wilmington, those on the Newport-Inglewood uplift, and several dozen smaller ones have produced more than 700 million metric tons (5,000 million barrels) of oil. They continue to be important producers, with nearly 140 million metric tons (1,000 million barrels) of reserves still in the ground.

Problems of Los Angeles Basin Geology

Several problems must be solved before an adequate history can be established for the area incorporating the Los Angeles Basin, the northern Peninsular Ranges, and the western Transverse Ranges. Among these problems are the following.

1. Distribution of the early middle Miocene rocks. In particular, the widely scattered occurrences and source areas of the San Onofre Breccia must be explained. It seems probable that present occurrences were once much closer to source areas containing glaucophane schist.

2. Juxtaposition of Catalina schist and granitic basement along the Newport-Ingle-

wood fault zone. When and how was this southwestern block emplaced, and where did it come from?

3. Development of deep marine basins in the Los Angeles area during the Miocene and how were these related to subduction of the East Pacific Rise and to the later block rotations that occurred.

4. How large and how deep are the rotated blocks and when did they rotate?

5. Establishment of the present geographic outlines of the area in Pliocene time.

6. Late Pleistocene uplift of the entire Los Angeles Basin, including Palos Verdes Hills and at least some of the offshore islands.

7. Relationships in time and origin of the right-slip and left-slip faults that cross the basin or terminate in unexplained patterns.

SAN BERNARDINO MOUNTAINS

The San Bernardino Mountains are the loftiest range south of the Sierra Nevada in either California or Baja California. The highest peak is Mount San Gorgonio (Grayback) at 3,508 meters (11,502 feet) (Figure 10-15). This high range has an extensive upland plateau of modest relief that is situated at 1,980 to 2,300 meters (6,500–7,500 feet) elevation. On this surface are the dams constructed to impound lakes Arrowhead and Big Bear, plus several smaller lakes.

The flanks of the range are cut by deep, narrow canyons similar to those in the Santa Gabriel Mountains. The San Gabriel Mountains, however, lack the distinctive rolling upland of the San Bernardino unit. If such a plateau ever did exist in the San Gabriels, it has been destroyed by headward stream erosion. The San Bernardinos are a larger and higher mountain mass and so perhaps were able to retain more of their rolling upland area. Whatever the case, headward erosion will eventually consign today's elevated plateau to oblivion.

The San Bernardino Mountains are about 105 kilometers (65 miles) long, with a maximum width of 48 kilometers (30 miles). They are bounded on the west and southwest by Cajon Pass and the San Andreas fault, on the north by the Mojave Desert, on the east by the Twentynine Palms Valley, and on the southeast by Morongo Valley. Morongo Valley lies along the Morongo Valley fault and separates the San Bernardino Mountains from the Little San Bernardino Mountains. Figure 10-4 shows the faults of the eastern Transverse Ranges.

On the south, the San Andreas fault branches near the city of San Bernardino. The north branch, which is closely related to if not identical with the Mission Creek and Mill Creek faults, crosses the south flank of the range along the deep canyon of Mill Creek, passes south of Mount San Gorgonio, and enters the Coachella Valley near Desert Hot Springs. To the southeast, near Indio, this branch joins the south branch of the San Andreas fault, which follows the north side of San Gorgonio Pass. This south branch of the San Andreas fault is also known as the Banning (or South Banning) fault. The Banning fault is probably responsible for the precipitous south face of the San Bernardino Mountains along San Gorgonio Pass; for a few kilometers east of Cabazon it has been mapped as a thrust.

Like the San Gabriels to the west, the San Bernardino Mountains are basically a horstlike block uplifted along bounding faults. Many smaller faults are present within the range. Several prominent Mojave Desert fault systems die away in the range or are lost where they enter the crystalline rocks.

FIGURE 10-15 View north across San Gorgonio Pass and the city of Beaumont, Riverside County. San Gorgonio Mountain (Grayback) is in the upper left corner. Its elevation is 3,507 meters (11,499 feet), the highest peak in the San Bernardino Mountains and the Transverse Ranges. (Photo by Spence Air Photos, courtesy Department of Geography, University of California, Los Angeles)

The distribution of rock types in the San Bernardino Mountains is as varied as in the San Gabriel chain, but differs in some important respects. Anorthosites are absent in the San Bernardino Mountains, but Proterozoic gneisses and schists are equally abundant in both ranges. These rocks are particularly prominent on the south flank of the San Bernardino Mountains and in Morongo Valley. Smaller patches of Proterozoic rocks occur in the central and northeastern parts of the range, and Ediacarian and Cambrian marine sedimentary rocks have been reported from west of the Cushenbury Grade.

Along the north flank and east and southeast of Big Bear Lake are extensive exposures of fossiliferous late Paleozoic rocks. These consist mostly of the Pennsylvanian Furnace Limestone, but include other limestones and quartzites from Ediacarian to Permian age. The total thickness is about 3,000 meters (10,000 feet).

Some geologists now believe that the San Bernardino Mountains are composed of two unrelated blocks of basement rocks, fortuitously juxtaposed along the North San Andreas (Mission Creek) fault. They note that the higher north block has a suite of intrusive and younger rocks closely akin to those of the Mojave Desert to the

FIGURE 10-16 Looking east toward San Bernardino Peak (3,248 meters or 10,649 feet), San Bernardino Mountains. The north branch of the San Andreas fault joins the south branch just to the left of the photo. The north branch passes through the lower foothills on the left and behind the dark ridge on the right. The south branch is at the base of the mountains. The north branch separates the two distinctive basement terranes discussed in the text. (Photo by Spence Air Photos, courtesy Department of Geography, University of California, Los Angeles)

north, but the southern block shares a similar basement with the San Gabriel range, including the Permo-Triassic Lowe granodiorite and Proterozoic gneisses (Figure 10-16).

At least 70 to 75 percent of the San Bernardino range is composed of a light-colored quartz monzonite dated between 70 and 85 million years (Late Cretaceous), called the Cactus Granite. About 18 kilometers (11 miles) southeast of Old Woman Springs, another quartz monzonite has been cut by pegmatite dikes tentatively dated as middle Jurassic (150 million years old), although the true age may be somewhat less. Some earlier plutonic rocks, named the Fawnskin Monzonite, occur in the eastern part of the range and have been dated at 214 million years (Triassic). Further, some quartz monzonites intrude probable Pelona Schist near Redlands, and may be Tertiary. Thus the granitic rocks in this range indicate several intrusive episodes from Triassic to Tertiary (?) time. Other than the probable Pelona Schist, no Mesozoic sedimentary rocks have been recognized to date.

The San Bernardino Mountains contain few Cenozoic deposits, although some Pliocene basaltic flows are exposed in the eastern end of the range, and Pliocene valley deposits that formed when the mountains were much lower are present on the south and west flanks (for example, at Barton Flats). There are, of course, also Quaternary lake beds, alluvial gravels, stream deposits in the larger canyons, and alluvial fans around the north and east flanks.

The presence of the Pliocene valley deposits and basalt flows, resting on surfaces of low relief, shows that the San Bernardinos have been elevated almost entirely during the Quaternary. Historical tectonic activity suggests that the mountains are still going up.

EASTERN BOUNDARY RANGES

Slightly southeast of the higher San Bernardino Mountains, the Little San Bernardino Mountains continue the east-west trend characteristic of the Transverse Ranges. The Little San Bernardino and adjoining Pinto, Hexie, and Eagle mountains are desert ranges and mark the eastern end of the Transverse Range province, although as noted previously, some include the Orocopia Mountains to the southeast. The northern edge of these eastern ranges lies along the Pinto Mountain fault zone, which merges on the west with the Morongo Valley fault. The southern margin of the Little San Bernardino Mountains is the Mission Creek fault, part of the San Andreas system. Eastward, into the Eagle Mountains, the southern boundary is less obvious; any faults that are present are concealed under large alluvial fans. The Mission Creek-San Andreas system turns south away from the ranges and strikes into the Salton Trough. The Pinto and Eagle mountains are separated by the Pinto Basin, a large upland alluviated valley across which the prominent Blue Cut fault extends in an east-west direction. This fault forms a prominent scarp along the south face of the Hexie Mountains in Joshua Tree National Monument (Figure 10-17).

Rocks of these ranges are about equally divided between the Proterozoic schists and gneisses and the granitic rock exposed on either side of Morongo Valley. The granitic rocks probably correlate with the Cactus Granite of the San Bernardino Mountains. These granitic rocks provide the striking boulder-pile mountains seen in Joshua Tree National Monument. Minor amounts of dolomite and other sedi-

FIGURE 10-17 The trace of the Blue Cut fault along the south base of the Hexie Mountains. View east in Pleasant Valley, Joshua Tree National Monument, Riverside County. (Photo by R. M. Norris)

mentary rocks of presumed early Paleozoic age occur in both the Pinto and Eagle mountains.

The Pinto and Eagle mountains both contain important iron ore bodies, though the principal mines are in the Eagle Mountains. The ores are located in contact metamorphic zones between the plutonic rocks and the older Paleozoic dolomitic limestones. The ore minerals are chiefly hematite and magnetite and carry from 50 to 53 percent iron. From 1948, the Eagle Mountain mine was the main source of ore for the Fontana steel plant until it closed in the 1980s.

STRUCTURAL FEATURES

The Transverse Range province is dominated by east-west trending folds, often faulted, and by faults that have east-west or northwest-southeast trends. Even the San Andreas fault is deflected across the province and assumes a more east-west direction, both east and west of where it slices obliquely across the Transverse Ranges at Cajon Pass. West of Cajon Pass, the San Andreas forms the northern boundary of the province for some distance. Similarly, east of Cajon Pass, the general trend of the fault is easterly and it forms the province boundary for some distance. The central axis of the Transverse Ranges is offset about 40 kilometers (25 miles) at Cajon Pass.

The western ranges and basins are primarily folded structures, although they are often bounded by steep faults that show both horizontal and vertical separation. The faults of the western Transverse Ranges are shown in Figure 10-4. The eastern ranges are chiefly elevated fault blocks; many of their bounding structures are reverse faults.

It was mentioned earlier that the Transverse Range province has been very active, tectonically, during the last half of the Tertiary and during the Quaternary. Rapid rates of uplift have been documented at a number of places. Near Ventura, for example, a rate of uplift of 75 centimeters (30 inches) per hundred years has been reported.

During the Miocene, clockwise rotation of about 90 degrees affected the central and western part of the province as demonstrated by paleomagnetic studies, though the exact size of the rotated blocks and their thickness remain uncertain. Rotation appears to have ceased by the end of the Miocene, but almost certainly was related to the arrival of the Pacific Plate and East Pacific Rise at the edge of the continent about 29 million years ago. As the continent overrode the East Pacific Rise, it was followed about 22 million years ago by considerable tectonic stretching, basin development, and block rotation. About 5.5 million years ago, when the San Andreas system was well established across the province, strong compression and uplift raised the modern Transverse Ranges. This historical development should be kept in mind as the following descriptions of some of the major faults in the province are read.

Big Pine Fault

The Big Pine fault has long been recognized as one of California's major left-slip faults. According to some interpretations, it forms the northern boundary of the Transverse Ranges in Ventura County. The fault is believed to extend about 80 kilometers (50 miles) from Frazier Mountain west into Santa Barbara County, to a belt of Franciscan rocks where it apparently merges with the northwest-trending faults of the southern Coast Ranges. Based on presumed lineations revealed in high-altitude photographs, it has been suggested that the Big Pine fault is somewhat offset to the north and continues west from the Franciscan belt to the coast south of Point Sal. Fieldwork to date provides little support for this idea, but some earthquake epicenters have been located along the supposed extension. If this extension is confirmed, it may be more logical to consider the Big Pine fault as the northern boundary of the western Transverse Ranges, rather than the Santa Ynez fault—which has been done in this book.

Because both the Big Pine and Garlock faults show prominent left slip, it has been inferred that they are part of the same system. As such, they would have to be subordinate to and older than the San Andreas, along which the western end of the Garlock is separated by 9.6 kilometers (6 miles) from the eastern end of the Big Pine. It may be, however, that the Big Pine is more logically related to faults in the Soledad Basin, from which it has been offset by the San Gabriel fault.

The earthquake of 1852 is usually attributed to movement on the eastern segment of the Big Pine fault, when ground breakage was observed in Lockwood Valley. There is another Lockwood Valley near Mission San Antonio in the southern Coast Ranges, however, and ground breakage may have occurred there instead. Investigations on the central and western parts of the fault suggest Quaternary horizontal stream offsets up to 900 meters (3,000 feet), and one fan may have been offset about $1\frac{1}{2}$ kilometers (1 mile). Much greater displacements are indicated since Miocene time, possibly $6\frac{1}{2}$ to 16 kilometers (4–10 miles). The preponderance of the evidence available indicates that much of the slip occurred in the Miocene with some later reactivation.

Santa Ynez Fault

The Santa Ynez fault is generally considered the northern boundary of the Transverse Ranges west of the Santa Barbara-Ventura county line. It is a relatively long fault, extending 140 kilometers (90 miles) west from the Agua Blanca thrust in central Ventura County to the coast (Figure 10-4). At Gaviota Pass, the fault divides into a south branch, which enters the Santa Barbara Channel, and a north branch, which continues west for about 19 kilometers (12 miles) and includes the Pacifico fault.

Matching the rock section north of the fault with that exposed to the south suggests a vertical separation of 2,900 meters (9,500 feet) with the south side up. The differences in the Tertiary rocks on opposite sides of the fault have been interpreted as evidence of considerable horizontal displacement. Investigators disagree regarding the direction of offset, however, although most support left slip. Furthermore, the amount of slip is undetermined; post-Eocene estimates range from $1\frac{1}{2}$ to $3\frac{1}{2}$ kilometers (1–2 miles) to as much as 60 kilometers (37 miles). Some recent interpretations suggest that slip increases from near zero at the eastern end of the fault toward the west, rather like a swinging beam.

Topographic features studied between 1957 and 1971 (including offset streams, sag ponds, offset terraces, and vertical changes in elevation) all suggest that the Santa Ynez fault is currently active. A very strong 1927 earthquake, with epicenter west of Point Arguello, has been attributed to movement on a presumed western extension of the fault.

Santa Cruz Island and Santa Rosa Island Faults

Santa Cruz Island and Santa Rosa Island are each cut by a prominent east-west trending fault. Whether these two faults are related is uncertain, but many geologists project them into the channel between the islands and also assume left-lateral offset on a younger northwest-trending sea floor fault.

The dissimilarity of rocks on either side of the two faults implies considerable displacement. For the Santa Cruz Island fault, vertical displacement of about 2,300 meters (7,500 feet) and horizontal separation of 1,500 meters (5,000 feet) has been proposed; for the Santa Rosa Island fault, a maximum of 16 kilometers (10 miles) of horizontal displacement has been suggested. Perhaps more important from a practical point of view is the recency of movement on the two faults. Streams and terraces have been displaced left laterally in both instances, showing that the faults have been active during the Quaternary.

It has been suggested that these island faults are merely the western portion of a major left-slip fault zone, which includes the Malibu Coast, Santa Monica, and Raymond Hill faults on the mainland to the east, although sea-floor profiling has not yet provided any convincing evidence that these faults are connected on the sea floor. If it proves to be a single fault zone, however, its potential for producing large earthquakes is much greater than would be the case if the faults are separate, shorter structures.

Santa Clara River Valley Faults

Two faults approximately outline the valley of the Santa Clara River, and their movement appears to be closing the valley as if it were caught in the jaws of an enormous vise (Figure 10-18). The river and its tributaries, of course, keep the valley open by erosion, and in historical terms the "jaws" are unlikely to close in the near future. Earthquakes are quite possible, however.

On the valley's north side is the San Cayetano thrust, with a sinuous trace from near Ojai to east of Piru. Its dip varies from 15 to 50 degrees north, which explains the irregular surface trace. Dip-slip displacement may be as much as 6,100 meters (20,000 feet), with the Topatopa mountain block riding up the fault plane in a southerly direction. East of Fillmore, the fault lies close to the base of these mountains, and to the west it is in their lower foothills. The fault can be seen fairly easily where it has brought Eocene rocks with bold outcrops and steep faces over softer, more gently dipping Miocene and Pliocene rocks. This fault is regarded as no older than Pleistocene.

On the south side of the valley is the poorly exposed Oak Ridge fault. This fault has been traced, through oil wells, from west of Piru to the mouth of the Santa Clara River and offshore into the Santa Barbara Channel. On land it is a high-angle reverse fault with south dips of 60 degrees or more. The anticlinal folds of South Mountain and Oak Ridge have moved upward on the fault, which has also offset Quaternary terraces.

San Gabriel Fault Zone

About 144 kilometers (90 miles) long, the San Gabriel fault is mapped as a part of the San Andreas system and like the San Andreas shows predominately right slip.

FIGURE 10-18 Looking east over the Santa Clara River valley, Ventura County. Oak Ridge is on the right and the Topatopa Mountains on the left. Sespe Creek enters the Santa Clara River from the left just below the city of Fillmore. (Photo by Spence Air Photos, courtesy Department of Geography, University of California, Los Angeles)

Moreover, it somewhat defies the general east-west trend of Transverse Range structure, as it strikes southeast from Frazier Mountain and enters the San Gabriel Mountains at their western end. North of Mount Wilson, the fault does become nearly east-west, however. It appears to be offset in San Antonio Canyon by north-south trending faults, with the eastern segment terminating against the San Jacinto fault. The San Gabriel's wide crushed zone has strongly affected topography and drainage, prompting the east and west forks of the San Gabriel River to follow the fault for most of their lengths (see Figure 10-11).

During the last 12 million years, the San Gabriel fault has undergone about 60 kilometers (40 miles) of right slip. Movement ceased for all practical purposes about 5 million years ago. This value has been deduced from displacement of Middle Miocene anorthosite-bearing conglomerates southwest of the fault from their only available source area, the western San Gabriel Mountains across the fault. The San Gabriel fault has also experienced varying vertical displacement, with the north side up in some places and down in others. Along the southwest side of the Ridge Basin, vertical displacement is as much as 4,250 meters (14,000 feet). The San Gabriel fault has experienced only minor activity in recent times, although it does show a few young breaks. Some Pleistocene deposits and stream courses were affected by movement on the fault, and there is even a little disturbance of Quaternary alluvial deposits.

In Big Tujunga Canyon, the San Gabriel fault bifurcates into the main east-west trending San Gabriel fault and a south branch that becomes the Sierra Madre fault. The Sierra Madre and its eastern counterpart, the Cucamonga, are steep north-dipping, range-front faults along which most uplift of the San Gabriel Mountains has occurred. Activity is quite recent, because the Cucamonga fault has offset Quaternary alluvium. In a broad sense, the Sierra Madre fault zone comprises all the range-front thrust and reverse faults from east of Arcadia to the arcuate, supposedly discontinuous faults in the western San Gabriel Mountains and the northern foothills facing the San Fernando Valley. The latter include the Santa Susana thrust, which cuts across the Santa Susana Mountains to join the Oak Ridge fault.

Seismologists have shown that the 1971 San Fernando earthquake defined a north-dipping reverse fault that corresponded to the surface breaks observed along segments of the Sierra Madre fault zone. Main motion below the epicenter was pinpointed at a depth of about 11 kilometers (7 miles), with no ground breakage directly above. A plane dipping northeast under the range at about 45 degrees includes both the focus of the quake and the surface offsets observed near San Fernando.

Soledad and Vincent Thrust Faults

The Soledad fault can be seen in Soledad Canyon, where it brings Proterozoic crystalline rocks into contact with the nonmarine middle and late Tertiary rocks of the Soledad Basin to the north. It is a normal fault.

On the north side of Mount San Antonio is the oldest major fault known in the San Gabriel Mountains. This is the Vincent thrust, a fault that dips south and west with an extremely irregular trace. As noted earlier, the Vincent thrust seems to have caused the northeast movement of plutonic rocks over the Pelona schist. Although it has influenced modern topography, the thrust is an old feature, formed about 60 million years ago, possibly in a subduction zone.

Malibu Coast—Santa Monica—Raymond Hill Fault System

The southern frontal faults of the Santa Monica and San Gabriel mountains constitute a fault system of substantial magnitude, with inferred left-slip of 64 kilometers (40 miles) or more. The Malibu Coast fault extends from Arroyo Sequit on the west to near the city of Santa Monica. It continues east as the Santa Monica fault, largely buried beneath the alluvium at the base of the Santa Monica Mountains. Near Glendale the Santa Monica fault seems to merge with the Raymond Hill fault, which has offset Quaternary terrace materials and is traced from South Pasadena east to near Monrovia.

The trend of this fault system aligns with the Santa Cruz Island fault, which it may well join. Such a connection is frequently assumed when accounting for the origin of the province and the distribution of such rocks as the Eocene Poway conglomerates and the Miocene San Onofre Breccia.

ORIGIN OF THE TRANSVERSE RANGES

The anomalous east-west trend of the Transverse Ranges has long puzzled geologists and led to a host of possible explanations. Thomas Dibblee, who has devoted much of his career to mapping the Transverse Ranges, has framed a number of questions that need to be answered before an acceptable history of the province can be developed. Some of these are:

1. Why are the ranges and valleys of the provinces oriented east-west, diagonal to the prevailing structural trends in California?
2. Are the east-west trends related to the east-west Murray fracture Zone in the Pacific?
3. How are the right slip San Andreas and the major left slip transverse faults in the province related?
4. Why is the San Andreas fault bent out of line as it crosses the province?
5. Why does the Transverse Range province extend about equal distances east and west on opposite sides of the San Gabriel and San Andreas faults?
6. Why isn't the province cut by the many northwest-trending right-slip faults apparent in the adjacent provinces, apart from the San Andreas and San Gabriel faults?

It is probably obvious that the San Andreas fault and its history are of major importance in understanding the origin of the Transverse Ranges. Although the San Andreas is probably older in other areas, in this part of California the fault system apparently developed no earlier than middle Miocene, about 12 million years ago. Since then, movement has been nearly continuous, usually reflecting right slip with perhaps a substantial vertical component. The amount of displacement is disputed, but it does not exceed 320 kilometers (200 miles) on the main trace of the San Andreas.

Plate tectonic theory identifies the San Andreas as a transform (fault) and as the shear boundary between a northwest-moving Pacific plate and the American plate. If this is correct, the Transverse Ranges must be relatively modern—at least post-late Oligocene (29 million years ago)—irrespective of the total offset on the fault. As far east as the San Gabriel Mountains the province definitely belongs to the Pacific plate, necessarily sharing its northward migration. The apparent eastward continuation of the province, which incorporates only about 40 kilometers (25 miles) of offset, must be just a coincidence if this interpretation of plate history is accepted.

A related point is that the enigmatic rock distribution of the Transverse and Peninsular ranges can be explained most satisfactorily by substantial strike-slip movement on the San Andreas and east-west Transverse Range faults. Present geography alone cannot account for the puzzling distribution of the widely scattered Poway-like conglomerates and the San Onofre Breccia. This is another indication that in former times little existed of the modern Transverse Range province.

In recent years Bruce Luyendyk and his students have been investigating the paleomagnetic history of the western Transverse Ranges as recorded mainly in its Miocene volcanic rocks. These studies suggest that the western ranges rotated 60 to 80 degrees in a clockwise direction from a pivot point near the eastern end of the Santa Monica Mountains, during the Miocene, and since the Miocene, the western ranges have moved northward the equivalent of about 10 degrees of latitude. They believe this resulted from the effects of right shear along the Pacific-American plate boundary (ancestral San Andreas fault), acting on a set of more or less north-south blocks like those of the Peninsular and Coast ranges (Figure 10-19).

Persuasive as the paleomagnetic evidence is, not all field observations are con-

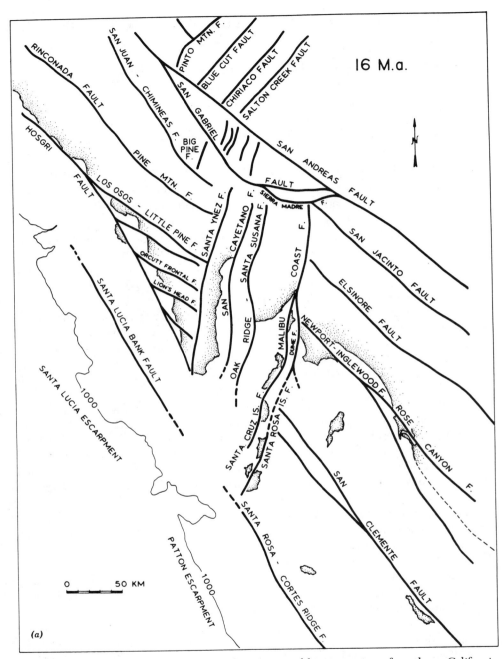

FIGURE 10-19a and 10-19b Diagrams showing possible orientation of southern California
faults and crustal blocks 16 million years ago (Middle Miocene), and 6 million years ago (Late
Miocene) following clockwise rotation of western Transverse Ranges. (Source, Hornafius,
Luyendyk, Terres, and Kamerling, Geological Society of America Bulletin, v. 97, pages 1484
and 1486, courtesy of Geological Society of America)

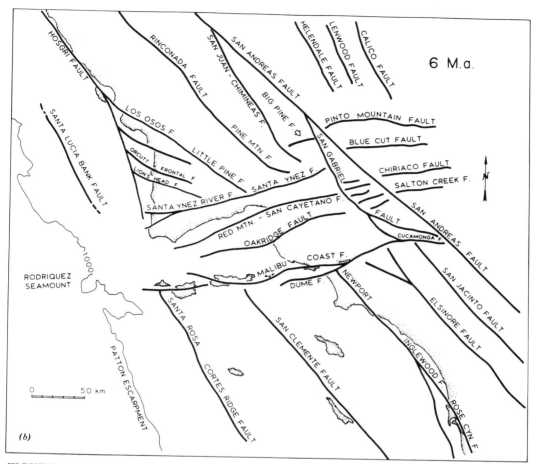

FIGURE 10-19 (*Continued*)

sistent with it. For example, the wedge of Miocene and younger sedimentary rock that occupies the triangular Santa Maria Basin between the western Santa Ynez range and the southern Coast Ranges suggests that the rocks were deposited in a triangular depression pulled open in early Miocene time by counterclockwise rotation of the western Santa Ynez Mountains away from the Coast Ranges. It is clear that we do not yet have all the evidence we need, or possibly we are incorrectly interpreting the evidence we do have.

Yet another view has been presented by A. O. Woodford and his colleagues. Their investigations call attention to substantial data to support the contention that the Transverse Ranges have persisted as a discrete block with an east-west trend since *before* the emplacement of the Cretaceous batholithic rocks. In particular, they cite east-west trends shown by the general petrology, crystalline rock patterns, chemical affinities, and structural features.

No matter how the bends in the San Andreas fault might have originated, it is likely that continued right slip on the present San Andreas system subjects the entire Transverse Range province to large-scale, north-south compression. This is doubtless related to the prominent, active east-west trending thrust faults that occur on both sides of the province as well as within it.

There is a lesson in this puzzling situation. Scientists must continually reexamine their assumptions when developing a hypothesis and must constantly check observed facts to see what they require of the hypothesis. Plate tectonic theory has been truly remarkable in providing a coherent explanation for a wide range of geologic phenomena. It is risky, however, to ignore geologic features that seemingly do not agree with the theory as applied to a specific region like the Transverse Ranges. When a theory has been as successful and revolutionary as Plate Tectonics, there is a tendency to force a recalcitrant fact to fit the grand plan, rather than reexamining the assumptions on which the particular application of the theory rests. It is equally necessary to recheck the validity of the so-called facts. In this regard, no application of plate tectonic theory to the west coast of North America can be entirely valid until it accords with the geologic realities of the Transverse Ranges.

SUBORDINATE FEATURES

Streams

The Transverse Range province has few sizable streams, but even though most of them are small, the streams are disproportionally important when one considers their modest discharges. This seeming paradox results from a combination of dry climate and the presence of about 40 percent of the state's population along the southern fringes of the Transverse Ranges.

The largest drainage system wholly within the province is that of the Santa Clara River (Figure 10-18), which drains most of central and southern Ventura County, the northwestern San Gabriel Mountains, Liebre Mountain, the Sierra Pelona, and the Soledad Basin. The river is more than 120 kilometers (75 miles) long, and although its surface flow is normally small, it provided almost all the domestic and agricultural water for its basin until the arrival of Feather River water. In addition, the riverbeds have been a valuable source of sand and gravel for the Los Angeles, Ventura, and Santa Barbara areas.

In the west, the Santa Ynez River, about 100 kilometers (60 miles) long, is the most important stream (Figure 10-20). It drains the northern Santa Ynez Mountains and the southernmost Coast Ranges and follows the Santa Ynez fault toward the sea. This river was dammed to provide water for the south coast of Santa Barbara County.

The San Gabriel and San Bernardino mountains are drained by several permanent streams that have been crucial to the communities of the region. On the western slope of the San Gabriel Mountains, streams drain into the San Fernando Valley and the Los Angeles River. The Los Angeles River is about 105 kilometers (65 miles) long and has a total drainage area of about 2,710 square kilometers (650 square miles). Moreover, although this river is the butt of many a local joke, it actually furnishes the city of Los Angeles with about 15 percent of its water supply from underflow and ground water.

Important streams to the east include the San Gabriel River, which drains the central San Gabriel Mountains. The San Gabriel forks downstream near Duarte into the San Gabriel River proper and the Rio Hondo, which joins the Los Angeles River near Downey, so that the two drainages intermingle. The Los Angeles River reaches the sea near Long Beach, and San Gabriel discharges near Seal Beach.

FIGURE 10-20 Santa Ynez River and its headwater drainage, north of the city of Santa Barbara. Gibraltar Reservoir is in the foreground. The east-west linearity of the ridge-valley topography is controlled primarily by faults. (Photo by Spence Air Photos, courtesy of Department of Geography, University of California, Los Angeles)

The Santa Ana River has the largest drainage basin in southern California. It drains both the high San Bernardino and the eastern San Gabriel mountains, plus the northern Peninsular Ranges. In the late 1930s, its limited supplies were augmented by Colorado River water. The interesting geomorphic history of this stream was considered in the preceding chapter.

Floods

Despite their normally modest discharges, like water courses in other arid and semiarid regions, the Transverse Range streams fluctuate greatly in volume. Every 20 or 25 years, heavy winter rains in the mountains cause widespread flooding in the lowlands. Owing to the high intensity of rainfall in such storms, small streams can accomplish tremendous erosion and transportation in short periods. In January 1934, one of these high-intensity storms struck the western San Gabriels, sweeping boulders weighing 80 metric tons (100 tons) from the mountains into the town of Montrose, with very destructive results. One creek was deepened 5.2 meters (16 feet) in a few hours (see Figure 1-19).

The Los Angeles River has been repeatedly subject to severe flooding, with the result that more than three-quarters of its channel has been lined with concrete. In several instances, the river even changed course. Prior to 1815, the Los Angeles River drained into the Long Beach area as it does today. The flood of 1815 caused the river to change its course westward, to join with Ballona Creek and empty into Santa Monica Bay. In 1825, flooding again forced the river to abandon the new course and cut the first well-defined channel south toward the Long Beach region.

FIGURE 10-21 Flooding in the Los Angeles Basin, March 1938. View of the San Gabriel River on its flood plain near Seal Beach. (Photo by Spence Air Photos, courtesy of Department of Geography, University of California, Los Angeles)

The flood of 1862 reportedly converted most of the coastal plain into a lake; at one place the Santa Ana River widened to 4.8 kilometers (3 miles). In 1884, all bridges over the Los Angeles River except one were washed out; this was the wettest season ever recorded in Los Angeles.

The flood of 1938 produced a discharge in excess of any previously recorded, although annual rainfall totals had been higher. Unfortunately, the ground had already been saturated by a series of heavy rains before the big storm arrived in late February. With rainfall of nearly 760 millimeters (30 inches) in the mountains and 250 to 380 millimeters (10–15 inches) in the lowlands in just four days, the streams in the Los Angeles region were unable to accommodate the volume of water. Large areas of the San Fernando Valley and the Los Angeles coastal plain were inundated (Figure 10-21). The raging Los Angeles River damaged more than 100 bridges, caused the loss of 43 lives, and resulted in perhaps $40 million in property damage. Total loss of life in southern California was 87 and property damage was about $78 million.

Palos Verdes Marine Terraces

Between Redondo Beach and San Pedro, the Palos Verdes Hills form a prominent headland that rises to 451 meters (1,480 feet) (Figure 10-22). From late Pleistocene until recently, the peninsula was probably an island, because land on the northwest

FIGURE 10-22 Elevated flight of marine terraces of Pleistocene age at Palos Verdes Hills, Los Angeles County. Photograph was taken in 1933 before urbanization of the peninsula had occurred. (Photo courtesy Map and Imagery Laboratory, Library, University of California, Santa Barbara)

is low and was very swampy until altered by humans. A 25-meter (80-foot) rise in sea level would make it an island again. The hills form cliffs and terraces along their seaward margin with some steps as high as 90 meters (300 feet), although the average is 30 to 45 meters (100–150 feet). Near San Pedro on the south and Redondo Beach on the north, cliff heights decrease to 15 meters (50 feet) or less.

Palos Verdes Peninsula is an anticline uplifted as a horst between faults on the northeast and southwest. The exposed rocks belong mostly to the Miocene Monterey formation, which covers a core of Franciscan Catalina Schist. Some Pliocene marine rocks are present in the northeast, and the marine terraces contain thin, late Pleistocene deposits.

The terraces are the landscape's most striking feature and can be seen best on the seaward side of the peninsula. Thirteen have been recognized, with elevations ranging from 30 to 400 meters (100–1,300 feet); it is likely that the subdued topography above 400 meters (1,300 feet) indicates complete late Pleistocene submergence. Relative ages of the terraces have not been determined, but the lower ones appear to be younger. Although many contain shelly marine sand, they cannot be distinguished paleontologically from one another. Elevations show that most of the terraces have been minimally deformed since they were cut, but near San Pedro the Gaffey anticline has gently warped the youngest terraces. Apparently deformation is continuing there, because recent alluvium has been deformed. Recent studies show that the Palos Verdes Peninsula has been rising since Pleistocene time at almost a meter per thousand years.

It should be remembered that the Palos Verdes and other California terraces are products both of local uplift and Pleistocene eustatic sea-level fluctuations that reflect worldwide cycles of glaciation. These two independent processes have so complicated local histories that correlation of the Palos Verdes terraces with any other terraces is difficult, except where datable materials are present on the terraces. Correlation by elevation alone is unsatisfactory.

SPECIAL INTEREST FEATURES

Landslides

Several Transverse Range landslides have achieved considerable notoriety, not because they are remarkable geologically, but because they have severely affected urbanized regions. Regrettably, some slides were prompted by activities of humans.

Portuguese Bend. Near Portuguese Bend, in the southwest Palos Verdes Peninsula, are a recently active large landslide and two smaller slides at higher elevations. Another slide, of moderate size lies near the coast west of White's Point. The large Portuguese Bend slide attained prominence in 1956, when it experienced renewed activity. Although the slide had been clearly mapped in a 1946 published report, building proceeded, and by 1956 many homes existed in the area (Figure 10-23). Several of these were destroyed or severely damaged when the slide began to move, sending porches downhill, developing faults in driveways and sags and cracks in walls, and jamming doors and windows. From time to time during the history of the slide, small closed depressions developed on its surface, and some of these persisted long enough to form small ponds. Fossil wood obtained from these former ponds has been dated by the radiocarbon method and shows that the slide has been present for at least 37,000 years.

Roads crossing the slide have been patched repeatedly, and eventually many were abandoned as segments moved downhill. Scarps 6 meters (20 feet) or more in height developed. Buried utility lines were excavated and laid on the surface; overhead wires required constant adjustment as distances between poles changed. Deep, undrained hollows and ominous bulges and buckles appeared within a few months. Near the toe of the slide, just offshore, a buckle arched a pier upward at its midpoint. The head of the slide was characterized by arcuate scarps and cracks, giving a steplike appearance.

Only the southeastern portion of this large slide moved during the most recent episode. Increasingly larger portions were involved, however, until the area affected was roughly three times as extensive as the area first deformed. The active portion continues to expand. Monitoring of movement showed that slippage was more rapid after winter rains, with a lag time of a month or two. Rates of movement have varied; a station near the head of the slide moved downward 3.6 meters (12 feet) in 600 days, a centrally located site subsided about 1.5 meters (5 feet) in a similar period, and upward movement near the slide's base was almost 1.2 meters (4 feet) in 388 days.

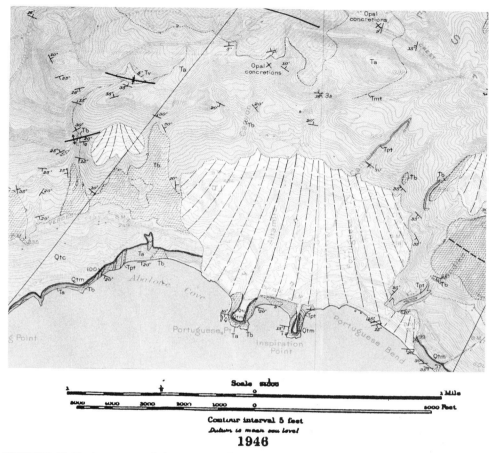

FIGURE 10-23 A portion of the 1946 geological map of the Palos Verdes Peninsula. The Portuguese Bend landslide area is indicated by the fan-shaped pattern of curved dashed lines. (Source: U.S. Geological Survey)

Causes of the Portuguese Bend landslide have been thoroughly investigated, with complex and controversial results. Apparently it was triggered by erosion of the seaward side of a synclinal fold, thus removing support from the dipping rocks. In addition, the area's rocks are thin-bedded and contain numerous beds of altered volcanic ash that becomes plastic when wet. So the combination of structure and lithology was initially unfavorable even under normal moisture conditions. In the few years preceding renewed movement, about 150 homes were built in the slide area, but none had municipal sewer service. This dependence upon septic tanks and cesspools introduced huge volumes of water into the slide, intensified by garden watering estimated at a minimum of 152,000 liters (40,000 gallons) per day. Another factor, alleged by some to be the chief cause, was emplacement of highway fill near the slide's head. Prevailing geologic opinion, however, was that introduction of water into the old slide was the critical factor, though the addition of load certainly exacerbated the situation. Recent studies show a very close relationship between moisture content in the slide and movement. Rainfalls are followed almost immediately by accelerated movement, and there is a strong seasonal effect as well—decreased movement during the dry summer months and increased movement in the rainy winter months.

Efforts to check movement were largely unsuccessful and consisted initially of installing concrete caissons in the slide. These were 1.2 meters (4 feet) in diameter and 6 meters (20 feet) long and extended 3 meters (10 feet) into the presumably stable material below the slide. Twenty-five of these "pins" were installed, but no appreciable change in slide movement resulted. Another plan entailed stabilizing the slide by placing a large fill on the toe, but this was never done. As homes were abandoned, the volume of water introduced into the slide decreased, and consequently the rate of movement slowed. Nonetheless, by 1968 the slide had caused nearly $10 million in damage to homes, utility lines, and roads. By 1986, damage had become very extensive and many of the remaining homes in the area were placed on temporary wood or steel foundations that could be releveled every few months as required. The main highway crossing the slide continues to be damaged and it must be repaved, regraded, or patched four to six times a year to permit traffic to cross the active part of the slide. In the fall of 1985, the slide moved 2.7 to 3.2 centimeters (1.1–1.2 inches) daily. One large depression containing a small pond known locally as Lake Ishibashi has developed in the farmlands next to the highway. This basin has steadily deepened and is now about 10–12 meters (30–36 feet) lower than the highway.

Point Fermin. In 1926, a landslide at the southeastern corner of the Palos Verdes Hills made a small segment of Point Fermin slip seaward (Figure 10-24). The landslide crack intersected the main waterfront street of San Pedro, rupturing streetcar tracks and damaging a few houses. There was no evidence that the slide was a reactivation of an older one, as in the Portuguese Bend case. This was an original slide, one of hundreds discernible along the coast of the Transverse Ranges. Like the Portuguese Bend slide, seaward-dipping beds undercut by wave action and garden irrigation that contributed to the softening of the expansible clay and ash layers in the bedrock, caused the Point Fermin slide.

Blackhawk. Certainly the largest slide in the Transverse Range province is the Blackhawk, on the north slope of the San Bernardino Mountains. This prehistoric slide is one of the largest known in North America. It was studied in detail by R. L. Shreve, who showed that the slide moved to its resting place on a cushion of compressed air. (This mechanism has since been recognized as applicable to other slides.) The end of the slide can be seen from State Highway 247 about 16 kilometers (10 miles) east of Lucerne Valley; the only satisfactory way to see the entire slide is from the air (Figure 10-25).

The Blackhawk slide is 8 kilometers (5 miles) long, about 3.2 kilometers (2 miles) wide, and 10 to 30 meters (30–100 feet) thick. It is a tonguelike sheet of brecciated Pennsylvanian Furnace Limestone derived from Blackhawk Mountain about 1,200 meters (4,000 feet) above. In the source area, the Furnace Limestone has been thrust northward over uncemented sandstone and weathered gneiss that subsequently were eroded away, leaving a precipitous slope.

FIGURE 10-24 Point Fermin landslide near San Pedro, Los Angeles County in 1984. Buckling and irregular settling are shown by the broken street, curb, and sidewalk pavement. The conical structure behind the left-hand palm tree was a manhole sewer access once flush with the ground surface. (Photo by R. M. Norris)

Once the softer rocks were undermined, presumably during a wet period about 17,000 years ago, a mass of limestone breccia collapsed and slipped rapidly into upper Blackhawk Canyon, forming a stream of rubble about 600 meters (2,000 feet) wide and 90 to 120 meters (300–400 feet) deep. As the slide moved down the canyon (at about 275 kmph or 170 mph), it passed over a resistant gneissic ridge that crosses the canyon, and was thus launched into the air—a geologic version of a flying carpet. Calculations indicate that the sheet of moving breccia was probably as high as 120 meters (400 feet) above the canyon floor immediately after becoming airborne, but that it settled quickly, compressing the air trapped beneath to a frictionless blanket less than a meter thick. While airborne, the slide possibly attained velocities of 435 kilometers per hour (270 mph), and the entire distance from launching point to resting place was covered in about 80 seconds. These values are based on a consideration of local geometry and are consistent with the behavior of similar slides observed during formation. As the slide spread over the desert floor, the air cushion became thinner, permitting the slide to settle.

A characteristic of such slides is the presence of large blocks that, although badly shattered, have fragments that retain their original orientation to one another—much like a jigsaw puzzle with pieces pulled slightly apart. This feature supports the view that carpetlike sheets of rock can be moved almost intact on cushions of compressed air.

Shortly after the slide occurred, small ponds developed in depressions on its surface and one of these has yielded fresh-water mollusk shells that give a radiocarbon

FIGURE 10-25 Blackhawk slide in the northern San Bernardino Mountains, with Mount San Gorgonio in the background. View to the south from the Mojave Desert. (Photo by John S. Shelton)

age of 17,400 years. Because the pond sediments are composed largely of materials pulverized during the slide and covered with different, probably windblown materials, the ponds are probably only slightly younger than the actual age of the slide.

Earthquakes

The Transverse Range province has sustained a number of severe earthquakes in historic times. Some are among the strongest observed in California, although at present the central Coast Ranges and Imperial Valley are experiencing earthquakes more frequently. The following earthquakes of magnitude 6.0 or greater have originated in the Transverse Ranges since 1800: Santa Barbara Channel (7.5?) in 1812, Big Pine (7.0?) in 1852, Fort Tejon (8.0?) in 1857, Santa Barbara (6.3) in 1925, Point Arguello (7.5) in 1927, and San Fernando (6.4) in 1971.

Seismologists have determined that a correlation exists between the length of a fault and the severity of earthquakes it may generate. Unfortunately, fault length cannot always be measured accurately because part may be concealed. Until the fault is revealed (by ground breakage or drilling, for example), mapped segments are often erroneously regarded as separate faults.

With its possible 1,600 kilometer (1,000 mile) length, the San Andreas is the most likely source of major earthquakes in the province, a probability confirmed by the historical record. The severest earthquake originating in the Transverse Ranges since Spanish settlement is the Fort Tejon earthquake, which occurred in 1857. Its magnitude is estimated at 8.0 ± 0.5 on the Richter scale, and about 320 kilometers (200 miles) of the San Andreas trace experienced ground breakage. Extensive rupture was accompanied by the greatest right-lateral offset yet observed on the San Andreas system, approximately 9 meters (30 feet). According to one account, a circular sheep corral astride the fault in eastern San Luis Obispo County was broken across the middle and converted to an open S-shaped figure. The earthquake was strongly felt on the southern California coast, severely damaging Santa Barbara and San Buenaventura missions and reportedly throwing the Los Angeles River from its channel.

During the past 50 years, repeated surveys have been made across the portion of the San Andreas affected in 1857. They have revealed no sign of creep, nor have any earthquakes been attributed to this segment since 1857—apart from one in 1916. Today, this part of the San Andreas is frequently cited as locked into position by its bend into the Transverse Ranges—while presumably strain energy accumulates for the next all-but-inevitable strong earthquake.

In analyzing the crustal movements associated with the 1971 San Fernando earthquake, it was discovered that a large area embracing the entire Transverse Range province and extending as far east as Arizona, had been uplifted, apparently first in the 1960s but continuing until about 1974, and involving a maximum of about 35 centimeters (12.8 inches) near Palmdale. Between 1974 and 1976, the *Palmdale Bulge* as it was named, rapidly relaxed and the elevation at Palmdale dropped 24 centimeters (9.4 inches) in just three years. The nature and indeed even the reality of the Palmdale Bulge was much disputed. To some it suggested a rapid increase in stress in the Big Bend area of the San Andreas fault. Such a compressional elevation is consistent with the structural features of the Transverse Ranges; some geologists thought it might be a precursor to a major earthquake, whereas others attributed the occurrence to measurement error. Since no such earthquake has occurred in the decade following these events, the significance of the bulge remains enigmatic.

The 1971 San Fernando earthquake was the most damaging in the Transverse Ranges this century. Consequences were severe even though the quake was of moderate magnitude (6.4), because the epicenter lay near the densely populated San Fernando Valley. The quake caused 64 deaths and property losses between $500 and $1000 million. The damage would have been much greater had the shock been longer or stronger, for the Van Norman Dam was perilously close to failure when the 60-second earthquake ended. Had the dam failed, a large residential area in San Fernando Valley would have been inundated. This quake was accompanied by the greatest ground accelerations ever recorded, to that time, mostly in the 0.5 to 0.75g range, but including a few peaks of more than 1.0g. Higher accelerations have since been measured in connection with other earthquakes and may well prove to be less remarkable than originally supposed.

On the positive side, many people survived because at the time of the shock (about 6 A.M.) they were in single-story, wood-frame homes that, owing to their flexibility, withstand earthquakes notably well (Figure 10-26). Furthermore, this shock was measured and studied more thoroughly than any previous earthquake,

FIGURE 10-26 House damaged by ground shortening (compression) during the February 1971 Sylmar earthquake. Note the buckling of the driveway to the left. Near Sylmar, Los Angeles County. (Photo by A. G. Sylvester)

which will contribute substantially to understanding earthquake forces and promoting the most suitable safeguards.

St. Francis Dam Disaster

A tributary of the Santa Clara River, San Francisquito Creek, was the site of the St. Francis Dam failure of March 12, 1928, which killed about 450 people and destroyed homes, bridges, several miles of highway and more than 4,000 hectares (10,000 acres) of field crops. Completed in May 1926, the dam had been built by the city of Los Angeles to be a reservoir for water from the Owens River and a source of electricity. It was 60 meters (200 feet) high and 210 meters (700 feet) long.

Unfortunately, the dam was planned without geologic advice. It was constructed in a narrow part of the canyon where the east wall is composed of thin-bedded Pelona Schist dipping down toward and underlying the canyon floor. The schist is fragile and breaks readily into small flakes. On the canyon's west wall the schist is separated from the younger Sespe conglomerate by the San Francisquito fault and a gouge zone a few centimeters to more than 1.5 meters (5 feet) thick. In this locality, the Sespe is not only badly sheared and fractured, but also has a dry crushing strength of 3,585 kilopascals (520 pounds per square inch or 14.6 kilograms per square centimeter). It was discovered, moreover, that a sample of this conglomerate placed in water almost immediately disintegrated to an incoherent pile of gravel and mud easily stirred with the finger! Even the pebbles broke up along tiny fractures. This simple test, regrettably not performed prior to construction of the dam, showed that the conglomerate was cemented only with thin films of clay and gypsum.

Ironically, upon learning of the proposed dam site and being familiar with the area's rocks and the location of the San Francisquito fault, several geologists recommended that the dam not be built. Their proffered advice was ignored, however.

The reservoir was first filled on March 5, 1928, and shortly afterward seepage was observed in the conglomerate. Later analyses of the seepage water showed a marked increase in dissolved calcium sulfate, which was derived from the solution of the gypsum in the conglomerate. The significance of the seepage was not fully appreciated by those on duty at the dam, and in the middle of the night of March 12, a week after the reservoir had been filled to a depth of 56 meters (185 feet), the dam failed.

It is likely that seepage increased rapidly in volume shortly before the dam's failure, washing out the soft conglomerate, undermining the west abutment of the dam, and permitting the concrete to crack into large blocks. Blocks up to 9,100 metric tons (10,000 tons) were swept down the canyon as far as 800 meters ($\frac{1}{2}$ mile). The intense swirling of the water probably undercut the schist on the east wall, allowing a huge block of concrete to slide down the dip of the rocks into the canyon. The torrent of water swept away every shred of vegetation and loose rock from the canyon to a height of 15 meters (50 feet) near the dam. The flood poured into the Santa Clara River near Castaic, sweeping people, houses, groves of trees, and bridges seaward. Some victims were not missed until their remains turned up during excavations for sand and gravel months, some even years later.

Rancho La Brea

In west Los Angeles, near the old Salt Lake oil field, are the tar seeps of Rancho La Brea. From this locality has come one of the most famous collections of Pleistocene animals in the world (Figure 10-27). The site consists of pools of viscous tar that oozed to the surface from deep petroleum reservoirs. In moving toward the surface crude oil tends to lose its more volatile constituents, becoming viscous and asphaltic. Normally, the tar pools were covered with thin sheets of slightly salty water, creating effective traps for the Pleistocene animals. As they waded into the pools seeking salt and water, the animals became mired in the sticky tar below.

Both extinct and living species have been collected. Included are 228 species of vertebrates made up of 59 species of mammals; 136 species of birds; a few snakes, toads, lizards, turtles, salamanders, and even fish; plus an assortment of land mollusks and arthropods. Of the large mammals, about 90 percent are extinct today. A few plant remains are preserved too, indicating that pine, oak, cypress, and manzanita grew in the area. Even humans were trapped by the tar, because at least one human skeleton 9,000 years old has been found. Interestingly, the ratio of carnivores and scavengers to herbivores is more than 9 to 1, reflecting the special nature of the deposit more than the actual ratio among animal inhabitants of the area. Many of these plants and animals are displayed in the Los Angeles County Museum.

Oil and Gas Fields

The Transverse Ranges contain more than 40 separate oil and gas fields, some long since abandoned and others still producing. The oldest producing field in California is in Pico Canyon near Newhall. Oil was collected from seeps here as early as 1850, and in 1869 several spring pole wells were sunk although regular production did not begin until 1875. Another pioneering "well" was a tunnel completed in 1866 on

FIGURE 10-27 Tar Pools at Rancho La Brea, Los Angeles. The pools are mostly water, but active oil seeps account for the shiny oil slick that can be seen on the surface. (Photo by Betty Crowell)

Sulphur Mountain near Santa Paula. This tunnel was only 25 meters (80 feet) long, and oil flowed by gravity to the tunnel's entrance. Later tunnels dug in the same area were as long as 490 meters (1,600 feet). Production was as high as 9,576 liters (60 barrels) a day, but usually less. Oil seeps in this area are still active; north of Santa Paula a seep produces a tarry oil that flows down the hillside and across State Highway 150.

Ventura Avenue. The Ventura Avenue field, just north of the city of Ventura, is the province's most productive field and ranks among the top producers in the state. Gas was first produced here in 1903. Early drilling was done with cable tools, which precluded control of the unusually high pressures encountered. As a result, great difficulties were experienced with cave-ins, large flows of water, and blowouts. Despite the field's small area (about 1,000 hectares or 2,500 acres), its oil-producing sands are so thick, sometimes more than 300 meters (1,000 feet), that the field had produced 124 million metric tons (865 million barrels) of oil by 1984. The field is still producing in 1989.

Conejo. One of the smallest fields in the province is perhaps the most interesting. This is the tiny Conejo field east of Camarillo at the base of the Conejo Grade. The field was unique in California because the reservoir rock was fractured Miocene Conejo volcanic rock rather than sedimentary rock, which normally provides oil and gas reservoirs.

First discovered in 1892, the oil was at shallow depth, generally between 18 and 25 meters (60–80 feet). For many years the wells were pumped in groups by little donkey engines that jerked cables fanned out on pulleys to the wells. Yield was

FIGURE 10-28 Windmills pumping oil, Conejo oil field. (Photo by Robert M. Norris)

never great, but the oil produced was highly desirable for lubricating oil stock at refineries.

The field was abandoned in the late 1940s, but an enterprising machinist subsequently leased the property and decided to let nature do the pumping for him. He set some old windmills to pumping six or eight of the wells (Figure 10-28). To his dismay, he got about 99 barrels of water for every barrel of oil. He then fitted each windmill with endless loops of chain that were turned by the windmill. Oil clung to the chains and the water dropped off. At the well head the chain passed a wiper, and the oil dripped into a collecting trough. At best this operation recovered only about 800 liters (5 barrels) of oil each week, so the field was once again abandoned.

Santa Barbara Channel. The Santa Barbara Channel has more than 20 oil fields, several produced from shore. The Summerland field, discovered in 1896, was the first offshore field developed in North America and at one time had several hundred wells producing from piers extending up to 200 meters (700 feet) from shore.

The most famous offshore field in this region is the Dos Cuadras field, first produced from a platform in 1968. It is a major field by American standards and is unusual in several respects. The highest point in the reservoir rock is only 90 meters (300 feet) below the sea floor, in contrast to the normal pattern of giant oil fields where 300 meters (1,000 feet) of rock may lie above the reservoir. Furthermore, at Dos Cuadras these intervening rocks are surprisingly permeable sandstones, siltstones, and clays. It was from Platform B of this field that the notorious Santa Barbara oil spill of January 18, 1969, occurred.

During drilling of the well that blew out, the casing had been cemented only to a depth of 70 meters (230 feet) below the sea floor when higher than expected pressures were encountered and a gaseous mist erupted from the well. The blowout preventer was closed, but the gas and oil were under high pressure and were able to move up the mostly uncemented hole into lower-pressure reservoirs that failed to contain them. In a few minutes the gaseous fluid boiled up from the sea floor, and within 24 hours oil and gas were exuding from fractures along a zone nearly 400 meters (1,300 feet) long. The flow was not checked until February 8, and it was months before the rate was reduced to about 1,600 liters (10 barrels) a day. In the meantime, beaches along the Santa Barbara coast were blackened with heavy oil, and many birds and sea mammals were killed.

Though the long-term effects of the oil spill are disputed, there is little evidence of the event today. Both the beaches and marine plants and animals have recovered substantially, perhaps chiefly because oil seeps have been part of the channel's natural environment for millenia.

Wilmington Field Subsidence. Subsidence of land over the Wilmington field was first observed in 1937, a year after the field's discovery. By 1941, an elliptical area near the east end of Terminal Island had sunk 0.36 meters (1.2 feet). By 1958, subsidence had increased to 7.5 meters (25 feet) and subsequently has reached about 10 meters (30 feet). After much litigation and geologic study, it was established that the removal of oil and gas had allowed the rock and mineral grains in the reservoir rocks to pack together more closely, reducing thickness of beds and encouraging subsidence. In a geologically active area, of course, it is always possible that subsidence is unrelated to the extraction of oil and gas. In the Wilmington case, however, the pattern of subsidence plotted on a map outlined the form of the oil reservoir almost perfectly, with maximum subsidence near the center of the field.

The subsidence caused severe problems in this heavily industrialized district. Bridges were jacked up and straightened; high dikes were built to protect a large power plant from encroachment by the ocean; the U.S. Navy dry dock was endangered by flooding; and numerous buildings were damaged or had to be raised. About the only advantage was the deepening of ship channels without dredging. Since 1958 the *rate* of subsidence has been greatly reduced by pumping salt water into the reservoirs to replace the oil and gas withdrawn. This nearly stabilizes the fluid pressure in the system, helps flush out more oil and gas, and reduces subsidence.

Water flooding itself may engender serious consequences, however. For example, it was the principal cause of renewed movements on small faults that ultimately provoked failure of the Baldwin Hills Dam in 1963. Nine hundred and fifty million liters (250 million gallons) of water were released into a residential neighborhood, killing 5 people, damaging 277 homes, resulting in property loss of at least $12 million.

Sea-cliff Retreat at Santa Barbara

Though nearly all sea cliffs are eroding landward, in only a few places in California has the rate of erosion been established. Santa Barbara is one of these places, and here the 50-year rates vary from 7.5 to 31 centimeters (3–12 inches) annually. Short-term rates may depart widely from these figures.

Santa Barbara sea-cliff erosion is attributable primarily to the following five processes, all well-known to geologists but hitherto less appreciated by homeowners, builders, and government officials.

1. Undercutting of the base of the cliff by direct wave attack.
2. Weathering resulting in the disintegration of rocks making up the cliff face.

FIGURE 10-29 Home threatened by active sea-cliff retreat west of Santa Barbara. (Photo by Robert M. Norris)

3. Emergence of underground water at the cliff face, weakening the rocks.
4. Rainwash on the face of the cliff.
5. Various kinds of landsliding.

Because most of these processes are slow, they generally are not appreciated by the public until some property is threatened or lost.

Development of coastal bluff property has been relatively intense in the Santa Barbara area, with the result that examples of all processes can be observed. Figure 10-29 shows a house being threatened chiefly by processes 1, 2, and 4.

REFERENCES

General

Dibblee, T. W., Jr., 1950. Geology of Southwestern Santa Barbara County. Calif. Div. Mines and Geology Bull. 150.

―――――, 1966. Geology of Central Santa Ynez Mountains, Santa Barbara County, California. Calif. Div. Mines and Geology Bull. 186.

―――――, 1970. Geology of the Transverse Ranges. Mineral Information Service (now California Geology), v. 23, pp. 35–37.

Elders, W. A., ed., 1971. Geological Excursions in Southern California. Univ. Calif. Riverside Museum Contrib. 1.

Fife, B. L., and J. A. Minch, eds., 1982. Geology and Mineral Wealth of the California Transverse Ranges. South Coast Geological Society, Santa Ana, Calif., 697 p.

Stout, Martin L., 1977. Radiocarbon Dating of Landslides in Southern California. Calif. Geology, v. 30, pp. 99–105.

Vedder, J. G., and others, 1969. Geology, Petroleum Development and Seismicity of the Santa Barbara Channel Region, California. U.S. Geological Survey Prof. Paper 679.

Yerkes, R. F., and others, 1965. Geology of the Los Angeles Basin, California—An Introduction. U.S. Geological Survey Prof. Paper 420A.

Special

Bennett, Jack, 1977. Palmdale "Bulge" Update. California Geology, v. 30, pp. 187–189.

Bryant, W. A., 1978. The Raymond Hill Fault. California Geology, v. 31, pp. 127–142.

Clements, Thomas, 1966. St. Francis Dam Failure of 1928. Assoc. Eng. Geol. Spec. Publ. pp. 90–91.

Crowell, John C., ed., 1975. San Andreas Fault in Southern California. Calif. Div. Mines and Geology Spec. Report 118.

_____, 1986. Geologic History of the San Gabriel Fault. Calif. Geology, v. 39, pp. 276–281.

_____, and M. H. Link, eds., 1982. Geologic History of Ridge Basin, Southern California. Pacific Sect., Soc. of Econ. Paleontologists and Mineralogists.

Hill, Mary, ed., 1971. San Fernando Earthquake. California Geology, v. 24, pp. 59–85.

Kiessling, Edmund, 1963. A Field Trip to Palos Verdes Hill. Mineral Information Service (now California Geology), v. 16, pp. 9–14.

Norris, Robert M., 1985. Southern Santa Barbara County, Gaviota Beach to Rincon Point, In Living with the California Coast, Griggs, G., and L. Savoy, eds., Duke Univ. Press, Durham, N.C., pp. 250–278.

Shaw, C. A., and J. P. Quinn, 1986. Rancho La Brea, a Look at Coastal Southern California's Past. Calif. Geology, v. 39, pp. 123–133.

Shreve, Ronald L., 1968. Geology of the Blackhawk Slide. Geol. Soc. America Spec. Paper 108.

Taylor, G. C., 1981. California's Diatomite Industry. Calif. Geology, v. 34, pp. 183–191.

Trent, D. D., 1984. Geology of the Joshua Tree National Monument. Calif. Geology, v. 37, pp. 75–86.

U.S. Geological Survey, 1971. San Fernando Earthquake of Feb. 9, 1971. Prof. Paper 733.

Weaver, Donald W., and others, 1969. Geology of the Northern Channel Islands. Amer. Assoc. Petrol. Geologists and Soc. of Econ. Paleontologists and Mineralogists (Pacific Sections) Spec. Publ.

eleven

COAST RANGES

> False facts are highly injurious to the progress of science, for they often endure long; but false views, if supported by some evidence do little harm, for everyone takes a salutary pleasure in their falseness.
>
> —*Charles Darwin*

Interpretations of Coast Range geology have been greatly affected by the theory of plate tectonics. As a result, much of the geologic history previously accepted for the province has now been substantially revised. In addition, studies of Coast Range rock relationships have contributed significantly to plate tectonic theory in general. Once mainly an enigma of local concern, the Coast Ranges are now a reference area of world importance.

The Coast Ranges stretch about 960 kilometers (600 miles) from the Oregon border to the Santa Ynez River and fall into two subprovinces: the ranges north of San Francisco Bay and those from the bay south to Santa Barbara County. This division is actually one of convenience rather than geologic distinction, for the ranges have more similarities than differences. The differences that do exist probably occur because the northern ranges lie east of the San Andreas fault zone, whereas most of the southern ranges are to the west. Moreover, the southern ranges are better known because of their oil and gas resources, easier accessibility, more intensive land development, and clearer rock exposures due to sparser vegetation from lower rainfall.

GEOGRAPHY

The province contains many elongate ranges and narrow valleys that are approximately parallel to the coast, although the coast usually shows a somewhat more northerly trend than do the ridges and valleys. Thus some valleys intersect the shore

359

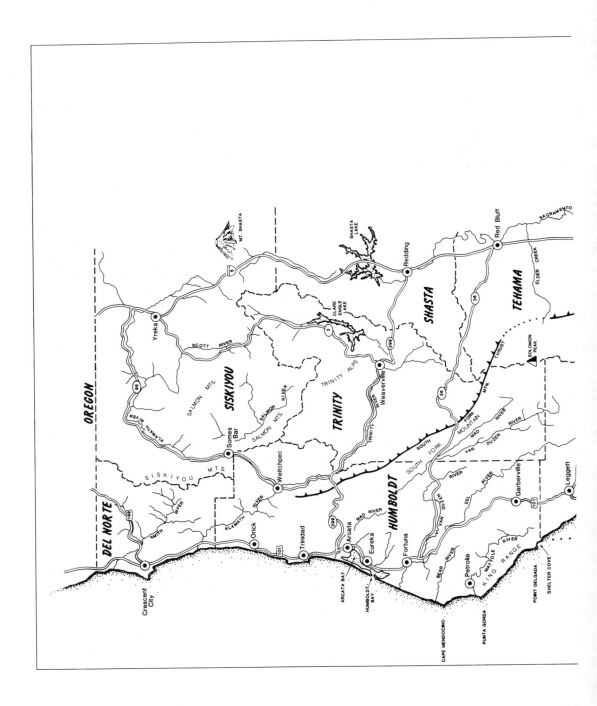

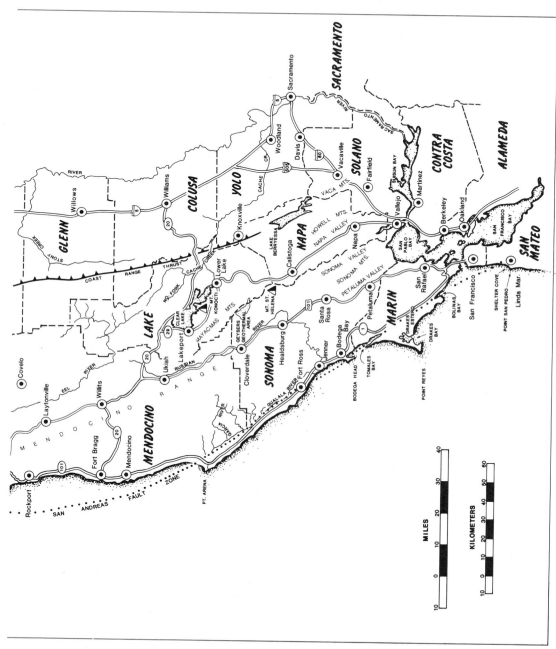

FIGURE 11-1 Place names, northern Coast Ranges.

361

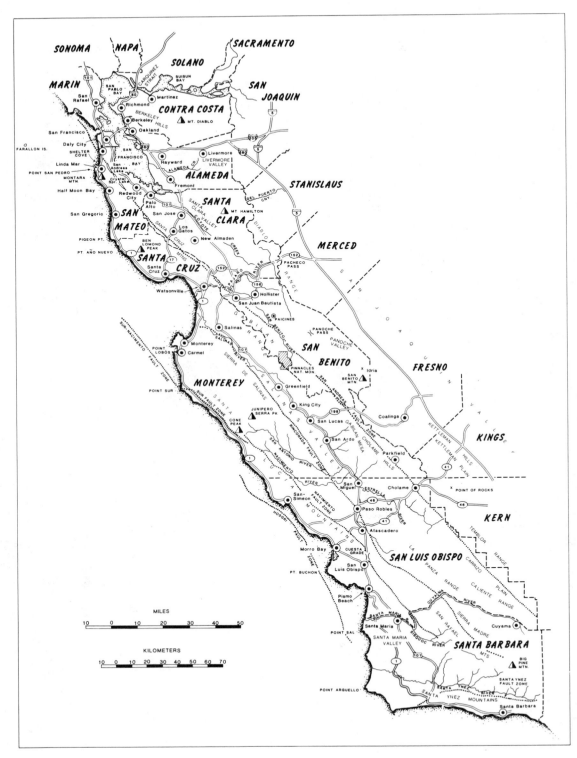

FIGURE 11-2 Place names, southern Coast Ranges.

at acute angles and some mountains terminate abruptly at the sea. Only minor streams enter the sea at right angles to the shore; most major streams flow many miles through inland valleys that roughly parallel the coast. Except at San Francisco Bay, where a pronounced gap separates the northern and southern Coast Ranges, travel in any direction in the province involves crossing range after range. Figures 11-1 and 11-2 show the locations of the main mountain units and valleys in the Coast Ranges.

Although elevations are moderate, relief is sometimes considerable. For example, within 1½ kilometers (1 mile) of the ocean are several peaks in the Santa Lucia Range that are more than 750 meters (2,500 feet) high; Cone Peak (1,572 meters or 5,155 feet) is only 6½ kilometers (4 miles) from the ocean. Travelers along the coastal highway from San Simeon to Monterey are almost always impressed with the precipitous seaward face of the Santa Lucia Range, for the road snakes along cliffs often 160 to 180 meters (500–600 feet) above the water. Highest elevation in the southern Coast Ranges is Big Pine Mountain (2,083 meters or 6,828 feet) in the San Rafael Mountains of Santa Barbara County. The northern Coast Ranges are higher, particularly in southern Trinity County where Solomon Peak rises to 2,312 meters (7,581 feet), the highest point anywhere in the California Coast Ranges.

DRAINAGE

Drainage is controlled primarily by structure. The large streams in the northern ranges, such as the lower parts of the Klamath, Mad, Eel, and Russian rivers, all follow the structural grain of faults or folds for much of their lengths. Of these, the Russian River has the most noteworthy drainage pattern. This river flows south in a normal way more than 64 kilometers (40 miles) to near Healdsburg where it abruptly turns west, crosses the Mendocino Range by a gorge up to 300 meters (1,000 feet) deep, and reaches the coast near Jenner. Because it appears simpler geologically for the river to continue south through the valley occupied by Santa Rosa and Petaluma and then into San Francisco Bay, geologists have long speculated about the origin of the river's lower course. Studies now indicate that the river established its initial channel on a more subdued Pliocene terrain. The river then flowed seaward across a blanket of sedimentary deposits, through which it subsequently cut downward into underlying Franciscan rocks. By the time the Santa Rosa-Petaluma lowland began sinking to its present elevation, the Russian River had already established the course that it retained even though the valley from Healdsburg to Santa Rosa and Petaluma continued to sink and the Coast Ranges continued to rise.

The southern ranges also contain drainages strongly controlled by faults and synclinal folds. The best example of a structurally controlled stream is the Salinas River, which lies in a synclinal trough for most of its course. Some faulting is involved in the lower valley, but even there folding seems to be the dominant structural control. Before joining the Salinas River, the San Antonio and Nacimiento rivers also follow linear systems of folds and faults.

Conversely, streams such as the Pajaro, Alameda Creek, and Santa Maria-Cuyama drain broad inland valleys and then follow gorges before emptying into the sea. Alameda Creek drains the Livermore Valley via Niles Canyon, a narrow gorge across the Diablo Range, and empties into southern San Francisco Bay. The Pajaro occupies a deep gorge between the Santa Cruz and Gabilan chains and, with its tributary the

San Benito River, drains the southern Santa Clara Valley. The Santa Maria River and a primary tributary, the Cuyama, have cut a zigzag course across the trend of the southern ranges, separating the Sierra Madre and San Rafael Mountains from the ranges of southern San Luis Obispo County. The upper Cuyama drains a broad structural depression between the Caliente Range and the Sierra Madre. The Sisquoc River, another principal tributary of the Santa Maria, follows the trend of faults and folds within the San Rafael Mountains.

Why the Pajaro and Santa Maria rivers behave as they do is not fully known, but there are several possibilities. Both rivers may be antecedent, with their present courses determined before folding and faulting outlined modern Coast Range structure; they may have been superimposed on sedimentary rocks that previously covered older Coast Range basement; streams eroding headward may have breached mountain barriers and captured upland drainages; or some combination of these events may have occurred. As with the Russian River, a complex explanation is likely, probably involving both antecedency and superimposition. In the case of Alameda Creek and Niles Canyon, it is clearer that the stream is antecedent because uplift is still in progress. First-order surveys through Niles Canyon, covering a 53-year period, show that the Diablo Range is being uplifted at a rate of about 1.5 to 2.0 millimeters (0.06–0.08 inch) per year with respect to the lowland on the east side of San Francisco Bay.

The discharge of southern Coast Range rivers, although important to the area, is no match for the discharge of the rivers draining the rainy north coast region. The largest of these northern rivers is the Klamath which, on average, accounts for about 18 percent of the total runoff from California, but the Klamath and its tributary the Trinity are mainly Klamath Mountain province streams; only a small part of their drainage basins lies within the Coast Ranges. On the other hand, the Mad, Van Duzen, and Eel Rivers are mainly Coast Range streams and all carry spectacular volumes of water during floods and substantial flows under normal conditions (Figure 11-3).

The Eel River is of special interest because it holds the record for the greatest average annual suspended load for any stream of its drainage area or larger in the United States; it exceeds both the Colorado and Mississippi in this respect! In tons of sediment per square mile of drainage basin, the Eel yields 4 times as much as the Colorado and 15 times as much as the Mississippi. One part of the Eel basin produced 1,079 metric tons of sediment per square kilometer (3,080 tons per square mile) per year. These very high rates are due to a combination of factors, including very high annual rainfall, soft, easily eroded sedimentary rocks in the basin, a multiplicity of landslides, and timber-harvesting practices.

ROCKS

Since the 1914 work of A. C. Lawson, it has been recognized that the Coast Ranges have two dissimilar core complexes (basement rocks) in contact along major longitudinal faults—a Franciscan subduction complex with oceanic crustal rocks and a granitic-metamorphic complex that includes the Sur Series and its equivalents. Present distribution of these rocks raises many questions that have not yet been answered satisfactorily.

FIGURE 11-3 Mouth of the Eel River just south of Eureka, Humboldt County. The town at the right edge of the photo is Loleta. This major north coast river carries about 3 percent of the total river discharge of California. (Photo by David Swanlund, courtesy Humboldt County Department of Public Works)

Franciscan Basement

The Franciscan subduction complex has been variously labeled a *series*, a *formation*, or an *assemblage*. Some portions are called a *melange*, a tectonic unit produced by fragmenting and mixing several rock types, presumably in a subduction zone. Lithologically, the Franciscan is dominated by greenish-gray graywackes (sandstones), generally in beds 0.3 to 3 meters (1–10 feet) thick. These graywackes were derived from rapid erosion of a volcanic highland and deposited in deep marine basins, usually by turbidity currents or submarine mudflows. The graywackes are composed mainly of quartz and plagioclase feldspar, with a chlorite mica matrix that confers the dark-greenish color. These rocks have immense volume and constitute 90 percent of the Franciscan. It is estimated that they average 7,600 meters (25,000 feet) in thickness and are exposed over 190,000 square kilometers (75,000 square miles) on both land and the sea floor. This gives an approximate volume of 1,500,000 cubic kilometers (350,000 cubic miles)—enough to cover all of California to a depth of 3,000 meters (10,000 feet) or all 48 contiguous states to a depth of 180 meters (600 feet).

The graywackes are interbedded with lesser amounts of dark shale and even occasional limestone. Sometimes, associated are thick accumulations of reddish radiolarian cherts, which are thought to represent organic deposition in marine waters possibly at least 3,000 meters (10,000 feet) deep. One limestone, the Calera, occurs discontinuously along the east side of the San Andreas fault from San Francisco to near Hollister. Another is the red Laytonville limestone found north of San Francisco Bay almost to Eureka; it has fewer and smaller outcrops than the Calera. Previously both limestones were thought to be chemical precipitates, because of their fine grain and lack of obvious fossils. Studies have shown, however, that these rocks contain siliceous radiolarians and planktonic calcareous foraminifers, with bulk composition resembling some modern deep-sea oozes. Furthermore, the dark color and bituminous character of some parts of the Calera show that deposition occurred in stagnant basins. Within this array of sedimentary rocks are some altered submarine volcanics, now mostly greenstones, and other metamorphic rocks such as the distinctive blue glaucophane schist and the more common green chlorite schist.

All these Franciscan rocks have been intruded by ultrabasic igneous rocks, now serpentinized peridotite (serpentinite). Sometimes the serpentinites have been injected as normal molten intrusives, but in other instances they occur in sill-like sheets that lack the thermal alteration of the enclosing rocks characteristic in most sills. In still other cases, these plastic serpentinites have squeezed up through the overlying rocks as plugs or diapirs. The prevailing view is that these serpentinized peridotites are altered masses derived from the upper mantle and transferred tectonically to the earth's surface.

Although the Franciscan is more than 15,000 meters (50,000 feet) thick, no recognizable top or bottom has yet been observed. This is surprising, because many rocks in adjacent provinces are much older. This curious record—like a book missing its first and last pages—has prompted the suggestion that the Franciscan sediments were deposited in a deep oceanic trench directly on mantle material or on a thin oceanic crust overlying the mantle. Supporting this contention is the presence of ophiolites in the Franciscan, which are distinctive assemblages of ultramafic rocks thought to represent typical oceanic crust. A complete ophiolite sequence includes: (top) pillow lavas often containing pockets of radiolarian chert; a mass of basaltic dikes and sills (the sheeted complex); gabbro and diorite; and (bottom) an ultramafic complex of serpentinites and dunites.

Fragments of ophiolite sequences occur in the Franciscan melanges, but in a number of places Great Valley beds of late Jurassic age rest in depositional contact on radiolarian cherts and typical Coast Range ophiolites of middle Jurassic age. Some of the best examples of ophiolite sequences occur at Point Sal in Santa Barbara County, in Del Puerto Canyon in the northern Diablo Range, and along the South Fork of Elder Creek in western Tehama County in the northern Coast Range. Many other partial sequences occur in the Coast Ranges, and other fine examples (previously mentioned) occur in the Sierra Nevada and Klamath Mountains.

Ophiolite sequences occur at many other places in the world; they are always associated with eugeoclinal sedimentary rocks similar to the Franciscan. Ophiolites are interpreted as masses of oceanic crust because of

1. Their lithologic similarity to samples dredged from oceanic fracture zones.
2. Their bulk chemical composition.
3. Their close association with pillow lavas and radiolarian cherts, indicating deep-sea volcanic extrusion.
4. Similarity of seismic characteristics measured in both ophiolites and ocean crust.

Glaucophane, which gives blueschist its characteristic color, jadeite, and lawsonite occur in some Franciscan melanges. These minerals are thought to form under low temperature (not over 300°C) and high pressure (15–30 kilometers or 9–18 miles of burial). Consequently, most geologists believe that Franciscan rocks were carried down rapidly from their depositional site on the deep-sea floor along a subduction zone beneath the edge of the continent, where they were subjected to high pressure. They were forced back up to the surface before becoming thoroughly heated, thus producing minerals that reflect both high pressure and low temperature. Not only is the mechanism that returned these glaucophane-bearing rocks to the surface still a mystery, but we have some evidence that suggests these rocks remained deeply buried in the subduction zone for perhaps 80 to 100 million years. It is difficult to imagine how these rocks could remain sufficiently cool for such a long interval.

Franciscan sedimentary rocks contain very few fossils, but widely scattered localities have yielded specimens ranging from late Jurassic to Eocene and possibly even younger. Radiometric dating of several associated ophiolite sequences has given a middle Jurassic age (153–165 million years) and blueschists have been dated at 150–155 million years.

The Franciscan complex includes some rocks that appear to have traveled thousands of kilometers to the subduction zone (exotic terranes).

Crystalline (Salinian) Basement

The second type of basement underlying the Coast Ranges occurs between the Nacimiento and San Andreas fault zones in the southern Coast Ranges and west of the San Andreas in the northern ranges. Known as the Salinian block, this basement consists of metamorphic rocks and granitic plutons, a common association in California. In modern plate tectonic terminology, this sort of rock association is known as a *magmatic arc*, in contrast with the Franciscan suite of rocks, which is designated a *subduction complex*.

The Salinian metamorphic rocks present in the Santa Lucia Range are known as the Sur Series and include gneiss, schist, quartzite, and marble. Lesser amounts of similar metamorphic rocks occur in the Santa Cruz, Gabilan, LaPanza chains, and in the Sierra de Salinas. The age of these rocks has not been firmly established, but poorly preserved fossils collected long ago and now missing were presumably from the Gabilan range. These were claimed to suggest a possible Paleozoic age, but there is a strong suspicion that these fossils actually came from the Sierra Nevada. Moreover, Sur series rocks, unlike the plutonic rocks with which they are intimately associated, do not resemble any of the metamorphic rocks known in either the Sierra Nevada or the Peninsular Ranges.

The granitic rocks and their associated metamorphic rocks are widespread and presumably underlie much of the Salinian block, although they are frequently concealed by younger sediments. The northernmost exposure of Salinian granitic rocks is at Bodega Head, west of Santa Rosa. Presumably these rocks underlie the northernmost slice of Salinian rocks near Point Arena, but if this is true, they are entirely concealed by the Cretaceous sedimentary rocks present there (Figure 11-4).

Composition of Salinian granitic rocks varies from granodiorite and quartz monzonite to quartz diorite, compositions much like those of the plutonic rocks of the

FIGURE 11-4 Salinian granite rock at Lovers Point, Pacific Grove, Monterey County. (Photo by R. M. Norris)

Sierra Nevada and Peninsular ranges. In contrast to the poorly dated metamorphic rocks, good radiometric dates exist for Salinian plutonics—from 69 to 110 million years (late Cretaceous). Generally the dates are younger on the west side of the block, suggesting that the western plutons cooled later, probably because they were deeper in the batholith. This in turn indicates that the western part of the Salinian block has been uplifted more than the eastern.

The Salinian granitic plutons are definitely younger than at least some of the Franciscan rocks. Furthermore, no contact metamorphism exists where Salinian granites are in contact with the Franciscan. This seems to indicate that the present proximity of the two basements results from large displacement along the San Andreas and Nacimiento fault zones. After the basement of the Salinian block was formed (wherever that might have been), this magmatic arc must have supplied large volumes of sediment to a *Forearc Basin*, just as the ancestral Sierra Nevada, also a magmatic arc, supplied sediment to form the thick pile of rocks called the Great Valley beds. These beds were deposited between the forearc basin and the subduction zone that normally lies about 200 kilometers (120 miles) to seaward. But all these forearc basin rocks derived from the Salinian granites are missing in the southern Coast Ranges. Instead, the Salinian magmatic arc rocks are in direct contact with the Franciscan subduction complex along the Sur-Nacimiento fault zone. What became of this voluminous pile of sediments is not known, though some of it may be preserved on the sea floor north of Point Sur where the western margin of the Salinian block is underwater; it remains a major mystery.

Other Jurassic and Cretaceous Sedimentary Sequences

Besides the Franciscan and the Sur series, the only major pre-Cenozoic sedimentary rocks in the Coast Ranges belong to the Great Valley sequence. This is an enormous thickness of miogeoclinal late Jurassic to late Cretaceous (some geologists claim it is

FIGURE 11-5 Lower Cretaceous Great Valley rocks near Beegum, Tehama County. The more resistant, light-colored beds are sandstones, and the softer, darker beds are shales. (Photo by R. M. Norris)

Paleocene in part) shale, sandstone, and conglomerate, generally quite unlike the contemporary Franciscan assemblage (Figure 11-5).

The lower part of the Great Valley sequence is the late Jurassic Knoxville Formation, a dark shale found mainly in a belt 170 kilometers (110 miles) long. These beds have an average thickness of about 4,900 meters (16,000 feet) along the western side of the Sacramento Valley. Similar rocks occur as far south as Kern County east of the San Andreas fault, and in the San Rafael Mountains and at Point Sal west of the Sur-Nacimiento fault zone. The Knoxville contains some graywacke and even some pillow lava and basalt-derived sandstone reminiscent of the Franciscan, but it is chiefly a dark, rhythmically bedded shale with minor sandstone beds. The base of the Knoxville, in many places, is a deep-water radiolarian chert resting on a ophiolite sequence generally known as the Coast Range Ophiolite.

Above the Knoxville are lower Cretaceous sandstones formerly called the Shasta Series. These rocks are as much as 10,500 meters (34,000 feet) thick and are associated with minor conglomerate and other sedimentary rocks. The lower Cretaceous appears only occasionally in the southern Coast Ranges. Shallow-water upper Cretaceous rocks (formerly designated the Chico Group) are widespread both east and west of the San Andreas fault however, and form a nearly continuous belt from northern

California to Kern County.[1] The names nevertheless are so thoroughly embedded in the vernacular of California geologists, that they remain useful if inexact terms.

The thickest section of late Cretaceous rocks occurs in the northern Coast Ranges, in the eastern Mendocino Range (4,550 meters or 15,000 feet) and in the eastern Diablo Range (8,500 meters or 28,000 feet).

Thus far there is no indication that material from the Salinian block contributed to the lower part of the Great Valley sequence. Instead, all materials seemingly came from the east. This absence of any materials derived from the now adjacent Salinian block is evidence that the block was not present during the late Jurassic and early Cretaceous. This is consistent with the late Cretaceous dates for the block's intrusives. Had the Knoxville and Shasta rocks been deposited on the Salinian block, they would have been intruded and metamorphosed by the younger granitic rocks and subsequently incorporated into the Sur series.

Latest Cretaceous marine sedimentary rocks are present on the Salinian block and on the Great Valley beds to the east. The two occurrences differ, however, and do not represent a continuous sedimentary sheet deposited across separate blocks situated in modern position. Latest Cretaceous beds east of the San Andreas fault and the Salinian block are the final members of the thick Great Valley sequence. Conversely, latest Cretaceous rocks on the Salinian block, west of the San Andreas, rest unconformably on granitic and metamorphic basement. These beds are almost 1,800 meters (6,000 feet) thick, and among them are rocks which, on the basis of paleomagnetic evidence, seem to have been deposited about 2,500 kilometers (1,560 miles) to the south. It seems evident that the San Andreas fault forms a major break between these two late Cretaceous sequences.

Cenozoic Sedimentation

Correlation Problems. By Cenozoic time, the sediments being deposited in the Coast Ranges were primarily of continental shelf origin. Although every Cenozoic epoch is represented, nowhere is there a complete section, which supports claims of repeated but localized tectonism. Since the gaps in the Cenozoic record occur at different points and in different localities, severe correlation problems exist. Ben Page, a Stanford geologist who has devoted much of his career to Coast Range studies, points out that correlating these Cenozoic units requires the incorporation of multiple approaches simultaneously, because any single method may succeed in one area only to fail in another. He suggests five approaches to the problem.

1. Study of megafossils, particularly mollusks.
2. Analysis of heavy mineral detrital grains.
3. Use of microfossils, chiefly foraminifers.
4. Evaluation of palynomorphs (spores and pollen grains).
5. Study of the relationship of critical floras and faunas to radiometric dates to avoid the confusion caused by the tendency of plant and animal groups to form communities in response to environmental constraints.

---

[1]Although the terms *Shasta Series*, *Chico Group*, and *Knoxville Formation* have been applied throughout California, recent studies discredit their use because they do not correspond to mappable units.

Correlation of pre-Cenozoic Coast Range rocks across structural boundaries has not yet been accomplished. The reason presumably is that movement on most fault zones has been extensive enough to bring very different pre-Cenozoic materials in contact with one another, sequences like the Franciscan and Great Valley beds. Conversely, Cenozoic rocks being younger and less extensive originally are not so widely separated across structural breaks and thus have been less affected by Coast Range tectonism. Moreover, they are unaffected by intrusion and metamorphism that altered many of the older rocks. Nevertheless, it is unlikely that a detailed Cenozoic history of the Coast Ranges will be available in the forseeable future; our understanding of the relevant lithologies and stratigraphic relationships is not yet clear enough to permit unequivocal interpretations.

For many years the differences in Tertiary beds on either side of the San Andreas were explained by supposed vertical uplift on the fault; first one side up and then the other. According to this view, a depressed block would receive thick deposits that might then be partially or totally removed when the block was reversed and became high standing. In such interpretations, opposing Tertiary sequences could be matched even though they involved notable differences in thickness or had un-matched gaps or extra units on one side of the fault. For example, if a basaltic flow west of the fault was absent in the corresponding eastern sequence, it was presumed that the eastern part of the flow had been eroded when the eastern block stood higher than the western block. Complicated explanations were often necessary to justify all observed differences, and yet large-scale horizontal slip was not generally accepted until about 1953.

Detailed mapping on the San Andreas fault zone has now established that several Cenozoic sequences on opposite sides of the fault match amazingly well, although they are separated by many tens of kilometers today. Mason L. Hill and Thomas W. Dibblee located an Eocene sequence near Palo Alto, which they believed closely matched one in the San Emigdio Mountains in southern Kern County on the opposite side of the San Andreas. This implied that post-Eocene separation on the San Andreas amounted to about 305 kilometers (190 miles). Skeptics pointed out, however, that the Eocene Butano sandstone, which was used for part of Hill and Dibblee's evidence, occurs on both sides of the fault near Palo Alto. Although the Palo Alto occurrences do differ in rock facies and stratigraphy, similar degrees of change exist within other pertinent formations when no faulting is involved, but it now appears that these occurrences were submarine fan deposits laid down in widely separated basins. Hill and Dibblee also matched a Miocene section across the fault and suggested that post-Miocene displacement amounted to 282 kilometers (175 miles). Ben Page points out that the 305-kilometer (190-mile) separation is still insufficient to bring the Salinian granites into alignment with the Sierra Nevada, and therefore suggests that there must have been about another 200 kilometers (120 miles) of pre-Butano offset on an ancestral San Andreas fault.

Miocene

The Monterey Formation. Miocene rocks are more widely distributed in the Coast Ranges than any other Cenozoic deposits. Most of this Miocene is marine and is characterized by organic deposits, silicic and phosphatic members. In addition, in the southern part of the province, Miocene rocks reflect more volcanic activity than any other Tertiary Series.

FIGURE 11-6 Monterey Formation of middle Miocene age exposed in the sea cliff at Shell Beach, San Luis Obispo County. The dark-colored bed at the top of the cliff is mainly shallow-water marine sands and gravels of Pleistocene age and rests by angular unconformity on the tilted Monterey Formation below, demonstrating recent uplift of this part of the coast. (Photo by R. M. Norris)

The Monterey Formation is perhaps the province's most distinctive Miocene sedimentary unit (Figure 11-6). Although it is preeminently a Coast Range rock, as noted previously it extends from Santa Rosa south into the Transverse and Peninsular ranges. The Monterey is characterized by abundant silica that was deposited organically as diatomites and inorganically as silicic ash beds. In many places the formation is 800 to 1,600 meters (2,600–5,200 feet) thick. M. N. Bramlette, who did a pioneering study on this formation, observed that the volume of sediment involved in the Monterey amounts to thousands of cubic kilometers, and noted that although its thickness and lithology vary, its silicification distinguishes it from other California rock units. The Monterey is easy to recognize in outcrop; it is pale buff to white, occurs in thin to very thin beds, and is often cherty where silicified and punky where rich in diatoms. It typically weathers to a dark, adobelike, clay-rich soil that normally supports grass rather than chapparal and trees.

California phosphatic deposits are most common in Miocene rocks, notably in the Monterey formation where phosphatic shale, pelletal sandstone, and phosphatic mudstone occur. The formation's phosphate and abundant diatoms both reflect deposition in a marine environment in which organic productivity was unusually high. It has been suggested that fluctuating temperatures played an important role. When sea water was cool, diatoms and silica were deposited; when temperatures rose, phosphates were deposited instead.

Pliocene and Pleistocene. At first Pliocene rocks were chiefly marine and more restricted areally than the Miocene units, but by the end of the epoch, the sea had

FIGURE 11-7 Mount Konocti (1,281 meters or 4,200 feet), a Quaternary dacite strato-volcano on the south shore of Clear Lake, Lake County. (Photo by Don Norris)

withdrawn from much of the province and widespread stream gravels and sand were deposited. Pliocene deposits are especially prominent on valley floors, as are Pleistocene deposits. In addition, the Pliocene often records continuing tectonic activity. Typically, the deposits make thick, conformable sequences with older valley rocks; thicknesses are often so great that continuing subsidence is indicated. Around valley margins important deformation has occurred, as shown by numerous unconformities that occur between late Cenozoic units.

Except along the coast and in the Sonoma and Clear Lake regions, Pliocene and Pleistocene strata are alluvial, with some lake-bed deposits. In several areas, these lake beds attest to the former presence of water bodies such as Lakes Merced and San Benito in Santa Clara Valley.

In the Sonoma area, volcanic activity began in the Pliocene and shifted north into the Clear Lake district during the Pleistocene. Chiefly lava flows and pyroclastics, the Sonoma volcanics are well displayed in the Sonoma and adjacent ranges and on either side of the Napa Valley. Mount St. Helena is a prominent peak composed mainly of fragmental Pliocene volcanic rocks. Near Clear Lake the volcanics are younger, with some of late Quaternary age, and include dacite, basalt, and obsidian. Mount Konocti is a nearly perfect, almost uneroded dacitic strato-volcano that rises 820 meters (2,700 feet) above the western shore of Clear Lake (Figure 11-7).

STRUCTURE

Certainly the dominant characteristic of the Coast Ranges is its division into elongate topographic and lithologic strips underlain by discrete basement rocks that are separated by profound structural discontinuities (Figure 11-8). The pattern extends east, and probably also west onto the sea floor. On the east, concealed beneath the Central Valley, is the enigmatic boundary between Sierra Nevada basement and the Coast Range Franciscan. Westward, the next major boundary is the San Andreas fault zone,

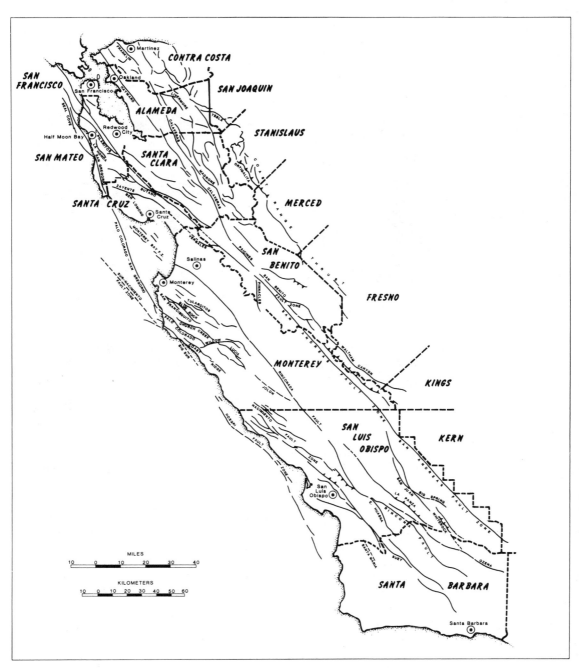

FIGURE 11-8 Major faults of the Coast Range province.

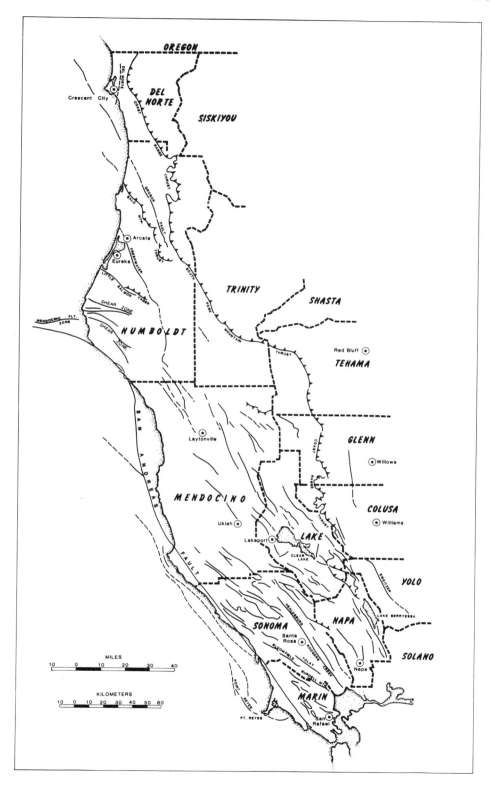

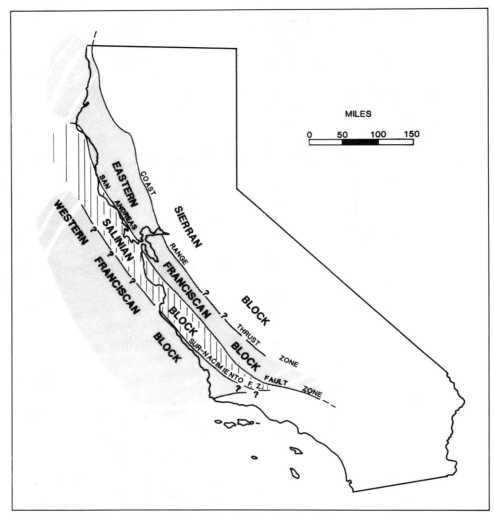

FIGURE 11-9 Major crustal blocks in the California Coast Ranges.

which separates Franciscan basement from the granitic-metamorphic basement of the Salinian block. South of Monterey, the Sur-Nacimiento fault zone separates Salinian rocks from more Franciscan basement to the southwest. Another boundary should occur farther west, offshore, where Franciscan basement is replaced by normal oceanic crust (Figure 11-9).

The Coast Range Thrust and Its Subdivisions

Most of the boundary between the Sierran and Franciscan basements lies beneath thousands of meters of late Mesozoic and Cenozoic sedimentary rocks in the San Joaquin and southern Sacramento valleys. North of Red Bluff, the boundary emerges as the South Fork Mountain thrust, separating the Klamath Mountains from the Coast Ranges.

Until fairly recently, this boundary was thought to lie between the ophiolite sequence and the Great Valley beds, but new studies indicate that the main tectonic boundary occurs between the Franciscan and the ophiolite instead. As was previously noted, beds of the Great Valley group sometimes lie in depositional contact on the ophiolite sequence.

Along the South Fork Mountain thrust, the Great Valley beds and the ophiolites on which they rest have been thrust westward over the Franciscan. This fault marks the boundary between the northern Coast Ranges and the Klamath Mountains and forms a prominent topographic feature for more than 150 kilometers (90 miles).

This same tectonic boundary follows a more southerly direction south of the latitude of Red Bluff as the Stony Creek thrust. The Stony Creek thrust is traceable as far south as Clear Lake and maybe beyond to the latitude of Sacramento. Most of it dips more steeply, and indeed is nearly vertical in places, than the flatter South Fork Mountain thrust.

Many geologists now agree that these two faults are parts of a major structural boundary called the Coast Range thrust, which some geologists think continues south on the eastern side of the southern Coast Ranges. One segment, approximately in the right location, is the Tesla-Ortigalita falt, and there is at least some evidence that Great Valley rocks were thrust on this or a related structure, over the top of the Diablo Range. Furthermore, in the Franciscan block west of the Sur-Nacimiento fault zone in the southern Coast Ranges, thrust faults bring Great Valley beds over Franciscan and some geologists think these thrusts are also part of the larger Coast Range thrust zone.

Whether all these thrust faults are truly parts of a single major structural system, what the cause of this thrusting is, and when it occurred are all problems waiting for solution.

Whatever the case, faults like the Stony Creek and Tesla-Ortigalita do separate Great Valley beds from the Franciscan, and no depositional contact between these units has yet been found. The Great Valley beds have been thrust as much as 80 kilometers (50 miles) westward along these faults. Age of this large-scale thrusting appears to be Paleocene or Eocene, and it must have accompanied a major orogenic event, evidently the late Jurassic to early Tertiary subduction.

San Andreas Fault

The San Andreas fault, the next major boundary to the west in the Coast Range province, brings into contact such contrasting basement rocks that large-scale movement is necessarily involved. The fault cannot be the simple demarcation between continental and oceanic crusts because the Salinian block possesses granitic, continental crust and lies seaward of the Franciscan basement, which itself contains some slices of oceanic crust. Any meaningful analysis of Coast Range structure must necessarily incorporate the role and history of the San Andreas fault. The fault itself is considered more fully in Chapter 13.

Sur-Nacimiento Fault

Disagreement exists concerning the name of this fault zone. Some call the northern end the Jolon-Rinconada fault; whereas others assign the name *Nacimiento* to several parallel faults in Santa Barbara and southern San Luis Obispo counties. For our purposes, however, the Sur-Nacimiento fault zone is taken as the western boundary of Salinian basement and is thought to extend from near Point Sur southeastward. It has also been suggested that this fault zone continues northward on the sea floor from the Point Sur area to near Point Arena in southern Mendocino County, and possibly south of the Transverse Ranges as the Newport-Inglewood fault zone.

Irrespective of whether or not these extensions have validity, the Sur-Nacimiento fault zone approximately parallels the San Andreas south to the Big Pine fault, which truncates or offsets the Sur-Nacimiento. Like the mysterious Sierran boundary fault beneath the Central Valley, Sur-Nacimiento brings granitic basement on the east into contact with Franciscan and ocean basement on the west. This has prompted several intriguing hypotheses.

1. The Sierran-Franciscan boundary fault and the Sur-Nacimiento faults are part of the same structure, which was broken and offset by right slip on the San Andreas.
2. The Franciscan rocks of the Santa Lucia Range west of the Sur-Nacimiento fault represent a huge remnant of the upper plate of a large thrust carried west across the Salinian block from east of the San Andreas.
3. Both the Sur-Nacimiento and the Sierran-Franciscan boundary are continental margin tectonic features that have been affected by large differential movements, possibly involving exotic terranes, which have been transported thousands of kilometers.

Unfortunately, the correct explanation for the puzzling basement pattern in the Coast Ranges is far from established. Further study of the ages of pertinent faults and a better understanding of Franciscan rocks and their relationship to the Great Valley sequence are still needed.

Hayward and Calaveras Faults

The Hayward and Calaveras faults are important parts of the San Andreas system in the San Francisco Bay region (Figure 11-8). The Hayward is currently active, but much of the Calaveras seems to be dormant. The Calaveras does offset some Plio-Pleistocene rocks, however, and a section of the fault presumably is responsible for offset curbs and sidewalks in Hollister.

The Hayward fault branches off the San Andreas south of Hollister. The junction itself is poorly exposed, and part of the Hayward is concealed beneath Quaternary alluvium. The fault extends north through the Santa Clara Valley, and along the western foot of the Berkeley Hills to San Pablo Bay, although some authorities trace it as far north as Petaluma.

The Calaveras fault branches off the Hayward east of San Jose, and the junction is not exposed at all. Some investigators extend the Calaveras south to Hollister and the San Andreas fault, but the U.S. Geological Survey now assigns this southern portion to the Hayward fault. Regardless of exact placement of its southern end, most geologists trace the Calaveras north as the Calaveras-Sunol fault, which disappears in the eastern Berkeley Hills. However, some recent work indicates that the Calaveras can be traced far to the north beyond Eureka where it forms the eastern boundary of a microplate.

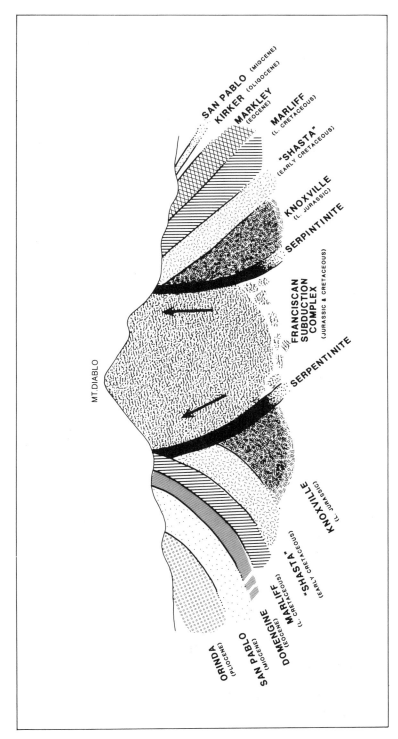

FIGURE 11-10 Cross section of serpentinite diapir at Mount Diablo, Contra Costa County. Note the way in which the intrusive plug has deformed the enclosing rocks. Because of the absence of metamorphism in the intruded rocks, this is an example of a cold intrusion.

Folding

Folding in the Coast Ranges is intense and widespread and involves all rocks. Fold axes tend to parallel faults, usually striking between N 40°W (320°) and N 60°W (300°). Like the faults, the fold axes frequently intersect the coastline at acute angles.

Tectonic activity has affected the province repeatedly, beginning in earliest Cretaceous time or perhaps earlier. Older structures have been refaulted and refolded, often with concomitant uplift and erosion, and unconformities are numerous and localized. This pattern has jumbled the geologic record, forcing investigators to figuratively "peel off" younger rocks and "iron out" structures so that the province's older features can be determined.

Often younger rocks resting on granitic basement are less intensely folded than rocks of similar age resting on Franciscan basement. It seems, therefore, that the granitic rocks tend to form tectonically resistant cores, such as that in the Gabilan Range, which protect younger Tertiary sediments from excessive deformation, though this does not mean granitic rocks are never folded.

Some ranges, like the Diablo, are crudely anticlinal, with younger strata dipping outward from the axis of the range. In the case of the Diablo Range, the archlike structure can be traced for almost 150 kilometers (90 miles), an unusually long distance in any Coast Range fold.

Among the intriguing structures of the Coast Ranges are the diapirs (cold intrusions) composed chiefly of Franciscan serpentinite (Figure 11-10). These intrusive bodies usually occur in anticlinal arches through which they have been forced upward as massive elongate plugs, a mechanism of emplacement first proposed by William Dickinson at Stanford. A well-known example occurs near Idria in the Diablo Range; it is 20 kilometers (13 miles) long and 8 kilometers (5 miles) wide, and is composed mainly of serpentinite. Another example is Mount Diablo, a plug of mixed Franciscan rocks including serpentinites and some sedimentary rock and volcanic rock. A few of these features may have been extruded as viscous surface flows, but most have only been unroofed by erosion. Examination of the surrounding younger sedimentary beds normally permits establishing the time of the intrusive body's first exposure. If the surrounding sedimentary rocks include little or no serpentinite detritus, the diapirs were probably covered when the sedimentary beds were deposited. If the sediments contain large amounts of clastic serpentinite, the plugs probably had breached the surface and were shedding detrital serpentinite. Field study shows that this unroofing first occurred in the middle Miocene. Some geologists have suggested that the source of these diapiric rocks—the serpentinites at least—is the upper mantle.

GEOLOGIC HISTORY

Four main orogenic events have been tentatively proposed by Ben Page for the history of the Coast Ranges.

1. Early (?) Cretaceous orogeny was accompanied by granitic intrusion in the Salinian block and by metamorphism of the older rocks to form the Sur Series.
2. Early Tertiary (Paleocene and Eocene) thrusting of the Great Valley beds over Franciscan rocks probably occurred.

3. This thrusting was followed by prolonged simultaneous strike-slip faulting on the San Andreas and related faults.
4. Late Pliocene and Pleistocene orogeny (the Coast Range orogeny) then occurred.

These events make a useful frame of reference when considering the following discussion of Coast Range history (Figure 11-11).

Pre-Cenozoic

No Pre-Phanerozoic or early Paleozoic rocks have been identified in the Coast Ranges, although Paleozoic rocks occur in the Klamath Mountains and Transverse Ranges. Consequently, there is little knowledge of Coast Range history prior to deposition of the sedimentary beds metamorphosed into the Sur Series. Confined to the Salinian block, Sur Series rocks are composed of gneisses, marbles, schists, quartzites, and granulites. They presumably were a marine shelf sequence originally, and were repeatedly metamorphosed and folded. This multiple metamorphism probably explains the scarcity of fossils, but even those that were reported—some crinoids and corals—are not certainly from the Coast Ranges, but may have come from the Sierra Nevada. The Sur series may include early Paleozoic and even Mesozoic rocks, but geologists are reasonably certain that it includes no Franciscan material.

The Sur series has undergone the effects of repeated plutonic intrusions, each imprinting folding and metamorphism. It is difficult to assign ages to these plutonic events, despite availability of potassium-argon dates. Some dates give *latest* Cretaceous ages for plutonic rocks, but these rocks are sometimes overlain by *late*, not latest Cretaceous sediments in depositional contact. The aberrant potassium-argon dates may reflect some post-intrusive event, perhaps further deformation.

At least partly contemporary with these plutonic events was the deposition of the Franciscan and Great Valley sediments. Relative locations of the depositional basins have not yet been precisely determined however, in relation either to one another or to the site in which the Salinian intrusives were forming.

During this same time interval, deposits were accumulating on the continental shelf, slope, and in an offshore subduction zone, and some of the deposits rested on oceanic crust in deep water. Mixed with land-derived materials delivered to offshore sites by turbidity flows were radiolarian oozes and submarine extrusive and intrusive volcanic rocks. The shelf and slope (Forearc basin) sediments became the Great Valley beds—the Knoxville and Shasta, and the subduction complex or trench deposits became the Franciscan. Deposition of the Franciscan was followed and accompanied by rapid downward transport, with underthrusting of both sedimentary and volcanic materials into the subduction zone beneath the continental margin. During this underthrusting, Franciscan beds were mixed tectonically and intruded by ultrabasic peridotitic rocks from the upper mantle. Subduction seems to have ceased rather abruptly, allowing the Franciscan to rise to the surface—as indicated by its high-pressure, low-temperature mineralogy. Contemporaneously, the Stony Creek, Tesla-Ortigalita, and other parts of the Coast Range thrust carried Great Valley deposits tens of kilometers westward over Franciscan terrain *or* the Franciscan eastward under the Great Valley beds. This zone of thrusting may have marked the boundary between oceanic and continental plates during subduction.

Composite Geologic Column, Southern Coast Ranges

HOLOCENE	Dunes, alluvium
PLEISTOCENE	Marine and nonmarine Terrace deposits Tulare (nonmarine) Orcutt (nonmarine), Paso Robles (nonmarine)
PLIOCENE	San Joaquin (marine), Careaga Sand (marine) Etchegoin (Jacalitos) (marine) Pancho Rico (marine)
MIOCENE	Reef Ridge (marine) Mclure, Pismo (marine) Santa Margarita (marine) Monterey (marine) Point Sal (marine) Obispo Tuff and miscellaneous volcanics Temblor (Rincon) (marine) Vaqueros (marine)
OLIGOCENE	Tumey (marine), Berry Conglomerate (nonmarine)
EOCENE	Kreyenhagen (marine), Point of Rocks (marine) Domengine, Yokut, Avenal (marine) Arroyo Hondo, Lodo (marine) Cantua (marine) Cerros (marine)
PALEOCENE	Dip Creek
CRETACEOUS	Moreno (marine) Panoche Group (many subdivisions) (marine) Toro, Marmalejo (marine) Gravelly Flat (marine)
JURA-CRETACEOUS	Franciscan, Salinian block granitic rocks
JURASSIC	Knoxville
PRE-JURASSIC	Sur Series, Gabilan Limestone

Composite Geologic Columns, San Francisco Bay Area, Central Coast Ranges

HOLOCENE	Dune sand, alluvium
PLEISTOCENE	Marine and nonmarine terrace deposits, Colma (marine) Alameda (nonmarine)
PLIO-PLEISTOCENE	Santa Clara, Cache, Glen Ellen (nonmarine) Tehama (nonmarine)

FIGURE 11-11 Geologic columns, Coast Ranges.

Composite Geologic Columns, San Francisco Bay Area, Central Coast Ranges

PLIOCENE	Merced (marine) Mulholland, Siesta, Moraga, Orinda, Petaluma, Wolfskill (nonmarine)	⎫ Leona rhyolite, ⎬ Bald Peak Basalt ⎭ Pinole Tuff
MIO-PLIOCENE	Purisima (marine)	
MIOCENE	San Pablo Group (Neroly, Cierbo, Briones) (marine) Monterey (marine) Sandholt (marine)	
OLIGOCENE	Kirker, San Ramon, San Lorenzo (marine)	
EOCENE	Butano, Markley, Tejon (marine)	
PALEOCENE	Martinez (marine)	
CRETACEOUS	"Chico" (marine) Oakland Conglomerate (marine) Shasta (Horsetown, Paskenta) (marine)	
JURA-CRETACEOUS	Franciscan, Franciscan metavolcanics, Granitic rocks of the Salinian block	
JURASSIC	Knoxville (Great Valley beds)	
PRE-JURASSIC	Sur Series metamorphics	

Composite Geologic Column, Northern Coast Ranges

HOLOCENE	Alluvium, dune sand	
PLEISTOCENE	Marine and nonmarine terrace deposits Hookton (Rohnerville) Carlotta	
PLIOCENE	Scotia Bluffs Rio Dell, St. George Eel River	Wildcat Group
MIOCENE	Pullen	
EOCENE	Capay, Meganos	
PALEOCENE	Martinez	
CRETACEOUS	Yager Unnamed graywacke	
JURA-CRETACEOUS	Franciscan, including serpentinites, gabbro	
PRE-JURASSIC	Kerr Ranch Schist and miscellaneous metamorphic rocks	

FIGURE 11-11 (*Continued*)

Early Tertiary

Paleocene and Eocene rocks rest on both Salinian and Franciscan materials. The Paleocene and Eocene rocks generally follow the pattern of late Cretaceous marine deposition, but were more restricted as seaways became smaller and shallower. Paleocene sandstones, conglomerates, and shales of submarine fan origin rest on the granites at Point Lobos and Point Reyes, on upper Cretaceous in the San Francisco Peninsula, on upper Cretaceous and Sur series in the Santa Lucia Range, and on upper Cretaceous in the Diablo Range. Reconciliation of large-scale subduction with continued continental shelf deposition from latest Cretaceous through the early Eocene is a major problem. Although available evidence indicates that these events were contemporary, the actual geography still eludes geologists.

The presence of coal beds, deeply weathered clays, and quartzose sandstones all suggest nearly tropical conditions during the Eocene. Minor unconformities between Eocene rock units testify to continued crustal unrest, but not to major orogenic activity.

From late Eocene to early Miocene, the sea withdrew from much of the Coast Ranges. The north particularly seems to have remained above sea level. In the south, shallow seas covered much of the Santa Cruz, Santa Lucia, and Diablo mountain units, but most of the extreme southern ranges remained above the sea.

Miocene

In early Miocene time, the sea once again spread over the Coast Ranges, forming numerous bays, straits, islands, and inlets. Deposits are characterized by notable facies and thickness changes over short distances, particularly in early Miocene beds, which were laid down in a complex archipelago. By the middle of the epoch, land areas were reduced, and deep elongate basins had developed. Volcanism (much of it submarine) became prominent, more so in the southern ranges than in the northern. The diatomaceous shales of the Monterey formation were deposited at this time, perhaps in a setting like the modern Gulf of California where deep elongate basins are receiving diatomaceous deposits.

By the close of the Miocene orogenic activity was again underway in the northern Coast Ranges. The lower Eel River valley became deeply depressed and flooded, a condition that persisted for most of the rest of the Tertiary (see Figure 1-15). Other embayments developed near Petrolia, along the Bear River north of Cape Mendocino, along the Mad River near Eureka, and at Crescent City. These downwarps and the intervening anticlinal arches all have northwesterly axial trends. Much of the marine Miocene and Pliocene in the northern Coast Ranges has been removed by late Cenozoic erosion. Small, downfaulted inliers still preserve isolated blocks of such rocks some kilometers inland, however, indicating that marine rocks were once more extensive.

Middle Miocene seas spread inland across the San Andreas fault into the San Joaquin Valley but near the end of the epoch, orogeny was renewed and seaways became more restricted and the sediments coarser. In the southern ranges, diatomaceous shales were replaced by the clean quartz sands of the Santa Margarita Formation, probably derived from the Sierran granitic rocks to the east.

Pliocene

The Pliocene sea persisted chiefly in the seaward ends of what are now prominent valleys, such as the Santa Maria Valley and in Half Moon and Bolinas bays. In addition, a long Pliocene embayment extended inland across the southern Gabilan Range and along both sides of the San Andreas fault from Kettleman Hills to the Diablo Range.

The orogenic episode begun in the late Miocene continued into the Pliocene, and by the end of that epoch most of the present ranges were dry land. In the southern ranges particularly, thick blankets of stream gravels were being deposited in the valleys. These sheets of gravel ultimately formed extensive, low-relief surfaces that almost covered some of the ranges. They generally rest by angular unconformity on the older rocks near the range margins, but are conformable with valley floor deposits. Almost all of the Cholame Hills and the northern Temblor Range from Paso Robles east are covered with these gravels. Patches of similar materials occur as far north as Eureka, but are generally absent in the northern Coast Ranges.

The Pliocene gravels have various local names. In the southern ranges they are known either as the Paso Robles or the Tulare Formation. In the central ranges, equivalent units are the Santa Clara Formation, the Livermore gravels, the Merced Formation, and the San Benito gravels. Near Eureka, the Packwood gravels are correlative.

Quaternary

Tectonism increased during the Quaternary and culminated, according to most opinions, in the middle Pleistocene Coast Range orogeny that produced today's topography. Much evidence for the orogeny comes from the southern ranges, where late Quaternary deposits often are unconformable on Pleistocene and older strata. These late Quaternary beds include some marine horizons and thus record a fluctuating sea level. In the San Francisco Bay region and in the central Salinas Valley, however, similar deposits are often conformable on Pleistocene beds.

This pattern of conformable relations between successive deposits in the valleys and unconformable relations between the same units in the foothills and ranges has persisted from the Miocene to the present. The pattern shows that since the middle Tertiary the basins and valleys have been intermittently depressed and nearby mountains and hills have been correspondingly uplifted. In several cases, sinking of the valleys has carried land-laid alluvial and near-shore deposits hundreds of meters below their former elevations. In the Sacramento Valley, some land-laid deposits now lie 900 meters (3,000 feet) below sea level, and in the Santa Clara Valley some freshwater deposits are 90 meters (300 feet) below sea level. The northern Coast Range record is less complete, owing to a scarcity of Pliocene and Quaternary deposits; where these are present, a history similar to that of the southern Coast Ranges is indicated.

Perhaps the most striking examples of Quaternary Coast Range tectonism are young faults and their accompanying belts of intensely deformed Quaternary strata. Movement on the major strike-slip faults has been substantial, although opinions differ regarding cumulative Quaternary slip on the San Andreas particularly.

Some suggestions for slip on the San Andreas since the middle Pliocene are listed here.

N.E.A. Hinds (1952)	Less than $1\frac{1}{2}$ kilometers (1 mile)
C.G. Higgins (1961)	$6\frac{1}{2}$ to 16 kilometers (4–10 miles)
L.F. Noble (1954)	32 kilometers (20 miles)
T.W. Dibblee (1966)	32 to 64 kilometers (20–40 miles)
T. Atwater and P. Molnar (1973)	140 kilometers (87 miles)

Elevated marine terraces along the shoreline also reflect Quaternary tectonism (Figure 11-6). Inevitably, however, eustatic changes associated with continental glaciation complicate the picture. Because maximum Quaternary sea level probably was not more than a few feet above the present level, virtually all elevated coastal terraces involve tectonic activity. Typically, higher terraces show greater deformation than younger, lower terraces. Further evidence of tectonism is provided by the disparity in elevation from one locality to another. Indeed, modern dating methods have shown that single terraces traceable continuously from one locality to another, may differ in elevation as much as 100 meters (330 feet) or more.

In the Coast Ranges, elevated terraces have been recognized as high as 275 meters (900 feet) above sea level. The terraces identified so far seem no older than Pleistocene, but considerable disagreement exists about whether they are middle or late Pleistocene. For a few localities, there is little argument. Near the mouth of the Santa Ynez River, for eample, marine terraces were cut into folded early and middle Pleistocene strata. These terraces are now more than 210 meters (700 feet) above sea level, and some have been dated radiometrically at about 100,000 years.

SUBORDINATE FEATURES

Santa Lucia Range

For dramatic coastal scenery the Santa Lucia Range has few equals in the coterminous 48 states. At several places along the Monterey coast, these mountains rise from the sea to significant heights less than 6.5 kilometers (4 miles) from shore (Figure 11-12). The range is about 225 kilometers (140 miles) long and extends from Monterey to the Cuyama River. For much of this distance, it is 32 to 40 kilometers (20–25 miles) wide and because it is so rugged north of Cambria, it is crossed by only one minor road. The highest point is Junipero Serra Peak (1,788 meters or 5,862 feet), west of King City.

Interestingly, the Santa Lucia Range, as a topographic feature, does not correspond closely to the underlying lithology and structure. Major faults, like the Sur-Nacimiento zone, cut diagonally across the range, resulting in the northern range being underlain by Salinian block rocks and the southern being developed on a Franciscan basement. Further, in many places in the Santa Lucia Range, fold axes have an even more westerly trend than the faults. These relationships suggest that the modern range has been—perhaps is being—elevated by crustal forces that are oriented differently than those responsible for the major folds and faults. This lack of a close match between present topography and structure is additional evidence for very

FIGURE 11-12 Looking north at the bold coastline of the Santa Lucia Range. (Photo by Robert M. Norris and David Doerner)

young orogenic activity in the southern Coast Ranges, and is not limited to the Santa Lucia Range.

Furthermore, the underlying cause of this Holocene tectonism is not at all clear, since it appears unrelated to both the faulting and to the Cenozoic subduction that ceased in the late Miocene.

The Salinian granitic rocks of the Santa Lucia Range are crucial to an understanding of the history of the Coast Range province and its prominent faults. Unlike the extensively metamorphosed Sur series, the granitic rocks can be dated, presumably precisely, by radiometric methods. Most dates from the Santa Lucia granites vary from 81 to 92 million years, making them younger than some Franciscan rocks. In the eastern part of the Salinian block, Franciscan sometimes is in fault contact with the granitic basement. Most geologists interpret this fault relationship as evidence for major strike-slip displacement of the granitic rocks. The younger granitic rocks have not been found in intrusive contact with the Franciscan.

Although geologists like to consider radiometric ages accurate, metamorphic or structural events ocurring after plutonic rocks cool can reset the radiometric locks, subsequently yielding ages too young for the original magmatic crystallization. Robert R. Compton of Stanford University has found Santa Lucia granitic rocks with radiometric ages of 69.6 and 75 million years overlain in *normal depositional contact* by unmetamorphosed fossiliferous Cretaceous strata 80 to 85 million years old. This dilemma has not yet been completely resolved, but Compton has suggested that the anomalous dates reflect post-granitic deformation. In reality then, the granitic rocks are older than their radiometric dates.

Apart from the Franciscan Formation, no pre-Cretaceous unmetamorphosed sedimentary rocks are known from the Santa Lucia Range. The thick section of Great Valley beds seen in the eastern ranges is missing here and apparently never deposited. Any late Jurassic or early Cretaceous beds once deposited on the Salinian block would undoubtedly be part of the Sur series.

The important Miocene Monterey Formation takes its name from extensive exposures south and east of the city of Monterey. The formation dominates the eastern half of the Santa Lucia Range, including the ridge adjoining the Salinas Valley from the town of Greenfield southward. The Sierra de Salinas forms the eastern wall of the Santa Lucia Range from Greenfield north to the lower end of the valley and is composed mostly of Sur series schists. The prominent ridge northeast of San Luis Obispo, with its oaks, grassy lower slopes, and pine-covered summits, is carved mainly from the Monterey, as is the ridge southwest of San Luis Obispo from Point Buchon to Pismo Beach (see Figure 11-6).

Salinas Valley

The structural history of the Salinas Valley is difficult to interpret. Steam-produced features are readily visible, but apparently folding has been more important in the development of the valley than either stream erosion or faulting.

The basement rocks exposed in the adjacent Santa Lucia and Gabilan ranges are either missing or concealed in the Salinas Valley. Although the basement surface may once have had considerable relief (which might account for the present valley), it probably was only slightly irregular when latest Cretaceous seas spread over the land and covered the granitic rocks. There are several reasons for inferring this. First, geophysical studies indicate that the basement's surface lies more than 1,500 meters (5,000 feet) below the valley floor in many places and as much as 3,000 meters (10,000 feet) in a few. The only logical explanation is that the basement under the valley has either been folded sharply downward or dropped downward on faults, or both. Second, detailed geologic mapping of the Salinas Valley has demonstrated that sedimentary rocks exposed on the flanks of the Gabilan and Santa Lucia ranges dip down toward the valley. Individual rock units thicken in the same direction, suggesting downward folding as the valley floor sank and sedimentary materials accumulated. Third, faulting is evident along the west. Moreover, the rocks have a pattern that is compatible with that in which a valley block had been down-dropped with respect to adjacent highlands.

Although the generally synclinal character of the Salinas Valley is not favorable for oil accumulation, there are several oil fields near San Ardo. Most of the oil is trapped by variations in rock permeability rather than by anticlinal or fault traps characteristic of most oil fields, which is why the San Ardo field was not discovered until 1947.

Diablo Range

The Diablo Range is a well-defined topographic feature 210 kilometers (130 miles) long and as much as 48 kilometers (30 miles) wide, extending southeast from Carquinez Strait along the west side of the San Joaquin Valley almost to Coalinga. San Benito Mountain (1,598 meters or 5,238 feet) is its highest point, but Mount Hamilton (1,284 meters or 4,209 feet) site of Lick Observatory, and Mount Diablo (1,174 meters or 3,849 feet) are better known (Figure 11-10). Some include the Berkeley Hills in the Diablo Range, although the Berkeley Hills are separated from the main range by the San Ramon Valley and the Calaveras-Sunol fault zone.

The upper part of the Great Valley sequence is well developed in the Diablo Range and rests on Franciscan basement along the west side of the San Joaquin Valley. The late Jurassic Knoxville formation accounts for only a small part of this sequence in the range, but the late Cretaceous portion is thick and widely distributed. These late Cretaceous marine rocks are usually divided into the Panoche group (mostly sandstone, shale, and minor conglomerate) and the overlying Moreno shale. They are notably developed in the Panoche Pass area where 8,500 meters (28,000 feet) are exposed.

In a broad sense, the southern Great Valley units, dominated by late Cretaceous beds, form large collars around the elliptical masses of Franciscan rocks that are the core of the Diablo Range. The Cretaceous rocks do not contain distinctive Franciscan detritus, but do incorporate some granitic materials and numerous clasts of a dark, fine-grained porphyry of unknown source. It has been suggested tht both the dark rock and the granitic detritus came from the Sierra Nevada. The absence of Franciscan clasts indicates that the extensive Franciscan exposures seen today were covered during late Cretaceous time and consequently were unavailable for deposition in the Diablo Range.

Younger sedimentary rocks vary from Paleocene to Pleistocene. They are widely distributed around the margins of the range and in some intermontane areas such as the Livermore, San Ramon, and Panoche valleys. Rocks of every Tertiary epoch appear in the Diablo Range, but nowhere is there a complete section.

Some particularly distinctive Cenozoic rocks are the middle Eocene nonmarine beds east of Mount Diablo, which contain clean, washed quartz sands and coal beds of the Domengine Formation plus clays suitable for ceramic use (Figure 11-11). These and similar beds elsewhere in California are thought to indicate almost tropical conditions, because the clays and quartz sands appear to be the end results of intense chemical weathering.

The Mount Diablo coal field west of Antioch produced most of California's commercial coal between 1860 and about 1920, with as many as six different mines operating near the turn of the century. Although considerable coal remains in place, the mining ceased because of competition with better coal from other western fields in such areas as Washington and Utah (Mount Diablo coal was all lignite, or brown coal, a low-grade variety), because of rising costs as mines were deepened, and because users found it more convenient to use oil and natural gas, both of which became readily available in California. However, the quartz sandstones in the Domengine Formation, which contained the coal, were mined for glass making after coal production ceased.

Miocene volcanic rocks occur at several places within the Diablo Range, but the best examples are the Quien Sabe volcanics east of Hollister. Flows, dikes, plugs, and agglomerates are included, some of which may be of submarine origin. Flows range from basalt through andesite to dacite; plugs are mostly andesites and rhyolites.

The lavas and agglomerates often produce steep cliffs and block-strewn surfaces that contrast sharply with the gentler topography developed on the underlying sedimentary rocks. In some places, the scenery developed on the agglomerates resembles that of Pinnacles National Monument in the Gabilan Range. The volcanics of both areas are of similar age, but have different composition.

The general structural pattern of the Diablo Range shows large anticlinal folds with Franciscan cores arranged en echelon and separated by synclinal folds containing younger rocks. Sometimes the crudely anticlinal features (antiforms) are diapirs composed of a mixture of serpentinite and other Franciscan volcanic and sedimentary rocks, which have been forced up along faults into and even through the younger rocks. The serpentinites, as previously noted, are thought to have been derived from the upper part of the mantle.

A large antiform with typical Franciscan core is the one dominating the range from Panoche Valley northwest to the Livermore Valley. It is about 144 kilometers (90 miles) long and averages 24 kilometers (15 miles) in width, with a maximum of more than 32 kilometers (20 miles) east of San Jose. Two notable intermontane synclinal folds are the Panoche Valley and a syncline between Panoche Valley and the New Idria mining district. Rings of Paleocene, Eocene, and Miocene rocks surround these valleys, which have floors filled by Pliocene and Pleistocene gravels.

The New Idria serpentinite diapir is about 22 kilometers (13   miles) long and up to 7 kilometers (4½ miles) wide, and is located mostly in southeastern San Benito County. We know that this diapir was first exposed to erosion in middle Miocene time, because debris from it first appears in the "Big Blue" member of the Temblor Formation. The greenish color of this distinctive rock is due to the presence of grains of Franciscan serpentine and the bright green chrome garnet, uvarovite. Incidentally, this diapir is also the source of California's state gem, the unique benitoite, a barium titano-silicate (see Figure 1-2).

Santa Cruz Mountains and the San Francisco Peninsula

The Santa Cruz Mountains and the San Francisco Peninsula are aspects of the same topographic unit. As in some of the Coast Ranges to the south, parts of the Salinian block and parts of the Franciscan block east of the San Andreas are incoporated with the Santa Cruz Mountains. The western Franciscan block, exposed in the southern Santa Lucia Range, presumably lies far offshore from the Santa Cruz Mountains.

The Santa Cruz Mountains extend from the San Francisco Peninsula 130 kilometers (80 miles) southeast to the Pajaro River, where they merge with the Gabilan Range. Generally less than 16 kilometers (10 miles) wide, between Santa Cruz and San Jose they widen to nearly 32 kilometers (20 miles). The chain inclines to more modest elevations than other ranges in the province; maximum is about 1,160 meters (3,800 feet) near New Almaden, and average summit elevation is only about 760 meters (2,500 feet).

Most of the range is developed on the Salinian block, but the southern portion is about equally divided between the Salinian and Franciscan blocks. Between Santa Cruz and Half Moon Bay, most of the range has Salinian basement, which is well exposed at Ben Lomond and Montara mountains. From Shelter Cove[2] north, the entire San Francisco Peninsula has Franciscan basement.

Numerous faults (most likely parts of the San Andreas system) parallel the range or slice obliquely across it. Near Redwood City the Pilarcitos fault branches off the San Andreas and follows a westerly course along Montara Mountain to Shelter Cove, where it strikes out to sea. This fault is apparently the easternmost demarcation of the Salinian block, for it separates the granitic rocks exposed in Montara Mountain from the Franciscan assemblage of the Peninsula. Some geologists have suggested that the Pilarcitos fault represents an older course for the San Andreas that was eventually abandoned in favor of the present trace. This view is corroborated by the Quaternary inactivity of the Pilarcitos.

The major crustal blocks of the Santa Cruz Mountains are subdivided into elongate slices separated by high-angle faults of substantial vertical movement. One of these faults, the Zayante, extends southeast from Ben Lomond Mountain and was the middle Tertiary boundary between a depressed block on the northeast and an elevated block on the southwest.[3] On the southwest, Salinian granitic rocks are exposed in and around Ben Lomond Mountain or are present at modest depth. In contrast, geophysical studies show that granitic rocks lie 1,800 to 2,750 meters (6,000–9,000 feet) below the surface northeast of the Zayante fault (Figure 11-8).

Other major faults include the Seal Cove-San Gregorio fault zone. This extends from Half Moon Bay across the bay, but parallel to the coast, returning to shore near San Gregorio and extending south to Año Nuevo Point where it again goes out to sea. The San Gregorio fault may join the San Simeon and Hosgri faults to the south. Thick Cretaceous sedimentary rocks occur in the narrow coastal block west of this fault, indicating that this slice of seacoast has been elevated with respect to the area on the east.

The San Andreas fault has long been active in the Santa Cruz Mountains. Normally it is marked by a distinct elongate valley that has been used for reservoir sites north of Redwood City (e.g., San Andreas Lake and Crystal Springs reservoir). Unfortunately the valley has also been used for intensive housing developments, especially in the Daly City area. This has happened despite displacement of this segment of the fault up to 3 meters (10 feet) during the 1906 San Francisco earthquake. Virtually all authorities regard this portion of the San Andreas as active today, ominously quiet only for the moment.

The oldest known rocks in the Santa Cruz Mountains are undated—probably Paleozoic—metasediments of the Ben Lomond district. They form small roof pendants in the granitic rocks and are probably equivalent to the Sur series of the Santa Lucia and Gabilan ranges.

---

[2]California has two Shelter Coves, both near the point where the San Andreas fault goes out to sea. One is in San Mateo County at Point San Pedro near Linda Mar; the other is at Point Delgada in southern Humboldt County.

[3]Elevation and depression refer here to the basement surface and not necessarily to the topographic surface. Even today, there is considerable difference in basement elevation on either side of the Zayante fault but little topographic difference.

Santa Cruz Mountains granitic compositions range from gabbro to granite, with most samples being quartz diorites. Well records have shown that these granitic rocks are continuous beneath the surface with those exposed in the Gabilan Range. In addition, quartz diorites almost identical with those of Montara Mountain occur on the bleak Farallon Islands, 45 kilometers (28 miles) west of San Francisco, and along the shelf edge north of the islands for about 48 kilometers (30 miles).

Northeast of the San Andreas and Pilarcitos faults, the San Francisco Peninsula has Franciscan basement with a typical array of graywackes, volcanic sills and dikes, deep-water red cherts, and ultrabasic intrusives with serpentine derivatives. The type locality for the Franciscan formation is the northern San Francisco Peninsula and was described by A. C. Lawson in 1895. The hills that dot the city of San Francisco are all exposures of various Franciscan rocks.

On the Salinian block, the only occurrences of Cretaceous marine sediments are those west of the San Gregorio fault, in the Pigeon Point area, and possibly north of Montara Mountain in the highly sheared rocks along the Pilarcitos fault. No Cretaceous marine rocks are known from well records in the central Santa Cruz Mountains. East of the San Andreas fault, late Cretaceous marine sedimentary rocks are exposed, primarily near the crest of the range. These rocks are in fault contact with the Franciscan, but are overlain by as much as 4,900 meters (16,000 feet) of clastic, mostly marine Cenozoic sediments.

Tertiary marine strata are present throughout the range and possess an aggregate thickness of more than 6,700 meters (22,000 feet). Considerable tectonic activity is indicated by the nature and distribution of the Tertiary, because some beds were deposited on land and others in deep marine waters. During the early Tertiary, deep-water conditions prevailed over much of the central Santa Cruz Mountain area. A small portion was uplifted during the Oligocene, however, forming a small island on which accumulated land-laid deposits were assigned to the Zayante sandstone. Marine conditions subsequently prevailed until at least the middle Pliocene.

Some of the Miocene rocks in the Santa Cruz Mountains are petroliferous. Some asphaltic sandstones were quarried west of Santa Cruz for a time to make paving materials, and three small oil fields once produced small amounts of petroleum and gas; all are now abandoned. One of these oil fields, the Moody Gulch area a few kilometers south of Los Gatos, produced a high-grade light oil from about 14 wells between about 1880 and 1930. Several small fields near Half Moon Bay produced a similar type of oil from 1867 until the 1940s. The Sargent field, the largest of the three, was located at the southern end of the Santa Cruz Mountains not far from San Juan Bautista. Production from this field totaled more than 8,000 metric tons (600,000 barrels), mainly between 1886 and 1945, with peak production in 1909.

San Francisco, San Pablo, and Suisun Bays

This system of bays, one of the world's finest harbors, is the only place where streams from interior California reach the sea. The system occupies a late Pliocene structural depression that has been flooded several times in response to Pleistocene glacial cycles.

The Pliocene date was determined in the San Francisco Bay area by stratigraphic studies. A thick Plio-Pleistocene deposit (the Merced Formation) occurs here, of which the lower 1,370 meters (4,500 feet) is marine and the upper 160 meters (500

FIGURE 11-13 Carquinez Strait, with San Pablo Bay in the distance. (Photo by Robert M. Norris and David Doerner)

feet) are mostly nonmarine. The lower portion contains heavy mineral grains, indicating locally derived sediment. About 30 meters (100 feet) above the marine-nonmarine transition, the mineral assemblage changes abruptly to one identical with that carried by the Sacramento River system and derived primarily from the Sierra Nevada. This mineral change reflects the initial establishment of the present drainage pattern.

San Francisco and San Pablo bays occupy part of the main structural depression that includes the Santa Clara Valley to the south. This depression extends from south of Hollister and northward beyond the bays. In the north section, it divides into the Petaluma, Sonoma, and Napa valleys. Suisun Bay to the east is separated from San Pablo Bay by the narrow, winding Carquinez Strait, thought to be the channel of a superimposed stream cut into bedrock by the Sacramento River (Figure 11-13). The stream channel is now as much as 60 meters (200 feet) below sea level. Its extension across the floors of San Pablo and San Francisco bays and out through the Golden Gate has not yet been traced, although the early river almost certainly followed such a course. Most bays of the San Francisco system are shallow. About 85 percent of the water area is less than 10 meters (30 feet) deep, sometimes less than $5\frac{1}{2}$ meters (18 feet). Sediment entering from the Sacramento-San Joaquin system will fill the bays soon (geologically), unless tectonic activity or a rise in sea level intervenes. Human activities have accelerated the rate of fill. Both miners (hydraulic mining and dredging) in the last century, and farmers (plowing and cultivating) have increased the sediment load in the Sacramento-San Joaquin system. Reclamation operations have developed such land additions as Treasure Island and San Francisco International Airport.

The San Francisco Bay area is of special geological significance because it was the first place in California for which a geologic map was prepared (Figure 11-14). Drawn in 1826, this historic map was not only a first for California, but was also probably the second geologic map for any part of North America. It appeared only 7 years

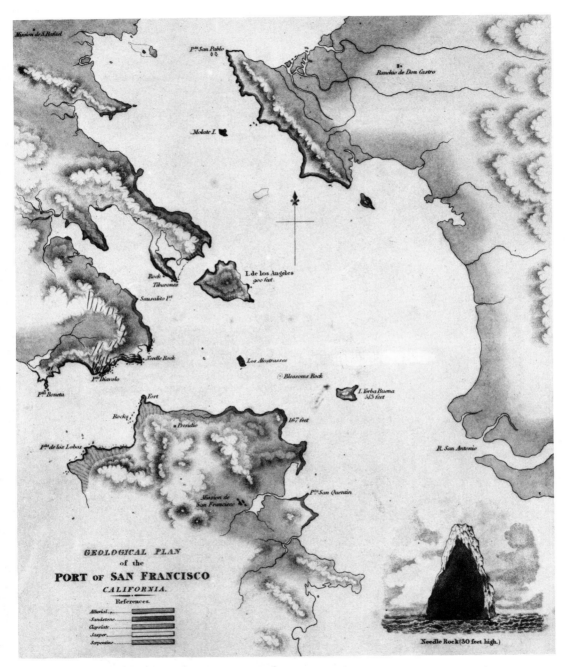

FIGURE 11-14 Geologic map of San Francisco area, 1826. (Courtesy of Bancroft Library, University of California, Berkeley)

after William Maclure's map of the eastern United States and only 11 years after William Smith (in England) drew the first geologic map for any region.

This pioneering map of San Francisco Bay was prepared by Edward Belcher, the surveyor, and Alex Collie, the surgeon, of HMS *Blossom*, a British sloop sent to meet the explorers John Franklin and William Parry in the Bering Strait. Belcher and Collie accurately represented the form of the bay, its islands, and the Carquinez Strait and also closely calculated the elevation of Mount Diablo. Furthermore, they correlated main rock types with different colors, recorded the trend of rock units, and listed a variety of mineral species present.

Mendocino Range

The northern units of the Coast Ranges are not as clearly defined as the southern units. They are also less well known geologically, although general distribution of rock types is now understood. The Mendocino Range, the main unit, extends 350 kilometers (215 miles) from San Francisco Bay north to Humboldt Bay. It lies between the coast and the elongate depression occupied by the Russian and Eel rivers. Most peaks are about 900 to 1,050 meters (3,000–3,500 feet) high, except west of Laytonville where some are a little more than 1,280 meters (4,200 feet) in height. The northern Coast Ranges are rugged, heavily wooded, and crossed by fewer roads than the drier southern ranges.

The Mendocino Range lies almost entirely on the Franciscan block, and its rocks resemble those of the Diablo and Santa Cruz ranges to the south. North of the Golden Gate the San Andreas fault hugs the coast, so here only headlands, promontories, and long narrow slices of coast lie on the granitic Salinian block. Possibly these selvages of coastal northern California do not belong to the Mendocino Range at all, but it is convenient to treat them together.

The main trace of the San Andreas leaves the coast at Mussel Rock and crosses the Golden Gate to Bolinas Bay, where it comes ashore. It has been established that the Pilarcitos fault, not the San Andreas, separates the Salinian from the Franciscan in San Mateo County, but it is not known how the Pilarcitos and San Andreas join on the sea bottom off the Golden Gate. In Marin County the San Andreas again forms the eastern edge of the Salinian block and granitic rocks are exposed west of the fault. The San Andreas occupies a well-defined valley from Bolinas to Tomales Bay, which is a long, narrow depression opening to the northwest and lying between the Bolinas and Inverness ridges (Figure 11-15).

A conspicuous geomorphic feature of this coastline is the Point Reyes headland. The landward side consists of the granitic Inverness Ridge, almost 450 meters (1,500 feet) high. To seaward, the ridge faces a tapering lowland more than 13 kilometers (8 miles) wide developed on synclinally folded Pliocene marine sedimentary rocks. Drakes Estero, a shallow lagoon open to the south, lies along the axis of this syncline.[4] Along the lowland's northern coast, northwesterly winds regularly drive beach sand ashore to form dunes that extend inland as much as 1,200 meters (4,000 feet). The

---

[4]It was here, allegedly, in 1579, that Sir Francis Drake formally took possession of the land for Queen Elizabeth of England.

FIGURE 11-15 Looking south along the San Andreas fault trace and Tomales Bay. Bolinas Ridge is on the left and Inverness Ridge on the right. (Photo by Robert E. Wallace, courtesy of U.S. Geological Survey)

lowland terminates on the southwest at the granitic ridge capped with Paleocene sedimentary rocks called Point Reyes (Figure 11-16).

A superificially similar but smaller feature occurs a few kilometers north at Bodega Head, the most northerly exposure of definitely Salinian rock. This granitic eminence is separated from the Mendocino Range by the low sandy area surrounding Bodega harbor (Figure 11-17). Bodega Head was once an island, but has been connected to the mainland by sand deposited by longshore currents. Once the connecting sand spit was established, winds and sand-loving vegetation created dunes that are prominently aligned with the northwest winds.

From near Fort Ross to Point Arena, the San Andreas fault zone is only a few kilometers inland, nearly paralleling the coast. The fault is again distinctively marked by a narrow, long, straight valley that is occupied by such streams as the Gualala and Garcia rivers. The slice of crust thus formed between the sea and the San Andreas is probably the northernmost land occurrence of the Salinian block. Neither granitic nor Franciscan basement is exposed in it, however, so there is some uncertainty about its affinities with Salinian rocks.

One of the subunits of the Mendocino Range is the rugged and almost roadless King Range bounded on the north and east by the Mattole River, and extending from Point Gorda south to Point Delgada. This range is underlain almost entirely by late Cretaceous rocks, mostly sandstones. The King Range presents a bold face

FIGURE 11-16 Point Reyes, Marin County. The rocks exposed here are marine Paleocene sandstones that rest unconformably on Salinian granitic basement rock. These sandstones correlate with the similar Carmelo Formation near Monterey Bay, about 180 kilometers (110 miles) to the southeast. (Photo by J. C. Clark)

FIGURE 11-17 Bodega Bay seen from Bodega Head, Sonoma County. The curving sandspit encloses the bay on the south. The north side of the bay is formed by a broad, dune-covered area that joins granitic Bodega Head to the mainland across the San Andreas fault zone. Bodega Head is the northernmost exposure of Salinian block granitic rocks. (Photo by G. R. Wheeler)

to the Pacific, much like the Santa Lucia Range in the southern Coast Ranges. For example, at 1,247 meters (4,087 feet), Kings Peak is only 5 kilometers (3 miles) from the coast.

The bulk of the Mendocino Range is made up of Franciscan basement overlain by Cretaceous sedimentary rocks of the Great Valley sequence. North of the gorge through which the Russian River crosses the range, Cretaceous beds and typical Franciscan graywacke and shale constitute bedrock in about equal amounts. Fossils indicate that here, as in the southern ranges, the Franciscan is late Jurassic to late Cretaceous.

The relationship between the Great Valley and Franciscan sequences poses unresolved problems in the northern Coast Ranges as it does elsewhere in the province. Great Valley beds of shelf and near-shore origin (the back arc basin) have the same ages as the Franciscan slope and deep-water beds (the subduction complex).

So far as is known, Great Valley beds are nowhere in depositional contact with the largely contemporaneous Franciscan, but are instead always in tectonic contact. As noted earlier, the Great Valley beds are sometimes in depositional contact with ophiolite sequences, including serpentinites, cherts, and greenstones, but these seem to be separated from the Franciscan by faults. Formerly, ophiolites had been included in the Franciscan.

The northern Mendocino Range includes two conspicuous downwarped basins, one near Garberville and another south of Cape Mendocino, where thick piles of Tertiary marine rocks are preserved. The range terminates near the mouth of the Eel River against a third, generally similar but larger basin that extends southeast from the coast for about 48 kilometers (30 miles). This large synclinal structure, the Eel River Basin, is outlined by marine strata of probable late Miocene age and is filled with Pliocene and Quaternary beds. (The smaller basins lack the Pliocene and most of the Quaternary.) Along the coast, the basin extends from about Cape Mendocino north almost to Eureka. Its Miocene beds are equivalent to such southern units as the Rincon and Santa Margarita formations; the Pliocene is referred to as the Wildcat group and is probably the age equivalent of the Pico and Repetto formations in the Los Angeles Basin. This Cenozoic section is more than 3,650 meters (12,000 feet) thick. It has been of particular interest for many years because of small oil seeps and evidence of natural gas (Tompkins Hill gas field), but only minor production has resulted. However, the small noncommercial wells at Petrolia on the lower Mattole River were among the first drilled in California, dating back to about 1865.

The folding and faulting of the Mendocino Range continues today. At Scotia Bluffs on the Eel River (see Figure 1-15), part of the Wildcat group has been uplifted 4 meters (13 feet) in the past 1,000 years. The 2 to 4 millimeter average uplift per year seems to be caused by folding, which is depressing the Eel River Basin and elevating the cliffs, related, no doubt, to the proximity of the Mendocino Triple Junction, where the Pacific, North American, and Juan de Fuca crustal plates meet.

Eastern and Northern Ranges

South of Clear Lake, the inland Coast Ranges are not only better known, but seemingly are also more complex than the ranges north and east of the Mendocino Range and Clear Lake. The latter are dominated by Great Valley and Franciscan rocks with little else present.

The significant volcanism that occurred in the southeastern ranges produced the Sonoma volcanics, which dominate the Sonoma, Mayacmas, and Howell mountains. This complex array of lava flows and tuffs rests mostly on Franciscan and Great Valley rocks, Eocene sediments, or the Pliocene Petaluma Formation. Because some Sonoma volcanics rest on Pliocene beds, the formation must be Pliocene or younger. The volcanics cover more than 900 square kilometers (350 square miles) from the Petaluma-Cotati lowland east to the Howell Mountains and from Suisun Bay north to Mount St. Helena. They originally formed a nearly continuous blanket over the whole region, but they have since been segmented by faulting, folding, and erosion. Andesites and andesitic tuffs are most common, but some basalts are present and rhyolites are prominent near Mount St. Helena.

The Clear Lake volcanic series is a belt of younger volcanic rocks stretching from near the north end of Lake Berryessa to Clear Lake. These rocks range from basaltic flows to rhyolites. On the southwestern shore of Clear Lake is Mount Konocti (1,280 meters or 4,200 feet), a nearly uneroded, prominent multiple cone composed mainly of dacitic to andesitic lava flows and interbedded pyroclastics (Figure 11-7). Adjacent mountains either have similar composition or are rhyolitic.

The volcanic area south of Clear Lake contains silicic volcanic rocks, including obsidian flows, layers of pumice, and numerous cinder cones. Ages range from Pleistocene to Holocene, and residual volcanic activity persists as hot springs, steam vents, and a borax-rich lake on the southeastern shore of Clear Lake (Figure 11-18). Sulfur and quicksilver (mercury) were mined in the Sulphur Bank area close to Borax Lake during the nineteenth century.

Clear Lake is the largest natural freshwater lake in the Coast Ranges and also the

FIGURE 11-18 Borax Lake, with Clear Lake in the distance. (Photo by Robert M. Norris and David Doerner)

largest landslide lake in California. Its drainage is unusual because it has two outlets: north into the Russian River and south into the Sacramento River. According to one view, its original eastern outlet was dammed by a small lava flow that raised the lake so it spilled west into the Russian River. A few centuries ago, a landslide blocked this western exit, causing another rise in lake level. This resulted in reestablishment of the eastern drainage across the lava flow into the Sacramento River. Erosion then cut a gorge about 18 meters (60 feet) deep into the lava flow, thereby reducing the lake level. Most of today's drainage is eastward, and only minor seepage occurs across the large landslide and into the Russian River. A control dam was built at the eastern outlet in the 1920s in order to regulate the lake level.

Most of the Coast Range province north of Clear Lake is developed on Franciscan rocks that may be as much as 15,000 meters (50,000 feet) thick. The eastern edge, however, is underlain by a very thick sequence of Jurassic and Cretaceous sedimentary rock that constitutes the type locality for the Great Valley sequence. The oldest part is the 6,100-meter (20,000-foot) thick late Jurassic Knoxville sequence composed of dark shales, thin-bedded sandstones, and minor conglomerate. The Knoxville grades imperceptibly into the overlying Cretaceous (Shasta) beds that aggregate at least 9,700 meters (32,000 feet) in thickness. The entire sequence is a continuum and can be subdivided satisfactorily only by fossils, not lithology.

Great Valley rocks are steeply tilted toward the Sacramento Valley, and their regular bedding and dip have produced long, parallel, hogback ridges. These mark

FIGURE 11-19 Strike ridges in Cretaceous rocks northwest of Williams. (Photo by Burt Amundson)

the eastern side of the Coast Ranges from west of Red Bluff south to the latitude of Sacramento (Figure 11-19). The Franciscan rocks west of the Stony Creek thrust are less regular and do not form distinctive topography.

At the latitude of Red Bluff, the Coast Ranges narrow sharply, veer west, and wrap around the seaward side of the Klamath Mountains. The provinces are separated by the South Fork Mountain thrust zone, which lies only about 8 kilometers (5 miles) east of the shoreline at Crescent City, the narrowest part of the northern Coast Ranges. Apart from scattered coastal Quaternary deposits, the northernmost part of the province is developed entirely on Franciscan terrain.

SPECIAL INTEREST FEATURES

Tertiary Intrusive Plugs near San Luis Obispo

Southeast of the city of San Luis Obispo is the first of 14 volcanic plugs that dot the landscape for 29 kilometers (18 miles) northwest to Morro Rock. The plugs form a string of picturesque hills with bold, rocky upper slopes, and grassy, oak-covered lower slopes. The most prominent are those between San Luis Obispo and Morro Bay. Morro Rock itself rises more than 160 meters (500 feet) above sea level (Figure 11-20).

These features are interpreted as eroded Miocene volcanic necks, probably emplaced along the West Huasna fault. The volcanic edifices themselves have long since been eroded away, leaving only the stumps of resistant lavas that once congealed in the throats of the vanished volcanoes. These volcanic centers are thought to be the sources of the Obispo tuff, a pyroclastic deposit widely distributed in the San Luis Obispo-Morro Bay area. Composition of rocks preserved in the volcanic necks varies from basalts and diabases to dacites. The larger peaks in the chain are generally andesite, but Morro Rock is a dacite.

Pinnacles National Monument

The Pinnacles are located about 56 kilometers (35 miles) south of Hollister, at the southern end of the Gabilan Range. This small area of rough and jagged topography is so conspicuous among the rolling, grass-covered hills characteristic of this part of the Coast Ranges that it was a popular attraction as early as the late eighteenth century. In 1794, George Vancouver took time from his coastal explorations to endure a mule trip to the Pinnacles. The region was proclaimed a National Monument by President Theodore Roosevelt in 1908, to a large degree because of the urging of Stanford University's naturalist president, David Starr Jordan (Figure 11-21).

The distinctive topography at Pinnacles, with its towering spires, numerous caves, and jumble of fallen blocks, is the result of weathering and erosion of a rhyolite breccia. Although the area includes other volcanic rocks, the most striking topography is sculptured on the breccia. All the rocks have been preserved in a downfaulted block about 10 kilometers (6 miles) long and 4 kilometers ($2\frac{1}{2}$ miles) wide.

The eastern half of the volcanic area is composed mainly of rhyolitic flows varying from gray to red and green and including large amounts of obsidian. The obsidian is concentrated either near old volcanic vents or wherever the lavas came in contact with the granitic basement through which they were intruded. Within the predom-

FIGURE 11-20 Morro Rock, a dacite volcanic plug at Morro Bay, San Luis Obispo County. Other similar plugs can be seen in the upper left corner of the photo. Morro Rock is a tombolo or land-tied island that resulted from beach and dune deposition on its landward side. (Photo by Spence Air Photos, courtesy Department of Geography, University of California, Los Angeles)

inantly rhyolitic flows are some andesitic and basaltic flows. The western half is composed of the slightly younger, thick pyroclastic volcanic breccias from which the Pinnacles themselves were eroded. Total volume of volcanic material originally erupted is estimated at between 25 and 40 cubic kilometers (6–10 cubic miles); all but 12½ cubic kilometers (3 cubic miles) of which has been removed by erosion. The volcanic rocks are considered Miocene because the first influx of volcanic detrital material occurs in Miocene sedimentary rocks deposited nearby. The rocks at Pinnacles are about the same age as those at Quien Sabe about 35 kilometers (22 miles) northeast in the Diablo Range, but they are of different composition. New evidence suggests the Pinnacles volcanic rocks are closely related to the Neenach volcanic rocks in the western Antelope Valley about 290 kilometers (180 miles) to the southeast, on the opposite side of the San Andreas fault.

The main events in the geologic history of the Pinnacles are as follows:

1. Masses of rhyolitic lava were erupted from at least five vents as flows on a slightly irregular erosional surface developed on the granitic rocks of the Gabilan Range.

FIGURE 11-21 Miocene rhyloite breccia at Pinnacles National Monument, Monterey County. (Photo by R. M. Norris and D. P. Doerner)

During the eruptive cycle, magma composition varied from andesitic and basaltic to rhyolitic. The possibility exists, of course, that the rhyolitic flows are the result of remelting of the underlying granitic rocks by the rising of less silicic magma.

2. Toward the end of the eruptive cycle, some vents became plugged, and periodic steam eruptions blasted out solidified or partly solidified material with new lava. This formed the breccias that subsequently were emplaced as avalanche deposits. There is little evidence to suggest that water transport was involved in forming these inclined, bedded rocks. They were deposited on steep slopes around the vents and most have remained in their original positions.

3. Faulting has preserved the volcanic rocks of the Pinnacles area by permitting the block on which they rest to be dropped down into the granitic rocks of the southern Gabilan Range. Most volcanic rocks that remained on the higher, unfaulted regions have been removed by erosion. The age of faulting usually appears to postdate volcanic activity, but it may well have begun before the eruptive cycle closed and perhaps was related to it.

4. Pliocene events are unrecorded, but by Pleistocene time, the volcanics were being eroded into today's distinctive landscape.

The Geysers

This unusual area of hot springs and steam vents is located in the Mayacmas Mountains. It occupies the northwestern end of a graben about 8.8 kilometers (5½ miles) long and 1½ kilometers (1 mile) wide, drained by Big Sulphur Creek. The entire region lies within Franciscan rocks, though Clear Lake rhyolites are exposed only

FIGURE 11-22 Steam wells at the Geysers Geothermal field, Sonoma County. (Photo courtesy Pacific Gas and Electric Company)

about 4.8 kilometers (3 miles) to the east, and the subterranean heat source is probably attributable to residual heat from magma closely associated with the volcanic rocks. The graben itself appears to be underlain mainly by Franciscan greenstones and serpentinites that have been downfaulted into the surrounding graywacke.

The main natural thermal area is along Geyser Creek, a tributary of Big Sulphur Creek, and is only about 400 meters (1,300 feet) long and 180 meters (600 feet) wide (Figure 11-22). Geyser Canyon contains numerous hot springs, with temperatures ranging from 50°C to boiling. Several small steam vents occur here, and modern geothermal wells are located just east of Geyser Creek.

The region is of particular interest because as of 1989 it is the largest geothermal power operation in the world. Power development began here, after a fashion, in 1921 when the first steam wells were drilled. By 1925, eight wells had been completed but there was little market for the power, so the project was abandoned. In 1955, more wells were drilled and generating equipment installed. Power was first produced in 1960, making the Geysers the third place in the world to use geothermal resources to produce commercial electric power.[5] The geothermal area at the Geysers is unusually favorable for power generation because it produces dry steam with few

_____

[5]Lardarello in Italy had been in operation since the early twentieth century, and Wairakei in New Zealand since 1950.

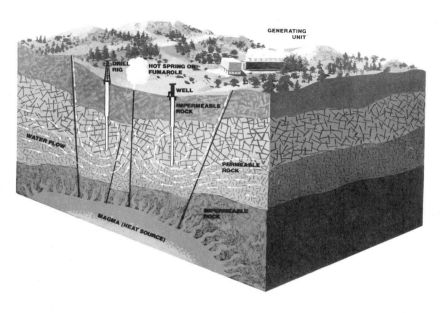

FIGURE 11-23 Block diagram of the Geysers Geothermal field showing the probable circulation of heated underground water. (Courtesy Pacific Gas and Electric Company)

corrosive products apart from some sulfur. This characteristic has greatly facilitated development, so the region now produces enough power to supply a city the size of San Francisco (Figure 11-23).

Landslides

Landsliding, of course, is not restricted to the Coast Ranges in California, but the phenomenon is so pervasive and important in the province that it deserves special attention. Because Franciscan rocks are more widely distributed in the Coast Ranges than any other group, and because their highly sheared serpentinites are so unstable, slides involve the Franciscan more than any other formation (Figure 11-24). In the northern ranges, where Franciscan is the dominant rock, slides are so prevalent that they probably account for more downslope transport of material than any other process, including streams.

An extensive study of geologic hazards in the nine counties around San Francisco Bay has pinpointed the location of hundreds of slides and provided an estimate of their direct and indirect costs. Slides cause direct losses by damage to houses, buildings, utility lines, and roads, and indirect losses by reducing property values, by tax forfeiture, and the need for emergency help from public agencies.

In one storm, from January 3 to January 5, 1982, 50 to 600 millimeters (2–24 inches) of rain fell in the San Francisco Bay area (the highest falls occurred in the Santa Cruz Mountains), which resulted in major damage by landsliding and flooding. Damage to private property was about $172 million and $108 million in public

FIGURE 11-24 Typical Coast Range landslide developed on Franciscan rocks, Russian River Valley, Mendocino County. The hummocky, grass-covered slopes are often dotted with massive knobs of peridotite or other resistant rocks around which the weaker rocks slowly flow. (Photo by R. M. Norris)

property; 26 deaths, 7 missing persons, and 477 injuries resulted; 231 homes were demolished and 6,295 were damaged. Twenty-four businesses were destroyed and 1,014 sustained appreciable damage. Nearly 12,000 people were forced to apply for some sort of relief. Unfortunately, events of this type are not unique and can be expected to recur every 20 years or so, with numerous less disastrous events in the intervening years.

Landsliding continually threatens and often forces the temporary closure of highways in the Coast Ranges, particularly in the northern ranges. The high cost of frequent landslides was an important factor in the decision in the early 1980s to abandon the only rail line from the San Francisco Bay area to Eureka.

Winter is the time of greatest slide activity in California, and wet winters produce more sliding than dry winters. Water is the crucial factor in activating landslides, because water reduces internal coherance of earth materials, adds weight to the unstable mass, and separates bedding surfaces. In addition, sliding is promoted by oversteepening of slopes caused by such activities as marine and stream erosion or earth moving and grading done by humans. Shearing and crushing associated with faulting also make sliding more likely; for example, landslides are particularly common along traces of active faults like the San Andreas and Hayward (Figure 11-25). Table 11-1 shows the widespread occurrence of landsliding in the Coast Ranges, plus some of the rock types involved.

Marine Terraces

Marine terraces occur at numerous places along the Pacific shore, all reflecting changes in relative sea level in the recent past. Although only a few have been

FIGURE 11-25 Devils Slide between Montara Mountain and Point San Pedro, San Mateo County. This area lies within a highly sheared zone just south of where the Pilarcitos fault strikes out to sea. California State Highway 1 crosses the slide and is frequently closed for repairs. (Photo by D. P. Doerner)

studied in detail, such as those at Santa Cruz, it is clear that the terraces are discontinuous, and terraces of the same age do not necessarily have the same elevation in different localities (see Figure 1-30). The picture has been complicated by the Pleistocene events that affected the California coast; it is difficult to distinguish eustatic changes in sea level from changes caused by diastrophism unless one has a well-dated Pleistocene sea level curve and can get precise dates on the terraces. These two requirements are seldom met unequivocally, but the picture is gradually coming into focus as worldwide sea level curves are developed and local terraces accurately dated.

No terrace has been dated earlier than Pleistocene, and as a rule the higher terraces are older and often show more deformation than the younger, lower terraces. The principal areas of well-developed terraces along the Coast Range shoreline are, from north to south: Crescent City, Trinidad, Cape Mendocino, from near Davenport to Santa Cruz, from Point Lobos to Point Sur, from Ragged Point to Cambria, from Morro Bay to Point Buchon, and the Shell Beach-Pismo Beach region.

At Santa Cruz, five conspicuous terraces have been recognized between elevations of 30 to 260 meters (100–850 feet). The 30-meter terrace is the lowest and most prominent and currently produces most of California's brussels sprouts. It is cut in the Santa Cruz Mudstone and in some places is almost $2\frac{1}{2}$ kilometers ($1\frac{1}{2}$ miles) wide.

TABLE 11-1 Coast Range Landsliding and Associated Rocks

Area	*Associated Rocks*
Berkeley Hills area (Alameda and Contra Costa counties)	Orinda Formation: weakly consolidated sands, silts, clays, and tuffs, Pliocene
Blue Lake area (Humboldt County)	Franciscan serpentinites; Falor Formation: soft clays and sands, Pliocene
Eel River area (Humboldt County)	Rio Dell Formation: mudstone; Yager Formation: sandstone; Wildcat Formation: mudstone, all Pliocene
Healdsburg area (Sonoma County)	Franciscan serpentinites
Lower Lake area (Lake County)	Franciscan serpentinites
Marin headlands (Marin County)	Franciscan cherts
Mount Hamilton area (Santa Clara County)	Alum rock slide: Franciscan serpentinites; Oak Ridge area: mainly Franciscan serpentinites; Santa Clara Formation: sands and gravels, Pliocene
Ortigalita Peak area (Merced County)	Franciscan serpentinites
Quien Sabe area (Merced and San Benito counties)	Large slide in Quien Sabe volcanics, Miocene. Franciscan shales and shaly sandstones.
San Juan Bautista area (San Benito County)	Purisima Formation, Pliocene
San Mateo coast area (San Mateo County)	Devils slide: faulted and sheared mass involving Franciscan, Cretaceous, and Paleocene rocks
Suisun Bay area, north side (Solano County)	Franciscan serpentinites. Sonoma volcanics, Pliocene. Petaluma sands and gravels, Pliocene.

It generally slopes seaward less than 1° and is typically covered with a thin layer of marine sand. Near the base of the old elevated sea cliff, the terrace and its thin veneer of marine sediment are often blanketed by thick nonmarine deposits. These have slumped and washed down from above in a pattern that is characteristic of marine terraces.

Gold and Quicksilver (Mercury) Mining

The California Coast Ranges have long been the leading American source of mercury, accounting for about 85 percent of the total past U.S. production, although activity has been minimal in recent years. Between 1850 and 1980, about $200 million worth of mercury was recovered, amounting to about 2.8 million 76-pound (34 kilogram) flasks. Mercury has been produced from the southernmost Coast Ranges in Santa Barbara County, where mercury was first discovered in 1796, to Lake County in the northern Coast Ranges. By far the bulk of production has come from just three areas: New Idria in southern San Benito County, New Almaden in Santa Clara County near San Jose, and from a group of mines in the Mayacmas Mountains near the junction of Lake, Sonoma, and Napa counties.

Most of the deposits are closely associated with young high-angle faulting, hot spring activity, and bodies of a somewhat unusual rock called silica-carbonate rock, which consists of an intimate association of chalcedony, quartz, opal, magnesite, dolomite, and calcite, often containing the bright red mercury mineral cinnabar, and finely divided gold. The silica-carbonate rock bodies often occur with serpentinites and are sometimes overlain by volcanic rocks.

The New Almaden mine was the first mercury-producing mine in California and was opened in 1824. This mine ultimately became the world's deepest mercury mine (747 meters or 2,450 feet) and by the 1970s had yielded $50 million worth of mercury. Even in 1861, this mine had reached a depth of 76 meters (250 feet) and was producing 3,000 flasks of mercury per month.

The New Idria mine went into production in 1859, and by 1861 was yielding 970 flasks per month. As recently as 1965, it was still the largest mercury mine in the United States and by then had yielded more than 500,000 flasks.

Long before the days of OSHA (Occupational Safety and Health Administration) and regulations protecting the health of miners, the hazards of mercury mining were fully recognized. William Brewer, a member of California's first geological survey — the Whitney Survey — gave the following account of conditions at New Idria in the summer of 1861, in his book, *Up and Down California*, 1974, Francis P. Farquhar, ed., Univ. Calif. Press, 583 p.).

The work at the furnaces is much more unhealthy and commands higher wages. Sulphurous acids, arsenic, vapors of mercury, etc., make a horrible atmosphere, which tells fearfully on the health of the workmen, but the wages always command men and there is no want of hands. The ore is roasted in furnaces and the vapors are condensed in great brick chambers, or "condensers." These have to be cleaned every year by workmen going into them, and many have their health ruined forever by the three or four days' labor, and all are injured; but the wages, twenty dollars a day, always bring victims. There are but few Americans, only the superintendent and one or two other officials: the rest are Mexicans, Chileans, Irish (a few), and Cornish miners.

I can hardly conceive of a place with fewer of the comforts of life than these mines have — a community by itself, 75 miles from the nearest town (San Juan) and 135 from the county seat, separated from the rest of the world by desert mountains, a fearfully hot climate where the temperature for months together ranges from 90° to 110°F, where all necessities of life have to be brought from a great distance in wagons in the hot sun. As might be expected, little besides the bare necessities of life is seen, and if any luxuries come in, it is only at an extravagant price.

Such is New Idria and by such toils and sufferings do capitalists increase their wealth.

Small amounts of gold have been recovered from the Coast Ranges for many years, but until recently the province was much less important than the Klamath Mountains, the Sierra Nevada, and the Mojave and Colorado deserts. In the past few years, however, the discovery of very finely divided gold associated with the mercury deposits in the Mayacmas Mountains, near the small settlement of Knoxville, has led to the development of a new gold mine, the McLaughlin. This deposit is expected to yield 3.2 million troy ounces of gold in roughly 20 years. The ore also contains mercury, arsenic, antimony, tungsten, and thallium. The McLaughlin mine was dedicated in 1985 and is already California's major gold producer.

REFERENCES
General

Bailey, Edgar H., and others, 1964. Franciscan and Related rocks and Their Significance in the Geology of Western California. Calif. Div. Mines and Geology Bull. 183.

Durham, David L., 1974. Geology of the Southern Salinas Valley Area, California. U.S. Geological Survey Prof. Paper 819.

Jenkins, Olaf P., ed., 1951. Geological Guidebook of the San Francisco Bay Counties. Calif. Div. Mines and Geology Bull. 154.

Oakeshott, Gordon B., 1970. Geology of the California Coast Ranges. Mineral Information Service (now California Geology), v. 23, pp. 7–10.

_____, 1980. Tectonic History of the Area North of San Francisco. California Geology, v. 33, pp. 266–273.

Page, Ben, 1981. The Southern Coast Ranges *in* Geotectonic Development of California, W. G. Ernst ed., Rubey Vol. 1, Englewood Cliffs, NJ: Prentice-Hall, pp. 329–417.

Special

Anonymous, 1979. The Geysers Heat Source. California Geology, v. 32, pp. 226–227.

Bedrossian, Trinda L., 1974a. Geology of the Marin Headlands. California Geology, v. 27, pp. 75–86.

_____, 1974b. Fossils of the "Merced" formation, Sebastopol region. California Geology, v. 27, pp. 175–182.

Burnett, John L., 1986. A New Type of Gold Lode Deposit—The McLaughlin Mine. California Geology, v. 39, pp. 15–16.

Carlson, Paul R., and others, 1970. The Floor of Central San Francisco Bay. Mineral Information Service (now California Geology), v. 23, pp. 97–107.

Chipping, David H., and others, 1982. Coastal Dune Complexes Pismo Beach and Monterey Bay. California Geology, v. 35, pp. 7–12.

Cleveland, George B., 1977. Rapid Erosion along the Eel River, Calif. California Geology, v. 30, pp. 204–211.

Farquhar, Francis P., ed. 1974. Up and Down California in 1860–64, The Journal of William H. Brewer, University of California Press, Berkeley and Los Angeles, 583 p.

Hinds, N.E.A., 1968. Pinnacles National Monument. Mineral Information Service (now California Geology), v. 21, pp. 119–121.

Hopson, Clifford A., and others, 1981. Coast Range Ophiolite, Western California *in* Geotectonic Development of California, W. G. Ernst, ed., Rubey Vol. 1, Englewood Cliffs, NJ: Prentice-Hall, pp. 418–510.

Jenkins, Olaf P., 1973. Pleistocene Lake San Benito. California Geology, v. 26, pp. 151–163.

Koenig, James B., 1963. The Geologic Setting of Bodega Head. Mineral Information Service (now California Geology), v. 16, pp. 1–9.

_____, 1969. The Geysers. Mineral Information Service (now California Geology), v. 22, pp. 123–128.

Morton, Douglas M., and Robert Streitz, 1967. Landslides. Mineral Information Service (now California Geology), v. 20, pp. 123–129, 135–140.

Oakeshott, Gordon B., 1951. Guide to the Geology of Pfeiffer-Big Sur State Park, Monterey County, California. Calif. Div. Mines and Geology Spec. Rept. 11.

————, and Clyde Wahrhaftig, 1966. A Walker's guide to the Geology of San Francisco. Mineral Information Service (now California Geology), v. 19, supplement.

Pestrong, Raymond, 1972. San Francisco Bay Tidelands. California Geology, v. 25, pp. 27–40.

Rogers, Thomas H., 1969a. Where does the Hayward fault go? Mineral Information Service (now California Geology), v. 22, pp. 55–60.

————, 1969b. A trip to an Active Fault in the City of Hollister. Mineral Information Service (now California Geology), v. 22, p. 159.

————, 1972. Santa Cruz Mountain study. California Geology v. 25, pp. 131–134.

Saul, Richard B., 1967. The Calaveras Fault Zone in Contra Costa County, Calif. Mineral Information Service (now California Geology), v. 20, pp. 35–37.

Smith, Theodore C., and others, 1982. Landslides and related Storm Damage, January 1982, San Francisco Bay Region. California Geology, v. 35, pp. 139–152.

Sullivan, Raymond, and others, 1977. Living in Earthquake Country. California Geology, v. 30, pp. 3–8.

————, 1980. Mount Diablo Coal Field, Contra Costa County, California. California Geology, v. 33, pp. 51–59.

Vredenburgh, L. M., 1982. Tertiary Gold-Bearing Mercury Deposits of the Coast Ranges of California. California Geology, v. 35, pp. 23–27.

Williams, J. W., and Trinda L. Bedrossian, 1977. Coastal zone Geology near Gualala, California. California Geology, v. 30, pp. 27–34.

twelve

GREAT VALLEY

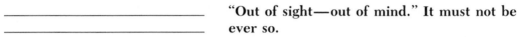

> "Out of sight—out of mind." It must not be
> ever so.
>
> —*Anonymous*

Between the Sierra Nevada and the Coast Ranges is the elongate lowland known as the Great Valley. About 640 kilometers (400 miles) long and 80 kilometers (50 miles) wide, this lowland rises from slightly below sea level to about 120 meters (400 feet) at its north and south ends. The valley is unusual for a lowland because it is a relatively undeformed basin surrounded by highly deformed rock units. The Sacramento River drains the northern portion and the San Joaquin River the southern part. The lowland is also referred to as the Central Valley, its northern segment as the Sacramento Valley, and its southern segment as the San Joaquin Valley. The locations of some major features of the province are shown in Figures 12-1 and 12-2.

GEOGRAPHY

The Great Valley is monotonous geologically, representing primarily the alluvial, flood, and delta plains of its two major rivers and their tributaries (Figure 12-3). The region persisted as a lowland or shallow marine embayment during the entire Cenozoic and at least the later Mesozoic. In the late Cenozoic, much of the area was occupied by shallow brackish and freshwater lakes. This was particularly true in the San Joaquin section, which has had interior drainage in its southern third since the Pliocene. Lake Corcoran, now extinct, spread over much of the northern San Joaquin Valley during the middle and late Pleistocene. Today the only outward drainage is through Carquinez Strait, into San Francisco Bay. The valley's most fertile lands lie at the head of this strait, where the deltas of the two main rivers converge.

412

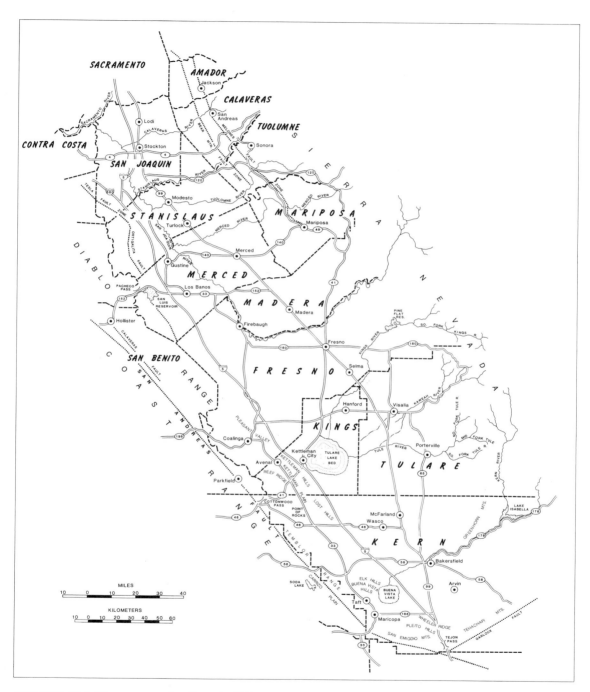

FIGURE 12-1 Place names, San Joaquin Valley.

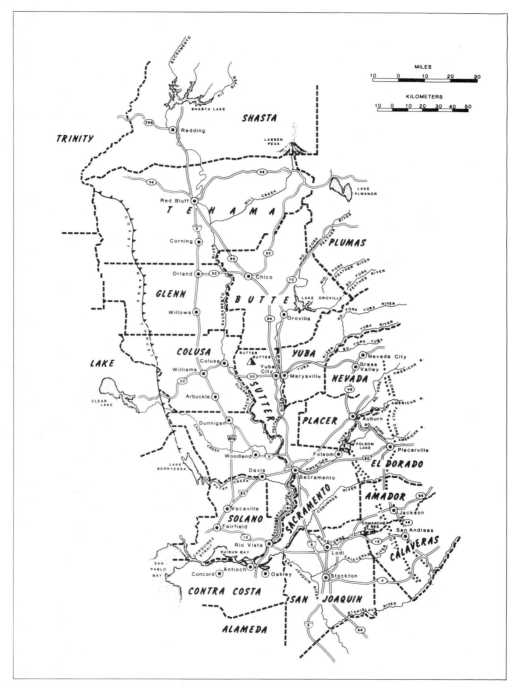

FIGURE 12-2 Place names, Sacramento Valley.

FIGURE 12-3 San Joaquin-Sacramento Delta region. The Sacramento River is the large channel on the upper left and the San Joaquin the one on the lower right. Montezuma Hills border the Sacramento River to the north and the city of Antioch is on the south bank of the San Joaquin to the left of the bridge. These two rivers together account for about 40 percent of the total river discharge in California. Of this, the Sacramento is responsible for 31 percent and the San Joaquin about 9 percent. (Photo by Ames Research Center, National Aeronautics and Space Administration)

Annual rainfall ranges from 120 to 500 millimeters (5–20 inches). The land is well watered, however, because its rivers are fed by the heavy rain and snowfall of the Sierra Nevada. In fact, during severe floods the valley's playas may fill and inundate thousands of acres of crops.

Only two topographic breaks occur on the flat lowland floor: Sutter (Marysville) Buttes in the Sacramento Valley and the Kettleman Hills and other anticlinal arches on the western and southern sides of San Joaquin Valley. Sutter Buttes reach 640 meters (2,100 feet) in elevation, but have an area of only a few square kilometers (Figure 12-4). The Kettleman, Elk, and Buena Vista Hills are outliers of the Coast Ranges and have elevations of about 550 meters (1,800 feet).

FIGURE 12-4 Sutter Buttes from the southwest. The highest peak is South Butte (646 meters or 2,117 feet). The Sacramento River is in the foreground. Sutter County. (Photo by John Burnett, courtesy California Division of Mines and Geology).

FIGURE 12-5 Wheeler Ridge south of Bakersfield, Kern County. Pliocene and Quaternary sedimentary rocks have been arched upward because of strong north-south crustal shortening on the buried Wheeler Ridge fault. The sharp line at the base of the hills in the upper part of the photo is the trace of the Pleito thrust fault. (Photo by John S. Shelton)

In the extreme southern end of the San Joaquin Valley are several very young fold features, most prominent of which is Wheeler Ridge (Figure 12-5). Compression probably associated with development of the Transverse Ranges is now elevating this ridge as well as several smaller nearby features.

ROCKS

The surface of the Central Valley is composed of unconsolidated Quaternary sediments. Where streams have cut channels into these sediments, lake beds are occasionally exposed that include clays, diatomites, and other rocks that can be correlated and mapped. Generally, rock sequences must be inferred from well records and by extension of formations that are exposed on valley margins and then dip beneath the valley floor. Fortunately, information from the thousands of oil and gas wells drilled has permitted a fairly accurate reconstruction of the valley's geologic history. Thousands of water wells have been drilled also, but they can help with only the latest geologic record; water wells rarely exceed depths of 450 to 600 meters (1,500–2,000 feet) and seldom penetrate any pre-Pliocene rocks.

The rock sequences of the Great Valley can be divided into two sections, a division supported by geophysical studies and well records. First is a belt along the west base of the Sierra Nevada; this includes minimally deformed alluvial fans and lake deposits that feather east onto the Sierran basement. Second is a linear belt from 16 to 32 kilometers (10–20 miles) wide at the east base of the Coast Ranges; this is composed of deformed Mesozoic and Cenozoic rocks that dip east beneath the valley. The sedimentary cover of gravels and sands is thinner in the western belt (Figure 12-6).

Composite Geologic Column, San Joaquin Valley

HOLOCENE	Alluvium, dune sand, lake beds
PLEISTOCENE	Corcoran (nonmarine) Tulare (nonmarine)
PLIOCENE	San Joaquin (marine) Etchegoin, Jacalitos (marine), Chanac (nonmarine)
MIOCENE	Santa Margarita (marine) Reef Ridge (marine) McLure (marine) Temblor (marine) Vaqueros, Media, Salt Creek, Leda (marine)
OLIGOCENE	Tumey (marine), Simmler (nonmarine)
EOCENE	Kreyenhagen, Point of Rocks, Domengine (marine) ⎱ Walker
PALEOCENE	Arroyo Hondo, Lodo, Cantua (marine) ⎰ (nonmarine)

FIGURE 12-6 Geologic columns, Great Valley province.

(*Continued*)

Composite Geologic Column, San Joaquin Valley (*Continued*)

CRETACEOUS	Moreno (marine)	
	Panoche Group	Brown Mountain (marine) Ragged Valley (marine) Joaquin Ridge (marine)

JURA-CRETACEOUS	Franciscan (West side basement), Sierran granitic rock, (east side basement)

Composite Geologic Column, Sacramento Valley

HOLOCENE	Alluvium, Modesto
PLEISTOCENE	Riverbank, Montezuma (nonmarine) Arroyo Seco, Red Bluff (nonmarine)
PLIO-PLEISTOCENE	Fair Oaks (nonmarine) Laguna (nonmarine)
PLIOCENE	Tehama, Red Bluff (nonmarine) Mehrten (nonmarine) } Sutter Buttes volcanics
MIOCENE	San Pablo (Cierbo, Neroly) (marine) Valley Springs (nonmarine)
EOCENE	Ione (nonmarine) Markley, Capay, Domengine (marine)
PALEOCENE	Martinez (marine)
CRETACEOUS	Chico (marine)
JURA-CRETACEOUS	Sierran granitic basement

FIGURE 12-6 (*Continued*)

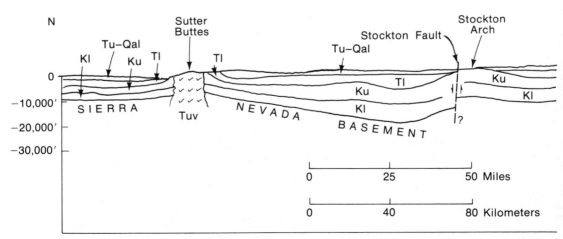

FIGURE 12-7 North-south diagrammatic section of the Great Valley. (Sources: California Division of Mines and Geology and Sacramento Geological Society)

Although generally concealed, the contact between the Sierran and Coast Range basements is presumed to be a fault or a subduction zone contact. Nearer the Coast Range margin, the contact between the two basement types has been definitely located and is exposed in the ranges west of Redding.

Establishing the relationship among the Great Valley rock sequences has been a major geologic detective project. A fairly complete record is now available, however, as indicated by Figures 12-7, 12-8, and 12-9.

STRUCTURE

The Great Valley is an asymmetrical synclinal trough with its axis off center to the west. It is interrupted by two major surface cross structures that had developed by the beginning of the Cenozoic: the Stockton fault in the Stockton arch, and the White Wolf fault in and south of the Bakersfield arch. In the Bakersfield arch, a subsidiary southward extension known as the Tejon embayment crosses the White Wolf fault, which is the approximate southern boundary of the Bakersfield arch. The surface trace of the White Wolf fault was extended nearly 27 kilometers (17 miles) by the 1952 Kern County (Arvin-Tehachapi) earthquake, although the subsurface existence of this segment had been documented previously by well records.

Other prominent faults in the southern San Joaquin Valley include the Kern Front fault north of Bakersfield, known to be slipping or creeping about 11 millimeters ($\frac{1}{2}$ inch) yearly since about 1968 when monitoring equipment was first installed. This fault marks the east side of the Kern Front oil field and is displayed by local low scarps in alluvial materials. The Buena Vista thrust, on the north side of the Buena Vista Hills, is another continuously active fault first discovered about 1930 when compressional movement on it forced oil pipelines out of the ground. Since monitoring began on this fault, it has been moving at about 25 millimeters (1 inch) a year, about double the rate observed on the Kern Front fault. Movement on this thrust has not only buckled buried pipelines out of the ground, but has produced low

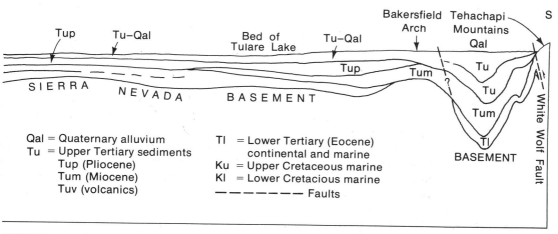

FIGURE 12-7 (*Continued*)

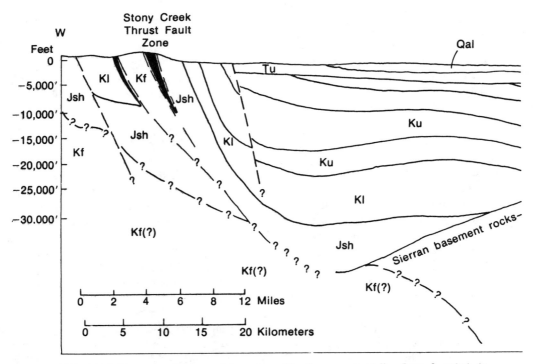

FIGURE 12-8 East-west diagrammatic section of the Sacramento Valley (latitude 39°N). (Sources: California Division of Mines and Geology and Sacramento Geological Society)

scarps, bulged roads, and sheared off several oil wells drilled through the fault plane (Figure 12-10).

Many major valley folds trend parallel to the Coast Range–Great Valley boundary and include some of the most famous oil-producing domes in North America: Elk Hills, Lost Hills, Kettleman Hills, Buena Vista Hills, McKittrick, and Wheeler Ridge. Less well-known are the more recently discovered gas-producing structures near Sutter Buttes, Willows, Dunnigan, Lodi, and Rio Vista.

Along the north flank of the San Emigdio Mountains at the southern end of the San Joaquin Valley there are numerous striking evidences of folding and faulting, which are raising and extending the mountains northward at the expense of the valley.

Anticlinal Wheeler Ridge is rising and growing eastward as shown by deformation of dated Holocene alluvial surfaces and by streams that have been shunted around the end of this rising barrier (see Figure 12-5).

About 20 kilometers (12 miles) west of San Emigdio Canyon, the range front facing the valley has migrated 5 kilometers (3 miles) valleyward in two jumps since the late Pleistocene. The Pleito thrust fault marks the oldest active zone, which is now quiet. At present, most of the activity is on the buried Wheeler Ridge fault—Wheeler Ridge is the surface manifestation of movement on this fault, but river terrace deposits in and near Wheeler Ridge show rapid deformation as well (Figure 12-11). Furthermore, another 3 kilometers (1.8 miles) or so to the north, a new fold is developing

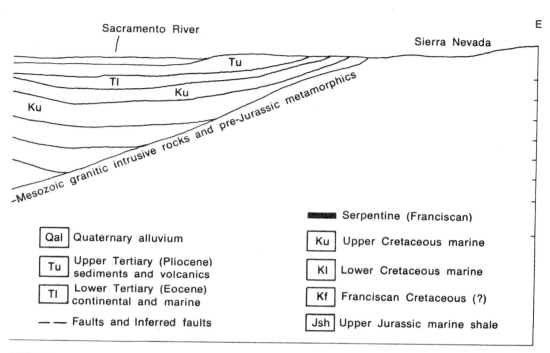

Serpentine (Franciscan)

Qal	Quaternary alluvium
Tu	Upper Tertiary (Pliocene) sediments and volcanics
Tl	Lower Tertiary (Eocene) continental and marine

— — Faults and Inferred faults

Ku	Upper Cretaceous marine
Kl	Lower Cretaceous marine
Kf	Franciscan Cretaceous (?)
Jsh	Upper Jurassic marine shale

FIGURE 12-8 (*Continued*)

in the alluvial fans, more or less parallel to the older belts of deformation. It is likely that these very young folds result from the development of a third, buried thrust fault, and that they will one day mark a new boundary between the San Emigdio Mountains and San Joaquin Valley.

Rates of deformation determined by comparing dated surfaces with elevation changes and with locations of the most active folding give surprising results. It appears that the San Emigdio Mountains have achieved much of their present elevation during the past 2 to 4 million years, having been raised at a rate of 2 to 4 meters (6–12 feet) per thousand years. During the same period, the range has widened northward at the expense of the valley at a rate of 5 to 10 meters (17–33 feet) per thousand years.

As noted in Chapter 11, this geologically recent and active compression in the Transverse Ranges appears to be related to the change in trend of the San Andreas fault where it crosses the province, but the complete story may prove to be much more complex.

The White Wolf fault was responsible for the strong 1952 Kern County earthquake (magnitude 7.6–7.7), the strongest California quake since the 1906 San Francisco event. The White Wolf fault is probably either an extension of the Pleito thrust or the Wheeler Ridge fault, though the connection, if any, is concealed beneath valley fill. Nevertheless, this earthquake and other evidence previously mentioned show that tectonism is very active in the southern San Joaquin Valley.

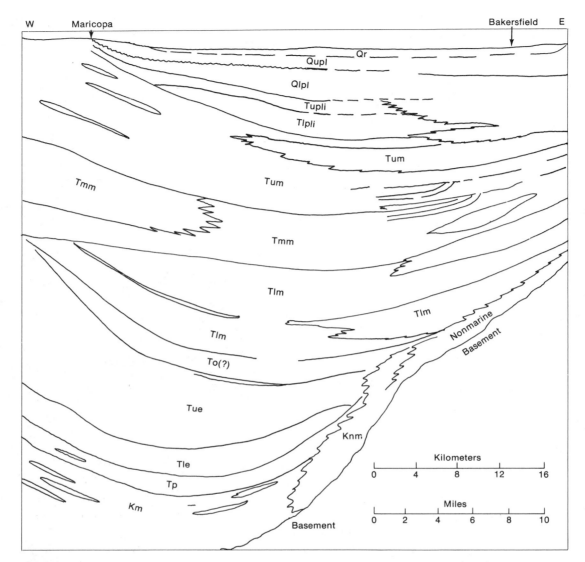

FIGURE 12-9

QUATERNARY
  Qr   = Holocene
  Qupl = Upper Pleistocene (Corcoran)
  Qlpl = Lower Pleistocene (Tulare)
TERTIARY
  Tupli = Upper Pliocene (San Joaquin)
  Tlpli = Lower Pliocene (Jacalitos, Etchegoin, Chanac)
  Tum = Upper Miocene (McLure, Reef Ridge)
  Tmm = Middle Miocene (Temblor)
  Tlm = Lower Miocene (Media, Salt Creek, Leda)
  To   = Oligocene (Tumey)
  Tue = Upper Eocene (Domengine, Kreyenhagen)

Tle = Lower Eocene (Cantua)
Tp   = Paleocene ("Martinez")
CRETACEOUS
  Km = Cretaceous marine
  Knm = Cretaceous nonmarine
Basement
Lenses in section
Interfingering contact,
sometimes with local
unconformities

422

Age	Lithology	Maximum Thickness (feet)
Quaternary	Clay, sand, and conglomerate; buff to gray and greenish-gray color, poorly cemented and poorly sorted; almost entirely alluvial-fan and lacustrine material.	8,000–10,000 (2,440–3,000 m)
Pliocene	Soft greenish-gray claystone and interbedded permeable sands; upper third nonmarine and marine; lower two-thirds marine, particularly in central basin areas; megafossil control.	8,000–9,000 (2,440–2,750 m)
Miocene	Brown and gray clay shale and hard siliceous shale, with numerous permeable sandstone and conglomeratic sandstone members; marine with foraminifers, diatoms, and megafossils, except uppermost and basal nonmarine members along eastern and southeastern borders; basaltic and andesitic flows and intrusions in lower part along southeastern borders.	12,000–13,000 (3,650–4,000 m)
"Oligocene" and Upper Eocene	Gray and brown shale and hard siliceous shale with some thin and thick permeable sands in local border areas; marine with foraminifers and megafossils, except for red and green nonmarine beds in "Oligocene" of eastern and southeastern border.	8,000–9,000 (2,440–2,750 m)
Lower Eocene and Paleocene	Gray shale with some sands that become thick and very permeable, particularly in Coalinga and southern border areas; marine, with foraminifers and megafossils.	5,000–6,000 (1,500–1,800 m)
Upper Cretaceous	Upper part weathered to purple, with dark gray siliceous and calcareous foraminiferal shale and clay shale with local sands; middle and lower parts massive thick concretionary sandstone, conglomerate, and dark shale with intercalated sandstone, marine.	25,000* (7,600* m)

FIGURE 12-9 (Continued)

FIGURE 12-10 Active crustal shortening on the Buena Vista thrust, Kern County. Movement on the thrust has forced the buried pipe to buckle and to be raised out of the ground. (Photo by John S. Shelton, *Geology Illustrated*, 1966, p. 413, courtesy W. H. Freeman and Company)

FIGURE 12-11 Deformed stream terraces, Los Lobos Canyon, looking south toward the San Emigdio Mountains, Kern County. A zone of active shortening crosses the stream at right angles at the dark patch of vegetation in the lower left corner of the photo. This compression or crustal shortening has caused the stream terraces to slope upward toward the observer. (Photo by R. M. Norris)

Persistent tectonic activity affected much of the Great Valley during the Cenozoic and is shown by the numerous unconformities that occur in the deposits that underlie the valley margins. In the Sacramento Valley, four or five erosional events separate Cenozoic deposits from one another and at least that many are known from the edges of the San Joaquin Valley. In this sense, the geologic history of the Great Valley is similar to the history of Coast Range valleys where deposition in the central parts of the valleys continued with little interruption during most of the Cenozoic, but where deposition was frequently separated by tectonic spasms and erosion along the margins of the bordering ranges.

GEOLOGIC HISTORY

Situated as it is between the Sierra with its two great orogenies and the presently active Coast Ranges, the Great Valley has a rather complex history. In some respects it is surprising that the Central Valley has persisted as a recognizable unit.

The valley's history begins with late Jurassic sedimentation from the east and northeast, derived from the rising Nevadan Mountains. Presumably restricted by the western shelf edge, the pattern of deposition was typically shallow water except for the absence of limestone. Concurrently, a deeper western basin—the slope and deep-sea floor—received great thicknesses of deposits, including volcaniclastics, volcanic flows, and intrusives, plus other clastic materials—the subduction complex. These are all represented today by the Franciscan Formation of the Coast Ranges. It is possible that some of the shallow-water Great Valley beds and the Coast Range Franciscan rocks were deposited in contiguous areas, but it seems likely that some of the Great Valley shelf deposits and the Franciscan slope and trench deposits, now separated only by a fault zone, originally were deposited hundreds of kilometers from one another and are now juxtaposed only because of long-continued strike-slip faulting.

Deposition continued, with Cretaceous shallow-water sediments accumulating from continued erosion of the Nevadan Mountains, especially in the northern regions. Cretaceous rocks are less evident in the south, possibly reflecting more distant or less well-developed source areas. The Mesozoic sedimentary rocks usually show great continuity and uniformity and are often easily recognized in the field even when separated by several kilometers.

The Cenozoic was initiated by deformation that was regional in the north and localized in the south. Intense local deformation apparently continued throughout the San Joaquin Valley region, where extremely thick marine sections accumulated during the Miocene. These sections were often highly localized, implying the presence of deep, narrow seaways (possibly block-faulted valleys or pull-apart basins) extending to the Pacific through embayments and straits across the site of the Coast Ranges. These linear structures may reflect extension of Basin and Range block faulting into the Coast and Transverse ranges during the middle and late Cenozoic, or they may have been produced by the sort of spreading center rifting, which is currently widening the Gulf of California.

The floor of the major basin sank and was concomitantly filled, although no significant deep-water areas ever developed. Nevertheless, localized nearly abyssal depths are indicated by some microfaunal associations in the southern San Joaquin

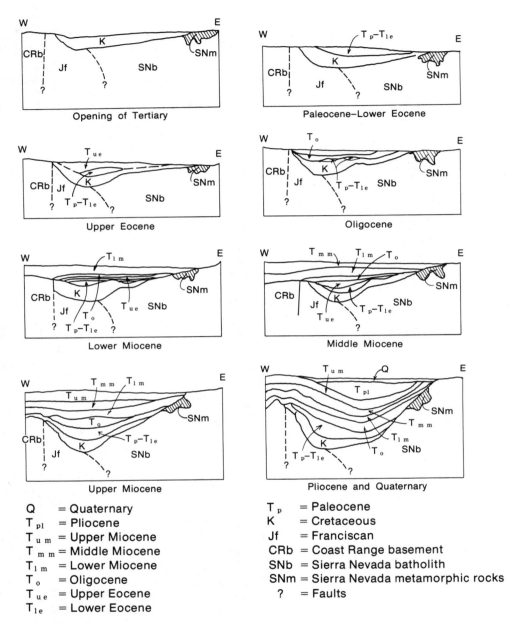

Q = Quaternary
T_pl = Pliocene
T_um = Upper Miocene
T_mm = Middle Miocene
T_lm = Lower Miocene
T_o = Oligocene
T_ue = Upper Eocene
T_le = Lower Eocene

T_p = Paleocene
K = Cretaceous
Jf = Franciscan
CRb = Coast Range basement
SNb = Sierra Nevada batholith
SNm = Sierra Nevada metamorphic rocks
? = Faults

FIGURE 12-12 Evolution of sedimentation in southern San Joaquin Valley. Thicknesses of sediment increased as the Tertiary advanced, with depression of basement and concomitant development of east-west arches and domes. Although the sections are for the southern San Joaquin, they represent conditions throughout the Great Valley. Note that the contact between Franciscan and plutonic basement is not defined, although it is presumed to be an east-dipping fault. (Source: California Division of Mines and Geology)

Valley. Simultaneously, deformation continued until marine waters were expelled completely, presumably because of the increasing intensity of the Coast Range orogenic activity. By Pliocene time, most of the valley's seas were drained, via Carquinez Strait. Brackish and freshwater lakes replaced marine waters, and the Central Valley assumed its present form.

Figure 12-12 gives the inferred profile of Great Valley basement throughout the Tertiary, showing a continually sinking basin being filled with sediment. Increased deformation is related to accelerating strike-slip movement on the San Andreas fault and compressional forces from the rapidly rising Coast and Transverse ranges.

SUBORDINATE FEATURES

Sutter (Marysville) Buttes

A prominent monolithic cluster of steep-sided pinnacle-like hills covering about 25 square kilometers (10 square miles) breaks the topographic monotony of the Sacramento Valley near Marysville (Figure 12-4). Rising over 600 meters (2,000 feet) from the valley floor, this pile of volcanic rocks is the only major igneous outcrop anywhere in the Great Valley.

What forces drilled these vents? Why this location? Why only one such volcanic center? The buttes are plug domes with a core of andesite porphyry, and an outer ring of intrusive rhyolite. In the earlier stages, magma was apparently injected slowly and in pasty form, although there are some breccias and tuffs that suggest mildly explosive activity. The andesite was subsequently intruded by the rhyolite. Late Cretaceous and Eocene sedimentary rocks are included as xenoliths in the volcanic rocks, and the Plio-Pleistocene Tehama Formation is upwarped around the buttes, showing that volcanic activity followed deposition of the Tehama Formation, ripping off chunks of the older sedimentary rocks buried below to form the xenoliths.

The position of these intrusions has long intrigued geologists, but explanations have been inconclusive thus far. The Tehama Formation underlies the Tuscan Formation, which is composed of extensive volcanic flows and volcaniclastics widely distributed across the northern end of the Sacramento Valley. Are Sutter Buttes in fact the southernmost of the Cascade volcanos? Or are they unrelated pulsations that coincidentally broke the valley floor? The buttes take on added importance because their volcanic domes and plugs have deformed the underlying sedimentary rocks, producing traps for considerable natural gas.

Kettleman Hills

The Kettleman Hills hug the western edge of the San Joaquin Valley, extending south from near Coalinga (Figure 12-13). They are usually considered an outlier of the southern Coast Ranges, but they also represent a clear interruption of the floor of the Great Valley. At their north end, they stand about 450 meters (1,500 feet) above the valley. From here they stretch south for 48 kilometers (30 miles), descending in elevation and finally merging with the valley floor. Almost a perfect elongate dome, the Kettleman Hills are divided into three segments, North, Central, and South domes. The hills are about 8 kilometers (5 miles) wide.

The Kettleman Hills lie in the driest part of the San Joaquin Valley. Consequently,

FIGURE 12-13 The Kettleman Hills anticlinal dome, Kings and Fresno counties. The city of Avenal is in the lower center. (Photo by Fairchild Aerial Surveys, courtesy Map and Imagery Laboratory, Library, University of California, Santa Barbara)

since their surface rocks are weakly consolidated, drainage patterns have produced some badland areas where the sparse vegetation has been destroyed. The hills are the largest of a series of folds in this area, nearly all of which have been important petroleum and gas producers. By 1986, North and Middle domes had produced a little more than 450 million barrels (64 million metric tons) of oil.

The anticlinal folds at Kettleman Hills are very young structural features. The freshwater Tulare Formation of Plio-Pleistocene age, together with the older Cenozoic deposits, participated in the folding. It thus follows that all the oil and gas they held must have migrated into the reservoir rocks in the last million years or so.

Oldest rocks exposed in the Kettleman Hills belong to the Pliocene Etchegoin Formation. This unit and the overlying San Joaquin Formation were deposited for the most part in shallow, near-shore waters and both are quite fossiliferous. Further, older rocks below the surface in the Kettleman Hills, but well exposed in the nearby Coast Ranges to the west, are also notably fossiliferous and include the early and middle Miocene Temblor Formation with its unique association of sea cow (*Desmostylus*) and horse (*Merychippus*) fossils, demonstrating that it was deposited very close to shore. The late Miocene Santa Margarita Formation also contains many

FIGURE 12-14 *Ostrea titan*, a very large late Miocene oyster from the Santa Margarita Formation, Diablo Range, Fresno County. (Photo by R. M. Norris)

fossils, among which are the reefs composed of the giant oyster (*Ostrea titan*), some specimens of which are more than 15 centimeters (6 inches) long. The Coalinga-Kettleman Hills area is an excellent place to collect Tertiary fossils, but extreme caution must be exercised because of the danger of contracting Valley Fever (Coccidioidomycosis) from the spores that are abundant in the dry, dusty soils of the area (Figure 12-14).

SPECIAL INTEREST FEATURES

Natural Gas Fields

The Great Valley has produced trillions of cubic meters of natural gas since gas was first consumed in Stockton in the 1850s. The Rio Vista gas field alone had produced more than 90 trillion cubic meters (3,200 trillion cubic feet) since its discovery in 1936 (1986 figures).

Natural gas is classified into two categories, wet or dry, with most oil fields producing wet gas. After its more valuable components like ethane, butane, and propane have been extracted, wet gas is often pumped back into the wells to recharge pressure and increase petroleum yield. Most gas thus recycled reappears with oil produced subsequently from the wells. Dry gas is produced from porous and permeable rock reservoirs where oil is absent. Dry gas is primarily methane and is usually too low in other gases to make further treatment profitable, so it is piped directly to customers.

Enormously important gas fields occur north of the Stockton arch and extend almost 320 kilometers (200 miles) to the northern end of the valley (Figure 12-15). First discovery of commercial dry gas was in 1933 at Sutter Buttes, followed by discovery of the McDonald Island and Rio Vista fields in 1936. Dry gas was produced

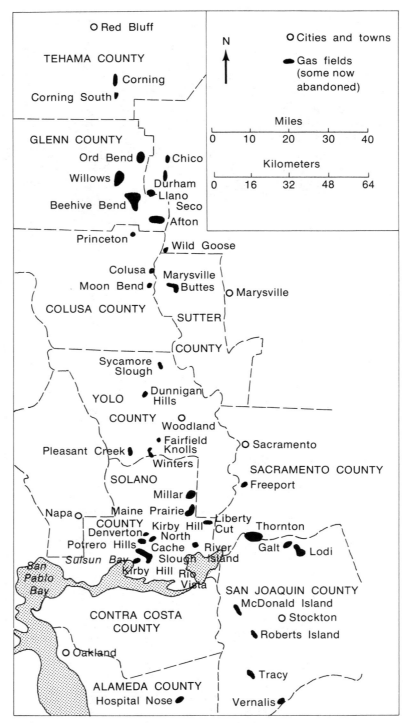

FIGURE 12-15 Gas fields of the northern Great Valley. (Source: California Division of Mines and Geology)

in the San Joaquin region as early as 1909, and some prolific fields were discovered at Elk Hills in 1919 and at Buttonwillow in 1927 (Figure 12-16). It is the Sacramento Valley, however, that dominates California in gas production.

The gas fields of the Sacramento differ from those of the San Joaquin Valley by abnormally high reservoir pressures, almost complete absence of associated petroleum, and high production from Cretaceous formations. The reasons for these discrepancies are not yet known, although differences in parent organic material may be the answer.

Subsidence in the Great Valley

Since 1940, agricultural development of the arid San Joaquin Valley has placed a considerable strain on the water resources. Although plentiful water actually enters the valley via Sierran rivers, inadequate distribution has necessitated several major irrigation projects. Each has been plagued with massive subsidence along its distributary canals, partially as a consequence of overpumping and overuse of ground water.

Subsiding regions include the San Joaquin-Sacramento deltas where agricultural development has occurred in peat lands, the overpumped artesian basins in the valley's western and southern sections, and in places compacted by wetting of moisture-deficient soils by irrigation. Subsidence as much as to 9 meters (28 feet) has occurred in some areas of overpumping, but 3 to 5 meters (10–15 feet) are typical of the delta. In one area on the west side of the San Joaquin River near Los Banos, the ground was subsiding at a rate of $\frac{1}{2}$ meter (1.8 feet) per year, but this rate has slowed with the importation of water from the north. Recharge of the underground reservoirs, where attempted, has not produced rebound toward original levels (Figure 12-17).

Subsidence creates major problems; several areas of 6 to 10 kilometers (4–6 miles) wide and from 30 to 110 kilometers (20–70 miles) long have been affected. A technique to help reduce subsidence has developed in the canal areas, however. Suspect parts of an excavated canal are divided into small segments by earth-fill barriers, and the depressions are then filled with water that is maintained at a certain level until stability is achieved. The trench is then regraded and the canal developed. In an extreme case, subsidence up to 5 meters (15 feet) occurred before stability was finally achieved 10 to 18 months later.

About 13,500 square kilometers (5,200 square miles) of the San Joaquin Valley have subsided more than 25 centimeters (1 foot) since the 1920s. Expressed as a volume, this subsidence is about equivalent to half the volume of the Great Salt Lake of Utah.

Pleistocene Lakes

The Pleistocene lakes of the San Joaquin Valley are all extinct and today form playas that have been temporarily reclaimed for agriculture. The lakes were concentrated south of the Stockton arch and generally occupied the western side of the valley floor.

Lake Corcoran. Lake Corcoran occupied approximately the western half of the San Joaquin Valley, from the Stockton arch south to the area where the San Joaquin

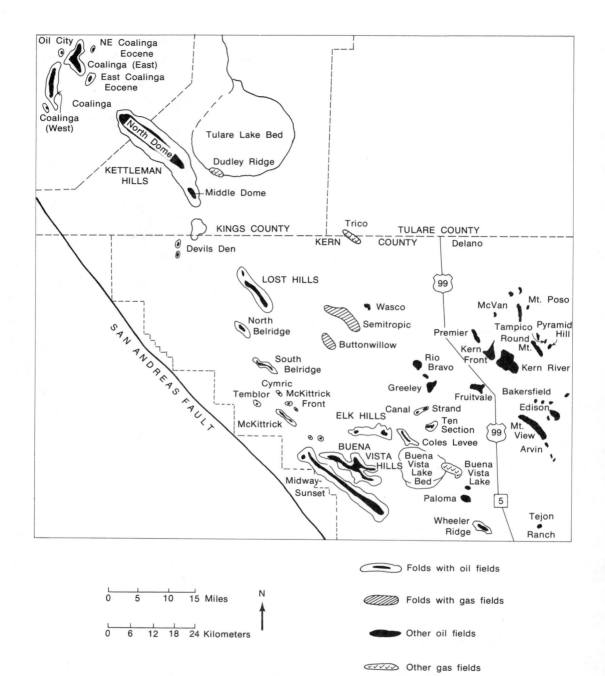

FIGURE 12-16 Some structures producing petroleum and natural gas in the San Joaquin Valley. (Source: California Division of Mines and Geology)

FIGURE 12-17 Ground subsidence causing roadway to sink and fracture caused by compaction of arid climate soils following irrigation, southwestern Kern County. (Photo by Stanley E. Karp)

River turns abruptly north (Figure 12-1). The lake had its high stand about 600,000 years ago and has left lake-bed clays, diatomite, and other deposits. Sand dunes are preserved in places along the old shoreline notably adjacent to Kettleman Hills. The extreme flatness of this part of the valley floor is due to the ancient lake bottom.

Lake Tulare. The floor of the valley south of the San Joaquin River is the playa of Lake Tulare. Were it not for irrigation diverting the waters of the Kings and other rivers, in years of heavy Sierran precipitation the playa would become an extensive, shallow body of water because its basin is the normal terminus for drainage of these rivers. In fact, until about 1920, Tulare Lake was the largest natural lake (in surface area) in California. At present, the playa is cultivated for agriculture, but every decade or so, in unusually wet years when runoff in the Kings River is very great, large-scale flooding of these farmlands occurs and Tulare Lake (at least briefly) is reincarnated.

Buena Vista and Kern Lakes. Buena Vista Lake was formerly the terminus of the Kern River. The Kern has one of the largest drainages in the Sierra, and the lake persisted as a permanent feature until the river was controlled by the dam at Isabella. This, plus irrigation and domestic diversion, dried Buena Vista Lake, and its playa too is now used for farming. Water rarely stands in the lake basin today except in years of heavy local rainfall. The drainage divide between the Tulare and Buena Vista basins is low, and occasionally the two lakes have formed a single shallow water body. Evidence suggests that late Pleistocene drainage was south into Buena Vista Lake when glacial meltwater entered Tulare Lake from the Kings River. Buena Vista, in turn, overflowed into smaller Kern Lake, which extended southeast from Buena Vista into the southernmost San Joaquin Valley.

The Delta. To most Californians, the Delta means the low, triangular area lying generally between Sacramento on the north, Stockton on the south, and Suisun Bay on the west. Nearly one half of the total river volume of California flows through the Delta. It is the area where the Sacramento and San Joaquin rivers join, and before reclamation and development of agricultural lands occurred early in this century, it was a vast malarial swampland of tules (reeds), interlacing river channels, and low islands underlain by thick deposits of peat and muck. Perhaps the earliest attempts at reclamation were made in 1868, when the railroad lines were pushed across the delta country using Chinese labor.

It may properly be considered a delta, though quite an atypical one. Unlike most deltas, it lies far inland from the coast and grows not seaward, but rather landward as sediments largely from the Sierra Nevada are deposited around its margins in times of high sea level, as at present. The Delta shows a pattern of growth and expansion during high sea levels and contraction during low ones.

Because of the proximity of the Sierra, the Delta is one of the few places in the world where glacially derived deposits merge with marine deltaic deposits, providing an unexcelled opportunity to relate glacial advances directly with sea-level changes.

As was noted, the peat lands in the Delta have experienced considerable subsidence since they were cleared for agricultural use. In a number of places the islands now lie as much as 7 meters (23 feet) below sea level, necessitating protective dikes to keep the adjacent streams from flooding these rich agricultural lands. Further, the levee system begun with Chinese labor in the last century is vulnerable to erosion, and some believe parts of the levees might fail in an earthquake. Such a failure would not only inundate farm lands, but it might break the Hetch Hetchy aqueduct that brings Sierran water to San Francisco, and would, of course, damage roads and rail lines as well. State Route 160 follows one of these dikes along the main channel of the Sacramento River from Antioch north to Sacramento, giving a fine view of the elevation difference between the river and the adjacent land.

Subsidence in these deltaic peat soils results from several factors. Oxidation of the peat results in a loss of volume, not fundamentally different than if the peats were set afire and burned, which has also happened. Second, winds pick up the fine materials and blow them away, and loss of water following reclamation has resulted in a volume decrease.

Though subsidence certainly imperils Delta lands, many levees have failed because of high water in the distributaries. These conditions usually result when periods of high runoff from the Sierra coincide with high tides in the ocean that act to back up the water in the distributary channels.

Recent studies have revealed that a series of buried and filled river channels lie below the present delta surface, demonstrating that the present pattern of islands and channels is only the most recent configuration. Earlier channels and islands related to Pleistocene high sea levels during interglacial times had a very different geometry.

These buried channels and interbedded glacial gravels and peat deposits all testify to a Quaternary history of subsidence—some of the peat deposits lie as much as 18 meters (60 feet) below the present land surface.

REFERENCES

General

Hackel, Otto, 1966. Summary of the Geology of the Great Valley. *In* Geology of Northern California. Calif. Div. Mines and Geology Bull. 190, pp. 217–238.

Poland, J. F., and R. E. Evenson, 1966. Hydrology and Land Subsidence, Great Central Valley, California. *In* Geology of Northern California. Calif. Div. Mines and Geology Bull. 190, pp. 239–247.

Smith, M. B., and F. J. Schamback, 1966. Petroleum and Natural Gas, *In* Mineral and Water Resources of California, Part 1, Committee on Interior and Insular Affairs, Eighty-ninth Congress, Second Session, pp. 291–328.

Woodring, W. P., and others, 1940. Geology of the Kettleman Hills Oil Field, California. U.S. Geological Survey Prof. Paper 195.

Special

Christian, Louis B., 1970. Ancient Windblown Terrains of Central California. Mineral Information Service (now California Geology), v. 23, pp. 175–179.

Finch, Michael, 1985. Earthquake Damage in the Sacramento-San Joaquin Delta. California Geology, v. 38, pp. 39–44.

Frink, J. W., and H. A. Kues, 1954. Corcoran Clay—A Pleistocene Lacustrine Deposit in San Joaquin Valley, California. Bull. Amer. Assoc. Petroleum Geologists, v. 38, pp. 2357–2371.

Manning, John C., 1973. Field Trip to Areas of Active Faulting and Shallow Subsidence in the Southern San Joaquin Valley. Far Western Sect. National Assoc. Geology Teachers Guidebook.

Meehan, J. F., 1973. Earthquakes and Faults Affecting Sacramento. California Geology, v. 26, pp. 32–36.

Newmarch, George, 1981. Subsidence of Organic Soils, Sacramento-San Joaquin Delta. California Geology, v. 34, pp. 135–141.

thirteen

SAN ANDREAS FAULT

Civilization exists by geological consent . . .
subject to change without notice.

—*Will Durant*

With a total length of more than 1,200 kilometers (740 miles), the San Andreas is the longest fault in California and one of the longer in North America (see Figure 1-1). It extends 960 kilometers (600 miles) through western California, cutting indiscriminately across rock and structural boundaries (Figure 13-1).

The San Andreas fault initially came to world attention because of the great 1906 San Francisco earthquake, which was caused by sudden right-slip movement on the fault of up to 4.9 meters (16 feet). Subsequent investigation of the San Andreas established California as an internationally famous center for the study of structural geology and seismology. Since 1906, dozens of earthquakes have occurred on the San Andreas. Moreover, sizable tremors are continually predicted, even though earthquake prediction is not yet possible in the sense apparently expected by the lay public. We can predict the location of impending earthquakes with reasonable success, however, but our ability to pinpoint the exact time is still lacking.

The population explosion and concentration of cities on the coastal block of western California make understanding of the San Andreas critically important. Furthermore, the San Andreas is just one of several faults with high potential for sudden slippage. The programs necessary to promote understanding of these faults must be implemented primarily at the state level, however. Federal legislators are likely to be unimpressed by requests for substantial appropriations for earthquake studies, especially when California and Alaska seem to be the principal beneficiaries. Nevertheless, a program was initiated after the 1964 Alaskan earthquake and is presently directed by the U.S. Geological Survey.

The San Andreas fault was not recognized as a continuous regional structure until after the 1906 San Francisco quake, although faulting as a cause for topographic

436

FIGURE 13-1 Looking southeast at the San Andreas fault (center) from near Palmdale. The San Gabriel Mountains are in the right background. East of the fault, in the far left, are Mount San Gorgonio and the San Bernardino Mountains. The San Jacinto fault crosses the San Gabriel Mountains through the first notch on the skyline west (right) of the San Andreas notch. Palmdale Reservoir is in the center foreground, with the Sierra Highway between the reservoir and a smaller pond across the road. These are both sag ponds on the San Andreas, but the volume of the larger one has been increased artificially from its original size. (Photo by Spence Air Photos, courtesy of Department of Geography, University of California, Los Angeles)

expression had been recognized in California as early as 1891. The name San Andreas was apparently first employed by A. C. Lawson in 1895, after other geologists had used various names for discrete sections of the single feature now considered the San Andreas fault. The California Earthquake Commission's 1908 report used the term *San Andreas rift*, and from this evolved the usage of San Andreas fault, San Andreas (fault) zone, and San Andreas system.

Generally, the San Andreas fault is defined as the principal surface rupture produced by recent movements within the San Andreas zone. This zone incorporates many nearly parallel features, few more than 18 kilometers (11 miles) long, and most confined to a belt a few kilometers wide. Many segments record movements older than those of the present fault. The system also contains faults that are subparallel, but not directly connected superficially to the San Andreas proper. Some examples are the San Jacinto and San Gabriel faults in southern California and the Hayward and Calaveras in northern California. Rift is normally used as a geomorphic term meaning *a narrow cleft or fissure*, but may also be used to encompass all the features of a fault zone, both erosional and structural.

Although important earthquakes have been associated with sudden slippage on the San Andreas fault, the record is incomplete and too short historically to have much geologic significance. California's first recorded earthquake was reported in 1769 by Gaspar de Portola and was felt while he camped near the Santa Ana River. The state's first seismographs were installed at the University of California, Berkeley, and at Lick Observatory in 1887; the 1906 earthquake was the first on the San Andreas fault for which a seismogram is available. Earthquakes on the Hayward fault were reported on June 10, 1836, October 8, 1865, and October 21, 1868, when a maximum of 1 meter (3 feet) of right slip was observed. Earthquakes on the San Andreas fault were reported in June 1838, on April 24, 1890, and on April 18, 1906 (the San Francisco disaster). The 1906 earthquake measured 8.25 on the Richter scale.

The Richter scale is a numerical, logarithmic scale devised by Charles F. Richter in 1935 to express the amplitude of the largest trace recorded by a standard seismograph 100 kilometers (60 miles) from the epicenter. Tables have been devised to express a magnitude of any earthquake from any seismogram. For a 1-unit increase in magnitude, there is an increase of about 30 times in released energy. The Richter scale defines earthquakes of 7.0 to 7.75 as major and more than 7.75 as great. Earthquakes of magnitude 2.0 may be felt, and those of 4 to 5.5 cause local damage. The San Francisco and Fort Tejon (1857) earthquakes were probably about the same magnitude. The 1872 Owens Valley earthquake (on the Sierra Nevada fault) may have been of higher magnitude, on the basis of the area affected, the intensity of its effects and possibly the amount of ground breakage, but no seismographic records were available to permit an accurate judgment.

In the Fort Tejon earthquake of 1857, the San Andreas fault showed surface rupture for a distance of at least 350 kilometers (220 miles). Major surface breaks occurred as far as 160 kilometers southeast and northwest of the fort. No other significant quakes have been recorded on the southern California section of the San Andreas proper, but many have occurred on faults that may yet be established as parts of the San Andreas system.

On land, the San Andreas fault extends south from Shelter Cove, Humboldt County, across the Golden Gate under western San Francisco, and southeastward through the Coast Ranges (where it is straight and narrowly confined) to the intersection with the Garlock fault near Frazier Mountain. Southeast of the Garlock fault, the San Andreas approximates the boundary between the Mojave Desert and Transverse Range provinces. The fault separates the San Gabriel and San Bernardino mountains at Cajon Pass, and as the southern boundary of the San Bernardino Mountains it forms the San Gorgonio Pass into Coachella and Imperial valleys.

One of the more remarkable characteristics of the San Andreas fault is the tendency for each new displacement to follow almost the exact trace of an earlier displacement. In some instances, the record shows that displacements have occurred repeatedly on the same breaks for periods of 3,000 years to perhaps as many as 10,000 years. But the most recently active zone of the San Andreas is seldom more than 0.8 kilometer ($\frac{1}{2}$ mile) wide and is often much less. An especially notable feature is the sharpness of the fault's definition (Figure 13-2). One cannot help but wonder how this tendency to reoccupy old breaks squares with the obvious fact that the fault

FIGURE 13-2 San Andreas fault on the Carrizo Plain. Almost every stream course in the Carrizo Plain shows right-lateral offset. The dark line on the left is made by tumbleweeds collected against a fence. (Photo by Robert E. Wallace, courtesy of U.S. Geological Survey)

zone includes many other strands in a belt sometimes several kilometers wide. If each new displacement occurs on the same strand, how and under what circumstances were the other strands developed?

GEOMETRY

Surface Geometry

Certain aspects of its plane geometry have been emphasized by describing the San Andreas as a seam, but the fault's planar features are actually far more complex. The San Andreas fault frequently consists of many closely spaced braided or anastomosing fractures, all previously surfaces of movement that often had slip different than today's. Various sections of the fault have features that are strikingly dissimilar from one another. At places this reflects the rocks involved, but at others the patterns seem to disregard rocks and preexistent structure completely.

Structural Knot or the Big Bend.

The Frazier Mountain. This intersection of the Garlock, Big Pine, San Gabriel, and San Andreas faults has been described as the Big Bend or structural knot of southern California. When the overall tectonics of the southern San Andreas fault are eventually worked out, key ingredients certainly will be found here accounting for the cause of the Big Bend and the two major left-slip faults (the Big Pine and Garlock) that terminate here. A better understanding of the other east-west trending segment of the San Andreas in the Cajon-San Gorgonio area and the left-slip east-west trending Malibu Coast-Raymond Hill-Cucamonga alignment also will figure prominently in the final structural interpretation of southern California (Figure 13-3).

The origin of the Big Bend has been related to left slip on the Big Pine and Garlock faults and to movement on the Frazier Mountain thrust and related faults. The Frazier thrust is a short fault that dips north and northwest beneath Frazier Mountain. It has forced Proterozoic banded gneiss into a high, small massif that cuts across folded Pleistocene nonmarine sedimentary rocks. Frazier Mountain is interpreted as a mountain without roots—a squeeze-up block of previously buried Proterozoic rock, a splinter that rose up and out from the San Andreas fault zone because of rotational compression. As the block moved south, it was thrust over the weak underlying younger formation. Similar thrust slices occur elsewhere along the San Andreas zone.

Mendocino Triple Junction. This feature, one of the earliest of its type recognized as the theory of plate tectonics was developed, is the locus of the junction of three crustal plates. These are the Juan de Fuca to the north, the North American to the

FIGURE 13-3 Frazier Mountain, Ventura County, seen from the northeast. The trace of the Garlock fault strikes toward the mountain from the lower center of the photo, intersecting the San Andreas fault that crosses the photo at the base of Frazier Mountain approximately along the stream valley. The San Gabriel fault strikes southeasterly from the far side of Frazier Mountain. Frazier Mountain lies in what has been characterized as the "structural knot" of southern California. (Photo by John C. Crowell)

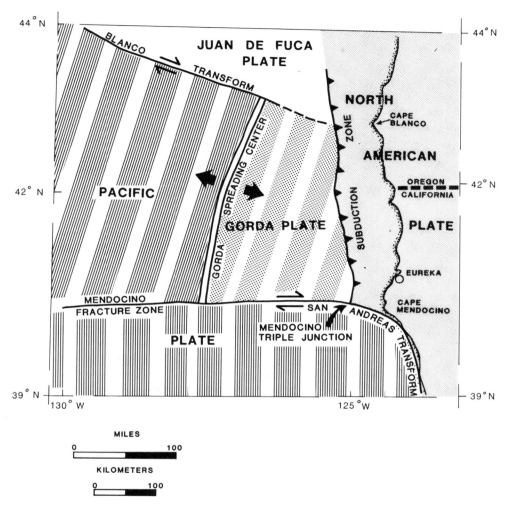

FIGURE 13-4 Diagrammatic representation of the Mendocino Triple Junction where the Pacific, Gorda, and North American crustal plates meet.

east, and the Pacific to the west (Figure 13-4). To the south the San Andreas fault is undergoing strike slip (transform displacement), and to the north, the Juan de Fuca plate appears to be moving obliquely southeast under the North American plate. In this interpretation, the San Andreas fault curves westward joining the great Mendocino fracture zone, which has been traced more than 1,600 kilometers (1,000 miles) westward.

The northward motion of the Pacific plate on the west side of the San Andreas is gradually affecting the southeast-moving Juan de Fuca plate over which it is riding. Faulting, forming ahead of the advancing Pacific plate is recognized as much as 120 kilometers (75 miles) northwest of the present junction. In addition, the subduction of the Juan de Fuca plate is regarded as the cause of volcanism in the Cascade Range from northern California to southern British Columbia.

In this interpretation, the San Andreas is a long transform, or a fault displacing two active spreading centers. One of these is north of Cape Mendocino and the other lies buried beneath the southern end of the Salton Sea. Shorter transforms offset these spreading centers both north and south of the ends of the San Andreas fault.

Gulf of California. The San Andreas fault disappears almost on the Riverside-Imperial county line near a railroad siding called Bertram. As was noted, the southern end of the San Andreas fault appears to terminate at a spreading center or divergent plate boundary beneath the southern end of the Salton Sea, although some older faults continue the trend of the San Andreas as far as Yuma. This spreading center is also the northernmost segment of the East Pacific Rise, segments of which, each offset by short transforms, extend southerly through the Gulf of California (Figure 13-5). Similarly, other southern California members of the San Andreas system, such as the Brawley, Imperial, San Jacinto, and Elsinore faults are mapped as transforms that offset the East Pacific Rise from the Salton Sea southward into the Gulf of California.

The northwesterly offset observed on these transforms played a role in the opening of the Gulf of California about 5 million years ago as Baja California and at least the Peninsular Ranges were rafted northwesterly away from the North American plate as a result of the underlying Pacific plate drift.

Vertical Geometry. Little is known about the San Andreas at depth. Only within the last 20 years have studies been initiated to define the San Andreas zone at depth. Defining the vertical geometry of any fault normally requires careful study of seismic records and possibly extensive drilling, maintenance of well logs, and collection of cores. Recent studies indicate that earthquakes are generated from as litle as 7 kilometers (4.3 miles) depth in the Imperial Valley to as many as 20 kilometers (12 miles) in the San Bernardino Mountains. An average depth would be about 14 kilometers ($8\frac{1}{2}$ miles). In the central Coast Ranges, for example, recent studies provide evidence that fault gouge extends downward to a depth of about 18 or 19 kilometers (11 – 12 miles). Accordingly, the San Andreas fault is customarily presented as a fault of dominately vertical dip. In a few places, such as in the San Emigdio Mountains of southern Kern County, deep erosion has provided a good cross section of the San Andreas fault, allowing us to examine it at greater depth than is usual elsewhere. This is in the Big Bend area where the fault zone has a strong east-west trend. This sort of deviation is often called a *restraining bend* because the normal northwesterly movement of the coastal slice of California west of the fault presses the westerly block firmly against the continental block on the opposite side (Figure 13-6); this greatly inhibits slip. Further, this is the region (as noted earlier) in which there is much evidence of strong crustal shortening on both sides of the fault.

Deep erosion along the fault zone reveals a fanlike series of rock slices converging downward. Compression across the fault has squeezed the slices upward, and the highest parts of some of the slices form nearly flat-lying thrusts that merge and steepen at depth. A mechanism such as this may account for the fact that most of the San Andreas fault is associated with hilly topography that stands above the surrounding terrain. Even where the fault crosses the relatively flat Carrizo Plain in eastern San Luis Obispo County, the fault is marked by a linear welt (see Figure 13-2).

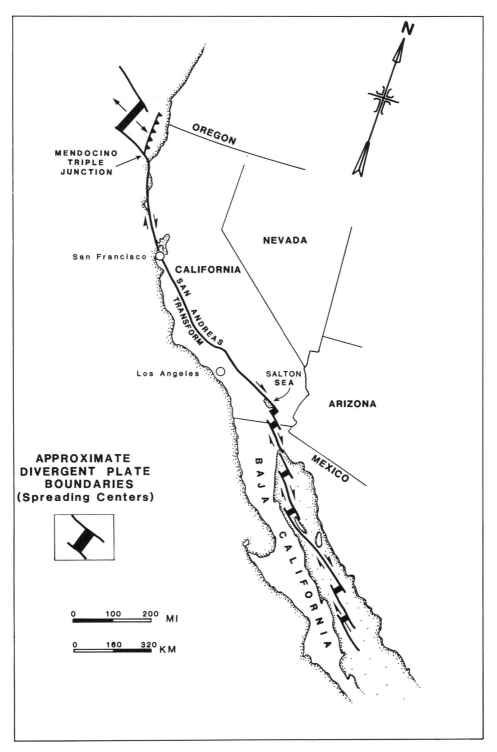

FIGURE 13-5 Relationship between the San Andreas transform (strike-slip) fault and the spreading centers in the Gulf of California and off northern California. The number and placement of the spreading centers shown are only approximate.

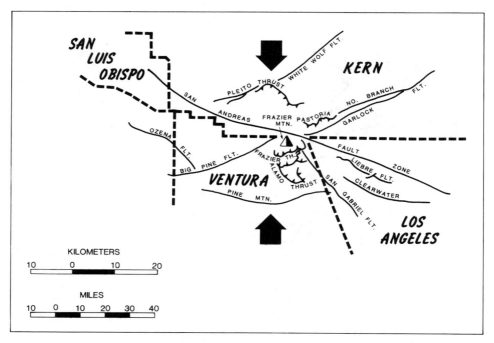

FIGURE 13-6 Faults associated with the restraining bend (Big Bend) in the San Andreas fault. Frazier Mountain occupies the center of what has been called "the structural knot" of California. Crustal shortening shown by the heavy arrows has produced the Frazier, Alamo, and other thrust faults south of Frazier Mountain, and the Pleito, Pastoria, and other thrusts to the north.

Further, in those places where the San Andreas fault changes direction, splays or branching faults may change from nearly vertical fractures to nearly flat-lying thrusts or normal faults as their trends deviate increasingly from the San Andreas.

Despite such local exceptions, the Cenozoic history of the San Andreas with its many kilometers of strike slip, discourages the acceptance of any geometry but a nearly vertical dip for a transcurrent fault of such magnitude. The question of how strike slip is accommodated in the Transverse Ranges where the San Andreas displays two marked deviations remains to be answered. One view is that the surface trace of the fault has been decoupled from the true plate boundary at depth in a "soft zone." Another hypothesis, consistent with the evidence of crustal shortening characteristic of the Transverse Ranges, is that some degree of subduction is occurring on either side of the restraining bend on the fault. But, to date, the nature of the deep structure where the San Andreas fault crosses the Transverse Ranges is very much an open question.

SEDIMENTATION

In cases where the fault direction veers right, the opposite of a restraining bend, or where a spreading center is involved, the crust is stretched rather than shortened, and a pull-apart basin develops into which sediment is poured (see Figure 9-12). Several good examples of such basins occur along the southern San Andreas.

One of these is the Ridge Basin, which formed mainly during the Miocene when the San Gabriel fault served as the main break from Frazier Mountain to near San Bernardino. The immense pile of Miocene and Pliocene sedimentary rock that accumulated in this pull-apart basin is described more fully in Chapter Ten on the Transverse Ranges. Another pull-apart basin still receiving sediment is the Salton Basin, which is believed to lie over several segments of the spreading center or diverging plate boundary that is responsible for forming the Gulf of California to the south.

In addition, there are a great many places where the San Andreas fault steps a short distance to the right. In each of these a pull-apart is formed, often only a few hundred meters across. These are called sag ponds, and each of them preserves a record of sedimentation that sometimes proves very useful in determining the earthquake history. For example, sediments accumulated in the sag can sometimes be dated by radiocarbon or other methods and a chronology established. In a number of cases deposits directly related to earthquake events such as *sand blows* (sandy layers formed by the liquefaction of water-saturated sands during shaking) can be identified and dated, yielding a measure of the recurrence interval for the larger earthquakes.

GEOMORPHOLOGY

At its two final land expressions, the San Andreas fault lies at sea level, but elevations as high as 2,100 meters (6,800 feet) occur along the fault in the San Gabriel Mountains. Relief as much as 500 meters (1,500 feet) is found where vertical offset results from a change in fault trend. Smaller scale topographic features are abundant along the San Andreas. At any scale on which the San Andreas fault is mapped in detail, it proves to consist of an en echelon set of breaks, which step either to the left or to the right. As noted earlier, when the fault steps right, a pull-apart results, often forming an undrained sag pond, with or without water. When the fault steps left, shortening occurs and an elongate ridge or squeeze-up is produced. These occur at all scales; perhaps the largest of the squeeze-ups is the Transverse Range province, where the fault steps left about 200 kilometers (125 miles). Pull-aparts range from large features like the Salton Basin to modest-sized sag ponds like Palmdale Reservoir and San Andreas Lake, to tiny little basins a few meters across (Figure 13-7).

Most types of landforms produced by faulting or subsequent erosion are found in the San Andreas zone. Although unspectacular when compared with Basin and Range faults, facets and truncated spurs abound on the San Andreas. Such features are usually considered evidence of vertical (dip-slip) movement, but this interpretation does not necessarily apply to their San Andreas occurrence. In strike-slip fault zones, localized but prominent facets and scarps are readily produced by horizontal shift where a hill or ridge is crossed by the fault. The lateral shift exposes offset facets that face each other. Such facets should not be confused with fault-line scarps, which are strictly erosional features developed along the trace of a fault.

R. P. Sharp divides geomorphic forms developed along faults into primary (movement) and secondary (erosional) features. Besides scarps, facets, scarplets, primary features along the San Andreas include offset streams, squeeze-up blocks, shutter ridges, sag ponds, fault valleys and troughs, gaps, saddles, kernbuts, and kerncols. Secondary features include, besides fault-line scarps and valleys, drainage derange-

FIGURE 13-7 Sag ponds on the San Andreas fault near Reyes Station, southeastern San Luis Obispo County. The dry pond in the distance has a salty crust because it lies above the seasonal water table. The nearer, one, slightly below the seasonal water table, remains moist and supports a green patch of vegetation contrasting strongly with the surrounding dry, grassy hills. (Photo by R. M. Morris)

FIGURE 13-8 Native desert palms (*Washingtonia filifera*) growing along the south branch of the San Andreas (Banning) fault. The fault plane acts as a dam for ground water moving toward the valley from the hills, forcing water upward toward the surface along the fault trace and producing a true oasis. Near Pushawalla Canyon, Indio Hills, Riverside County. (Photo by R. M. Morris)

ments and springs or oases that often result from ground-water barriers incident to faulting (Figure 13-8).

Detailed analyses of surface stream derangements and fault patterns have been conducted by Robert E. Wallace, and John Crowell and his students. Figure 13-9 summarizes patterns in the Carrizo Plain. Edward A. Keller and his students have investigated a beautifully displayed set of fault-generated landforms in the Indio Hills where two major branches of the San Andreas fault converge. Although such landforms are particularly clear in the Carrizo Plain and in the Indio Hills, they are by no means limited to these two areas.

Geomorphic forms are often heavily influenced by the rock types involved. The San Andreas fault is special because almost all of California's rock types are found somewhere along the fault. They may occur in small quantities, and as granulated, mylonitized breccias, or simply unaltered material. In southern California, banded gneiss, anorthosite, and other Proterozoic crystalline rocks are comingled with Pleistocene alluvium, lake beds, and Holocene deposits. Slivers of early Cenozoic sedimentary strata, Mesozoic granite, and occasional Paleozoic metasedimentary rocks are all found together. North of the structural knot, the rocks are less varied. They include Mesozoic Franciscan and granitic rocks and often thick sections of post-Jurassic sedimentary rocks and many Quaternary marine and nonmarine units.

MOVEMENT

Recent hypotheses about displacement on the San Andreas fault differ significantly from earlier views. Although no single interpretation is currently acceptable, consensus seems to exist on the following points.

1. The San Andreas has had two main periods of activity separated by quiescence.
2. Different parts of the fault have been active at different times.
3. No single section of the zone has had a unique history.
4. Irregularity of movement over time has produced variety in rock deformation and geomorphology.
5. The conventional idea of a fault as an identifiable structure with similar characteristics throughout does not apply to the San Andreas fault.

A major redirection in study of the San Andreas fault occurred in 1953, when Mason L. Hill and Thomas Dibblee advanced the view that the San Andreas had been active geologically since Jurassic time and that it had experienced 560 kilometers (350 miles) of displacement since inception. Their hypothesis was based on suggested matching of similar rock types, fossil sequences, and structures on opposing sides of the fault. Moreover, it once again directed attention to the San Andreas questions of how much, when, how, and why. Although geologists had accepted 1.6 kilometers (1 mile) of displacement and some even as much as 40 kilometers (25 miles), Hill and Dibblee's hypothesis was nothing less than outrageous to many investigators. Similarly, a second major redirection in study occurred after 1965, when J. Tuzo Wilson introduced the concept of transform faults as one type of lithospheric plate boundary. He proposed that the San Andreas fault was an example of a transform, an idea that enjoys almost universal acceptance today.

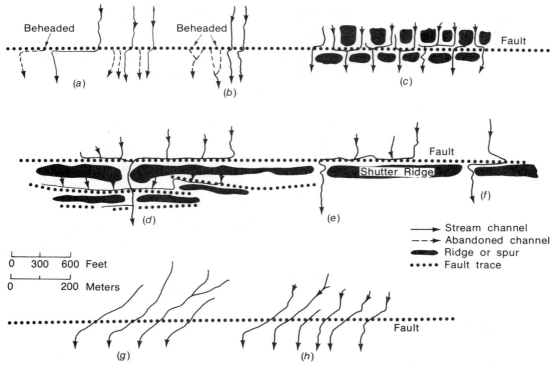

FIGURE 13-9 Patterns of fault-related stream channels found in the Carrizo Plain area. OFFSET
CHANNELS: (a) Misalignment of single channels directly related to amount of fault displacement
and age of channel. No ridge on downslope side of fault. Beheading is common. (b) Paired
stream channels misaligned. COMBINATION OF OFFSET AND DEFLECTION: (c) Compound offsets
of ridge spurs and offset and deflection of channels. Both right and left deflection. (d) Trellis
drainage produced by multiple fault strands, sliver ridges, and shutter ridges. (e) Offset plus
deflection by shutter ridge may produce exaggerated or reversed apparent offset. (f) Capture
by adjacent channel followed by right slip may produce "Z" pattern. FALSE OFFSETS: (g)
Differential uplift may deflect streams to produce fast offset. (h) En echelon fractures over
fault zone followed by subsequent streams produce false offset. (Source: California Division
of Mines and Geology)

Vertical Components—Dip Slip

Prior to the advent of the Hill-Dibblee thesis, the San Andreas was believed to have
extensive vertical displacement, particularly south of the Big Bend. Relative to the
Mojave block, the San Gabriel Mountains have risen at least 1,200 meters (4,000
feet) on faults that are subordinate but parallel to the San Andreas. The San Bernardino Mountains rise more than 2,150 meters (7,000 feet) above San Gorgonio
Pass along faults in the San Andreas zone. The San Jacinto Mountains rise 3,000
meters (10,000 feet) relative to the Coachella Valley, another impressive evidence
of a vertical component of movement. In the San Gorgonio Pass, strands of the San
Andreas appear to be the cause of that depressed block between the San Bernardino
and San Jacinto mountains. Studies of comparative thicknesses of sediment in the

central Coast Ranges suggest that the coastal block moved up a maximum of 1,800 meters (6,000 feet) during middle Cenozoic time. In northern California, the vertical component of displacement is less.

Apparently the San Andreas fault zone has experienced major vertical movement either at intervals or concurrently with the horizontal displacement generally characterizing the fault. Although pictorially represented as a line, a shear zone like a transform is actually a wide belt in which vertical displacement may occur simultaneously with strike-slip displacement. Both regional basins of sedimentation and regionally uplifted blocks have been produced in and along the splintered margins of the Pacific and North American plates. Concomitantly, where rotation of segments with the fault zone has occurred, crustal shortening may cause large units to be uplifted. This might be seen as a shift in tectonic patterns when in fact the vertical component of displacement is compatible with the activity on the transform. For example, it has been suggested that large crustal slabs are squeezed and uplifted where major strands of the San Andreas fault system converge. Such a feature is the high San Jacinto Mountain block where the San Andreas and San Jacinto faults converge. Similarly, large depressions tend to develop where branches of the fault diverge, an example of which is San Francisco Bay where the Hayward and San Andreas faults diverge, or perhaps the western Antelope Valley between the diverging San Andreas and Garlock faults. Thus large landscape features with considerable vertical relief can be produced either by strike slip on the San Andreas as a result of its en echelon character, or because of strike slip on associated branching faults.

Horizontal Components—Strike Slip

Right slip as the major movement on the San Andreas fault was first deduced from topography. Horizontal slip, at least in the central Coast Ranges, gradually was deemphasized as a result of studies by N. L. Taliaferro, who called attention to widely exposed Salinian granitic basement west of the fault and its absence east of the fault. He assumed the Franciscan rocks on the east were deposited on a similar, deeply buried granitic basement. On the basis of this erroneous assumption, it was reasonable to postulate considerable vertical uplift of the Salinian block along the San Andreas fault. Most of the geomorphic features requiring recent strike-slip displacement were young and apparently did not require much strike slip to account for them. For these reasons, Taliaferro proposed that the San Andreas was a major structure of dominately vertical movement, cut at a low angle of about 10 to 15 degrees by a younger strike-slip fault, the present-day active trace, for which Taliaferro conceded up to 1.6 kilometers (1 mile) of strike-slip displacement since the Pleistocene epoch.

The earlier suggestions of up to 48 kilometers (30 miles) of right slip in southern California and 1.6 kilometers (1 mile) in the Coast Ranges are understandable, but it is now widely agreed that the total slip on the San Andreas and its main branches in southern California is about 330 kilometers (205 miles), composed of 300 kilometers (186 miles) on the San Andreas proper, including the now inactive San Gabriel fault, plus 24 kilometers (15 miles) on the San Jacinto and 40 kilometers (25 miles) on the Elsinore. If right slip on the borderland faults offshore is included, the total slip would exceed 330 kilometers (205 miles) by a substantial amount. These additions may prove to be justified as the earlier history of the San Andreas becomes better understood.

North of the Big Bend, according to J. C. Crowell, who has long been interested in the displacement history of the San Andreas fault, the evidence now available suggests a total of about 600 kilometers (370 miles) of displacement divided into a late Cretaceous to early Paleocene episode during which about 300 kilometers (185 miles) of displacement occurred, and a late Miocene to Holocene slip of an additional 330 kilometers, separated by a period of quiescence during the Eocene and Oligocene epochs.

How one reconciles the 600 kilometers (370 miles) total north of the Big Bend with the 330 kilometers (205 miles) south of that area is still a problem, but the answer may be found in the history of the Newport-Inglewood fault and related borderland faults to the southwest.

Although displacements observed during earthquakes and those affecting geomorphic features of Quaternary age provide obvious and unequivocal evidences of strike-slip movement, determination of total offset since birth of the fault is much more difficult. A water main, when sheared off during a displacement, forms a perfect *piercing point* on which one can measure the amount of strike slip as well as any dip slip. A vertical volcanic dike cut by the fault would be useful for measuring strike slip, but less helpful in establishing vertical offset. Most geologic reference planes are horizontal or nearly so, however, and these are the least useful types for establishing amounts of strike slip. Unfortunately, there is a total lack of Miocene water mains, so one must find other sorts of piercing points.

Fossil shorelines and intrusive contacts as well as various volcanic features have provided some of the better correlations on which the history and amount of displacement on the San Andreas is based. For the northern section, displacements are found to increase with age back as far as middle Miocene time. Eocene and Oligocene piercing points show no differences, whereas displacement of contacts increases again from the early Paleocene epoch back to the late Cretaceous period. All contacts older than late Cretaceous show the same total displacement. Different views regarding the validity of any piercing point account for much of the range of offsets one finds proposed in the scientific literature (Figure 13-10).

Creep

Records of progressive horizontal movements along the San Andreas have been systematically maintained since 1930 through measurements of triangulation networks installed by the U.S. Coast and Geodetic Survey and its successor organizations. Creep was first documented on the San Andreas fault at the Almaden Winery near San Juan Bautista, and then on the Calaveras fault at Hollister (see Figure 1-14). In the 1960s, continuous creep was demonstrated along the Hayward fault in the San Francisco Bay region.

Specific rates of creep vary considerably, although 1966 reports show all movements are nearly horizontal. Average creep rate is summarized in Table 13-1. Very recent studies suggest that little or no creep is occurring on the San Andreas north of San Juan Bautista, but it has been documented on other bay area faults such as the Hayward and Calaveras.

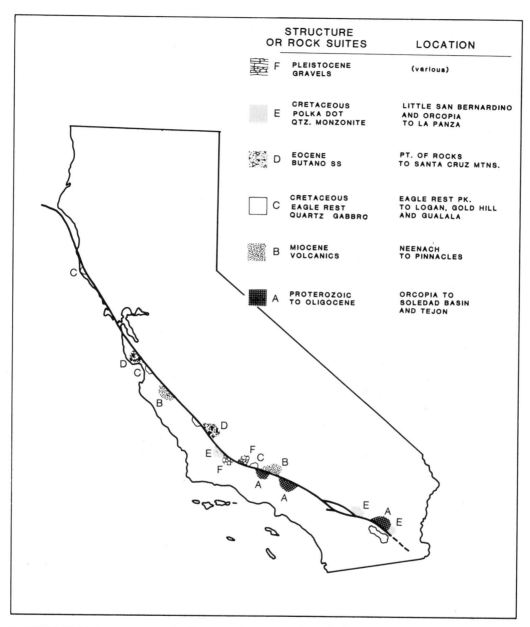

FIGURE 13-10 Some proposed offsets on the San Andreas fault system. These offsets are based on a suggested match of rock suites, structural features, or both. In general the amount of offset is less in the younger rocks and structures, but the pattern is not entirely consistent. Anomalous patterns such as example C may result from small scraps being left behind along the fault as the main portion of the slice is carried beyond. If so, the Logan and Gold Hill exposures of the Eagle Rest Quartz Gabbro may be such examples, whereas the Gualala exposure represents the true measure of total offset. But all these examples depend on the basic validity of the proposed matches, and all are subject to revision.

TABLE 13-1 Creep on the San Andreas Fault

Area	Years	Rate of Creep
Point Reyes to Petaluma	1930–1938	2 cm/yr
	1938–1960	1 cm/yr
Vicinity of Monterey Bay	1930–1962	1.6 cm/yr
San Juan Bautista	1968–1977	1.4 cm/yr
Vicinity of Hollister	1959–1963	1.5 cm/yr
	1963–1966	1.7 cm/yr
Pinnacles National Monument	1972–1977	2.3 cm/yr
Bitterwater Valley	1967–1971	3.3 cm/yr
Parkfield	1968–1979	1.5 cm/yr
Cholame	1966–1979	0.4 cm/yr
Maricopa to Cajon Pass	to 1962	none
Imperial Valley	1941–1954	3.0 cm/yr
	1975–1979	1–2 cm/yr

Sources: California Division of Mines and Geology, 1967; Burford and Harsh, 1980, Bull. Seismological Soc. Am., v. 70, pp. 1233–1262; Cohn, Allen, Gilman, and Goulty, USGS Prof. Paper 254, 1982.

Creep has not yet been unequivocally related to the major causal mechanism of earthquakes—sudden slip. What are the relations? Are zones of creep on faults less prone to sudden slip, and adjacent areas therefore less prone to earthquakes? Or are the two processes unrelated? Is creep a subtle assurance of greater immunity from earthquakes of high Richter magnitude? Could the opposite be true?

If one examines the history of slip on the San Andreas fault over the past few million years, it is found that long-term slip rates (not necessarily creep) may be determined. If they are compared with measured rates, perhaps some sense of earthquake risk can be obtained. Table 13-1 shows no measured creep on the San Andreas fault from Maricopa to Cajon Pass, yet the long-term geologic evidence reveals that approximately $5\frac{1}{2}$ centimeters (2 inches) of slip per year is necessary to account for displaced geologic features. If this segment has been locked into place since the last great earthquake in 1857, as is generally believed, then the rocks had by 1988 accumulated about 7.2 meters (24 feet) of strain. Because the 1857 event was accompanied by perhaps as much as 10 meters (30 feet) of displacement, one might be forgiven for suggesting that the next earthquake is imminent and that it is likely to be a big one.

Conversely, in the Hollister area in San Benito County, where the fault is straighter, moderate earthquakes and creep are common, and it appears that the strain is released in a less damaging and dramatic way before it builds to such high levels as are believed to exist in the Maricopa to Cajon Pass segment.

A number of workers, beginning with Kerry Sieh in the late 1970s, have established a prehistoric chronology of earthquakes on the locked segment of the San Andreas fault between Maricopa and Cajon Pass. This chronology was deduced from detailed descriptions of such features as offset beds, fractures, sand blow deposits, and filled crevices, all dated by radiocarbon methods and displayed in the walls of trenches purposely cut across the trace of the San Andreas fault. These studies suggest that major earthquakes have been releasing accumulated strain on this part of the fault about every 130 to 185 years for at least the past 1,000 years or so. Displacements of 8 to 12 meters (26–40 feet) were typical. Presumably, each of these events would generate an earthquake of 7.5 to 8.5 Richter magnitude.

In contrast, a straighter part of the San Andreas fault near Parkfield in extreme southeastern Monterey County has, during the last century or so, generated moderate earthquakes about every 22 years. The most recent was in 1966 and was about magnitude 5.5. It seems to be quite possible that the next earthquake in this area will take place in 1988 or 1989.

Curiously, creep on faults is not a common phenomenon. Apart from the North Anatolian fault in Turkey, the central and southern parts of the San Andreas system are the only places where active fault creep has so far been observed. The reasons for this are not understood.

ORIGIN

Sixty years ago, geologists were more confident that they understood the origin of faults than they are today. Faults were simply breaks where rocks had failed. They were thought to be major boundaries of mountain units, on which the mountains were lifted relative to the adjacent lowland. Movement was therefore dip slip. Strike-slip movement was considered theoretically possible, but it was recognized as subordinate to the vertical responsible for major uplifts.

Related to the standard dip slip expected on faults was the widely accepted (and still valid) theory of isostasy. According to this theory, rock failure on the margins of erosional-depositional boundaries results in fracturing and vertical movement. Isostasy produces greater instability on the continental margins, which are usually sites of mountain ranges. Thus the San Andreas was a major fault of strong vertical displacement along the Pacific margin.

In global tectonic theory, however, the San Andreas is a boundary on which the eastern edge of the Pacific Plate is drifting north against the North American plate to the east. As has already been noted, changes in trend on the San Andreas fault appear to be responsible for producing such notably mobile units as the Transverse Ranges.

Many questions fundamental to an adequate understanding of the San Andreas system remain to be answered. A few of them are:

1. Why is creep so prominent on parts of the San Andreas system and so rare on other faults around the world?

2. How is movement on the San Andreas related to such apparent swellings as the Palmdale Bulge, which may occur from time to time?

3. What is the mechanism that leads to the en echelon strands of the San Andreas fault zone?

4. If most breaks occur on existing traces, how is a new fault strand ever born?

5. Why was the San Andreas system inactive during the Eocene and Oligocene?

6. What is the cause or causes of the two bends in the San Andreas fault?

7. How does the surface trace of the San Andreas fault relate to the plate boundary at depth?
8. Just how is the San Andreas related to the major left slip faults like the Garlock and Big Pine?

TRANSFORM FAULTS AND GLOBAL TECTONICS

Because there is considerable evidence that large-scale northward transport of large crustal blocks affected the Pacific margin in late Mesozoic time, it follows that the area was dominated by transform faulting. Hence, it is evident that California was situated near a plate boundary at least as far back as Cretaceous time, and that movements of the Pacific Ocean plates caused at least the northern part of the San Andreas to evolve into a trench-linked transform by the end of the Cretaceous period.

California's persistent location near a plate boundary doubtlessly provides a general mechanism for producing most of the state's major faults and many of its smaller ones as well. Thus, in a sense, it is likely that the same forces that have created and maintained the San Andreas fault can be used to explain many of the state's major topographic features. It is likely that most right-slip faults are related to the San Andreas system. Several left-slip faults also may result from the same forces that generated movement along the San Andreas. The same may be true of other major faults that are not obviously connected to the San Andreas system. Further, as the North American plate continues to ride west, new faults are likely to develop. In recent years evidence has been found that suggests the present plate boundary from San Francisco northward to the Mendocino Triple Junction is being replaced by one that lies to the east and continues northwestward beyond the Mendocino fracture zone.

It is appropriate to mention again two previously discussed geomorphic lineaments in California and western Nevada. First is the Furnace Creek fault, which cuts indiscriminately across rock and topographic boundaries and may be a major strike-slip fault extending from Fish Lake Valley southeast to the Colorado River near Blythe. It is subparallel to the San Andreas. A second trend appears from southeast of Las Vegas striking northwest, and extends about 320 kilometers (200 miles). Known as the Walker Lane or Las Vegas shear zone, this lineament is closely parallel to the San Andreas. In addition, it has been proposed that a lineament similar to the Walker pattern extends more than 750 kilometers (450 miles) from central Oregon southeast into central Nevada. They all appear to result from the same forces that initiated the San Andreas system.

Faults similar to the San Andreas are known elsewhere in the world. An example is the major alignment of the Denali, Fairweather, and Queen Charlotte faults in southeastern Alaska and British Columbia. The great Alpine fault, which cuts across the west side of New Zealand's South Island and is traced northward into two or three prominent North Island faults, may be about 1,600 kilometers (1,000 miles) long. In addition to vertical displacement, the Alpine fault shows as much as 560 kilometers (350 miles) of strike-slip displacement. In Asia Minor, major faults of San Andreas-type trend east-west and intersect similar northwest-southeast trending faults that extend several hundred kilometers south into the Jordan Valley and Sinai Desert.

The Dead Sea is marked by a large strike-slip structure. All these features are interpreted as transforms in areas where the crustal plates are being broken.

HUMAN INVOLVEMENT

In an unstable region with large population and continued growth, maximum safety precautions to offset earthquake hazards are extremely important. In California, about 75 percent of the population resides in the areas of greatest instability. It is difficult to implement earthquake precautions, however. They are expensive, and because earthquakes may never affect the majority, appropriate precautions are often ignored. Building for earthquake resistance may increase costs by 10 percent or more. In the past, incorporating earthquake safety into structures was hampered because geologic science could not provide adequate data for structural engineers, architects, and construction firms. Today better information is available.

An interesting example of this problem relates to the development and deployment of strong-motion seismographs. For many years, seismologists had been most interested in developing extremely sensitive instruments that would allow scientists working at places like Berkeley or Pasadena to record and analyze in detail the nature of distant earthquakes. Nearby strong shocks were expected either to disable or deactivate these sensitive instruments. Structural engineers, on the other hand, were much less concerned with the nature of a quake in Kamchatka, for example, and instead wanted to know what forces local earthquakes imposed on buildings and other structures. Beginning about 1965, strong-motion instruments were perfected and placed in buildings, on dams, bridges, and other structures. These began to yield data that surprised nearly everyone. During the moderately strong 1971 San Fernando earthquake (magnitude 6.4), accelerations equal to or slightly greater than gravity were measured. This meant that objects not tied down tended to float when the acceleration exceeded gravity, just as they do in spacecraft; rocks tossed in the air often landed upside down, for example. These high accelerations forced considerable review of the adequacy of existing construction standards and design. Such large accelerations were thought to have been extremely rare, but it is now clear that they are not.

Establishing adequate safety standards assumes willingness to recognize earthquake hazards, expenditure of large sums of money required, and counteracting the apathy that arises from the "it can't happen to me" philosophy. This last obstacle might be overcome quickly if many continuing, sufficiently strong but not disastrous earthquakes were to occur. Instead, infrequent, localized, but unfortunately sometimes severe tremors affect comparatively few people at a time. For example, the 1987 Whittier Narrows earthquake (magnitude 5.9, $400 million damage) was termed a "wake-up call" to the Los Angeles metropolitan area which, at present, is bracing for "the big one."

Engineering

It is now feasible to construct modern earthquake-resistant (not earthquake-proof) buildings. Since the locally disastrous 1933 Long Beach earthquake (magnitude 6.3), the Field Act has required that all California public school buildings be earthquake resistant. Moreover, the uniform building codes adopted in the 1950s have applied Field Act standards to virtually all buildings. Yet damage sustained in the 1971 San Fernando quake showed that even these standards were not completely adequate (see Figure 10-26). Nonetheless, dams, waterworks, highways, and utility structures may all be built with reasonable safety provided certain precautions are followed and provided that they are not built directly on faults or on unstable ground subject to liquefaction or sliding.

Public lack of awareness and failure to demand reasonable precautions are the main impediments to earthquake safety, although some hazards will always remain. In the 1971 San Fernando earthquake, damage to newly constructed freeways (built with reasonable safeguards) was quite extensive. The quake's surprisingly large forces led some seismologists to assert, however, that the San Fernando case can reasonably be expected to occur once every 5,000 years. Most seismologists are now much less sure that these high accelerations are rare, and on the basis of what is known about the setting and history of the San Fernando fault system it is now estimated that the recurrence interval is between 100 and 300 years. Should huge expenditures be allocated to safeguard structures, roads, and bridges that have an estimated life of 50 years or so? Reasonable precautions are required, but some degree of risk must be accepted when people choose to live in an unstable, earthquake-prone region.

Psychological Factors

The psychological discord experienced when an earthquake strikes can be severe for a few people. There is often a tendency to accept rumors about earthquakes without regard for the facts. The psychological consequences of other destructive natural phenomena such as tornados, hurricanes, and floods are also extremely traumatic, but they do not seem to be accompanied by false rumors to the same degree as earthquakes. Perhaps other natural disaster-producing phenomena are less damaging to the psyche because they can be seen—people see them arrive and see them leave—something rarely possible with earthquakes. Storms, floods, landslides, windstorms, and fires take far larger tolls of life and property in the United States than do earthquakes. Furthermore, safeguards against earthquakes are no more costly than those required to prevent damage from other natural phenomena.

PREDICTION OF MOVEMENT

Prediction of earthquake-producing movement for the San Andreas fault is not new. It has usually involved both geologic pronouncements and the ruminations of soothsayers, mediums, astrologers, and clairvoyants.

In the geologic category are specific predictions like the one attributed to an eminent California geologist about 1920. This authority allegedly predicted that a major earthquake would occur in the Los Angeles Basin from movement on the San Andreas fault. He theorized that the movement would occur closer to 5 years than 10, and closer to 15 than 20. In 1940, a national news service was said to have telephoned the geologist, reminding him that it was the twentieth anniversary of his prediction and no earthquake had yet occurred from movement on the San Andreas. The reporter supposedly asked, "Have you any statement that you would care to make?" "Oh, yes," responded the geologist. "It isn't every geologist who is privileged to live long enough to have his predictions disproved!" Regardless of its accuracy, this story makes a point: Specific predictions about slippage on any fault can as yet be based only on statistical probabilities, although the geologic basis for our statistics is vastly better than it was even a decade ago.

Studies in Japan, the United States, China, and the U.S.S.R. suggest that definitive earthquake prediction may someday be possible, at least for some shocks at some times and in some places. Long-range predictions on the order of a year or so will probably come first and include the cautiously predicted Parkfield quake in 1988 or 1989. Short-range forecasts are considerably more valuable, and although there have been a few notable successes such as the moderate shock predicted in California in 1974, and the major Haicheng earthquake in China in February 1974, other quakes occur without warning. Chinese success in 1974 was followed by the unexpected, disastrous Tanshan quake of 1976, which killed more than 650,000 people, and was very likely the second most lethal earthquake in recorded history (the most disastrous earthquake was the one of January 24, 1556, which killed 830,000 in Shaanxi Province, China). Although the Chinese did not predict the Tanshan quake, they had observed *most* of the signs recognized as warnings, but not *all* of them, so they made no announcement.

In the U.S.S.R., investigations suggested that the ratios of the velocities of P and S waves measured in small shocks sometimes dropped significantly, followed by a sharp increase in the ratios just before a major shock. Subsequently, U.S. studies showed that this was not a reliable criterion. Anomalous ground tilt in regions of active faulting may signal impending sudden movement as was thought possible in the mid-1970s when the Palmdale Bulge appeared to be developing. Porosity and dilantancy changes in critical areas adjacent to faults and shifts in electrical conductivity of rocks are being explored as possible avenues for prediction. The discharge pattern of radon gas from ground waters seems to be related to earthquakes and is being monitored at a number of stations. The meaning of small shock patterns both before and after known major quakes is being restudied in conjunction with changes in rock properties when the quakes occur. Among these properties are porosity, density, and electrical characteristics. It is possible that the rocks along faults undergo detectable strain before an earthquake is triggered. Apparently, the effects of strain in small earthquakes are measurable for a few days prior to the actual shocks; in the case of large earthquakes, it may be possible to make predictions three or four years in advance.

Any predictions that do not specify the time frame and the approximate epicentral location do not deserve serious consideration by the public, because earthquakes are frequent in California and strong shakes are almost an annual phenomenon somewhere in the state.

REFERENCES

Allen, C. R., 1981. The Modern San Andreas fault. *In* The Geotectonic Development of California, W. G. Ernst, ed., Rubey vol. 1, Englewood Cliffs, NJ, Prentice-Hall, pp. 511–534.

Crowell, John C., 1962. Displacement along the San Andreas fault, California. Geol. Soc. Amer. Spec. Paper 21.

———, ed., 1975. San Andreas Fault in Southern California. Calif. Div. Mines and Geology Spec. Report 118.

———, 1981. An Outline of the Tectonic History of Southeastern California. *In* Geotectonic Development of California, W. C. Ernst, ed., Rubey vol. 1, Englewood Cliffs, NJ, Prentice-Hall, pp. 583–600.

Dibblee, T. W., Jr., 1966. Evidence for Cumulative Offset on the San Andreas Fault in Central and Northern California. *In* Geology of Northern California. Calif. Div. Mines and Geology Bull. 190, pp. 375–384.

Dickinson, William R., and Arthur Grantz, eds., 1968. Proceedings of the Conference on Geologic problems of the San Andreas Fault System. Stanford Univ. Publ. Geol. Sci., v. 11.

Hill, M. L., 1981. San Andreas Fault: History of Concepts. Bull. Geol. Soc. America, v. 92, part 1, pp, 112–131.

Iacopi, Robert, 1964. Earthquake Country, Menlo Park, CA, Lane Book Co. 192 p.

Kovach, Robert L., and Amos Nur, eds., 1973. Proceedings of the Conference on Tectonic Problems of the San Andreas Fault System. Stanford Univ. Publ. Geol. Sci., v. 13.

Lawson, A. C., and others, 1908. The California Earthquake of April 18, 1906. Rept. of State Earthquake Investigation Commission, Washington, D.C., Carnegie Institution Publ. 87, part 1, v. 1.

Sherburne, Roger W., 1988. Ground Shaking and Engineering Studies Near the San Andreas Fault Zone, Parkfield, California. Calif. Geology, v. 41, pp. 27–32.

Sieh, Kerry E., 1978. Earthquake Recurrence Intervals, San Andreas Fault, Palmdale, California. Calif. Geology, v. 31, pp. 143–145.

Sullivan, Raymond, and David A. Mustart, 1977. Living in Earthquake Country. California Geology, v. 30, pp. 3–8.

Veek, Eugene B., 1981. Earthquake Prediction. California Geology, v. 34, pp. 154–155.

Weber, F. Harold, 1986. Geologic Relationships Between San Gabriel and San Andreas Faults. California Geology, v. 39, pp. 5–14.

Wootton, Tom, 1980. Strong Motion Instrument Program. California Geology, v. 33, pp. 215–218.

Ziony, J. I., ed., 1985. Evaluating Earthquake Hazards in the Los Angeles Region—An Earth Science Perspective. U.S. Geol. Surv. Prof. Paper 1360.

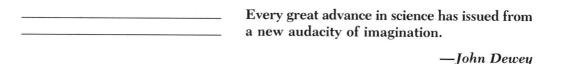

fourteen

OFFSHORE

Every great advance in science has issued from
a new audacity of imagination.

—*John Dewey*

Offshore California is related to three discrete land provinces. Evaluating it as a separate entity, however, usually leads to the most accurate understanding of offshore geology. One onshore counterpart is the Coast Range province, north of Point Arguello, where the Offshore includes continental shelf and slope that are probably submerged extensions of the Coast Ranges. The other related land provinces are the Transverse Ranges and Peninsular Ranges, south of Point Arguello. A distinctive topography occurs in the southern part of Offshore California that is quite unlike either the steep continental slopes or the gently outward-sloping shelves seen elsewhere in the world. Francis P. Shepard named this section the Continental Borderland.

NORTHERN SHELF AND SLOPE

Usually the northern shelf is only a few kilometers wide, but off the Golden Gate it widens to about 48 kilometers (30 miles). It widens again north of Cape Mendocino, probably in response to the westward bend of the San Andreas fault as it joins the west-trending Mendocino Fracture Zone. This bend is accompanied by the marked narrowing of the northern Coast Ranges onshore to only 13 kilometers (8 miles).

Although the sedimentary cover on the northern shelf is reasonably well known, few bedrock exposures are available. Apart from rocky islets and sea stacks close to shore, the only known exposures are on the Farallon Islands off San Francisco. Most information has been deduced from geophysical studies, which include sub-bottom profiling and seismic surveys of deeper structures. Echo soundings made in the last 50 years, coupled with earlier wire soundings, have provided a fairly accurate and

459

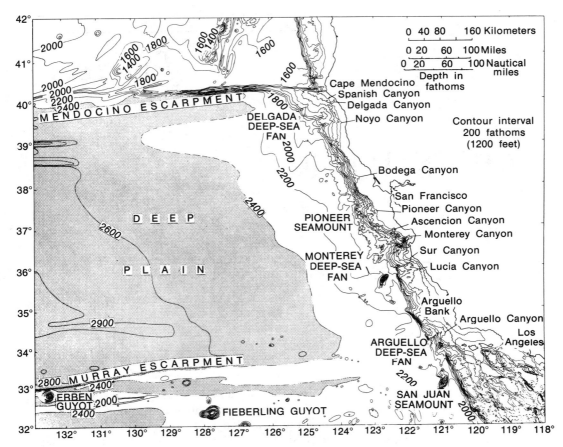

FIGURE 14-1 Topographic features: northern California shelf. (Source: California Division of Mines and Geology)

detailed picture of shelf topography. Recently, development of side-scan sonar has revealed much additional detail (Figure 14-1). All investigations have been hampered, however, by the persistently rough seas that pound the northern California coast.

The most striking topographic features of the northern shelf are the submarine canyons that cut both the shelf and the continental slope beyond. Several head within a kilometer or so of the beaches and continue as distinctive topographic features to the deep ocean floor. By far the largest is Monterey Canyon and its branches. Not only is this the largest submarine canyon on the California coast, but it is also believed to be the largest canyon on the Pacific side of the Americas (see Figure 1-33).

Monterey Canyon extends to within a kilometer (0.6 mile) of the shore and continues as a recognizable feature to depths of 3,650 meters (12,000 feet). It follows a winding course and occupies a V-shaped gorge, at least in its headward reaches. Two main branches join the Monterey in its upper reaches: Soquel Canyon from the north and Carmel Canyon from the south. Although no bedrock is found exposed in the main Monterey Canyon until a depth of 550 meters (1,800 feet), borings along the beach show that a connecting canyon 1,500 meters (5,000 feet) deep, extending inland into the lower Salinas River Valley, was once cut in bedrock but was later filled with sediment. Although not as long, Monterey Canyon equals the depth and width of the Grand Canyon of the Colorado River. The Monterey's seaward extension below a depth of 1,550 meters (5,100 feet), becomes less V-shaped and more trough-like with distinct levees up to 24 meters (75 feet) high. These eventually disappear near the outer margin of the huge submarine fan at the mouth of the canyon. This submarine delta is almost 100 times the volume of rock that was removed from the canyon above, establishing that the canyon has long been a channelway conveying sediment from beach and near-shore shelf into the deep ocean. The submerged deltaic material is both mineral and organic, derived from beach and adjacent shallow-water areas.

Other submarine canyons include Delgada Canyon south of Cape Mendocino, Noyo Canyon about 65 kilometers (40 miles) south of Delgada Canyon, Bodega Canyon off the mouth of Tomales Bay, Sur and Lucia canyons off the Monterey coast south of Carmel, and Arguello Canyon well off Point Arguello.

Rocks

The northern outer shelf and upper slope appear to be dominated by an elongate ridge of basement rock from near Cape Mendocino to about 96 kilometers (60 miles) south of Monterey. Composition of this buried bedrock ridge is uncertain because geophysical measurements have not clearly distinguished between granitic (Salinian) and Franciscan (subduction complex) basements. Nevertheless, available evidence does suggest that the ridge is chiefly granitic from Monterey north to Cape Mendocino.

Although the surface of the northern shelf has a nearly regular outward slope, geophysical studies have shown that a prism of sedimentary rock lies between the present shoreline and the basement ridge. Both features are blanketed with recent sediments. It has been suggested that the bedrock ridge represents a slice of continental crust detached from the Coast Ranges, with the intervening elongate basin subsequently filled with sediments. This may be analogous to the Gulf of California, which is generally regarded as a pull-apart structure. Furthermore, gravity measurements on the shelf between the Golden Gate and the Farallon Islands have revealed a distinct negative gravity anomaly that could reflect either a thick section of low-density sedimentary rocks or stretching and thinning of the underlying crystalline crustal rocks (Figure 14-2).

Sediments. Much of the northern shelf is covered with sands and muddy sands. The few banks and higher places that exist are usually overlain by shelly materials often mixed with the clay mineral glauconite (also known as greensand) plus phosphorite. (Both minerals are chemical precipitates or alteration products of other minerals or organic materials that accumulate where detrital sediments from land are scarce.) Sedimentary patterns were affected by extensive exposure to wave action during lower stands of sea level in the Pleistocene. Several relict near-shore deposits are preserved well below the reach of modern wave action.

An unusual sedimentary feature is the submerged lunate sandbar off the mouth of the Golden Gate. This sandbar has induced many cases of seasickness because its presence is often marked by a patch of higher waves and rougher sea. It is formed

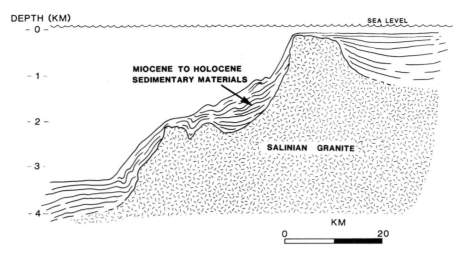

FIGURE 14-2 Probable cross section of northern shelf north of the Farallon Islands. Vertical exaggeration 10 times. The folded sedimentary materials are known to include some of Miocene age and younger and may well include other sedimentary rock as old as Cretaceous. (Source: Curray, California Division of Mines and Geology)

from river sand carried from the Golden Gate and spread out as a low crescentic ridge on the sea floor. A delicate balance is maintained with the tidal currents that sweep back and forth through the Golden Gate. Previously, the bar was somewhat farther offshore, but as sediment from mining and farming activities in interior California was carried into San Francisco Bay, the bay became shallower and the volume and velocity of tidal water moving back and forth through the Golden Gate were reduced. This effect was augmented by man-made fills added to the San Francisco foreshore. Weakening of the tidal currents subsequently allowed wave action in the open Pacific to shift the bar about 100 meters shoreward.

Structure

Knowledge of the structure of the northern shelf is incomplete. Although it is clear that many land structures extend onto the shelf, important but puzzling changes seem to occur there. For example, the San Andreas and Pilarcitos faults diverge near Redwood City and continue onto the shelf as separate entities. These faults almost certainly rejoin somewhere off the Golden Gate because the San Andreas reappears on the shore alone, but no details have yet been established about their offshore reunion. Investigations in the Monterey Bay area, however, permit fairly confident extension of faults from the Santa Cruz Mountains north of the bay across the shelf to the Santa Lucia Mountains on the bay's southern shore (Figure 14-3).

Near Cape Mendocino is the Mendocino triple junction, a place where three major crustal plates meet (see Figure 13-4). West of the San Andreas fault is the Pacific plate, east is the North American plate, and north of the Mendocino Fracture zone into which the San Andreas fault merges, is the Juan de Fuca plate, which is currently being subducted beneath northernmost California, Oregon, and Washington. The spreading center (divergent plate boundary) that separates the Juan de

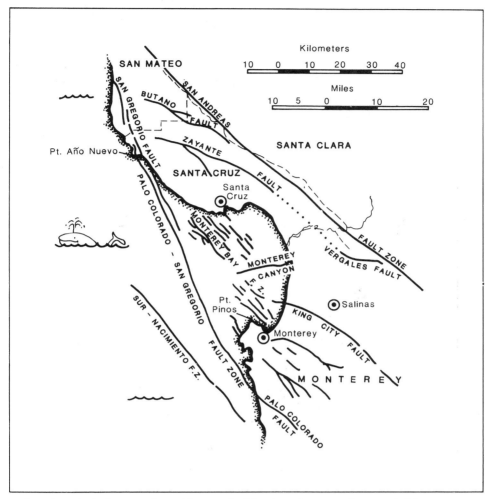

FIGURE 14-3 Faulting on the sea floor in the Monterey Bay area. (Source: California Division of Mines and Geology)

Fuca plate from the Pacific plate lies about 100 kilometers (65 miles) west of Cape Mendocino.

The Mendocino Fracture zone seems to curve south to join the San Andreas fault and does not extend east into the Coast Ranges or Klamath Mountains. As was noted in the previous chapter, evidence has recently been found suggesting that the northern part of the San Andreas fault is beginning to be replaced by a new break east of the present fault. This new break appears to extend northwest perhaps 100 kilometers or so onto the shelf from near the mouth of the Mad River north of Humboldt Bay.

In contrast, far to the south, the Murray Fracture zone dies out on the sea floor west of Point Arguello where the east-west trending Transverse Ranges begin. Many geologists have sought to relate the 3,050-kilometer (1,900-mile)-long Murray Fracture zone to the Transverse Ranges because of their similar east-west orientation. But to date no really persuasive evidence has been found to corroborate such a connection.

Construction of the Diablo Canyon nuclear power plant on the coast near San Luis Obispo spurred further investigations of some faults previously mapped on the nearby sea floor. The most prominent of these, the Hosgri fault, is now regarded as a portion of a major structural trend called the San Gregorio-San Simeon-Hosgri fault

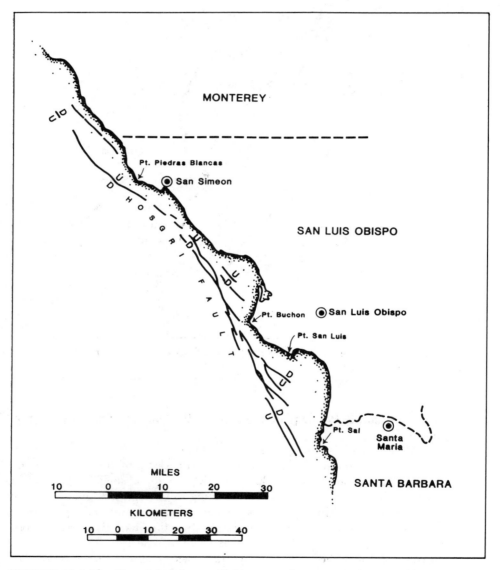

FIGURE 14-4 The Hosgri fault zone off the coast of San Luis Obispo and Santa Barbara counties. (Source: California Division of Mines and Geology)

zone. At least the Hosgri portion of this zone seems to have experienced between 80 and 115 kilometers (50 to 70 miles) of right slip since late Miocene time. The fault zone extends from the Santa Cruz Mountains southward to near Point Sal in Santa Barbara County, a distance of about 350 kilometers (217 miles) (Figure 14-4). The Hosgri fault proper is nowhere exposed on land, though branches of it appear to extend onshore in both San Luis Obispo and Santa Barbara counties. The San Simeon fault, however, is well exposed near Point Piedras Blancas.

Although the three faults that collectively comprise the San Gregorio-San Simeon-Hosgri zone are not known to be joined at the surface, very distinctive suites of rocks—one east of the Hosgri fault at Point Sal in Santa Barbara County, and the other west of the San Simeon fault in San Luis Obispo County—are now separated by about 100 kilometers (65 miles). If these suites of rock were once adjacent, as their similarities suggest, this fault zone may well be one of the more important units of the greater San Andreas system of faults.

CONTINENTAL BORDERLAND

The section of Offshore California south of Point Conception, the Continental Borderland, is far better known than the northern portion (Figure 14-5). In fact, it is among the most thoroughly investigated areas of sea floor anywhere in the world. Despite this, our knowledge of the area is still far from complete.

The Continental Borderland is composed of elevated blocks and ridges, sometimes with islands above the marine datum and sometimes with deep, often enclosed basins. The seaward edge, known as the Patton Escarpment, approximately connects Point Conception with Punta Banda in Baja California. Seaward, a typical continental slope with relief up to 4,000 meters (13,000 feet) descends to the deep floor of the Pacific. The landward margin of the borderland is a narrow continental shelf seldom more than 8 kilometers (5 miles) wide. In the Continental Borderland maximum relief is about 2,600 meters (8,500 feet), although the sedimentary fill in the basins suggests that the relief once may have been much greater.

Shelves as wide or wider than the Continental Borderland are known throughout the world, but in only a few localities do they depart from the typical pattern of almost flat, gently sloping platforms interrupted by low banks and occasional canyons or channels. The southern California-northern Baja California region is certainly neither typical shelf nor slope, although narrow shelves do extend along the mainland California coast and around the islands. Some steep declivities reminiscent of slopes also occur, but these normally lead into deep, closed basins. Only at the outer edge of this unusual region is there a well-defined slope leading into the deep sea.

Topography

Apart from the Santa Barbara Basin and the four Channel Islands, which follow Transverse Range trends, most of the basins and intervening banks and ridges trend distinctly northwest. Depths of basin floors vary somewhat systematically. Those closest to shore have the shallowest depths, flattest floors, and thickest sedimentary fillings. Partially filled inshore basins are Santa Monica, San Pedro, San Diego, and Catalina. As noted in Chapter 10, the Offshore region once included much of the

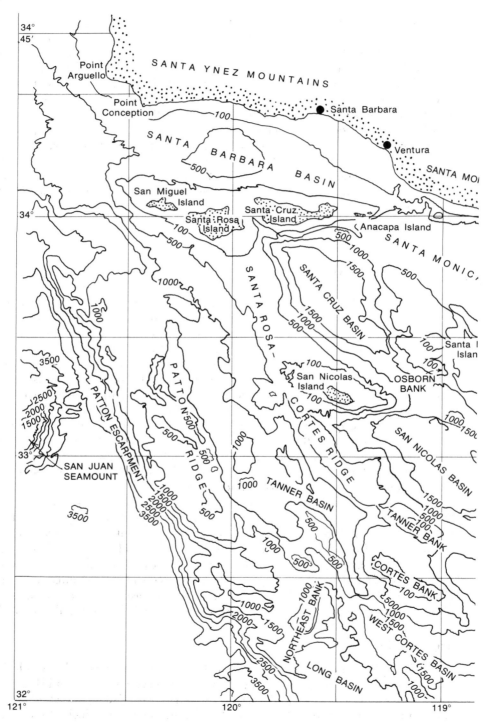

FIGURE 14-5 Topographic features: southern California shelf. (Source: U.S. Geological Survey)

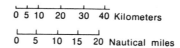

Bathymetric contours in meters

0 5 10 20   30   40 Kilometers

0   5   10   15   20 Nautical miles

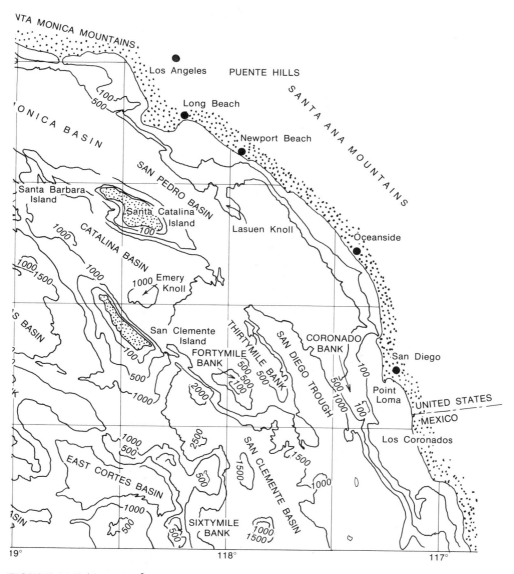

FIGURE 14-5 (Continued)

TABLE 14-1 Characteristics of Continental Borderland Basins

	Number of Sills	Bottom Depth		Lowest Sill Depth		Closure		Area of Flat Floor	
		(m)	(ft)	(m)	(ft)	(m)	(ft)	(km²)	(mi²)
Santa Barbara	1	627	2,056	476	1,560	151	496	1,090	420
Santa Monica	1	939	3,078	737	2,418	212	696	2,030	780
San Pedro	2	913	2,994	737	2,418	176	576	700	270
San Diego	1	900–1,400	3,000–4,500	1,400	4,500	0	0	1,770	680
Santa Cruz	1	1,967	6,450	1,085	3,558	882	2,892	730	280
Catalina	2	1,358	4,452	983	3,222	375	1,230	1,790	690
San Clemente	1	2,108	6,912	1,817	5,958	291	954	1,250	480
San Nicolas	1	1,834	6,012	1,107	3,630	727	2,382	960	370
East Cortes	1	1,980	6,492	1,416	4,644	564	1,848	620	240
Tanner	1	1,552	5,088	1,166	3,822	386	1,266	520	200
West Cortes	5	1,797	5,892	1,363	4,470	434	1,422	650	250
Long	2	1,940	6,360	1,698	5,568	242	792	550	210

Source: After K. O. Emery, 1960.

Los Angeles area, but being closest to the copious sedimentary supply, the former marine basins were filled and converted to dry land.

North of 32 degrees latitude, the Continental Borderland contains 12 distinct basins. Several others exist to the south, but they are outside California waters and will not be considered here. K. O. Emery has assembled considerable data on these basins. Table 14-1 incorporates some of this information, and Table 14-2 shows the relative importance of the varied topographic features of the area.

The boundaries between relief features are by necessity somewhat arbitrary, but it is still evident that the basins and their surrounding slopes constitute a major portion of the province. Admittedly a minor part of the total, the islands, submerged banks, and ridges are of particular interest because they have yielded most of the bedrock samples. Furthermore, they may prove to contain oil reservoirs. Several of the banks are shallow enough to have been emergent during the Pleistocene low sea levels. For example, a drop in sea level of about 100 meters (330 feet), which almost certainly occurred during the last Pleistocene glacial stage, would unite the four northern islands—Anacapa, Santa Cruz, Santa Rosa, and San Miguel—into a single large island. Similarly, such a drop would at least quadruple the area of San Nicolas Island, and would produce three or four new islands along the Santa Rosa-Cortes Ridge. Evidence suggests that there were indeed more islands during the Pleistocene. Submerged wave-cut benches or terraces have been recognized by their topography and by accumulations of wave-rounded pebbles in water that is now too deep for these materials to be exposed to wave action.

Although a substantial drop in sea level today would recreate some of these probable Pleistocene islands, what actually happened was considerably more complex, and there may have been fewer or more islands than at present during different parts of the Pleistocene. Presently submerged marine terraces show that *relative sea level* has risen since their formation, but elevated marine terraces of Pleistocene age exposed on today's islands and mainland coast show that *relative sea level* has dropped. Some of these elevated marine terraces occur at elevations about 620 meters (1,900 feet) on San Clemente Island, at 425 meters (1,300 feet) on Palos Verdes Peninsula, a former island, and 295 meters (900 feet) on San Nicolas Island. It is evident that all three of these presently elevated features were submerged or nearly submerged during the Pleistocene. If all these Pleistocene marine benches could be attributed to a simple rise in sea level, we would expect the elevations of the benches to match from island to island and to agree with elevated marine platforms along the mainland coast and with similar features elsewhere in the world. They do not match, however, and we have no worldwide evidence for Pleistocene high sea levels of 300 meters (1,000 feet) or more above present sea level.

Hence, the evidence is overwhelming that vertical tectonic movements differed in magnitude and timing from place to place on the borderland and that, in combination with the well-established Pleistocene sea-level fluctuations have led to an exceedingly complex history, which will not be fully resolved until virtually every southern California marine terrace is accurately dated.

This independence of block movement seems to be shown by the differences among Palos Verdes Peninsula, Santa Catalina, and San Clemente islands, which lie in a north-south line separated from one another by about 40 kilometers (25 miles). Palos Verdes, at the northern end, has a very well-developed flight of marine terraces of Pleistocene age (see Figure 10-22). Santa Catalina in the middle has few or no

TABLE 14-2 Areas of Continental Borderland Features

Feature	Areas (km²)	(mi²)	Percent of Total Area
Islands	880	340	1.1
Mainland shelf	4,890	1,890	6.2
Insular shelves	3,580	1,390	4.6
Banks	6,270	2,420	8.0
Basin and trough slopes	49,700	19,210	63.0
Basin and trough floors	13,260	5,120	16.8

Source: After K. O. Emery, 1960.

recognizable terraces and in fact shows some evidence of Pleistocene submergence, and San Clemente at the south has perhaps the best developed set of elevated marine terraces in California. Two of these islands show clear evidence of post-Pleistocene uplift and the third appears to show submergence.

Like the northern Offshore area, the southern region also contains submarine canyons. There are at least 32, of which 19 are named. Some are as well known and surveyed as any in the world, although none is so large as Monterey Canyon. In addition, much of the investigation into the origin of such features generally has been conducted in this region. Apparently some canyons have been produced by presently operating sea-floor processes, whereas others have resulted from processes that are similar but currently inactive. Six prominent canyons, presumably related to the modern shoreline, occur along the mainland coast. Several others are cut into the insular shelves. Still others appear to reflect the shoreline and lower sea levels of the Pleistocene epoch.

Rocks

Most submarine rock samples recovered from the Continental Borderland have come from islands, banks, knolls, ridge crests, and steep slopes because most other sections are thickly blanketed with Quaternary sediments. To date, sampling has revealed a complex distribution of basement rocks, but because virtually all the Borderland lies west of the Newport-Inglewood fault zone (now believed to be continuous with the Rose Canyon fault near San Diego), the basement seems to represent various phases of a subduction complex. Examples are exposed on Palos Verdes Peninsula and Santa Catalina island, where the Franciscan-like Catalina Schist (with some gabbro) is present. Somewhat different schists and associated plutonic rocks, including diorite and tonalite, are exposed on Santa Cruz Island, but these too are now considered to be a subduction complex suite. East of the Newport-Inglewood fault, the basement is granitic, much like that exposed in the Sierra Nevada and Peninsular Ranges.

Granitic basement has not been found anywhere in the Borderland, though it is certainly present in the Transverse and Peninsular ranges. Conversely, rocks closely resembling the Franciscan have been dredged from a sea knoll east of Sixtymile Bank, a low ridge northwest of Santa Rosa Island, and the Patton escarpment. Other rocks of Franciscan type have been obtained from several places, including the Patton Ridge, the saddle between Santa Barbara and San Clemente islands, southeast of

San Nicolas Island, and a knoll in the Tanner Basin. Since Franciscan basement is so widely distributed, it seems likely that it underlies the entire Borderland.

Upper Cretaceous sedimentary rocks are abundant in coastal southern California and are exposed on San Miguel Island, and are present below the surface on Santa Cruz Island as well as beneath the Santa Barbara Channel. Rocks of this age are likely to occur at depth throughout the area. Much of the Upper Cretaceous sandstone is relatively unfossiliferous, though, and is not easily dated.

Paleocene rocks have been found on San Miguel and Santa Cruz islands, but none have been recovered elsewhere off southern California. The most likely sites for additional occurrences are the Santa Barbara Channel and the Santa Rosa-Cortes Ridge. These areas contain known Eocene rocks that could easily conceal Paleocene rocks. Eocene rocks are 1,200 meters (4,000 feet) thick on San Miguel, Santa Rosa, and Santa Cruz islands and at least 1,050 meters (3,500 feet) thick on San Nicolas Island.

Among the more enigmatic Eocene rocks are the Poway-type conglomerates mentioned in Chapter 9, whose lithology and general character suggest deposition in a delta-submarine fan setting. Recent studies of the Poway suggest a source east of the Elsinore fault, perhaps in northwest Sonora in Mexico where suitable source rocks are exposed. The occurrence of Poway-type conglomerates on San Nicolas and all the northern islands except Anacapa seems to require considerable post-Eocene strike slip on such faults as the San Andreas, San Jacinto, Elsinore, Newport-Inglewood, in addition to similar movement on some offshore faults and appreciable block rotation of the sort described in Chapter 10.

Because the California Oligocene-Miocene time boundary is disputed, it is difficult to describe Oligocene rock distribution without creating some confusion. If one restricts the Oligocene to Sespe Formation-equivalent rocks, however, it is accurate to say that the Oligocene is so far known only from the Transverse Range part of the Borderland and that there it is entirely nonmarine.

Of all Tertiary rocks, those from the Miocene are most widely distributed both onshore and offshore. They include a wide variety of sedimentary and volcanic rocks, some so distinctive that they often can be identified even when quite unfossiliferous. Middle Miocene rocks are particularly widespread and are distinctive lithologically. They represent three different facies: organic-siliceous shales similar to the Monterey Formation; feldspathic sandstone equivalent to the Topanga Formation in the Transverse Ranges; and blueschist breccias like the San Onofre Breccia of the Peninsular Ranges.

The Monterey facies is the most widely distributed. It is missing in only a few places, chiefly where older rocks are exposed. It is likely that almost all of the Borderland was a relatively deep marine environment during the middle Miocene and that little or no land was present.

Distribution of the San Onofre Breccia raises questions much like those raised by the occurrence of the Eocene Poway rocks. Recent studies seem to suggest, however, that the scattered exposures of San Onofre Breccia did not all come from a single source, but instead were derived from fairly extensive basement ridges now submerged or concealed beneath younger rocks.

Similar-appearing exposures of San Onofre Breccia occur on Santa Cruz, Anacapa, and Santa Catalina islands and on the small Los Coronados group off Tijuana in northwestern Mexico. Mainland exposures include the type locality at San Onofre in northern San Diego County, several exposures near Point Dume on the Malibu coast, at Laguna Beach in Orange County, and some near Tijuana.

This scattered distribution has led geologists to infer the now concealed presence of several Catalina Schist ridges trending northwest-southeast, one of which probably paralleled the coast of Orange and San Diego counties.

Volcanic rocks, primarily of middle Miocene age, are also widespread and occur on most of the islands; flows, sills, dikes, and ash beds are present. Some are land-deposited, but most are submarine. Anacapa and Santa Barbara islands are composed almost entirely of such volcanic rocks. Similar rocks are likewise present on San Clemente, Santa Catalina, and Santa Cruz islands.

Distribution of Miocene rocks is largely independent of present topography. The implication is strong, therefore, that the system of basins, ridges, and islands distinguishing the region today came into existence only after Miocene strata were deposited. This view is confirmed by the distribution of marine Pliocene strata, which are confined almost entirely to existing basins and their margins. Pliocene strata reached phenomenal thickness in the near-shore and onshore basins, as demonstrated by well records. It is likely that thick Pliocene strata are present in many offshore basins as well, because geophysical measurements indicate at least 3,000 meters (10,000 feet) of sedimentary fill in such basins as San Nicolas. However, the Pliocene portion of this fill can only be estimated from geophysical data.

Quaternary basin deposits are undoubtedly often continuations of Pliocene patterns, although Quaternary thicknesses have been established at only a few places such as the Santa Barbara Channel, where 490 meters (1,600 feet) of Holocene have been identified. On the islands, deposits of this age are mainly thin, near-shore shelf and terrace materials.

Special Characteristics of the Quaternary

Despite considerable uncertainty about thickness, Quaternary sediments are widespread and varied. Generallly basins and slopes are blanketed with fine-grained, often greenish muds. Exceptions occur near the mouths of submarine canyons, where coarser-grained sandy deposits have been carried down from beach and shelf environments (turbidites). K. O. Emery has divided shelf sediments into five main categories.

1. Detrital materials derived from stream and cliff erosion along shorelines. This is the dominant type.
2. Relict sediments not now in equilibrium with their environment. These are mostly late Pleistocene coarse-grained deposits submerged by the rising postglacial sea level.
3. Residual organic debris left from in-place weathering of outcrops.
4. Organic sediments, mostly shell gravels.
5. Chemical precipitates, mainly glauconite and phosphorite.

The higher parts of the sea bottom (ridges, banks, and knolls) contain authigenic deposits. These are produced in place by either direct chemical precipitation or alteration of materials already present. Authigenic deposits are commonly associated with organic materials that have accumulated in those places where they were not overwhelmed by detrital materials from land. In contrast, on shelves and in near-shore basins, detrital materials greatly dilute authigenic sediments.

Fine-grained materials are usually absent from topographic highs, mostly because of the strength of tidal currents that sweep across these eminences. Fine deposits can accumulate only in sheltered spots between loose rocks or in cracks and crevices. It is in such environments that much of the authigenic material develops. Two minerals predominate: glauconite (not to be confused with the blue metamorphic mineral glaucophane) and phosphorite.

Glauconite tends to form in oxygen-deficient microenvironments like the interiors of foraminiferal tests and between and beneath the grains of shell gravels. Phosphorite forms more effectively on the better-exposed upper surfaces of bank tops and ridges where oxygenated water is present. Both minerals have potential economic value, though neither is being exploited at present. Glauconite is a low-grade source of potassium suitable for some fertilizers because the potassium is released slowly during weathering. Phosphorite is also a potential fertilizer and a source of phosphorus. Nodules, lumps, and crusts of marine phosphorite probably cover at least 15,600 square kilometers (6,000 square miles) of the Borderland, primarily between depths of 30 and 300 meters (100–1,000 feet).

Structure

Until quite recently our understanding of the structure of the southern Offshore province was based on island exposures, distribution of earthquake epicenters, and sea-floor topography. The rather striking similarity between Offshore basins and those of the Basin and Range province led to the widely-held view that the deep basins were the same kind of down-dropped fault blocks that occur in the Basin and Range province. Distribution of earthquake epicenters seemed to support this hypothesis.

During the 1960s and 1970s, the U.S. Geological Survey conducted a detailed study of the California Borderland. As a result, many basins and ridges are now thought to be the result of large-scale synclinal and anticlinal folding rather than faulting. Because of the great structural relief, this conclusion was unexpected.[1] Maximum submerged structural relief recognized so far occurs between Santa Catalina Island and the San Nicolas Basin—about 6,100 meters (20,000 feet). On land, the Ventura Basin has structural relief of up to 18,000 meters (60,000 feet), and the Los Angeles Basin shows at least 12,200 meters (40,000 feet).

Folds. Several ridges, such as the Santa Rosa-Cortes Ridge, are antiformal in general plan. They probably are not simple anticlines, but vary in size and orientation. A subordinate anticline well exposed on San Nicolas Island is a broad, nearly symmetrical fold quite unlike the tight, asymmetrical folds occurring on land in such places as the Ventura and Los Angeles basins. There is a general tendency for the folds to become broader and more symmetrical away from shore.

---

[1]In this case, structural relief means the difference in elevation between a reference surface under the basin floor and the adjacent ridge or island.

Faults. Despite the discovery that it is less important than formerly thought, faulting is still considered significant. The prominent strike-slip faults exposed in the northern islands undoubtedly continue on the adjacent sea floor and may connect with faults on the mainland. Some faults appear to be long, continuous structures such as the San Clemente Island fault along the east side of the island. Other reasonably established faults are found along the southwestern side of Santa Catalina Island and on the north and east sides of the San Nicolas Basin.

South of the northern islands, faults and folds have a northwesterly trend, and most are believed to be right-slip faults like the San Andreas. Because a number of these faults converge, geophysicists have suggested that movement on these converging structures may be the cause of much of the folding. This is the same sort of process discussed in the previous chapter in connection with the San Andreas fault and its main branches.

According to our present understanding of the structural history of the Borderland, it appears that oblique subduction was in progress off the California coast by the end of the Oligocene epoch. This subduction continued into the Miocene, and resulted in stretching of the Borderland and probably the Basin and Range province as well. This stretching was followed by widespread Miocene volcanism, which decreased as the oblique subduction was replaced by right slip on the San Andreas system of faults, perhaps by mid-Miocene time. By the end of the Miocene, much of the strike-slip movement and rotation of crustal blocks had been completed, so that the geography at the opening of the Pliocene was similar to what we see today, though subsequent vertical movements have modified the shoreline and other details of the scene.

Gravity Measurements

Because gravity profiles must be interpreted, investigators do not always reach the same conclusions from the same data. For example, a negative gravity anomaly indicates a deficiency of mass that some consider a downbuckle of less dense sedimentary material into more dense basement rock. Others may interpret the negative gravity anomaly as a thinning of the underlying crustal material. Deep drilling or some other sort of independent evidence may be needed to settle the question.

Negative anomalies occur a short distance from shore, between La Jolla and Laguna Beach; in the San Pedro, Santa Barbara, and Catalina basins, in the San Diego trough, southern San Clemente Basin, and southwest of San Clemente Island; and in Arguello Canyon. Gravity highs, generally believed to indicate basement or crystalline rocks close to or at the surface, are found seaward from Point Loma and Del Mar; northwest and southeast of Palos Verdes Hills and in outer San Pedro Bay; and southwest and west of San Miguel Island and south of Santa Cruz Island.

SUBORDINATE FEATURES

Coastal Sand Transport and Harbor Problems

As humans have attempted to manage the shoreline and build harbors, breakwaters, jetties, and so on, they unfortunately have learned about some of nature's processes at great cost to property and financial resources. Perhaps the most depressing aspect of this situation is the apparent reluctance to learn from past mistakes or to seek and heed sound advice when it is available. Earlier mistakes are certainly understandable, but recent ones are much more difficult to rationalize. In considering the matter, southern California examples are used primarily because this region of the state has experienced the most intensive coastal development, and most of the studies have been conducted there. However, many equally good examples could have been used from central and northern California; southern California has no monopoly on what might be called "coastal foolishness."

Santa Barbara. Early in the century, construction of a breakwater and harbor at Santa Barbara were repeatedly urged. Santa Barbara had only an unprotected coast with a wooden wharf, which was impossible to use in heavy weather. Moreover, there was no sheltered archorage for boats. Three proposals were reviewed by federal agencies, and each time Santa Barbara was advised not to construct a permanent breakwater. Despite this advice, local interests prevailed and private funding was arranged.

In the late 1920s, an open-ended breakwater was constructed offshore. Sand subsequently accumulated on the protected beach inside the breakwater and inad-

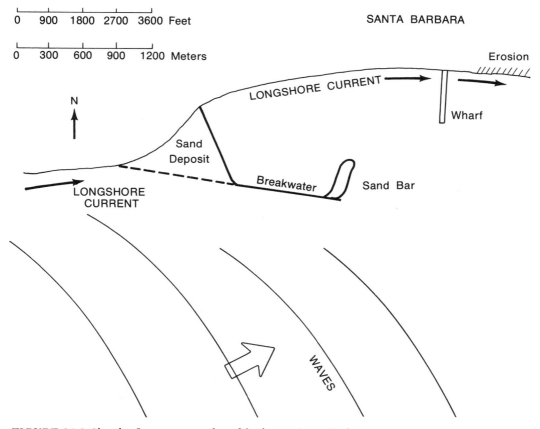

FIGURE 14-6 Sketch of wave approach and harbor at Santa Barbara.

equate protection from wind also became evident. The breakwater's western end was extended to shore to keep sand from entering the harbor and to provide additional wave protection; this greatly improved the harbor—for a while. The longshore current moving toward the east along the shore was forced to deposit its load of sand on the western side of the breakwater (Figures 14-6, 14-7, 14-8). This triangular wedge of sand accumulated for about 5 years until sand began to pass along the outer arm of the breakwater and enter the harbor at the east. While the sand was being trapped, the longshore current continued its activities unabated east of Santa Barbara. Denied its usual load of sand from the west, it obtained a fresh supply by severely eroding the beaches for 16 kilometers (10 miles) east of Santa Barbara. During a particularly stormy winter in the late 1930s, some beaches were cut away as much as 45 meters (150 feet), damaging buildings and narrowing the beaches east of Santa Barbara.

In the middle 1930s, dredging was initiated to counteract the sand moving around the east end of the breakwater and into the harbor. The spoil from this first dredging was carried east of the wharf and dumped in $5\frac{1}{2}$ to 6 meters (18–20 feet) of water, forming a submarine ridge about 300 meters (1,000 feet) long. It was expected that the sand from this ridge would be picked up by the longshore current or waves and be carried east to the depleted beaches. This did not happen, however, and erosion of the eastern beaches continued for another few years until further dredging of the

FIGURE 14-7 Santa Barbara harbor under construction, before sand filling, 1929. (Photo by Fairchild Aerial Surveys, courtesy of Department of Geography, University of California, Los Angeles)

FIGURE 14-8 Santa Barbara harbor after sand filling, 1935. (Photo by Fairchild Aerial Surveys, courtesy of Department of Geography, University of California, Los Angeles)

harbor became necessary. By this time, the U.S. Army Corps of Engineers had begun detailed studies that resulted in landmark reports on the nature of southern California beach processes.

The 15 or more dredgings that occurred subsequently have placed the dredged sand on the beach east of Santa Barbara. From here the waves and longshore current can move the sand along the shore, at least stabilizing if not improving the depleted beaches to the east. The ridge from the first dredging remains today, almost unaltered, and in water too deep and far from shore to augment the sand carried by the currents along the beaches.

The studies made by the Corps of Engineers showed that the currents at Santa Barbara moved an average of 585 cubic meters (770 cubic yards) of sand into the harbor daily. This is a volume about equal to 100 fully loaded dump trucks. This amount has risen to as much as 3,200 cubic meters (4,200 cubic yards) daily during stormy winters and has dropped to as few as 230 to 300 cubic meters (300–400 cubic yards) during quiet summer periods. Studies elsewhere in southern California have shown that the volume of sand moving along the Santa Barbara coast is not particularly high; some stretches of coast, such as that between Ventura and Port Hueneme, experience about three times as much sand transport.

Santa Monica. In the early 1930s, a breakwater was built at Santa Monica, a site as unsuitable for a harbor as Santa Barbara. The breakwater was left open at each end, on the presumption that the longshore current could move the sand through the sheltered area rather than impounding it on the upcurrent side of an attached breakwater. It was not realized, however, that the longshore current and associated beach drifting required normal wave and surf action at the beach. The breakwater was built to eliminate or reduce vigorous wave activity, but it thereby slowed or stopped the current behind the breakwater. Sand was promptly deposited in the sheltered area and by the early 1940s threatened to fill the harbor and join the breakwater to the land. A dredging program was instituted and is still required to keep the harbor open.

Natural Beach Processes. Under normal conditions, what happens to beach sand when it routinely moves along shorelines? Approximate beach equilibrium is maintained by a balance between loss and supply (principally stream and cliff erosion). Sometimes winds move sand inshore to locations beyond the reach of waves. Particularly during storms, waves and currents carry sand into deeper waters, again beyond the reach of normal transporting mechanisms. In addition, sand deposited into the heads of submarine canyons will be carried into deeper waters by turbidity flows and slumping.

Humans disturb any of these processes at considerable cost; for instance, trapping sand behind inland dams will eventually starve the beaches normally nourished by this supply. Similarly, detention of sand caused by breakwater construction results in down-coast beach erosion, often with associated property loss.

Douglas Inman, one of California's leading authorities on beach processes, recognized some years ago that the coastline is divided into more or less self-contained cells, which are often bounded at their upcurrent ends by headlands around which little or no sand is transported. At the down-current ends submarine canyons frequently are present. Beach sand is supplied to the cell chiefly by streams reaching the shore within the cell. Lesser amounts are supplied by wave erosion of sea cliffs. Surplus sand is carried down submarine canyons into deeper water, some is lost offshore during storm activity, and in some instances winds blow beach sand inland to form coastal dunes.

One example of such a cell is the coast from Point Conception in western Santa Barbara County to Port Hueneme in Ventura County. Little or no sand comes around Point Conception from the north, and the beaches of Santa Barbara County rely on supplies from the short streams that drain the south face of the Santa Ynez range. The beaches of Ventura County are more generously supplied by streams like the Ventura and Santa Clara rivers, which drain much larger areas than the short streams of Santa Barbara. The excess sand is lost down Hueneme Submarine Canyon or is blown inshore to form the low coastal dunes between the Santa Clara River mouth and Point Hueneme.

Southern California Submarine Canyons

Eight of the 13 main southern California submarine canyons cross most of the mainland shelf; the other five cross the outer parts of the shelf. These five (and possibly others yet undiscovered) may have once extended all the way across the shelf almost to the beach, but they have been partially or totally filled with sediment since formation.

Most of the more prominent canyons approach shore where longshore currents are deflected seaward either by the orientation of the coast or by the position of headlands. The pattern is so consistent that it can hardly be accidental. As noted previously, longshore currents are known to transport considerable sediment along the coast and near shore, and shallow-water sediments often move down submarine canyons into deeper water. Evidence for this includes the following elements.

1. Presence of near-shore sediments and the remains of shallow-water organisms in the deep-water fan deposits at the mouths of submarine canyons. One of the very first documentations of this process came from a study of the submarine fan in the San Diego trough fed by La Jolla submarine canyon.
2. Direct observations of underwater streams of sand flowing down canyons.
3. Surveys of sedimentary fills in the heads of canyons that prove these materials periodically move down the canyons.
4. Undersea fans that possess levees along the channels crossing the fans.
5. Direct observation of the effects of erosion and abrasion on the canyon walls.
6. Volumes of fans at canyon mouths are characteristically larger than canyon volumes. This indicates tha fan sediments cannot be derived exclusively from erosion of the canyons but require additional material.

For many years it was widely supposed that submarine canyons were cut by land rivers and then drowned by a relative rise of sea level. As these features became better known and their courses were traced to depths of as much as 3,000 meters (10,000 feet) or more, it became difficult to attribute their existence to sea-level change. Furthermore, other considerations apparently contradicted the possibility of such major sea-level shifts. The close association of canyon heads, present sea level, and coastal configuration seemed to demand an origin related to presently active marine processes.

Although parts of some canyons may reflect the location of land streams during lowered Pleistocene sea levels, most canyons are not clearly associated with land streams. Most are eroded and kept open by periodic flows of turbid water-sediment mixtures that carry near-shore materials into deeper water, build fans and levees, and scour canyon walls. Canyons with heads some distance from the present shore apparently formed when lowered sea level changed the coastline. This permitted longshore currents to transport sediments into what are now fossil canyons.

The main obstacle to a satisfactory explanation of submarine canyons has been the inability of scientists to observe what happens in these environments. Although turbidity currents have been studied in the laboratory for many years, and their existence in submarine canyons was suspected, their erosional competence had not been clearly demonstrated. Consequently, many workers were reluctant to acknowledge that a process with questionable erosional efficacy could account for features the size of the Grand Canyon. Stream erosion, however, was well documented and understood. It was therefore considered the most likely agent, despite the serious problems associated with major sea-level changes. On the one hand, geologists knew of an effective process that would only work if the sea level could be disregarded. Conversely, a likely process appeared to exist, but its presence on the sea floor was doubted and its erosional competence was speculative at best. Nevertheless, as a cause of submarine canyons, it did not require any troublesome change in sea level. The turbidity current origin is obvious today, now that so much is known about events and processes on the sea bottom. The choice certainly was not so obvious even three decades ago, when neither the river-cut nor the turbidity current adherents could present invincible arguments. This is a vignette of the way that geology advances, and the same pattern of dispute and discovery of new evidence can be expected to solve many of today's problems.

Several of the world's most intensively studied submarine canyons occur off the California coast. The best known is La Jolla Canyon, with its two branches — Scripps and Sumner canyons. Francis P. Shepard long supervised regular surveying and monitoring of La Jolla Canyon and its sediments. It has been established that La Jolla Canyon is one of the longer canyons off southern California, extending 40 kilometers (25 miles) from the beach to the floor of San Diego Trough. Its outer end is about 1,000 meters (3,300 feet) deep.

The deepest southern California canyon is Redondo Canyon in Santa Monica Bay. Although it can be traced only 16 kilometers (10 miles) to its terminus on the floor of the Santa Monica Basin, Redondo Canyon reaches depths of 600 meters (2,000 feet). Like many others, the head of Redondo Canyon has repeatedly and gradually filled with sediment, only to be abruptly deepened as sediment suddenly moved down the canyon into the basin below. These abrupt flushings have occurred in various types of weather and at irregular intervals. It is not known what triggers these outward flows of sediment.

Oil and Natural Gas on the Continental Shelf

During the last 25 years or so, the oil and natural gas resources of offshore southern California have been subject to intensive development, particularly the portions lying off Santa Barbara, Ventura, and Los Angeles counties. But the great majority of oil activity has been located off the coast of Santa Barbara County — much of it in the Santa Barbara Channel. Beginning about 1980, the west-facing coast from Point Arguello northward toward San Luis Obispo was being explored. The major Point Arguello field was discovered in 1981 and it may prove to be a giant oil field, which means that it contains more than 500 million barrels of recoverable oil; a super-giant field is one containing more than 5,000 million barrels.

Other fields have been found off Point Sal and the mouth of the Santa Maria River, and additional fields may well exist farther north off the coast of San Luis Obispo County (Figure 14-9). Indeed, at this writing, exploration of the central and northern California coast is proceeding apace, and one area thought to be especially promising lies offshore from Eureka in Humboldt County. Past experience on shore suggests that thick accumulations of Tertiary sedimentary rocks offer the most favorable sites for petroleum, though considerable natural gas has come from Cretaceous rocks in the Sacramento Valley. Some future shelf areas likely to be tested when and if environmental concerns can be allayed are from Monterey to Half Moon Bay on the central California coast.

Offshore oil development has pitted a number of constituencies against one another. Many coastal residents dislike the appearance of offshore oil platforms, are concerned about air pollution from processing the oil and gas, and fear the consequences of another major oil spill such as those that occurred off Santa Barbara in January 1969 and at Valdez, Alaska, in March 1989. Conversely other residents are interested in exploiting and profiting from this important natural resource, and cor-

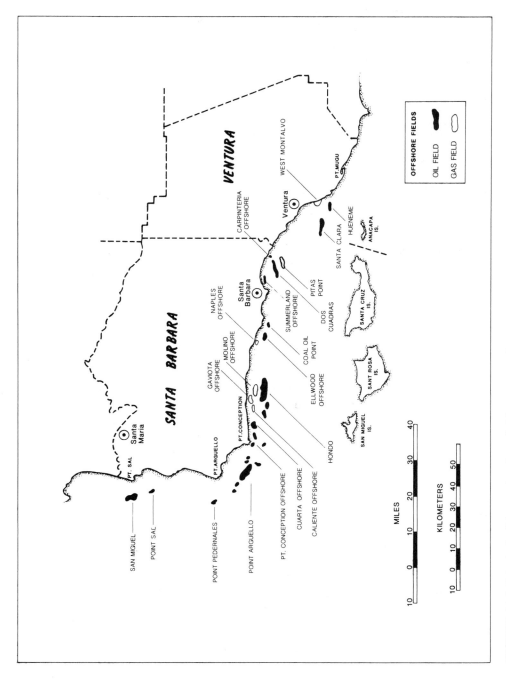

FIGURE 14-9 Principal offshore oil and gas fields, Santa Barbara-Ventura area. Some other offshore fields occur near Long Beach in Los Angeles County.

FIGURE 14-10 Small active tar seep spreading out onto the beach at Carpinteria, Santa Barbara County. (Photo by R. M. Norris)

rectly call attention to increasing American dependence on foreign supplies, many of which are in politically unstable areas. It is ironic that exploitation of California's offshore petroleum resources has simultaneously made important contributions to the local and national economy and given enormous impetus to a heightened environmental concern in California, the United States, and even the world.

Oil and gas have played a role in southern California's coastal region throughout history. Forty or more active oil and gas seeps are known from the Santa Barbara Channel alone. Oil from the offshore seeps as well as from those along the beach have long produced the tarry blobs beach users complain about (Figure 14-10). The Indians used this tar to caulk boats, to seal baskets and abalone shells for carrying water, and for various kinds of adhesives or other purposes.

The first written account of submarine oil seeps in the Santa Barbara Channel appears to be that of the English explorer, George Vancouver who wrote in 1792:

> The surface of the sea, which was perfectly smooth and tranquil, was covered with a thick slimy substance, which when separated or disturbed by a little agitation, became very luminous. Whilst the light breeze, which came principally from the shore, brought with it a strong smell of tar, or some such resinous substance. The next morning the sea had the appearance of dissolved tar floating on its surface, which covered the sea in all directions within the limits of our view . . .

Tar on the beach, often blamed by today's beach habitués on tankers pumping bilges, or sometimes correctly on offshore oil operations, also has a long history. Except for rare disasters such as the January 1969 oil spill, the tar on the beach comes mostly from offshore oil seeps. In 1776, Padre Pedro Font, while near Goleta, wrote:

> . . . much tar which the sea throws up is found on the shores, sticking to stones and dry. Little balls of fresh tar are also found. Perhaps there are springs of it which flow out into the sea, because yesterday on the way, the odor of it was perceptible, and today the scent was as strong as that perceived in a ship or in a store of tarred ship tackle and ropes . . .

Although there are some giant oil fields off the California coast, most are not large by world standards. These fields probably will not produce as much oil and gas during their entire productive lives as is used in United States in a few months' time. The immense capital cost of developing California's offshore oil means that only the very largest offshore fields can be exploited unless notable changes occur in national policy, the price of oil, or taxation. Deep-water platforms of the type required cost as much as $300 million or more, and each well drilled from the platform may cost an additional $5 million. This high cost of offshore drilling means that a substantial resource may never be developed, or may remain untouched for many years.

REFERENCES

Anonymous, 1959. Offshore Geology and Oil Resources. Mineral Information Service (now California Geology), v. 12, p. 7.

Dupras, Don, 1987. A Thumbnail Sketch, California's Marine Geology. California Geology, v. 40, pp. 171–180.

Emery, K. O., 1960. The Sea Off Southern California. New York, John Wiley and Sons, Inc.

Graham, S. A., and W. R. Dickinson, 1978. Evidence for 115 Kilometers of Right-Slip on the San Gregorio-Hosgri Fault Trend. Science, v. 199, pp. 179–181.

Greene, H. G. et al., 1973. Faults and Earthquakes in the Monterey Bay Region, California, next accompanying U.S. Geological Survey Map MF-518.

Hanna, G. D., 1951. Geology of the Farallon Islands. California Division of Mines and Geology Bulletin 154, pp. 301–310.

Howell, D. G., ed., 1976. Aspects of the Geologic History of the California Continental Borderland. Pac. Sect. Amer. Assoc. Petroleum Geologists Spec. Publ. 24, 561 p.

Rusnak, Gene A., 1966. The Continental Margin of Northern and Central California. In Geology of Northern California. Calif. Div. Mines and Geology Bull. 190, pp. 325–335.

Shepard, F. P., and K. O. Emery, 1941. Submarine Topography off the California Coast— Canyons and Tectonic Interpretations. Geol. Soc. America Spec. Paper 31.

Vedder, J. G. et al., 1969. Geology, Petroleum Development and Seismicity of the Santa Barbara Channel Region. U.S. Geological Surv. Prof. Paper 679.

_____, 1974. Preliminary Report on the Geology of the Continental Borderland of Southern California. U.S. Geological Surv. Map MF-624 and accompanying text.

GLOSSARY

Acid Igneous rocks containing a high percentage of silica.

Aggradation Process of building up a surface by deposition.

Allochthonous Formed in another area than that in which it is presently located.

Alluvial fan Low, cone-shaped stream-deposit at the base of a mountain range.

Alluvial plain Plain resulting from the deposition of alluvium by water.

Alluvium Detrital deposits of modern rivers and streams.

Amphibole Group name for common rock-forming silicate minerals high in iron and magnesium.

Andesite Volcanic rock composed essentially of plagioclase and one or more mafic constituents. Quartz is often present.

Anorthosite Plutonic rock composed almost wholly of plagioclase.

Antecedent stream A stream whose course was maintained despite the uplift of a barrier in its path by tectonic processes.

Anticline Strata that dip in opposite directions from a common ridge or axis with younger rocks on the flanks and older rocks in the core.

Antiform An anticline-like structure in which the stratigraphic sequence is not known.

Aquiclude A body or layer of relatively impervious rock that forms a cap or a base to a water-bearing aquifer.

Aquifer A formation or group of formations that are water bearing.

Armored mudball A rounded pebble or boulder originally composed of soft mud that became coated or studded with small pebbles during stream transport.

Artesian water Ground water that is under sufficient pressure to rise above the level at which it is encountered in a well. It does not necessarily rise to or above the surface of the ground.

Ash Fine volcanic material with particles less than 4 millimeters across. The term is usually applied to unconsolidated material but may be used for the consolidated equivalent (tuff).

Attitude General term to describe the relation of some directional feature in a rock to north or to a horizontal plane.

Augite A common rock-forming silicate mineral belonging to the pyroxene group.

Autochthonous Formed or occurring in the place where it was formed.

Autolith A fragment of igneous rock enclosed in another igneous rock of later consolidation, each being regarded as a derivative of a common parent magma.

Badlands Region nearly devoid of vegetation where erosion has cut a large number of closely spaced gullies and channels. Typical or arid or semiarid regions and soft, impermeable rocks.

Bajada An apron formed of coalescing alluvial fans at the base of a desert mountain range.

Barchan A crescent-shaped sand dune that migrates in the direction the horns point. Usually indicative of consistent wind direction.

Basal conglomerate Coarse, usually well-sorted and well-rounded lithologically homogeneous sedimentary deposit found just above an erosional break.

Basalt An extrusive rock composed primarily of calcic plagioclase and pyroxene, with or without olivine.

Base level The level below which a land surface cannot be reduced by running water.

Basement Underlying rock complex, usually crystalline, that behaves as a unit mass.

Base surge A ring-shaped cloud of gas and suspended particles that moves outward from the base of a vertical column accompanying a volcanic eruption.

Basic Refers to igneous rocks comparatively low in silica.

Batholith A massive body of intrusive (plutonic) igneous rock that includes many related but independent intrusives (plutons) that developed in sequence often over long periods of time.

Bed Smallest division of a stratified series, marked by a more or less well-defined plane from its neighbors above and below.

Bedrock Any solid rock underlying unconsolidated surficial deposits such as alluvium.

Bergschrund A crack, often crescentic, at the head of a mountain glacier, produced as the ice moving down the valley separates from ice remaining attached to the rock headwall.

Biotite A common rock-forming silicate mineral of the mica group, characterized by flexible cleavage plates and black color.

Bituminous Containing much organic or carbonaceous matter, usually in the form of tarry hydrocarbons described as bitumen.

Block A chunk of solid material, between 5 centimeters and several meters in diameter, erupted from a volcano.

Block faulting Process by which rock is divided into small or large units by faulting.

Block mountains Mountains carved by erosion of large uplifted blocks bounded on at least one side by fault scarps.

Bolson An alluvial-floored desert valley without a surface outlet.

Brachiopod A member of a phylum of marine, shelled animals with two unequal valves (shells), each of which is bilaterally symmetrical.

Brackish Waters with saline content intermediate between sea water and fresh water.

Breccia A coarse-grained sedimentary rock composed of angular fragments.

Brine Water strongly impregnated with various salts.

Butte Conspicuous isolated hill or small mountain, especially one with steep sides or a turretlike form. Typical of badlands.

Caldera Large basinlike volcanic depression, more or less circular in plan, often produced by collapse following a violent eruption.

Carbonatite A plutonic igneous rock consisting mainly of calcium or magnesium carbonate of uncertain origin.

Chert A compact, siliceous rock formed of chalcedony or opaline silica.

Cienaga springs Small springs or seeps occurring near the base of alluvial fans in desert or semiarid settings.

Cinder cone A conical edifice formed of fragmental material, mostly of pea to orange-size erupted from a volcanic vent, generally of basaltic composition.

Cirque A deep, steep-walled recess or hollow formed at the head of a glaciated mountain valley.

Clast A sedimentary rock fragment.

Clastic A textural term applied to rocks formed from fragments of preexisting rocks.

Clay A fine-grained sediment that can be deformed plastically. *Also*, a group of closely related sheetlike micaceous silicate minerals.

Climatic firn limit An average line or elevation above which some snow lies on the ground throughout the year.

Climatic optimum A period between 9,000 and about 2,500 years ago during which the earth's climate, particularly at high latitudes, was warmer than at present (see also Hypsithermal interval).

Coal A dark brown or black organic rock composed by the accumulation and alteration of plant material.

Coastal plain A plain with its margins on a large body of water, usually the sea. Often represents a strip of recently emerged sea bottom.

Colluvium Soil and other unconsolidated rock material on hill slopes; not transported by water.

Columnar jointing Variety of jointing that produces columns, usually forming a hexagonal pattern. Most characteristic of basalts and andesites. Generally regarded as a consequence of shrinkage during cooling of a lava flow.

Columnar section A graphic expression of the sequence and stratigraphic relations of rock units in a given region.

Comminuted ash Finely divided volcanic ash.

Concretion A nodular or irregular concentration of certain mineral constituents in a sedimentary rock or tuff; developed from localized deposition of material from solution.

Conglomerate Cemented clastic, coarse-grained rock composed of rounded particles from pebble to boulder-size.

Connate water Water trapped in a sedimentary rock at the time of deposition. Often fossil sea water.

Conodont A small, generally $\frac{1}{2}$ to 3 mm long toothlike structure of apatite of uncertain organic origin used for stratigraphic purposes.

Contact Place or surface where two kinds of rock come together.

Continental deposits Sedimentary deposits laid down on dry land, in streams or in lakes, contrasted with marine deposits formed in the sea.

Continental drift The slow movement of continental crustal blocks presumably driven by convective movements in the mantle. One of the processes of plate tectonics.

Continental shelf The submerged margin of continents and continental islands extending to a depth of about 100 meters (350 feet), the outer edge of which is usually marked by a steep descent to the deep ocean floor (continental slope).

Continental slope The steep declivity at the outer edge of the continental shelf, marking the boundary zone between continental and oceanic crustal blocks or plates.

Convection Process of mass movement of portions of any medium in a gravitational field as a consequence of different temperatures or densities in the medium.

Convoluted bedding Twisted and contorted bedding resulting from disturbance of sedimentary materials after deposition but before consolidation.

Coulee A short, stubby, steep-fronted lava flow, usually of rhyolitic composition.

Creep Slow, more or less continuous deformation of soil, rock, or ice under gravitational or tectonic stress.

Crystal In mineralogy, the regular polyhedral form bounded by plane surfaces (faces), that is the outward expression of an orderly internal arrangement of atoms. In petrology, a crystalline grain with or without crystal faces.

Crystallization Process through which crystal phases separate from a fluid, viscous, or dispersed state.

Current structures Forms produced by currents of wind, waves, or streams in sediments during deposition.

Cyanobacteria Photosynthetic single-celled aquatic organisms often called blue-green algae.

Cyclic salt Salt transported inland from the sea by winds blowing onshore.

Dacite A fine-grained extrusive or volcanic rock having roughly the same composition as andesite but with more calcic feldspar.

Deflation Erosion and transport by wind.

Deformation Any change in the volume or form of rock masses produced by tectonic forces.

Delta Deposit of sediment in a body of water at the mouth of a river.

Dendritic drainage A branching tree-like pattern of stream channels developed generally on rocks of uniform texture and structure.

Desert pavement A mosaic of pebbles forming flat surfaces of varying extent between desert water courses. Formed by one or more processes, including deflation of finer grains, upward movement of pebbles through a clay substrate, or by sheetwash.

Desert varnish A surface stain of clay and oxides of iron and manganese deposited on exposed surfaces in arid regions by the action of bacteria.

Detachment fault A structural feature produced by crustal extension or stretching in which the overlying block behaves independently of the underlying block. Listric normal faults may merge at depth to form a detachment fault or zone.

Detritus Material that results from the disintegration and erosion of minerals and rocks.

Diapir (piercing fold) An antiform in which a mobile core has injected the more brittle enclosing rocks.

Diastrophism Processes by which the crust of the earth is deformed.

Diatomite A sedimentary rock composed of the siliceous remains (frustules) of one-celled green plants called diatoms.

Differentiation Process by which different types of igneous rock are derived from a parent magma.

Dike Tabular body of rock, usually igneous, that cuts across the structure of adjacent rocks or cuts massive rocks.

Dilatancy An increase in volume during deformation. Results from a change in close-packed structure to a more open structure with an increase in pore space.

Diorite A plutonic igneous rock composed of soda-rich feldspar and hornblende, biotite, or pyroxene.

Dip Angle at which a stratum or any planar structure is inclined from the horizontal.

Dip slip Component of slip (on a fault) parallel with the dip.

Divergent plate boundary A spreading center where basaltic magma rises by convective upwelling. Associated with lateral spreading at a rate of 1 to 10 centimeters per year.

Dolomite Common rock-forming carbonate mineral. Rocks that approximate the mineral dolomite (calcium, magnesium carbonate) in composition.

Dome
> *Structural* A roughly symmetrical upfold of strata with beds dipping in all directions, more or less equally, from a point.
> *Volcanic* Steep-sided protrusion of viscous lava forming a more or less dome-shaped or bulbous mass over and around a volcanic vent.

Drainage basin Part of the land surface that is occupied by a drainage system or contributes surface water to that system.

Dripstone A collective term describing limestone cave features formed by water dripping from the ceiling.

Dry lake A playa lake or an ephemeral desert lake only occasionally containing water. The dry bed of such a lake.

Dunite Peridotite, a plutonic igneous rock consisting almost wholly of the mineral olivine with accessory pyroxene or chromite.

Earthquake Groups of elastic waves propagating in the earth, set up by a transient disturbance of the elastic equilibrium of a portion of the earth. Vibration resulting from waves produced by sudden slippage along a fault.

Efflorescence A whitish crystalline crust left on rocks by water brought to the surface by capillary action and then evaporated.

Elastic rebound theory Faulting arises from the sudden release of elastic energy that has slowly accumulated in the earth. Just before rupture, the energy released by faulting is potential energy stored as elastic strain in the rocks. At the time of rupture, the rocks on either side of the fault rebound to positions of little or no strain.

En echelon An overlapping, staggered arrangement, particularly of faults.

Epicenter Point on the earth's surface directly above the focus or hypocenter of an earthquake.

Erratic An ice-transported rock carried a varying distance from its source.

Escarpment Cliff or relatively steep slope separating gently sloping or level tracts.

Estuary Drowned river valley, usually along the seacoast, forming an inlet or narrow bay.

Eugeocline A subsiding marine trough, often characterized by the accumulation of sediments of volcanic origin and often lying between an island arc and a more stable continental margin. Formerly *eugeosyncline*.

Eustatic Real change in sea level resulting from variation in the volume of seawater. Often caused by continental glaciation.

Exfoliation A weathering process by which thin shells are successively detached from a larger rock mass. Particularly common in plutonic igneous rocks.

Exotic terrane A small plate fragment or microplate attached to a larger plate, but of distant origin. May have been transported by continental drift or transform faulting.

External drainage Drainage to the sea in contrast to internal or inland drainage.

Extrusive rocks Igneous rocks derived from magmatic material ejected at the earth's surface by volcanic activity. Contrasts with intrusive or plutonic rocks

that cool or solidify at depth in the earth's crust.

Facet Nearly plane surface abraded on rocks or a rock fragment; polished surface of a cut gemstone; fault facet on the ridge ends along a fault scarp.

Fanglomerate Heterogeneous sediment originally deposited as an alluvial fan, but subsequently cemented into solid rock.

Fault Fracture or fracture zone along which there has been displacement of rocks on either side relative to one another.

Fault block A body of rock bounded by one or more faults.

Fault-line scarp Scarp that is the result of differential erosion along a fault rather than the direct result of movement along the fault.

Fault-line valley Valley excavated by erosion along a fault.

Fault scarp A cliff formed by a fault.

Fault system Two or more faults or groups of faults that are related in space, movement type, or time.

Fault valley Valley formed by faulting.

Fault zone Belt tens or hundreds of meters wide consisting of numerous interlacing or branching small faults and associated zones of gouge.

Fauna The animals collectively of any given region or time.

Feldspar Group of abundant silicate rock-forming minerals, containing high proportions of potassium, sodium, or calcium, plus aluminum.

Fenster A window or hole through a thrust fault upper plate exposing the rocks below the thrust.

Flood plain Portion of low-lying land adjacent to a river that is periodically flooded by the stream.

Flora The plants collectively of any given region or time.

Flows Extrusions of fluid lava.

Fluid pressure The pressure exerted by fluids in the pore spaces of a rock.

Focus True center of an earthquake, within which strain energy is first converted to elastic wave energy. Hypocenter.

Fold A bend in strata or in any planar structure.

Fold axis Trend of the crest of an anticline or trough of a syncline.

Foliation Laminated structure resulting from the segregation of different minerals into layers parallel to schistosity.

Foraminifer Member of a subdivision of the Phylum Protozoa; shells (tests) usually microscopic in size, commonly composed of calcium carbonate and more rarely of sand or other mineral particles.

Forearc basin An accumulation of marine clastic sediments mainly of volcanic origin deposited between an island arc and a trench or subduction zone.

Formation Lithologically distinct product of essentially continuous sedimentation selected from a local succession of strata as a convenient unit for mapping, description, or reference.

Fossil Animal or plant remains or traces preserved in the earth's crust by natural methods. Excludes organisms buried since the beginning of historic time.

Frustule The skeleton secreted by diatoms, generally composed of opaline silica.

Fumarole Hole or vent in a volcanic area that emits fumes or vapors.

Gabbro Plutonic rock consisting of calcic plagioclase and clinopyroxene, with or without orthoclase and olivine.

Gastropod Member of the Phylum Mollusca, class Gastropoda, usually with an asymmetrically coiled calcareous exoskeleton.

Geomorphology Branch of science that treats surface features, their form, nature, origin, and development.

Geocline Geosyncline. Large trough filled with sediments as much as 10,000 meters thick, or regional extent and a long history of subsidence, mostly under marine conditions.

Geyser Intermittent eruptive hot spring in which discharge occurs at more or less frequent and regular intervals, caused by the expansive force of highly heated steam.

Glacier Mass of ice with definite lateral limits and motion in a definite direction, originating from the accumulation and compaction of snow.

Glass-sand Relatively pure silica sand suitable for making glass or pottery.

Glauconite A green micaceous mineral composed of hydrous potassium, iron silicate. Commonly occurs in rocks of marine origin; a rock rich in glauconite.

Glaucophane Blue metamorphic mineral of the amphibole group containing sodium. Indicative of metamorphism at high pressure and relatively low temperature.

Gneiss Coarse-grained metamorphic rock in which bands of granular minerals alternate with bands of schistose minerals.

Graben Block that has been down-thrown along faults relative to the blocks on either side.

Graded beds Stratification in which each layer displays a gradation in grain size from coarse below to fine above.

Granite A plutonic igneous rock composed chiefly of quartz and potash feldspar.

Granitization A metamorphic process that produces a granite from sediments or other rocks without melting.

Granodiorite Plutonic rock consisting of quartz, plagioclase, and orthoclase, with biotite, hornblende, or pyroxene as mafic constituents; intermediate between quartz diorite and quartz monzonite.

Granulite A metamorphic rock composed of even-sized, interlocking, granular grains.

Gravel A coarse-grained sediment in which many grains are coarser than sand.

Graywacke A dark-colored sandstone with abundant plagioclase and various dark minerals or rock fragments. Often derived from volcanic source rocks.

Ground water Water contained in the pore spaces in soils and rocks.

Hogback A ridge formed from steeply-dipping layered rocks.

Horizontal component Amount of movement reflecting displacement in the horizontal plane when movement on a fault occurs in an oblique direction.

Hornblende Commonest member of the amphibole minerals.

Horst Block that has been uplifted along faults relative to the adjoining blocks.

Hypsithermal The climatic optimum, a warmer than average period between about 9,000 and 2,500 years ago.

Hypocenter The focus of an earth-quake. The point where strain energy is converted to elastic wave energy.

Igneous Class of rock formed by the solidification of molten or partly molten parent matter.

Ignimbrite A welded tuff, composed of pumiceous volcanic ash erupted as ash flows or nuées ardentes.

Inclusion A fragment of an older rock included in an igneous rock to which it may or may not be related.

Inlier Approximately circular or elliptical area of older rocks surrounded by younger strata.

Internal drainage Drainage into an enclosed basin with no outlet to the sea.

Intrusive rocks Igneous rocks that, while fluid, penetrate into or between other rocks, but solidify before reaching the surface.

Isocline (isoclinal fold) Anticline or syncline so closely folded that beds on the two sides of the fold have the same dip.

Isostasy A condition of equilibrium in which crustal rocks appear to float on a plastic substratum.

Isotopes Elements having an identical number of protons in their nuclei, but differing in their number of neutrons; isotopes have the same atomic number, different atomic weights, and almost the same chemical properties.

Joint A fracture or parting in a rock without displacement.

Jökulhlaup A glacial outburst flood, usually due to volcanic activity beneath an ice sheet.

Kernbut Buttelike hill or buttress on a canyon wall or an outer ridgelike edge to a fault terrace or bench; separated from the main hillside by a sag (kerncol).

Kerncol Sag between a kernbut and the adjoining hillside.

Klippe An isolated remnant of a thrust sheet detached from the main part of the sheet. Underlain by a thrust fault. Sometimes referred to as a mountain without roots.

Laccolith A concordant intrusive body that has domed up the overlying rocks, but usually has a flat floor.

Lahar A volcanic mudflow.

Landslide A gravity-propelled down-slope movement of a mass of soil or rock.

Lapilli Fragmental volcanic ejecta ranging from 1 to 64 millimeters in grain size.

Lateral moraine A deposit of glacially transported material along the sides of a mountain glacier.

Lava Fluid rock (magma) that issues from a volcano or fissure at the earth's surface; the same material solidified by cooling.

Left lateral, left slip, left separation Occurs where the motion on a fault is such that an observer standing on one block would see evidence that the opposite block had shifted to his left.

Limestone: Bedded sedimentary rock consisting chiefly of calcium carbonate.

Lineation Parallel orientation of structural features in rocks that are in lines rather than in planes.

Listric fault A type of fault in which the fault surface is concave upward. Results from crustal extension. Listric faults may merge at depth with detachment surfaces or faults.

Lithology Study of rocks based on a megascopic examination; used loosely to mean the composition and texture of rocks.

Load cast Mark consisting of a swelling on the bottom of a sedimentary layer that extends downward into the under-lying layer.

Lode Several veins closely spaced such that all together, with the intervening rock can be mined as a unit.

Louderback cap A remnant of a lava flow on a tilted fault block.

Maar Relatively shallow, flat-floored explosion crater with walls that consist largely of fragments of country rock and only partly of magmatic ejecta.

Mafic Pertaining to rocks composed dominately of iron and magnesium silicates.

Magma Naturally-occurring molten rock material generated within the earth and capable of intrusion or extrusion.

Magmatic arc Regions of roughly ar-cuate plan in which volcanism and seis-micity are intense. Where the sub-ducted plate lies below oceanic crust, an island magmatic arc results; where it lies below continental crust, a conti-nental magmatic arc occurs. The gran-itic rocks of the Sierra Nevada were formed as a part of a continental mag-matic arc, for example.

Magnetic chronology The time record of changes in the polarity of the earth's magnetic field.

Magnetic dip The vertical component of the earth's magnetic field at any given geographic location.

Magnetic reversal The time at which the earth's magnetic field reverses polarity.

Mantle Segment of the earth's crust that lies between the Mohorovičić discontinuity and the core (Gutenberg discontinuity); regolith originating from weathering of bedrock.

Mass wasting Processes by which large masses of earth material are moved by gravity.

Meander A regular, looplike bend in a stream course, developed when a stream flows at grade and erodes its channel on the outsides of the bends.

Medial moraine A deposit of glacial debris formed by the merging of two lateral moraines where ice streams join.

Melange A heterogeneous mixture of rock materials, often pervasively sheared, of angular and variable-sized blocks, often from local or distant sources, developed in a subduction zone.

Mesa Flat-topped mountain or other elevation bounded on at least one side by a steep cliff.

Metamorphic rock Rock formed in the solid state in response to pronounced changes in the temperature, pressure, or chemical environment.

Metasedimentary rock A metamorphic rock derived from sedimentary materials.

Metavolcanic rock A metamorphic rock derived from volcanic materials.

Mica A silicate mineral group with sheetlike structure.

Microcline A potassium feldspar, common in granitic rocks.

Microplate A relatively small fragment of a major crustal plate, usually isolated by transport away from the parent plate.

Mineral A homogeneous inorganic, naturally-occurring crystalline material of definite chemical composition.

Mineralizer Mineralizing agent: substance that, when present in magmatic solutions, lowers temperature and viscosity, aids crystallization, and permits the formation of minerals.

Miogeocline (miogeosyncline) A depositional environment near the edge of the stable continental interior (craton) with little volcanic material. Approximating the continental shelf depositional environment.

Mohorovičić discontinuity The boundary zone between the base of the crust and the top of the mantle.

Mollusks Members of the Phylum Mollusca of invertebrate, generally shell-bearing animals, including gastropods, pelecypods, and cephalopods.

Moraine Accumulation of drift material built chiefly by direct deposition of glacial ice.

Mudflow Flow of heterogeneous debris mixed with water and usually following a stream course.

Mudstone A sedimentary rock with poor or indistinct bedding formed from deposition of fine-grained materials (less than silt-size).

Mylonite Fine-grained, laminated rock formed by extreme micro-brecciation and milling of rocks during fault movement.

Natural gas Mixture of naturally-occurring hydrocarbon gases, frequently associated with petroleum deposits.

Normal fault A dipping fault in which the upper block moves downward with respect to the underlying block.

Nose (anticlinal) The end of an elongate fold.

Obsidian Volcanic glass.

Olistrostrome A heterogeneous sedimentary deposit formed by gravity sliding of a semifluid mixture.

Olivine Important rock-forming iron and magnesium silicate mineral, especially in mafic or ultramafic rocks.

Ooze A fine-grained pelagic deposit that is more than 70 percent organically derived.

Ophiolite A group of basic and ultrabasic igneous rocks, ranging from spilite and basalt to gabbro and peridotite, including rocks rich in serpentine, albite, chlorite, and epidote. Regarded as a sample of oceanic crust and often associated with a subduction zone.

Orogeny Process of mountain building, particularly by folding and thrusting.

Orthoclase A potassium feldspar, common in granitic rocks.

Outcrop Exposure of unaltered rock above the surface of the ground.

Overthrust Crustal shortening in which field evidence indicates that one block was shoved over a relatively immobile block.

Paleontology Study of fossil remains, both animal and vegetable.

Paternoster lakes A string or series of lakes occupying glacially scoured rock basins, often at different elevations.

Pecten A scallop, a member of the pelecypod class of mollusks with two shells like a clam.

Pegmatite Coarse-grained igneous rock, usually found as a dike, and associated with a large mass of finer-grained plutonic rock.

Pelecypod A class of bivalve mollusks with shells asymmetrical mirror images of each other.

Peneplain Land surface worn down by erosion to a nearly flat or undulating plain irrespective of underlying rock resistance.

Percolation Movement under hydrostatic pressure, of water through interstices of soil or rock.

Peridotite Plutonic igneous rock consisting chiefly of olivine with or without other mafic minerals.

Periglacial A climatic environment close to, but not glacial.

Permeability The degree to which pore spaces in soils or rocks are interconnected.

Petrifaction Process of converting organic matter into rock.

Petrogenesis Origin of rocks, particularly igneous rocks.

Petrology Study of the natural history of rocks.

Phenocryst Relatively large and conspicuous crystals of early formation in a porphyry.

Phyllite Argillaceous rock intermediate in grade between a slate and a schist.

Piercing point A reference line perpendicular to a fault plane that can be used to measure the amount of vertical and horizontal slip.

Pillar A column formed by joining of stalagmites and stalactites in limestone caverns.

Pillow lava Lava structure consisting of an agglomeration of rounded masses that resemble pillows; occurs in basic volcanic rocks erupted under water.

Placer Alluvial or glacial deposit, usually of sand or gravel, containing particles of a valuable mineral.

Plagioclase A group of closely related soda-lime feldspars. Common in igneous and metamorphic rocks.

Planation Widening of valleys through lateral erosion of streams after the streams achieve grade and begin to meander, including the formation of flood plains.

Plate A major unit of the earth's crust, capable of movement as a discrete entity.

Playa Dry lake bed in the bottom of an enclosed desert valley.

Pluton Mappable body of igneous rock formed beneath the earth's surface by consolidation of magma.

Plutonic A class of igneous rocks that has crystallized at depth and developed a granitic texture.

Porosity The amount of pore space in a rock.

Porphyry Textural igneous rock type characterized by conspicuous crystals (phenocrysts) set in a finer-grained matrix.

Pothole A cylindrical hole drilled in a stream bottom by pebbles rotated by the current and aided by finer-grained abrasive material such as sand.

Protolith The unmetamorphosed parent rock from which a given metamorphic rock was derived.

Pull-apart basin A depression in the crust produced as a result of movement on strike-slip faults that step either right or left.

Pumice Cellular, lightweight, glassy lava usually of rhyolitic composition.

Punky Spongy, woodlike texture.

P waves Elastic waves in the earth in which displacements are in the direction of wave propagation. Compressional waves. May pass through both solids and fluids.

Pyroclastic Detrital volcanic material explosively ejected from a volcanic vent.

Pyroxene An important group of rock-forming iron and magnesium silicate minerals.

Quartz Silicon dioxide; the commonest rock-forming mineral.

Quartzite Granulose metamorphic rock consisting essentially of quartz; a metamorphosed sandstone.

Quartz monzonite Plutonic rock containing quartz and approximately equal amounts of plagioclase and orthoclase feldspar. Intermediate between granodiorite and granite.

Radial drainage Drainage pattern in which streams radiate from a common center, like the spokes of a wheel.

Radioactive decay Change of one element into another by the emission of charged particles from the nuclei of its atoms.

Radiocarbon dating Determination of the age of a material by measuring the proportion of the isotope ^{14}C (radiocarbon) in the total carbon present in the material.

Radiolaria Free-living marine protozoans that secrete minute shells of opaline silica or occasionally other substances.

Reentrant Indentation or alcove in a landform, often angular.

Rejuvenation Development of youthful topographic features in an area previously worn down to an old, subdued landscape.

Relative time Geologic time based on the sequence of oldest to youngest, without regard for the specific number of years involved.

Relief Elevation or inequalities, collectively, of a land surface; difference in elevation between high and low points on a given landscape.

Regolith All the loose or fragmental material blanketing the earth's surface, of whatever origin.

Replacement Process by which a new mineral of partly or wholly different composition may grow in the body of an old mineral or mineral aggregate.

Restraining bend A bend in a strike-slip fault such that motion on the fault is inhibited by an increase in friction or sliding resistance.

Resurgent dome Development of domes or uplifts within an area such as a caldera, which was formed by subsidence following an eruption.

Reverse fault A dipping fault in which compression or shortening has caused one block to ride up over the other.

Rhyolite Extrusive equivalent of a granite.

Richter scale A logarithmic scale used for measuring the magnitude of earthquakes. Derived from seismograms of earthquakes.

Rift An elongate depressed feature formed either by faulting or erosion.

Right slip, right lateral, right separation Occurs where the horizontal separation on a fault is such that an observer standing on one block would see that the opposite block has moved to his right.

Rip-up Sedimentary structures formed by shale clasts torn up from semiconsolidated deposits by stream erosion and transported to a new depositional site.

Rock Any consolidated or unconsolidated mass of crustal material.

Roof pendant Older cover rocks projecting downward into a batholith or pluton.

Rubble Accumulation of loose angular fragments, not water-worn or rounded like gravel.

Sag pond Pond occupying a depression or pull-apart basin along an active fault, particularly a strike-slip fault.

Sandstone Compacted or cemented detrital sediment composed mainly of sand-sized grains.

Schist Medum- to coarse-grained metamorphic rock with subparallel orientation of its micaceous minerals.

Schollendome A detached elongate low ridge of lava, often with a crestal crack, rafted away from the leading edge of a flow by residual fluid lava.

Scoria Pyroclastic volcanic ejecta, usually of basic composition.

Screen An irregular sheet of material, often derived from roof rocks, that separates individual plutons in a batholith. See: **Septum.**

Sea-floor spreading The widening of the oceanic sea floor on either side of a spreading center or mid-ocean rise, accompanied by the addition of new ocean floor crust at the spreading center.

Sea knoll Submarine hill or elevation of the deep sea floor. Usually less prominent than a seamount.

Section Natural or artificial rock cut; the diagrammatic representation of such a feature.

Sedimentary Refers to rock formed from sediment, by transportation of fragments from their sources to a depositional site or by precipitation from solution.

Sedimentation Portion of a rock cycle from separation of particles from their parent rock to and including their consolidation into another rock.

Seif dune An elongate ridge of dune sand, often with an undulating crest.

Seismology Study of earthquakes and the measurement of the elastic properties of the earth.

Septa Singular *septum:* An irregular sheet of material, often derived from roof rocks, separating adjacent plutons in a batholith. See **screen.**

Serpentinite Rock consisting almost wholly of serpentine minerals, derived from the alteration of previously existing mafic minerals.

Shale Laminated sedimentary rock in which the constituent particles are predominantly clay, or fine silt.

Sheet joints Parallel jointing, usually in plutonic rocks, that simulates bedded rocks.

Sheet wash Runoff not confined to well-defined channels, usually seen in arid regions following thunderstorms.

Shutterridges Ridges that by horizontal or oblique fault slip tend to block canyons of stream crossing the fault.

Silica Silicon dioxide, usually quartz.

Sill A tabular intrusive body of igneous rock parallel to bedding of the enclosing sedimentary rock. Relatively thin compared to its lateral extent.

Sillimanite An aluminum silicate mineral formed under metamorphic conditions of high temperature and high pressure.

Silt Unconsolidated clastic sediment of a size grade that lies between sand and clay.

Slate Fine-grained metamorphic rock possessing well-developed fissility (slaty cleavage).

Slip Actual displacement of formerly adjacent points on the opposite sides of a fault. Three dimensional.

Slip face The leeward side of a sand dune where grains rest at the angle of repose, usually about 33 or 34 degrees.

Soil All unconsolidated material above bedrock that has been in any way weathered or altered. Usually contains appreciable organic material.

Spatter (driblet) cone Low, steep-sided hill or mound of spatter built by lava fountains along a fissure or central vent.

Specific time Measurement of rock ages in specific years.

Speleothem Any mineral deposit that is formed in a cave by the action of water.

Spreading center Convective upwelling of magma at a mid-oceanic ridge, accompanied by lateral spreading of the plates on either side of the ridge. See also Divergent plate boundary.

Squeeze-up An elevated block or slice between two en echelon strands of a strike-slip fault.

Stalactite A conical or cylindrical deposit that hangs from the roof of a cave. Usually with a central tube.

Stalagmite A conical deposit that develops upward from a cave floor by the action of dripping water. Often formed directly below a stalactite.

Stratigraphy Study of the formation, composition, sequence, and correlation of the stratified rocks of the earth's crust.

Strato-volcano (composite cone) Volcanic cone, usually of large size, built of alternating layers of lava and pyroclastic material. Often andesitic in composition.

Stream capture (beheading) Diversion of the upper part of a stream by headward growth of another stream.

Striation A short, narrow mark, often a few millimeters deep and many centimeters long, on rock surfaces, produced by abrasion; frequently the result of glacial action.

Strike Direction or bearing of an inclined bed or structure, measured in a horizontal plane and perpendicular to the direction of dip.

Strike-slip fault Fault in which most of the net slip is in the direction of fault strike.

Stromatolite A laminated, often domal algal structure, developed in shallow waters by the action of cyanobacteria (blue-green algae).

Subduction complex A mixture of rock types associated with a subduction zone, typically including ophiolites, deep-sea cherts, turbidite sandstones, and various sea-floor volcanic rocks.

Subduction zone In plate tectonic theory, a belt along which one crustal block, usually oceanic, descends beneath an adjacent crustal block or plate, usually continental.

Superposition Rule that if a stratified series of rocks is in its original relationship, the underlying beds are older than the overlying beds, a governing principle of sedimentation.

Suspect terrane A microplate or plate fragment thought to be of exotic origin or unrelated to the other plates with which it is in contact.

S waves Transverse or distortional waves that travel through an elastic medium.

They pass through solids but not through fluids.

Syncline A fold in rocks in which the strata on both sides dip inward. Younger rocks are found in the center and older ones on the margins.

Synclinorium A regional syncline on which smaller folds are superimposed.

Talus Coarse waste at the foot of a cliff or steep slope, transported largely by gravity.

Tarn A rock-floored lake basin scoured out of bedrock by glacial action.

Tear fault A strike-slip fault that trends across the strike of the deformed rocks.

Tectonism Crustal instability, structural behavior, and deformation of the earth's crust during and between major cycles of sedimentation.

Tephra A general term for all types of pyroclastic materials erupted by volcanic activity.

Terminal moraine A ridge of glacial material dumped by melting ice at the point of greatest glacial advance.

Terrace Relatively flat horizontal or gently inclined surface bounded by a steeper ascending slope on one side and a steeper descending one on the other.

Test The shell of single-celled protozoan foraminifera.

Thrust fault A low-angle fault developed in areas of crustal compression or shortening, in which one block is forced over another.

Till Unsorted, unstratified sediment carried or deposited by a glacier.

Tilted fault block Block rotated so that one side is relatively uplifted and the other depressed.

Tombolo A near-shore island tied to the mainland by deposition of a low sand bar.

Topographic inversion An area in which former valleys have been filled and become ridges in a later cycle of erosion.

Topography Physical features of a district or region, especially relief and contour.

Trace elements Elements present in minor amounts in a given rock, mineral, or crustal unit.

Transform fault A strike-slip fault that offsets oceanic ridges.

Transgressive unconformity Progressive pinching out, toward the margins of a depositional basin, of sedimentary units in a comformable series.

Trap (gas, oil, water) Reservoir rock completely surrounded by impervious rock.

Tremor Small earthquake.

Triangular facet A truncated ridge end along a fault or sometimes along the sides of a glacial or stream valley.

Trilobite A primitive extinct crustacean occurring throughout the Paleozoic, characterized by a segmented body divided into three parts.

Triple junction Point on the earth's crust where three crustal plates meet.

Tufa A sedimentary rock composed of calcium carbonate or silica deposited from solution and often by the action of cyanobacteria in a spring or lake or from percolating ground water.

Tuff Rock formed from compacted volcanic ash or pumice fragments.

Turbidity current A relatively dense suspension of sediment in water that moves along the bottom slope of a body of standing water. May occur in the sea, lakes, or reservoirs. Dust storms sometimes form turbidity currents as well.

Ubehebe A low volcanic cone formed by explosive activity resulting presumably from magma contacting ground water.

Ultramafic Applies to rocks containing less than 44 percent silica, but rich in magnesium and iron minerals.

Unconformity Contact between rock units that was once a surface of erosion or nondeposition.

Underthrust A thrust fault in which the field evidence suggests that the upper plate was relatively immobile and the lower plate active.

Vein Crack or fissure filled with mineral matter deposited by underground water solutions.

Vent Nearly vertical outlet from within the crust. Usually refers to volcanic openings.

Vertical component Portion of a vector that is perpendicular to a horizontal or level plane.

Vesicle Small cavity in a fine-grained igneous rock produced by the expansion of a bubble of gas or steam during solidification of the rock.

Volcanic bomb Mass of magma tossed into the air during an eruption. Often develops a spindle shape as a result of rotation in flight.

Volcanic glass (obsidian) Glass produced from magma when cooling is too rapid to permit crystallization.

Volcaniclastic Sedimentary material produced from materials ejected from a volcanic source.

Volcanic neck Solidifed lava in the feeder pipe of a volcano, left standing after the more easily eroded parts of the volcanic edifice are eroded away.

Wall rock Country rock into which magma is intruded.

Water table The upper surface of a zone in soil or rock in which all the pore spaces are filled with water.

Weathering The sum total of all proc-esses that act to break down rocks in place, chemical, physical or mechani-cal, but that do not involve transport.

Welded tuff A rock composed of vol-canic ash or pumice in which the grains were plastic at the time of deposition and acted to consolidate the rock.

Xenolith Rock fragment that is foreign to the body of igneous rock in which it occurs; an inclusion.

appendix two

COMMON MINERALS
AND ROCKS

COMMON MINERALS

The chemical elements found in minerals are thought to compose all materials of the earth, plus the materials of the other planets in our solar system. Lunar rock studies have so far confirmed this premise, since minerals discovered in lunar samples are similar or identical to minerals found on earth. Nearly 3000 minerals are known, but only about 60 are widely distributed in the earth and only about 25 are at all common.

Minerals are the building blocks of most rocks. In contrast to rocks, however, minerals have specifically organized sets of atomic relationships called *space lattices*. These lattices govern the external forms of mineral crystals and determine, within limited ranges, the physical, optical, and chemical properties of all minerals. Minerals are produced only by nature, and even though many identical chemical compounds can be made in the laboratory, such duplications are not minerals. In addition, materials formed through organic processes, like petroleum and coal, are rocks, not minerals. Organic materials have variable compositions, but minerals have definite compositions within specified limits.

Minerals are most accurately identified by methods that reflect their lattice structures, but involve instruments of limited availability. The lay investigator must rely on external characteristics that depend primarily on the composition of the mineral in question. It is possible to recognize 25 to 50 common minerals by making a few simple tests like those for hardness and color and by noting other distinctive characteristics.

Mineral Recognition It is helpful to keep the following points in mind when trying to recognize minerals.

1. The geology of the region in which specimens are found can provide important clues regarding what minerals will *not* be found in the area.

501

Table A-1 (continued)

Luster	Hardness ($>$ = greater than, $<$ = less than)	Mineral (and Color)	Other Characteristics
NONMETALLIC Colored minerals	< fingernail	Azurite (blue)	Blue streak; fizzes
		Talc (green, gray)	Soapy or greasy feel and look; very soft; white streak
		Hematite (red, red brown)	Usually harder; wipes brick red on fingers
		Limonite (yellow, yellow brown)	Usually harder; wipes rust brown on fingers
		Malachite (grass green)	Green streak; fizzes
		Graphite (black)	Usually metallic or submetallic luster; marks paper black
		Pyrolusite (black)	Often sooty; usually submetallic luster
		Cinnabar (bright red)	Very heavy; very bright luster
	> fingernail < copper coin	Biotite (black, brown, yellow)	Mica; sheets and flakes flexible and elastic
		Chlorite (black, green)	Sheets and flakes flexible and inelastic
		Serpentine (green, gray green)	Usually harder; greasy look
		Asbestos	Fibrous
	> copper coin < knife	Calcite (brown, gray)	Fizzes; good cleavage (3 ways)
		Dolomite (gray, brownish)	Does not fizz; good cleavage (3 ways)
		Lepidolite (lavender, purple)	Mica; sheets and flakes flexible and elastic
		Sulfur (yellow)	Burns with match (very hot) with odor
		Fluorite (blue, green, yellow, purple)	Does not fizz; good cleavage (4 ways)
		Serpentine (green, gray green)	Greasy look

Table A-1 (continued)

Luster	Hardness $\left(\begin{array}{l} > = \text{greater than} \\ < = \text{less than} \end{array}\right)$	Mineral (and Color)	Other Characteristics
		Asbestos	Fibrous
		Azurite (blue)	Blue streak; fizzes
		Malachite (grass green, gray green)	Green streak; fizzes
		Chrysocolla (blue green, aqua)	Almost colorless streak; no fizz; enamel-like appearance
		Sphalerite (yellow, dark brown)	Resinous; heavy to feel; looks metallic; much cleavage (6 directions)
		Hematite (red, red brown)	Usually submetallic; heavy
	> knife	Amphibole (green, black)	Cleavage very good and visible; hard to separate from pyroxene
		Pyroxene (green, black)	Cleavage very good, and often not visible; hard to separate from amphibole (amphibole is commoner than pyroxene)
		Chrysocolla (blue green, aqua)	Colorless streak; Enamel-like appearance; often mixed with quartz and therefore tests as though it is very hard
		Epidote (yellow green, pea green)	Hard to identify but a very common mineral in crystalline mats and as coatings, veinlets, and alterations in igneous (and other rocks)
		Olivine (green, black green)	Granular; very common in volcanic rocks
		Garnet (multicolored)	Often in good crystals; massive; heavy for a nonmetallic mineral
		Quartz (multicolored)	Amethyst is purple; rose quartz; citrine quartz (yellow)

(Continued)

Table A-1 (continued)

Luster	Hardness ($>$ = greater than, $<$ = less than)	Mineral (and Color)	Other Characteristics
		Jasper (brown, red, dark green)	Massive
		Chalcedony (multicolored but usually white)	Fine-grained quartz
		Tourmaline	
		Pink: rubellite	Gem variety; uncommon but found in southern California
		Black: schorl	Brittle; common; striated with vertical lines on crystal faces
White minerals	$<$ fingernail	Talc	Very soft; feels soapy
		Borax	Powder, coating; sweetish taste
		Kaolinite	"Clay"; odor of earth when moist (breathe on it); compact; very common mineral
		Bauxite	Kaolinite with spherical nodes in it; chief ore of aluminum
		Gypsum	Cleaves readily in two directions unless massive; looks like mica but not "flaky"; very common mineral
		Barite	Usually harder, but noticeably heavy for a white mineral
	$>$ fingernail $<$ knife	Kernite	Uncommon mineral but spread widely from world-famous occurrence at Boron, Calif.; like gypsum, but cleavage is different— breaks into brittle fibers
		Halite	Salt (taste it); often in cubical crystals
		Calcite	Fizzes in acid; 3 directions of perfect cleavage

Table A-1 (continued)

Luster	Hardness (> = greater than, < = less than)	Mineral (and Color)	Other Characteristics
		Dolomite	Does not fizz; 3 directions of perfect cleavage
		Barite	Glassy luster; does not fizz; heavy to feel for a white mineral
		Fluorite	High glassy luster, usually transparent, but sometimes variegated in color
		Cerussite (Anglesite)	These minerals cannot be separated except by chemical analysis; very high (resinous) luster; very heavy to feel for a white mineral; common ores of lead; associated with galena.
		Scheelite	Commonest tungsten ore; very heavy for white mineral; high luster, hard to distinguish from lead minerals; associated with garnet, epidote, calcite
	> knife	Opal	Opalescence
		Feldspar	
		Orthoclase Microcline	Softer than quartz; often dusted with clay alteration; common rock-forming mineral; widespread in granites; has cleavage; separation of these two not possible by physical properties
		Plagioclase	Plagioclase separated from other feldspars by fine striations (lines) due to twinning of crystals

(Continued)

Table A-1 (continued)

Luster	Hardness ($>$ = greater than, $<$ = less than)	Mineral (and Color)	Other Characteristics
		Quartz	Commonest and most widely spread mineral in rocks; sand is usually quartz; looks like glass; crystals are striated across faces
		Chalcedony	Chalcedony is finely crystalline quartz in which grains are invisible to eye or lens
METALLIC	$<$ fingernail	Graphite (black)	Submetallic luster; marks paper black
		Pyrolusite (black)	Often sooty; submetallic luster; dendrites are often pyrolusite
		Molybdenite (gray)	Metallic, bright; marks paper gray; heavy; in granites; chief ore of molybdenum
	$>$ fingernail $<$ knife	Stibnite (light gray)	Tarnishes to darker gray; one cleavage; blades and needles; lighter weight and color than galena; chief ore of antimony
		Galena (dark gray)	Very heavy; cleavage (3 directions, forming cubes); chief ore of lead
		Chalcocite (black or steel gray)	No cleavage; strong fracture; brittle not so heavy as galena; chief ore of copper
		Bornite (peacock blue)	Tarnishes to many colors; associated with chalcocite and chalcopyrite; ore of copper
		Sphalerite (black to brown)	Much cleavage; high luster, chief ore of zinc; not so heavy as lead or copper minerals
		Chalcopyrite (brass yellow)	Softer and yellower than pyrite

Table A-1 (continued)

Luster	Hardness (> = greater than < = less than)	Mineral (and Color)	Other Characteristics
	> knife	Chromite (black, brownish black)	Brown streak; nonmagnetic; softer than ilmenite and magnetite
		Magnetite (black)	Highly magnetic; important ore of iron
		Ilmenite (bluish black)	Less magnetic than magnetite
		Pyrite (pale brass yellow)	Fool's gold; lighter in color than chalcopyrite; harder than chalcopyrite; common mineral
		Hematite (steel gray)	High luster; common iron ore; red brown streak

2. Most common minerals are white, but white minerals can be difficult to recognize because dozens of uncommon minerals are also white. Frequently occurring white minerals are: quartz and the common varieties of the quartz group (agate, jasper, chalcedony, chert, flint, rock crystal, and onyx), feldspar, calcite, gypsum, kaolinite, barite, and fluorite.

3. The most frequently found minerals are physically hard and durable and usually chemically stable. Quartz and feldspar are good examples.

4. About 100 chemical elements compose the minerals of the earth. Eleven elements make up 95 percent of the crust, with the result that the common minerals tend to be chemical compounds involving these elements. The dominant elements are aluminum, calcium, hydrogen, iron, magnesium, nitrogen, oxygen, potassium, silicon, sodium, and carbon.

5. The common minerals are mostly silicates, compounds involving silicon and oxygen.

6. All physical, chemical, optical, and other properties of minerals are constant within prescribed ranges. Although there are some limitations, this constancy permits the use of physical properties to recognize common minerals. Luster, color, hardness, cleavage, and specific gravity are the properties most often considered; occurrence and habitat are also important.

Items normally used to help determine the physical properties of minerals are a knife blade, magnifier, magnet, weak acid, copper coin, and an unglazed porcelain plate. Procedures are as follows.

1. Examine specimen with eye and magnifier. Note texture and uniformity. If it is nonuniform, it may be more than one mineral, or a rock. Note color and color variation. Is color throughout or a coating? Is it a white or colorless mineral, stained?

2. Determine luster (reflection) of mineral on a fresh surface: metallic (like a metal) or nonmetallic.

3. Determine hardness (always examine with magnifier to see whether in fact a scratch has been made): harder than knife blade—over 5.5; harder than copper coin, but softer than knife—between 3 and 5.5; harder than fingernail, but softer than copper coin—2.5 to 3; softer than fingernail—2.5 or less.

4. Put small drop of weak acid on specimen. Is there a reaction (fizz)?

5. Locate mineral in recognition chart and name specimen.

COMMON ROCKS

Rocks are composed primarily (but not entirely) of minerals, and it is the minerals of a rock on which the name of the rock is usually based. Since rock names are numerous and have been assigned over many years by many petrologists, precise

TABLE A-2 Recognition of Rocks: Their Properties

	Appearance	Minerals Visible	Name	Variety
IGNEOUS* (from cooling of magma)	Glassy (very rapid cooling)	None	Obsidian (gray to black)	Basalt glass (shiny cinders on volcanos)
	Fine grained (rapidly cooled)	Feldspar Quartz	Rhyolite (lava; pink to white)	None
		Feldspar Dark minerals: (1 or more) Biotite Amphibole Olivine Pyroxene No quartz	Basalt (lava; black to brick red)	None
	Coarse grained (slowly cooled)	Feldspar Quartz	Granite (pepper and salt appearance; sugary texture)	None
		Feldspar Dark minerals No quartz	Gabbro or diorite (dark and colored minerals)	None

*Notes on igneous rocks: Pumice is froth from obsidian—lightweight, porous, and light-colored. Porphyry is an igneous rock with discrete large grains (crystals) in a massive or glassy matrix. Dark minerals are often called "ferromagnesian" because of their iron and magnesium content.

TABLE A-2 *(continued)*

	Appearance	Minerals Visible	Name	Variety

		CLASTIC: Made of rounded particles of other rocks and minerals		
	Material	Coarse Grain	Medium Grain	Fine Grain	Very Fine Grain
	Unconsolidated material	Gravel	Sand	Silt	Mud (clay)
	Consolidated material	Conglomerate	Sandstone	_____Shale_____ (Siltstone) (Mudstone)	

		ORGANIC: Made of organic materials	
	Material	Rock Name	
	Shells	Coquina (fizzes with acid)	
	Shells and limy mud	Shelly limestone (fizzes with acid; chalk, ooze)	
	Diatoms	Diatomite (does not fizz)	
	Plants	Coal	

		CHEMICAL: Formed from water solutions	
	Material	Rock Name	
	Salt	Rock salt (salty taste)	
	Limy mud (as on inside of tea kettle)	Limestone (fizzes with acid; dolomitic or calcareous)	

Left margin for above sections: SEDIMENTARY (from erosion, precipitation, accumulations of other rocks, minerals, and organic matter)

Original Rock	Metamorphic Rock	Properties
Shale	Slate	Many colors; splits easily
Sandstone	Quartzite	Very hard, tough; rock grains hard to see; many colors; makes pebbles
Slate	Schist	Splits easily; rich in mica flakes
Granite	Gneiss	Looks like granite, but minerals arranged in crude bands
Limestone	Marble	Coarse grained; many colors; fizzes with acid; knife scratches it; when dark colored, and hit with hammer, gives off odor
Basalt	Serpentinite	Green to black; shiny, slippery feel; soft

Left margin for above section: METAMORPHIC (rocks changed from original nature by heat, pressure, chemical action)

naming is often complicated and can be resolved only by microscopical study. The problem may be somewhat simplified for the general student, however, if some basic guidelines are followed. First, recognize the process by which the rock of interest originated and, second, note the minerals in the specimen (most rocks are composed of more than one mineral).

TABLE A-3 Igneous Rock Classification (primarily for hand specimen recognition)

	CHIEF FELDSPAR ALKALI		ALKALI and SODA-LIME		CHIEF FELDSPAR SODA-LIME		FELDSPARS ABSENT	
	Potash Feldspars: Microcline or Orthoclase		Orthoclase-Plagioclase		Plagioclase			
Characteristic (essential) minerals	Quartz: 0–5%	Quartz: greater than 5%		No quartz	Quartz	No quartz	No quartz ±Olivine	No quartz / Olivine / Other Fe-Mg minerals
Percentage of silica	66–55%	75–65%	75–65%	65–50%	70–62%	65–50%	60–45%	50–30%
Family or clan	Syenite	Granite	Quartz monzonite	Monzonite	Quartz diorite	Diorite	Gabbro	Peridotite
Characteristic texture								
PLUTONIC — 1. Granitic (megacrystalline)	Syenite	Granite, Alaskite	Quartz monzonite, Granodiorite	Monzonite	Quartz diorite	Diorite	Gabbro, Anorthosite, Norite	Peridotite, Dunite
HYPABYSSAL — 2. Porphyritic: Ground mass a. megacrystalline b. microcrystalline / Phenocrysts	Syenite porphyry (a. mega-)	Rhyolite porphyry (b. micro-)	←———— Porphyries ————→					
VOLCANIC — 3. Felsitic	Trachyte	Rhyolite	Quartz Latite	Latite	Dacite	Andesite (less than 50% dark minerals)	Basalt (greater than 50% dark minerals)	Various terms in wide use

Pyroclastic Materials

Unconsolidated: Blocks, bombs; lapilli; ash
Consolidated: Volcanic breccia; agglomerate; tuff

Glassy Rocks

Obsidian: Bright vitreous luster; usually black
Pitchstone: Dull, pitch-like obsidian
Perlite: Gray, black, pearly, with small and large spherical modes layered like an onion
Scoria: Vesicular basaltic lava
Pumice: Froth of obsidian; strong cellular structure

Igneous Environments Some igneous rocks are formed when volcanos erupt molten material (magma) that congeals on contact with the atmosphere. Such congealed products may take many forms, but all are volcanic igneous rocks. Magma congealed within the earth's crust forms plutonic or hypabyssal (intermediate in depth) igneous rocks. The specific names assigned depend on cooling rate and on the original minerals that crystallized from the magma.

Sedimentary Environments Once a rock is exposed on the surface of the earth, the atmosphere begins to change it by alteration (weathering) and transportation (erosion). Eroded products ultimately reach the sea as rounded fragments, gravel, sand, or mud, which are then consolidated into conglomerates, sandstones, shales, and mudstones. All these belong to the large class known as sedimentary rocks, the environment of formation being that of sedimentation. The source of such sediments can be any material, including previously formed sedimentary rocks, igneous rocks, or recycled metamorphic rocks.

Metamorphic Environments Under the impact of pressure, heat, and chemical change, any material buried in the crust of the earth can become a new (metamorphic) rock type. By definition, however, the heat involved cannot be great enough to change the metamorphic rock to magma. Metamorphic rocks have complex mineralogy, but fortunately they are named according to external characteristics. Five common types are recognized: gneiss (often metamorphosed plutonic rock), schist (often metamorphosed shale or volcanic rock), slate (metamorphic rock with good rock cleavage), quartzite (metamorphosed sandstone), and marble (metamorphosed—"crystalline"—limestone).

appendix three

GEOLOGIC SEQUENCE AND TIME

Geologists use two approaches when considering the vast dimension of geologic time: relative or qualitative and specific or quantitative. Relative measure is not a true measure, but rather a sequence, since a given event occurs before or after another event. Specific measure is a true although not exact measure. Geologists think primarily in terms of a sequence of events, each event being relative to every other event, sequentially from oldest to youngest. Specific ages are in some ways incidental to the basic task of the geologist. Hence geologic events are arranged in chronologic sequence, based on stratigraphic and paleontologic data. The magnitude of each known event is estimated in both time and space; relative magnitudes are acknowledged by assigning designations with varying degrees of emphasis. Thus the term *era* designates major time units, separated from one another by major events; *period*, *epoch*, and *age* refer to time spans between events of progressively lesser magnitudes.

A standardized set of labels for recording geologic history is the geologic time scale, which has been accepted by international agreement. Continental and regional designations may vary, however, particularly those of lesser magnitude. Estimates of specific time are commonly inserted.

Current techniques for establishing specific measurement of geologic time involve pairs of chemical elements that have "mother-daughter" relationships. Elements such as uranium-lead, rubidium-strontium, and potassium-argon occur widely, though in exceedingly small proportions, in some of the commonest rock-forming minerals. Age determinations from the elements in these minerals may yield measurements up to many millions of years. The date of mineral crystallization from magma or even the date of later metamorphism can often be determined. In general, igneous rock-forming minerals furnish the best results. Carbon-14 measurements are only valid to 50,000 years before the present, so this method is of limited geologic use.

Table A-4 summarizes the isotopes currently used for age dating. For example,

TABLE A-4 Common Isotopes Used in Radiometric Age Dating

| Materials Used | Isotopes | | Half-life of Original (years) | Maximum Dating Range (years) |
	Original (parent)	Decay Product (daughter)		
Uraninite, pitchblende, zircon in rocks	Uranium-238 Uranium-235	Lead-206 Lead-207 (not terrestrial)	4.5 billion >10 million	10 million to 4.6 billion
Micas (muscovite and biotite); feldspar; microcline; some metamorphic rocks	Rubidium-87	Strontium-87	47 billion	10 million to 4.6 billion
Micas (muscovite and biotite); amphibole (hornblende); some volcanic rock	Potassium-40	Argon-40 Calcium-40	1.3 billion	100,000 to 4.6 billion
Wood, peat, and various plant materials; charcoal, some bones, cloth, shells; waters, both ground and ocean; deposits like stalactites and stalagmites	Carbon-14	Nitrogen-14	5730 ±	100 to 50,000

potassium breaks down into calcium and argon at a known rate. Thus, if the amounts of each of these elements in a certain mineral is known, the time involved in the change can be calculated.

Geochronologic techniques appear to suggest solutions to some fundamental problems. Present applications provide: (1) an independent test of the relative time scale, particularly in the post-Precambrian; (2) greater sophistication in determining rates of sea-floor spreading and continental drift; (3) refinements of dating reversals in the earth's magnetic field; (4) increasing definition and division in the tremendous span of Precambrian time; and (5) assistance in meteoritic studies, which promise insight into pregeologic history of the earth.

In another use of the radioactive decay process, the rubidium[87]-strontium[86] system has proved especially helpful in establishing rock origins. Rubidium is known to be concentrated in the continental crust, where rubidium-strontium ratios are higher by a factor of at least 10 than they are in rocks derived from the upper mantle. Furthermore, rubidium-strontium ratios for rocks derived from the lower mantle contrast with those derived from the continent or upper mantle.

appendix four

SOME THEORIES PERTINENT TO CALIFORNIA GEOLOGY

Within 15 years, the wide acceptance of continental drift made the formerly conventional view of stable continental and oceanic masses untenable to most geologists. Corollaries of continental drift include sea-floor spreading, polar wandering, and plate tectonics. California's position at the edge of a continent confers on the state a geologically important position as a testing ground for many new and exciting ideas in the earth sciences.

GEOSYNCLINES AND GEOSYNCLINAL THEORY

Continental bodies are subject to periodic marine invasions that develop elongate basins, primarily along the edges of the continents. The seas thus established have seldom been deeper than 1,500 feet (450 meters), but they frequently accrued great thicknesses of marine sediments as the basin floors sank under the loads of deposition. *Geosyncline* is the term for the large depression (fold) that sedimentary deposits create as they settle on the underlying crust. The geosynclinal theory of mountain building contends that marine waters are eventually expelled from the geosyncline by compressive or vertically activated forces that typically produce a mountain chain of parallel folds. The Appalachian chain is a classic North American example.

The geosynclinal theory has substantially influenced North American geology and California geology in particular. Prior to development of plate tectonic theory, the standard explanation for the Sierra Nevada and Coast Ranges involved their evolution from continentally based geosynclines. About 1940, however, it was suggested that geosynclines are primarily marginal to continents rather than continentally based.

The geosynclinal setting is now understood to embrace several aspects or depositional settings that may succeed one another in time. For example, the California continental margin from near the beginning of the Phanerozoic, about 700 million years, to late in the Devonian about 370 million years ago, was a miogeoclinal setting (Atlantic-type margin) in which a wedge of carbonate and other sedimentary materials thickening toward the ancient ocean were deposited. Beginning late in the Devonian, this was replaced by a volcanic island arc setting like that seen today in Japan. The continental edge was separated from the volcanic islands by a marginal sea in which volcanic sediments played a more important role than heretofore. Near the end of the Triassic, about 200 million years ago, as subduction of the oceanic crust brought the island arc in contact with the continental edge, sediments deposited in the forearc basin (the eugeoclinal setting) seaward of the island arc were formed. The seaward end of these deposits was mixed in part with deep-water sediments and oceanic crust in the subduction zone to form a subduction complex now represented by the Franciscan Formation. The landward part of the Forearc basis sediments represent the Great Valley beds.

ISOSTASY

Isostasy has been accepted as a fundamental premise of earth science for more than 100 years. Initially it was asserted that high-standing rock masses like the Himalayas and the Andes deflected the plumb bob (in measurements of gravity) less in amount than the mathematical calculation of gravitational variance required for the supposed mass of the mountain block in question. We now know that this anomaly occurs because mountains have deep roots and a mountain mass with roots has a lower density (mass) than the adjacent, crustal units. Constant shifting is required to maintain balance between high-standing (mountain) blocks and low-standing (basin) blocks as erosion transfers material from the mountains to the lowlands.

The principle of isostasy applies to small and large areas alike, although because of the general rigidity of rocks the results may not be measurable if the area is small. Larger areas may show deformation appropriate to the volume (mass) transfer on the surface. Isostasy has been frequency suggested as the causal mechanism of mountains, both those originating from geosynclines and those formed by block faulting. A variety of geologic events could trigger isostatic adjustment: widespread lava outpourings like those of the Columbia and Modoc plateaus, the formation of deltas or continental ice masses, and response to erosional and depositional processes in general. Changes in sea level often reflect isostasy, for epicontinental seas may be created or removed through isostatic adjustment. Large vertical movements such as those that produced the Tibetan and Colorado plateaus have been attributed to isostasy.

OVERTHRUST FAULTS

Horizontal movement represented by overthrusting on flat-lying fault planes was once considered an impossibility by most physicists, who questioned the concept that blocks of rigid crust rode upward and outward across nearly horizontal planes.

Physicists generally maintained that forces could not move bodies of such rigidity across miles of underlying, also rigid bodies without substantially shattering overlying rock masses. Nevertheless, geophysicists M. King Hubbert and W. W. Rubey demonstrated that the supporting pressure of fluids in pore spaces of rocks could permit rock masses to ride miles across underlying units with minimal friction and without causing the severe shattering predicted for such brittle bodies. In fact, it appears that vertical movements may well be incidental to the worldwide horizontal movement of rock bodies called for by plate tectonics. Although horizontal displacement on strike-slip faults like the San Andreas was traditionally assumed to have occurred on vertical or nearly vertical planes, the cause of such horizontal shift was difficult to explain until plate tectonic models introduced the transform fault as an element in plate boundaries.

ROCK MAGNETISM AND PALEOMAGNETISM

It is generally accepted that the earth's changing magnetic field has impressed its past orientations on the iron minerals found in rocks. Accordingly, plotting of magnetic polar positions from orientation of minerals in ancient rocks supports the contention that continents have changed their positions with respect to the magnetic poles. Furthermore, we now know that magnetic field reversals are common over both short and long periods of geologic time; there is evidence that reversals have occurred frequently in the past 160 million years or so, for which a polarity calendar has been established. The geomagnetic polarity time scale is well controlled for the past 60 million years and further refinements can be expected as well as an extension of the scale backward in time.

CONTINENTAL DRIFT

The idea of drifting continents was first advanced seriously in 1910 by F. B. Taylor, who did not pursue the concept. It remained for Alfred Wegener to articulate and develop the theory with his first paper in 1912. Wegener's ideas were eventually synthesized in his 1915 treatise entitled *The Origins of Continents and Oceans*. His view was that an original single continent became slowly but continuously dissociated by horizontal shift of its adjacent parts, creating suture lines and rifts or trenches between adjacent masses. The discrete parts evolved by slow drift into the major continents we know today. Wegener presumed that the continents were supported in a substrate of mantle or subcrustal material. Although American geologists were slow to accept Wegener's hypothesis, today drift is accepted by almost all earth scientists as the basic explanation for shape and position of the continents.

SEA-FLOOR SPREADING

In the early 1960s, the view was advanced that midoceanic ridges like the Atlantic ridge are the places from which magmas rise from the mantle and spread laterally to produce new crust and, eventually, new continental plates. As plates migrate laterally, they collide. The deep trenches existing in many ocean basins (often impinging on continental borders) are now thought to be the result of the more dense oceanic plates riding under the lighter continental bodies.

Evidence from polarity reversals in deep-sea sediments seems to support the view that new crust develops by mantle extrusion from midoceanic spreading centers. It is of particular interest for California geology that spreading centers along the East Pacific rise may extend into the Gulf of California, and possibly beneath the Imperial Valley and even farther north (with the concomitant development of the San Andreas fault).

PLATE TECTONICS

In 1967, it was first proposed that the earth's crust is divided into immense plates, five or six of which are rigid, relatively thin continental plates. These plates are thought to be moving continually, primarily horizontally with lesser vertical components of movement. Downward movements tend to occur along plate margins, where spreading centers cause rifting or where oceanic plates impinge on the forward edges of continental plates in downsweeping subduction zones. Upward movements tend to occur where the forward edges of continental plates float up over oceanic units, or where continental plates collide or impinge on each other. According to plate tectonic theory, California's mountains have been produced chiefly by the collision of the east-moving Pacific plate with the west-drifting North American plate. Moving plates presumably have at least three kinds of margins: (1) sutures like the mid-Atlantic ridge, where new material is being added to the trailing edge of a plate; (2) boundaries where moving plates impinge on one another, usually with one plate moving below the other (subduction); and (3) boundaries in which two plates slide past one another (shearing), rather than one passing beneath the other.

TRANSFORM FAULTS

Transform faults, or simply transforms, are special cases of strike-slip faulting that separate major structural elements of tectonic plates. A North American example would be the interpretation of the San Andreas fault as a partial boundary between the Pacific and North American plates. The transform was originally proposed to describe the relative motion between discrete units of oceanic crust upwelling at oceanic rises, where magma repeatedly fills a central suture and shoves the blocks of oceanic crust away from the rise. The upwelling zones in the central rise are themselves offset, so the blocks on either side of the faults in some places move in the same direction and in other places move in opposite directions, with normal strike-slip offset. These are ridge-ridge transforms, in contrast to the boundaries where plates collide or move past one another by shearing. Both are correctly termed strike-slip faults, however.

EXOTIC AND SUSPECT TERRANES

As the rocks formed in subduction zone settings have been examined more thoroughly, some units occasionally tens or even hundreds of kilometers long have been found to have one or more characteristics that indicate the block (or microplate if quite large) formed at some distance from its present location. Indeed, the evidence occasionally indicates these exotic blocks may have moved thousands of kilometers from their place of origin. The degree to which an exotic origin is established determines whether the block is labeled a Suspect Terrane or an Exotic Terrane, although no exact standard for such a distinction has yet been agreed upon.

Types of evidence used to establish a foreign origin for the blocks includes, but is not limited to (1) Paleontologic evidence, which may show that the block was, for example, deposited in an Asian setting with an Asian fauna distinct from the North American fauna of the same age; (2) Paleomagnetic dip measured in suitable rocks in the exotic terrane may demonstrate that the block formed at a very different latitude than its present setting, or (3) petrologic or isotopic evidence that indicates a much closer affinity with a distant suite of rocks than with those with which they are at present associated.

These exotic blocks are believed to have been transported by transform faulting, as in the case of the Salinian granitic rocks of the Coast Ranges, or by long-distance rafting on a spreading sea floor as is thought to be the case for examples in the Sierran foothills and the Klamath Mountains.

LISTRIC FAULTS AND DETACHMENTS

For many years flat-lying faults have been known in the eastern Mojave Desert, the Basin and Range, and in the Colorado Desert and were interpreted as thrust faults that developed during crustal shortening or compression. Some of these faults are still believed to be true thrusts, but in recent years as evidence for stretching or extension of the crust in these areas became more and more firmly established, a different model was proposed to account both for the high-angle range-front faults of the Basin and Range and some of the flat lying faults as well. It was proposed that these high-angle normal faults curved and flattened at depth forming what are known as listric faults or listric normal faults.

Geophysical evidence as well as surface exposures in more deeply eroded areas has convinced many geologists today that listric faults define the many Basin and Range block-faulted mountains and that these faults merge at depth to form a zone called a Detachment or detachment fault. This model allows the crust above the detachment to behave independently from the crust below. In an extensional regimen, this might allow the upper part of the crust to break up in a brittle manner in response to crustal stretching, and the underlying crust to respond by stretching and thinning.

index

Note: References to figures are italicized